Starter cultures in food production

Starter cultures in food production

EDITED BY

Barbara Speranza
University of Foggia, Italy

Antonio Bevilacqua
University of Foggia, Italy

Maria Rosaria Corbo
University of Foggia, Italy

Milena Sinigaglia
University of Foggia, Italy

WILEY Blackwell

Registered Office
John Wiley & Sons, Ltd, The Atrium, Southern Gate, Chichester, West Sussex, PO19 8SQ, UK

Editorial Offices
9600 Garsington Road, Oxford, OX4 2DQ, UK
The Atrium, Southern Gate, Chichester, West Sussex, PO19 8SQ, UK
111 River Street, Hoboken, NJ 07030-5774, USA

For details of our global editorial offices, for customer services and for information about how to apply for permission to reuse the copyright material in this book please see our website at www.wiley.com/wiley-blackwell.

Library of Congress Cataloging-in-Publication Data

Names: Speranza, Barbara, editor.
Title: Starter cultures in food production / edited by Barbara Speranza,University of Foggia, Italy, Antonio Bevilacqua, University of Foggia, Italy, Maria Rosaria Corbo, University of Foggia, Italy, Prof Milena Sinigaglia, University of Foggia, Italy.
Description: Chichester, West Sussex, UK ; Hoboken, NJ : John Wiley & Sons, Inc., [2017] | Includes bibliographical references and index.
Identifiers: LCCN 2016045438 (print) | LCCN 2016049868 (ebook) | ISBN 9781118933763 (cloth) | ISBN 9781118933770 (pdf) | ISBN 9781118933787 (epub)
Subjects: LCSH: Bacterial starter cultures. | Fermented foods.
Classification: LCC TP456.B32 S73 2017 (print) | LCC TP456.B32 (ebook) | DDC 664/.024–dc23
LC record available at https://lccn.loc.gov/2016045438

A catalogue record for this book is available from the British Library.

Wiley also publishes its books in a variety of electronic formats. Some content that appears in print may not be available in electronic books.

Cover image: Background Image: jcarroll-images/Gettyimages

Largest image on front cover: Science Photo Library – STEVE GSCHMEISSNER./Gettyimages

Clockwise, starting from the image beside the editors name:

1) Tobik/Shutterstock
2) Diane Diederich/Gettyimages
3) macida/Gettyimages
4) Genotar/Shutterstock
5) Petr Jilek/Shutterstock
6) Mariusz Szczygiel/Shutterstock
7) VeraLubimova/Gettyimages
8) Hong Vo/Shutterstock
9) Lukas Gojda/Shutterstock
10) Lauri Patterson/Gettyimages
11) Pavel L Photo and Video/Shutterstock
12) BSIP SA/Alamy Stock Photo

Set in 9.5/13pt Meridien by SPi Global, Pondicherry, India

Printed and bound in Malaysia by Vivar Printing Sdn Bhd

10 9 8 7 6 5 4 3 2 1

Contents

List of contributors

Clelia Altieri
Department of the Science of Agriculture, Food and Environment,
University of Foggia, Italy

Ersilia Arace
Department of the Science of Agriculture, Food and Environment,
University of Foggia, Italy

Francisco Noé Arroyo-López
Food Biotechnology Department, Instituto de la Grasa (CSIC), Spain

Salvatore Augello
Department of the Science of Agriculture, Food and Environment,
University of Foggia, Italy

Antonietta Baiano
Department of the Science of Agriculture, Food and Environment,
University of Foggia, Italy

Antonio Bevilacqua
Department of the Science of Agriculture, Food and Environment,
University of Foggia, Italy

Pietro Buzzini
Department of Agricultural, Food and Environmental Science,
Industrial Yeasts Collection DBVPG, University of Perugia, Italy

Giuseppe Calò
Department of the Science of Agriculture, Food and Environment,
University of Foggia, Italy

Daniela Campaniello
Department of the Science of Agriculture, Food and Environment,
University of Foggia, Italy

Angela Capece
Scuola di Scienze Agrarie, Forestali, Alimentari ed Ambientali,
Università degli Studi della Basilicata, Italy

Vittorio Capozzi
Department of the Science of Agriculture, Food and Environment,
University of Foggia, Italy

Emanuela Ciuffreda
Department of the Science of Agriculture, Food and Environment,
University of Foggia, Italy

Maria Rosaria Corbo
Department of the Science of Agriculture, Food and Environment,
University of Foggia, Italy

Roniele P. Cordeiro
Department of Food Science, University of Manitoba, Canada

Philippe Dantigny
Université de Brest, LUBEM, ESIAB, France

Fabio de Stefano
Department of the Science of Agriculture, Food and Environment,
University of Foggia, Italy

Barbara Di Maggio
Department of the Science of Agriculture, Food and Environment,
University of Foggia, Italy

Simone Di Mauro
Department of Agricultural, Food and Environmental Science,
Industrial Yeasts Collection DBVPG, University of Perugia, Italy

Francesca Fuccio
Department of the Science of Agriculture, Food and Environment,
University of Foggia, Italy

Mariangela Gallo
Department of the Science of Agriculture, Food and Environment,
University of Foggia, Italy

Pilar García
Instituto de Productos Lácteos de Asturias–Consejo Superior de Investigaciones
Científicas (IPLA-CSIC), Spain

Antonio Garrido-Fernández
Food Biotechnology Department, Instituto de la Grasa (CSIC), Spain

Maria Clara Iorio
Department of the Science of Agriculture, Food and Environment,
University of Foggia, Italy

Rufino Jiménez-Díaz
Food Biotechnology Department, Instituto de la Grasa (CSIC), Spain

Takashi Kuda
Department of Food Science and Technology, Tokyo University of Marine Science and Technology, Japan

Martina Loi
Department of the Science of Agriculture, Food and Environment,
University of Foggia, Italy

Fernando B. Luciano
School of Agricultural Sciences and Veterinary Medicine,
Pontifícia Universidade Católica do Paraná, Brazil

Renata E.F. Macedo
School of Agricultural Sciences and Veterinary Medicine,
Pontifícia Universidade Católica do Paraná, Brazil

Beatriz Martínez
Instituto de Productos Lácteos de Asturias–Consejo Superior de Investigaciones Científicas (IPLA-CSIC), Spain

Noemi Monacis
Department of the Science of Agriculture, Food and Environment,
University of Foggia, Italy

Labia Irène I. Ouoba
Consultant – Senior Research Scientist, London, UK

Marianne Perricone
Department of the Science of Agriculture, Food and Environment,
University of Foggia, Italy

Leonardo Petruzzi
Department of the Science of Agriculture, Food and Environment,
University of Foggia, Italy

Stefano Pignatiello
Department of the Science of Agriculture, Food and Environment,
University of Foggia, Italy

Angela Racioppo
Department of the Science of Agriculture, Food and Environment,
University of Foggia, Italy

Ana Rodríguez
Instituto de Productos Lácteos de Asturias–Consejo Superior de Investigaciones Científicas (IPLA-CSIC), Spain

Patrizia Romano
Scuola di Scienze Agrarie, Forestali, Alimentari ed Ambientali, Università degli Studi della Basilicata, Italy

Patricia Ruas-Madiedo
Instituto de Productos Lácteos de Asturias–Consejo Superior de Investigaciones Científicas (IPLA-CSIC), Spain

Pasquale Russo
Department of the Science of Agriculture, Food and Environment,
University of Foggia, Italy

Borja Sánchez
Instituto de Productos Lacteos de Asturias–Consejo Superior de Investigaciones Cientificas (IPLA-CSIC), Spain

Luca Settanni
Department of Agricultural and Forest Sciences, University of Palermo, Italy

Milena Sinigaglia
Department of the Science of Agriculture, Food and Environment,
University of Foggia, Italy

Giuseppe Spano
Department of the Science of Agriculture, Food and Environment,
University of Foggia, Italy

Barbara Speranza
Department of the Science of Agriculture, Food and Environment,
University of Foggia, Italy

Sanna Taskila
Chemical Process Engineering, University of Oulu, Finland

Benedetta Turchetti
Department of Agricultural, Food and Environmental Science, Industrial Yeasts Collection DBVPG, University of Perugia, Italy

Chibuike C. Udenigwe
Department of Environmental Sciences, Dalhousie University, Canada

Preface

As classically defined, starter cultures are living microorganisms or defined combinations that are deliberately used for the fermentation of raw material and applied to elicit specific changes in the chemical composition and sensorial properties of the substrate.

Due to their vital role in the manufacturing, flavour and texture development of fermented foods, the awareness that starter cultures are of great industrial significance is a well-established fact. Once mainly used in the dairy industry, nowadays the addition of selected starter cultures has spread to all fermented food products (meat, sourdough, vegetables, wine, fish), where their use ensures a correct and predictable process and avoids fermentation arrests or the production of undesired metabolites. Depending on the type of action and the product to be obtained, a starter should fit some predetermined selection criteria. In the last 20 years, the selection of starter cultures for food has been an emerging topic; the main issue has been the evaluation of the technological traits of autochthonous strains, with the main aim of selecting some biotypes adapted to the different raw materials. Many papers can be easily found in the literature dealing with these topics; namely, with the quali-quantitative composition of the lactic microflora from dairy products, vegetables, meat, sourdough and so on. These reports clearly underline the industrial importance of starter cultures (mainly lactic acid bacteria) for the manufacture of fermented food products, and different selection protocols are described.

Over the last decade new concepts have emerged, including the use of functional starter cultures, the use of genomic approaches to select promising starter cultures, the use of new kinds of starter (like fungi) and the use of microorganisms as non-conventional starters to manage the waste from the food industry. These emerging ideas could be the future as well as a tentative practical application of starter cultures in the food industry, as they could offer a solution to the increasing demand for new ways to give functional/added value to some traditional food products.

Therefore, the main goal of this book is to describe the most recent insights around this topic, through 19 chapters covering all new concepts related to this issue. For example, advances in genetics and molecular biology have recently provided opportunities for genomic studies of starter cultures, aimed to design and improve industrially useful strains. The selection of new starter cultures is beginning to take advantage of pangenomic, based on a comparison of the complete genome sequences of a number of members of the same species; pangenomic does in fact open up an array of new opportunities for understanding and improving industrial starter cultures and probiotics. These include understanding the formation of texture and flavour in food products; understanding the functionality of probiotics; and providing information that can be used for strain screening, strain improvement, safety assessment and process improvement.

Another growing issue is starter attenuation through physical methods. Attenuated starters are lactic acid bacteria that do not have the ability to produce acid during fermentation, but contain enzymes that can influence food quality (for example, during cheese ripening). Besides heat treatment, different methods to achieve attenuation have been studied, including freezing and thawing, freeze or spray drying, lysozyme treatment, high-pressure treatment, use of solvents, and natural and induced genetic modification. To the best of our knowledge, little information is actually available about both pangenomic and starter attenuation, so an overview of what has been done and what can be done could help the scientific and academic community.

Moreover, even if starter microorganisms have mainly useful and positive aspects, could they negatively affect human health and well-being? Some starter cultures can produce both biogenic amines and other toxic compounds; this aspect is often overlooked and we have devoted a chapter to this lesser-known issue.

Lactic acid bacteria are the main microorganisms responsible for fermentation and are consequently used as starter cultures by definition; surprisingly, fungal starters have also been reported as a promising means in some fermentations and appear to survive, and even grow, in stressful environments. However, neither their role nor the mechanism facilitating their survival and growth under these conditions is completely understood. A special focus on this new concept of starter cultures could be appreciated, especially if applied to the management of wastes from the food industry.

In this book we have tried to update and collate information and research carried out on various aspects of these innovative features. We have also devoted an entire second section to analysing and describing what has been done and what is known about different fermented food products: sourdough and cereal-based foods, table olives and vegetables, dairy and meat products, fish, wine and ethnic foods. One special focus is the selection of functional *Bacillus* starter cultures for alkaline fermentation.

We are grateful to all the contributing authors who accepted our invitation to write this book. We are happy to bring numerous foreign authors on board, and offer our thanks to Francisco Noé Arroyo-Lopez, Philippe Dantigny, Takashi Kuda, Renata E.F. Macedo, Labia Irène I. Ouoba, Ana Rodriguez, Patricia Ruas-Madiedo and Sanna Taskila and their colleagues, who have given an international dimension to this project. We are also grateful to our Italian colleagues Clelia Altieri, Pietro Buzzini, Angela Capece, Vittorio Capozzi, Leonardo Petruzzi and Luca Settanni, and to everyone who collaborated with them.

We also want to thank the editorial staff of John Wiley & Sons for their guidance in all the aspects that made the publication of this book possible.

We hope the book will be utilized by researchers, students, teachers, food entrepreneurs, agriculturalists, ethnologists, sociologists and people in general who are interested in fermented foods and starter cultures.

The editors
June 2016

CHAPTER 1
Lactic acid bacteria as starter cultures

Clelia Altieri, Emanuela Ciuffreda, Barbara Di Maggio and Milena Sinigaglia
Department of the Science of Agriculture, Food and Environment, University of Foggia, Italy

Introduction

Starter cultures have a basic role: to drive the fermentation process. Concomitantly, they contribute to all the characteristics of products, as well as to their sensorial and safety characteristics. Therefore, the introduction of starter cultures has undoubtedly improved the quality of products and the standardization of the industrial process.

A very important aspect is to have a good knowledge of the metabolic properties required to improve a specific product and to select useful microbial strains. Nevertheless, the limited number of already selected and studied strains that are also able to possess highly technological properties, as well as the constant risk of bacteriophage attacks, are stimulating research into new starter strains, in order to obtain higher quality and product diversification, in response to more and more aware consumers.

General aspects of starter cultures

The production of fermented foods today is based on the use of starter cultures, for example lactic acid bacteria (LAB), which initiate fast acidification of raw material. The great advantage of starter cultures is that they can provide controlled and predictable fermentation.

Starter cultures of LAB can contribute to microbial safety or offer one or more technological, organoleptic, nutritional or health advantages. Examples are LAB that produce antimicrobial substances, sugar polymers, sweeteners, aromatic compounds, vitamins or useful enzymes, or that have probiotic properties (Leroy and De Vuyst 2004).

While starter cultures, chosen on the basis of their good safety and 'functional' characteristics, can benefit the consumer, they must first be able to be

Starter Cultures in Food Production, First Edition. Edited by Barbara Speranza, Antonio Bevilacqua, Maria Rosaria Corbo and Milena Sinigaglia.

manufactured under industrial conditions (Saarela *et al.* 2000). Safety aspects of LAB include specifications such as origin, non-pathogenicity, certain metabolic activities (e.g. deconjugation of bile salts), toxin production, haemolytic potential, side effects in human studies (i.e. systemic infections, deleterious metabolic activities, excessive immune stimulation in susceptible individuals and gene transfer) and epidemiological surveillance of adverse incidents in consumers (post-market). Functional aspects can be related to viability and persistence in the gastrointestinal (GI) tract, survival at low and high pH and in the presence of bile salts, hydrophobic properties, antibiotic resistance patterns, immunomodulation, and antagonistic and antimutagenic properties. Technological aspects concern growth at different sodium chloride (NaCl) amounts, temperatures, pH values, acidifying ability and metabolism (arginin deamination, esculin hydrolysis, acetoin production) and the ability to produce adequate flavour/texture.

With regard to the effect of salting, the addition of NaCl is a common practice in most fermented dairy foods, and also affects the growth of starter bacteria. Most LAB are partially or fully inhibited by levels of NaCl higher than 5%. However, it is evident that salt tolerance is a strain-dependent characteristic, thus this criterion is important in starter selection (Powell *et al.* 2011).

LAB starters are primarily used because of their ability to produce lactic acid from lactose and for consequent pH reduction, leading also to important effects like inhibition of undesirable organisms, improvement of sensorial and textural properties, as well as contribution to health benefits. A major role of starter cultures in dairy production is the degradation of peptides generated by the coagulant to small peptides and amino acids. Starter cultures are also capable of degrading caseins and converting amino acids to a range of flavour compounds. However, since many of the proteolytic enzymes are intracellular, flavour development in maturing cheese also depends on the release of the enzymes from starter cultures into the cheese matrix through cell lysis. Cell lysis, and the consequent release into the cheese matrix of intracellular enzymes, particularly peptidases and amino acid-degrading enzymes, is an important characteristic for both general protein degradation and also the control of bitterness. Autolysis results from the enzymatic degradation of the bacterial cell wall by indigenous peptidoglycan hydrolases released into the growth medium, although it is still unclear how this process is controlled in the cell. The process is highly strain dependent and is also influenced by factors such as the nutrient status of the growth medium and environmental conditions (Lortal and Chapot-Chartier 2005).

Generally, in maturing cheese there is a positive relationship between the period of starter culture autolysis and the flavour-forming reactions, involving not only proteolysis but also lipolysis. Consequently, various screening assays using buffers or model cheese and milk solutions have been proposed to select highly autolytic strains for use in cheese manufacture. Lysis positively influences the ripening and flavour of the cheese, but the type of peptidases is also very important, in particular since low peptidase activities and low lytic properties produce bitter cheese. One of the most successful strategies to counteract this defect involves the use of LAB with high peptidase activities, particularly Pep N.

For these reasons, the use of good starter cultures can ensure the safety, quality and acceptability of both traditional and innovative fermented dairy products.

Types of starter cultures

In practice starter cultures may be categorized as mesophilic or thermophilic, according to the incubation and manufacturing temperatures under which they are used. Mesophilic cultures grow and produce lactic acid at optimal levels, at a moderate temperature (about 30 °C), whereas thermophilic cultures optimally function at a higher temperature (about 42 °C). Examples of mesophilic dairy starter cultures are the species *Lactococcus lactis* subsp. *lactis*, *Lactococcus lactis* subsp. *cremoris*, *Leuconostoc mesenteroides* subsp. *cremoris* and *Leuconostoc lactis*. On the other hand, the most thermophilic LAB species are *Streptococcus thermophilus*, *Lactobacillus delbrueckii* and *Lactobacillus helveticus*.

Nevertheless, the most common classification of starter cultures is based on the complexity of the culture and the way it is reproduced. All starter cultures available today are derived in one way or another from **natural starters** (or artisanal starters) of undefined composition (i.e. containing an undefined mixture of different strains and/or species). For some types of products, natural starters have been replaced by commercial **mixed-strain starters** (MSS), derived from the 'best' natural starters and reproduced under controlled conditions by specialized institutions and commercial starter companies, then distributed to the industries that use them to build up bulk starter or for direct vat inoculation. Natural starter cultures and commercial MSS, because of their long history, are called **traditional starters** (Limsowtin *et al.* 1996) as opposed to **defined strain starters** (DSS). DSS are usually composed of only a small number of selected strains and allow greater control over the composition and properties of the cultures. Table 1.1 shows a summary of culture types.

Traditional cultures contain many strains of many microbial species, sometimes including yeasts and moulds as well as bacteria; they all contribute biochemically to the complexity (and the variability) of the final product (Powell *et al.* 2011). Therefore, traditional starter preparation methods are still in use for some particular or traditional products, and have been adapted to a limited industrial scale. Industrial-scale production requires starters that give reproducible performance and are free of undesirable organisms. These goals are difficult to achieve using traditional methods. Thus, DSS have replaced traditional starters in industrial-scale production because of their optimized, highly reproducible performance and their high phage resistance.

Table 1.1 Culture types and their preparation.

Types of starter cultures		Description
Traditional starters	Natural starters	Low cost. Undefined composition. Highly variable composition and performance. Prone to undesirable contamination; microbiologically hazardous
Traditional starters	Mixed-strain starters (MSS)	Undefined composition. Variable composition and performance. With careful handling and some quality control testing, these are still in limited use, but have largely been replaced by laboratory-maintained cultures
Defined strain starters (DSS)		Defined composition, usually composed of only a small number of strains. This gives a high degree of control over starter performance parameters and product properties, as long as strains are carefully selected and managed

Traditional starters: Natural starters

The production of natural starters is derived from the ancient practice of backslopping (the use of an old batch of a fermented product to inoculate a new one) and/or by application of selective pressures (heat treatment, incubation temperature, low pH) (Carminati *et al.* 2010). No special precautions are used to prevent contamination from the environment, and the control media and culture conditions during starter reproduction are very limited. As a result, natural starters are continuously evolving as undefined mixtures composed of several strains and/or species (Carminati *et al.* 2010).

Natural starters are an extremely valuable source of strains with desirable technological properties (antimicrobials, aroma production); for example, they are considered to be highly tolerant to phage infection because they are reproduced in the presence of phages, which leads to the dominance of resistant or tolerant strains (Carminati *et al.* 2010). Also they seem to be advantaged by microbial interactions; in fact, many strains show limited acid-production ability when cultivated as pure cultures (Parente and Cogan 2004).

Traditional starters: Mixed-strain starters (MSS)

MSS, obtained by careful selection of natural starters, are maintained, propagated and distributed by starter companies and research institutions (Parente and Cogan 2004). Like artisanal starters, MSS contain an undefined mixture of strains that differ in their physiological and technological properties (Parente and Cogan 2004).

When undefined cultures are propagated under controlled conditions with a minimum of subcultures, the stability of their composition and performance is greatly improved in comparison to natural strains (Stadhouders and Leenders 1984).

The composition of MSS is undefined, but their reproduction under controlled conditions reduces the intrinsic variability associated with the use of natural starters (Limsowtin *et al.* 1996).

The traditional method for reproduction of MSS, which requires several transfers to build up the bulk starter by using small amounts of stock cultures, has been replaced by the use of concentrated cultures for the inoculation of the bulk starter tank, thus minimizing the need for transfers within the factory and the risk of fluctuations in starter composition and activity (Carminati *et al.* 2010).

Defined strain starters (DSS)

DSS are composed of one or more strains (the dominant species of the traditional product) and are selected, maintained, produced and distributed by specialized companies. Since the strains and/or species ratio in DSS is defined, their technological performance is extremely reproducible and this is a desirable property. In fact, in recent years DSS have replaced traditional starters (Carminati *et al.* 2010). However, as a consequence of the limited number of strains used, a phage infection may cause disruption of lactic acid fermentation. Furthermore, with the subsequent loss of natural microbial diversity, maintenance of the typical features is difficult. Nevertheless, examination of the key properties of each strain (i.e. genetic or biochemical features, growth and acid-production characteristics) can lead to the rational mixing of strains, in order to formulate a culture with a desirable set of properties (Carminati *et al.* 2010).

DSS usually have no defects of flavour, and have a distinctive trait of 'cleaner' aroma and flavour. In order to increase control over their nature and attain a flavour as close as possible to the traditional one, industrial companies are making increasing use of flavour-enhancing adjunct cultures; DSS cultures are added at low levels to the starter, and may themselves be defined or undefined (Powell *et al.* 2011).

Metabolism of lactic acid bacteria

LAB are important in many food fermentations because they contribute to sensory characteristics and preservative effects (Holzapfel 1995) with their physiological features such as substrate utilization and metabolic capabilities. Some LAB are homofermentative and produce lactic acid as the main product of glucose fermentation, while others are heterofermentative and produce carbon dioxide and ethanol in addition to lactic acid (Blandino *et al.* 2003).

It is clear that LAB adapt to various conditions and change their metabolism accordingly. This may lead to significantly different end product patterns, thus LAB metabolism is essential to study when selecting new starter strains.

Lactose metabolism

Lactose, a disaccharide composed of glucose and galactose, is the only free-form sugar present in milk (45–50 g/L). The main pathways for lactose metabolism are shown in Figure 1.1.

The transport of lactose into a cell requires energy. In the lactococci, this energy is sourced via energy-rich phosphoenolpyruvate (PEP), an intermediate of the glycolytic pathway. This is part of a transport mechanism referred to as the phosphoenolpyruvate phosphotransferase system (PEP-PTS), in which the lactose is phosphorylated as it is transported across the cell membrane. Once inside the cell, phosphorylated lactose is hydrolysed by the enzyme phospho-β-galactosidase to glucose and galactose-6-phosphate. The glucose moiety enters the glycolytic pathway, and galactose-6-P is converted into tagatose-6-phosphate via the tagatose pathway. Both sugars are cleaved by specific aldolases into triose phosphates, which are converted to pyruvic acid at the expense of nicotinamide adenine dinucleotide (NAD^+). For continued energy production, NAD^+ must be regenerated. This is usually accomplished by reducing pyruvic acid to lactic acid (Poolman 1993).

In other dairy starter bacteria, including *Strep. thermophilus*, leuconostocs, lactobacilli and bifidobacteria, lactose transport appears to be via a specific protein (a permease) that translocates the lactose into the cell without modification,

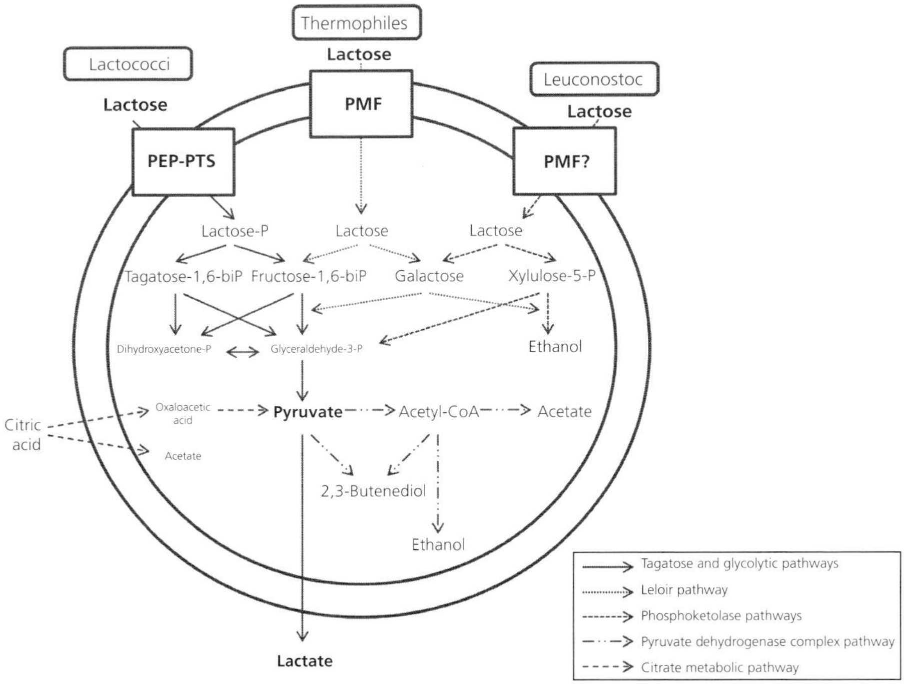

Figure 1.1 General pathways for carbohydrate catabolism by lactic acid bacteria.

although in many of these organisms the exact nature of the system used is still unclear. The lactose is then hydrolysed by β-galactosidase to glucose and galactose (Powell *et al.* 2011). The glucose moiety enters the glycolytic pathway, but galactose is excreted from the cells and accumulates in milk or cheese. Thermophilic lactobacilli that do not excrete galactose and *Lb. helveticus* strains utilize the Leloir pathway to metabolize galactose, while *Lb. delbrueckii* subsp. *bulgaricus* and most strains of *Strep. thermophilus* cannot metabolize galactose. This is a problem in cheese manufacture, since residual sugar can be metabolized heterofermentatively by other bacteria.

It is not known how lactose is transported in cells by *Leuconostoc* species or heterofermentative lactobacilli; however, lactose is known to be hydrolysed by β-galactosidase (Huang *et al.* 1995).

The galactose moiety is transformed into glucose-6-phosphate (Leloir pathway) and, together with glucose, is metabolized through the phosphoketolase pathway.

Lactic acid and ethanol, respectively, are formed during this metabolism to regenerate NAD^+; however, where lactococci are fermenting galactose or lactose at growth-limiting rates, products other than lactic acid can be formed from pyruvate. The enzyme pyruvate formate lyase is able to convert pyruvate to formate, acetate, acetaldehyde and ethanol under anaerobic conditions and at high pH (>7.0). Under aerobic conditions and at pH 5.5–6.5, pyruvate can be converted to acetate, acetaldehyde, ethanol and the minor products acetoin, diacetyl and 2,3-butanediol via the multienzyme pyruvate dehydrogenase complex.

Citrate metabolism

Citrate metabolism in LAB has been reviewed by Hugenholtz (1993). Milk contains 0.15–0.2% citric acid, but not all LAB can metabolize it. However, *Leuconostoc* species, Cit^+ *Lb. lactis* subsp. *lactis* and facultative heterofermentative lactobacilli do metabolize citric acid (Palles *et al.* 1998).

Many LAB use citrate as a substrate for cometabolism with sugars like glucose, fructose, lactose or xylose, providing NADH (citrate + 2 [H]/lactate + acetate + CO_2) (Hache *et al.* 1999) not directly as an electron acceptor, but as a precursor of acetate and oxaloacetate, which will be the final electron acceptor after being decarboxilated. Citrate metabolism is important in *Lc. lactis* and *Ln. mesenteroides* strains, which are often used in the dairy industry.

The latter organism was called *Streptococcus diacetylactis* in the old literature and more recently *Lc. lactis* subsp. *lactis* biovar *diacetylactis*. This name has no taxonomic status and the correct way to refer to it is citrate-utilizing (Cit^+) *Lc. lactis* subsp. *lactis*. Cit^+ strains of *Lc. lactis* differ from non-citrate-utilizing (Cit^-) strains because they contain a plasmid that encodes the transport of citrate. *Leuconostoc* species and Cit^+ *Lc. lactis* subsp. *lactis* strains utilize citric acid and lactose simultaneously and under certain conditions can derive energy via metabolism of citric acid.

Citric acid is transported into the cell by a citric acid permease, which is plasmid encoded in lactococci and *Leuconostoc* (Vaughan *et al.* 1995), and metabolized to pyruvic acid without generation of NADH. The result is an excess of pyruvic acid, which can be used to produce lactic acid to regenerate NAD^+, or in other reactions that regenerate NAD^+ and/or $NADP^+$.

The enzymes involved in these reactions are inducible and their expression is influenced by sugar concentrations and pH; in fact, a low amount of sugar and low pH favour diacetyl/acetoin formation.

Historically, there was a debate on which pathway was the most important. Evidence now clearly prefers the route via α-acetolactate, since α-acetolactate can be detected as an intermediate in cultures producing diacetyl and an α-acetolactate synthase has been identified in several LAB (Hugenholtz 1993).

Diacetyl contributes to typical yoghurt flavours and is produced by chemical decomposition of α-acetolactate (non-enzymatic). This reaction is favoured by aeration and low pH. Acetoin and/or 2,3-butanediol is produced in much larger amounts than diacetyl, but does not contribute to the aroma (Marshall 1987). Hugenholtz (1993) describes the use of genetic engineering to construct strains of lactococci able to produce high levels of diacetyl.

Nitrogen metabolism

Nitrogen metabolism by starters has an enormous impact on their activity and on cheese quality. LAB are fastidious microorganisms and are unable to synthesize many amino acids, vitamins and nucleic acid bases. Depending on the species and the strain, LAB require from 6 to 14 different amino acids (Kunji *et al.* 1996).

The proteolytic system of LAB is very complex and consists of three major components: a cell-wall bound proteinase that promotes extracellular casein degradation into oligopeptides, then peptide transporters that move peptides into the cytoplasm, where finally there are various intracellular peptidases that degrade peptides into smaller molecules and amino acids (Liu *et al.* 2010).

Proteolysis is a major event in cheese ripening: the proteolytic system of primary starter and secondary microflora contributes to the production of hundreds of flavour compounds through the synthesis of low-molecular-weight peptides and amino acids and their subsequent catabolism.

Free amino acids and peptides in cheese can contribute to flavour either directly or indirectly and with positive or negative effects. Cheese flavour development has been the subject of a comprehensive review (Smit *et al.* 2005). A major negative effect of proteolytic products is bitterness, which is believed to be caused by hydrophobic peptides ranging in length from 3 to 27 amino residues (Lemieux and Simard 1992). These peptides are believed to be generated from casein principally by the joint action of chymosin and LAB proteinases (Broadbent *et al.* 1998) and can be hydrolysed to non-bitter peptides and amino acids by LAB peptidases. In particular, the enzymatic degradation of proteins (caseins) leads to the formation of key flavour components, which contribute to the sensory perception of dairy products.

LAB can catalyse reactions such as deamination, transamination and decarboxylation, and metabolism of their amino acids also contributes to the flavour. As an example, same strains of importance in bakery production convert glutamine to glutamate during sourdough fermentation, imparting taste to the bread (Gänzle *et al.* 2007). The expression of the arginine deaminase pathway in *Lactobacillus* spp. promotes higher production of ornithine, and thus enhances the formation of 2-acetyl pyrroline, which is responsible for the roasty note of wheat bread crumb (Gänzle *et al.* 2007).

The proteolytic activity is also important for other mechanisms; several antihypertensive peptides produced during milk fermentation have a strong activity against angiotensin I-converting enzyme (ACE), a dipeptidyl carboxypeptidase that plays a major role in the regulation of blood pressure within the renine angiotensin system (Riordan 2003), inducing blood pressure increase. *In vivo* studies evidenced a reduction of blood pressure after consumption of fermented milks (Pina and Roque 2008). Moreover, *in vitro* ACE inhibitory (ACEI) activity of different traditional fermented milks has been reported in the literature (Chaves-López *et al.* 2011). Thus, selection of microorganisms to be used in fermented products is gaining in importance, due to the inherent variations in their ability to produce bioactive peptides, particularly those with specific health claims (Ramchandran and Shan 2008).

Recently, LAB-induced proteolysis has been suggested as an efficient method for decreasing the toxicity of wheat and rye flours. Gliadins are among the most affected proteins by food fermentation and the extent of hydrolysis of monomeric gliadins (α-, β-, γ-, ω-gliadins) is strain specific (Di Cagno *et al.* 2002). Di Cagno *et al.* (2002) showed that selected proteolytic LAB could efficiently hydrolyse the 31-43 fragment of the toxic peptide A-gliadin. On the basis of these results, the same authors showed that selected LAB could completely hydrolyse the highly toxic 33-mer peptide over prolonged (12–24 h) and semi-liquid fermentation of a mixture of wheat and non-toxic flours. Breads produced with 12-hour sourdough fermentation retained acceptable quality and when consumed by coeliac individuals, no alterations in the baseline values could be observed. The selected LAB were also successfully used for the detoxification of other fermented foods (De Angelis *et al.* 2006).

A variety of fermented foods, especially protein-rich foods, may contain biogenic amines (BAs). During the fermentation process protein breakdown products, peptides and amino acids, used by spoilage and also by the fermentation microorganisms, represent precursors for BA formation (Bodmer *et al.* 1999). The consumption of foods with high concentrations of BAs can induce adverse reactions such as nausea, headaches, rashes and changes in blood pressure (Ladero *et al.* 2010). Microorganisms suitable for food fermentation have been examined with regard to their potential to degrade histamine and tyramine (Fadda *et al.* 2001). A low potential for histamine and tyramine degradation among lactobacilli was noticed. In 35 well-known species with a practical function for the fermentation of dairy products and wine, Straub *et al.* (1995) observed a potential to form BAs only for a few strains.

Lipases and esterases

The lipolytic and esterolytic systems of LAB remain poorly characterized. Esterases from lactic acid bacteria may be involved in the development of fruity flavours in foods, and pregastric lipase and esterases are essential for the development of taste perception and typical flavour in Italian cheese. Microbial lipases and esterases may improve quality or accelerate the maturation of cheeses, cured bacon and fermented sausages. However, except for Parmigiano Reggiano, Pecorino and related Italian cheeses and blue cheeses, limited lipolysis occurs in cheese during ripening.

Lipolysis results in the formation of free fatty acids, which can be precursors of flavour compounds such as methylketones, secondary alcohols, esters and lactones. Generally, the role of LAB in lipolysis is less significant, but additional cultures, such as moulds in the case of surface-ripened cheeses, are often highly active in fat conversion. Flavours derived from the conversion of fat are particularly important in soft cheeses like Camembert and Roquefort (Smit *et al.* 2005).

Lipases are chemically defined as glycerol ester hydrolases (EC 3.1.1.3) that hydrolyse tri-, di- and monoglycerides present at an oil–water interface. Esterases (EC 3.1.1.6) hydrolyse esters in solution and may also hydrolyse tri- and especially di- and monoglycerides containing short-chain fatty acids (Medina *et al.* 2004). Esterases have been purified from several starter and LAB, including *Lc. lactis* (Chich *et al.* 1997), *Strep. thermophilus* (Liu *et al.* 2001) and *Lb. plantarum* (Gobbetti *et al.* 1997). All of them are serine enzymes that preferentially hydrolyse butyrate esters and are optimally active at pH 7. Some of them have no activity at pH 5.0; nevertheless, a very small amount of activity over a long time could result in significant hydrolysis of fat during cheese ripening. The major tributyrin esterase of *Lc. lactis* has been cloned, overexpressed and characterized (Fernandez *et al.* 2000).

Some probiotic strains of LAB can hydrolyse triglycerides, releasing most short- and medium-chain and essential fatty acids, which are valuable to today's health-conscious consumer. Medium-chain fatty acids (C6-C14), in particular, have become an accepted treatment for patients with malabsorption symptoms, a variety of metabolic disorders, cholesterol problems and infant malnutrition. These probiotic bacteria could alleviate lipase deficiency in the digestive tract during digestion (Medina *et al.* 2004).

Bacteriocins production

Bacteriocins are peptides produced by various bacteria that inhibit the growth of other bacteria. They could ensure the stability of fermented products, reduce microbial contamination during fermentation, inhibit the growth of moulds and prolong the microbiological spoilage time of baked goods (Juodeikiene *et al.* 2009).

In recent years, interest in starter/probiotic LAB has also grown substantially due to their potential usefulness as a natural substitute for food preservatives in the production of fermented foods with an enhanced shelf life and/or safety. *Lactobacillus* and *Lactococcus* include main strains with probiotic activity (Fuller 1989), producing bacteriocins (Altuntas *et al.* 2010).

The inhibitory host range and the molecular mass can be either large or small. Bacteriocins produced by LAB are divided into three classes: lantibiotics, small heat-stable non-lantibiotics and large heat-stable bacteriocins (Nes *et al.* 1996). Nisin, the best-known bacteriocin, is a lantibiotic that is produced by some strains of *Lc. lactis* and is used commercially in more than 50 countries as a food preservative to control the growth of spoilage and pathogenic bacteria.

Homofermentative *Pediococcus acidilactici* were isolated from spontaneous rye sourdoughs and characterized as producing pediocin Ac807 with antimicrobial activity against *Bacillus subtilis* (Narbutaite *et al.* 2008).

Exopolysaccharide production

Many food-grade microorganisms produce exopolysaccharides (EPS) (De Vuyst and Degeest 1999). EPS act as biothickeners and can be added to a variety of food products, where they serve as viscosifying, stabilizing, emulsifying or gelling agents (Tieking and Gänzle 2005).

They are divided into two classes: homopolysaccharides (HoPS), mainly glucan or fructans polymers; and heteropolysaccharides, with (ir)regular repeating units (De Vuyst and Degeest 1999). Heteropolysaccharide production is an important characteristic of many LAB involved in the production of fermented milks.

Lactic acid bacteria produce either homopolysaccharides, containing fructose or glucose residue, or heteropolysaccharides. composed of repeating units of several different sugars including glucose, galactose, fructose and rhamnose (De Vuyst *et al.* 2001). They may be involved in a wide variety of biological functions, including prevention of desiccation, protection from environmental stresses, adherence to different surfaces, pathogenesis and symbioses (Jolly *et al.* 2002). EPS-producing cultures have also been used to increase the moisture and improve the yield of low-fat Mozzarella cheese (Perry *et al.* 1998).

Glucan and fructans produced by fermenting LAB can strongly influence the quality of wheat bread in terms of bread volume and crumb firmness (Di Cagno *et al.* 2006). In particular, the production of EPS *in situ* is more effective than their addition (Brandt *et al.* 2003).

LAB can also produce gluco- or fructo-oligosaccharides (FOS), among which FOS, together with the fructan inulin, have been well described for their prebiotic effects (Biedrzycka and Bielecka 2004). In addition, the levan produced by *Lactobacillus sanfranciscensis* was proved to stimulate bifidobacterial growth *in vitro* (Dal Bello *et al.* 2001). In sourdough, *Lactobacillus reuteri, Lactobacillus acidophilus* and *Lb. sanfranciscensis* showed the ability to produce the prebiotic FOS 1-kestose (Tieking and Gänzle 2005).

Conclusion

The use of industrial starters has reduced the biodiversity and the organoleptic properties of fermented products. This phenomenon may be explained because the commercial availability of new, interesting starter cultures is very limited. Therefore, the selection of promising and wild strains from raw materials could be an interesting way forward. We can suggest at least three hot topics in selecting new LAB cultures: genome sequencing; interaction with natural microbiota; and functionality (Figure 1.2).

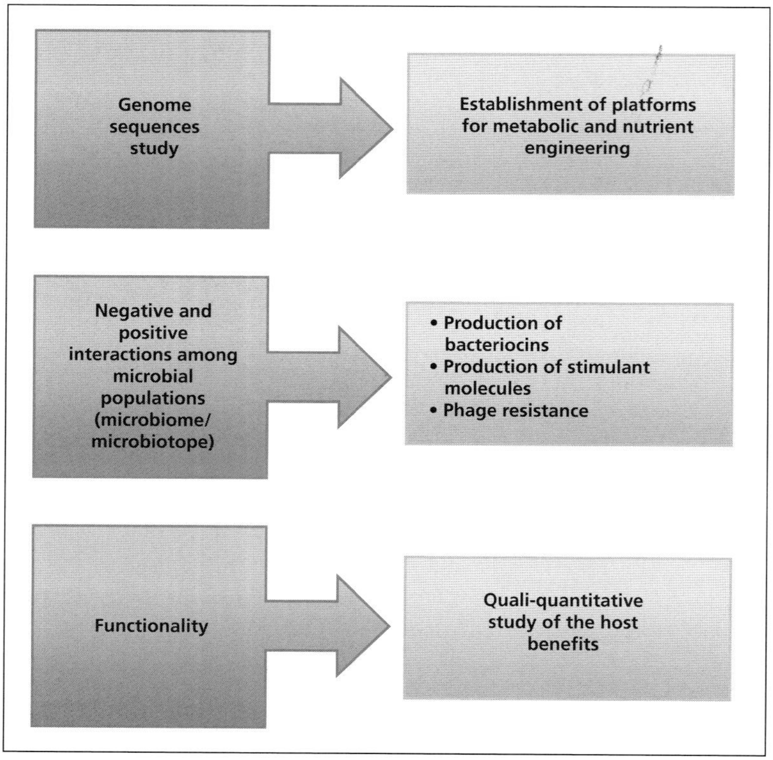

Figure 1.2 A new approach in the selection of microorganisms for innovative food purposes.

References

Altuntas, E.G., Cosansu, S. and Ayhan, K. (2010) Some growth parameters and antimicrobial activity of a bacteriocin-producing strain *Pediococcus acidilactici* 13. *International Journal of Food Microbiology*, **141**, 28–31.

Biedrzycka, E. and Bielecka, M. (2004) Prebiotic effectiveness of fructans of different degrees of polymerization. *Trends in Food Science and Technology*, **15**, 170–175.

Blandino, A., Al-Aseeri, M.E., Pandiella, S., Cantero, D. and Webb, C. (2003) Cereal based fermented foods and beverages. *Food Research International*, **36**, 527–543.

Bodmer, S., Imark, C. and Kneubűűhl, M. (1999) Biogenic amines in foods: Histamine and food processing. *Inflammation Research*, **48**, 296–300.

Brandt, M.J., Roth, K. and Hammes, W.P. (2003) Effect of an exopolysaccharides produced by *Lactobacillus sanfranciscensis* LHT 1729 on dough and bread quality. In L. De Vuyst (ed.), *Sourdough from Fundamentals to Application* (p. 80), Brussels: Vrije Universiteit Brussels IMDO.

Broadbent, J.R., Strickland, M., Weimer, B.C., Johnson, M.E. and Steele, J.L. (1998) Small peptide accumulation and bitterness in Cheddar cheese made from single strain *Lactococcus lactis* starters with distinct proteinase specific cities. *Journal of Dairy Science*, **81**, 327–337.

Carminati, D., Giraffa, G., Quiberoni, A., Binetti, A., Suarez, V., and Reinhemer, J. (2010) Advances and trends in starter culture for dairy fermentation. In F. Mozzi, R.R. Raya and G.M. Vignolo (eds), *Biotechnology of Lactic Acid Bacteria: Novel Applications* (pp. 177–192), Oxford: Blackwell.

Chaves-López, C., Serio, A., Martuscelli, M., Paparella, A., Osorio-Cadavid, E. and Suzzi, G. (2011) Microbiological characteristics of kumis, a traditional fermented Colombian milk, with particular emphasis on enterococci population. *Food Microbiology*, **28**, 1041–1047.

Chich, J.E, Marchesseau, K. and Gripon, J.C. (1997) Intracellular esterase from *Lactococcus lactis* subsp, *lactis* NDCO 763: Purification and characterization. *International Dairy Journal*, **7**, 169–174.

Dal Bello, F., Walter, J., Hertel, C. and Hammes, W.P. (2001) In vitro study of prebiotic properties of levan-type exopolysaccharides from lactobacilli and non-digestible carbohydrates using denaturing gradient gel electrophoresis. *Systematic and Applied Microbiology*, **24**, 232–237.

De Angelis, M., Coda, R., Silano, M. *et al.* (2006) Fermentation by selected sourdough lactic acid bacteria to decrease coeliac intolerance to rye flour. *Journal of Cereal Science*, **43**, 301–314.

De Vuyst, L. and Degeest, B. (1999) Heteropolysaccharides from lactic acid bacteria. *FEMS Microbiology Reviews*, **23**, 153–177.

De Vuyst, L., de Vin, F., Vaningelgem, E. and Degeest, B. (2001) Recent developments in the biosynthesis and applications of heteropolysaccharides from lactic acid bacteria. *International Dairy Journal*, **11**, 687–707.

Di Cagno, R., De Angelis, M., Lavermicocca, P. *et al.* (2002) Proteolysis by sourdough lactic acid bacteria: Effects on wheat flour protein fractions and gliadin peptides involved in human cereal intolerance. *Applied and Environmental Microbiology*, **68**, 623–633.

Di Cagno, R., De Angelis, M., Limitone, A. *et al.* (2006) Glucan and fructan production by sourdough *Weissella cibaria* and *Lactobacillus plantarum*. *Journal of Agricultural and Food Chemistry*, **54**, 9873–9881.

Fadda, S., Vignolo, G. and Oliver, G. (2001) Tyramine degradation and tyramine/histamine production by lactic acid bacteria and *Kocuria* strains. *Biotechnology Letters*, **23**, 2015–2019.

Fernandez, L., Beerthuyzen, M.M., Brown, J. *et al.* (2000) Cloning, characterization, controlled overexpression and inactivation of the major tributyrin exterase gene of *Lactococcus lactis*. *Applied and Environmental Microbiology*, **66**, 1360–1368.

Fuller, R. (1989) Probiotics in man and animals. *Journal of Applied Bacteriology*, **66**, 365–378.

Gänzle, M.G., Vermeulen, N. and Vogel, R.F. (2007) Carbohydrate, peptide and lipid metabolism of lactic acid bacteria in sourdough. *Food Microbiology*, **24**, 128–138.

Gobbetti, M., Fox, R.E. and Stepaniak, L. (1997) Isolation and characterization of tributyrin esterase from *Lactobacillus plantarum* 2739. *Journal of Dairy Science*, **80**, 3099–3106.

Hache, C., Cachon, R., Wache, Y. *et al.* (1999) Influence of lactose-citrate co-metabolism on the differences of growth and energetics in *Leuconostoc lactis, Leuconostoc mesenteroides* spp. *Mesenteroides* and *Leuconostoc mesenteroides* ssp. *cremoris*. *Systematic and Applied Microbiology*, **22**, 507–513.

Holzapfel, W. (1995) Use of starter cultures in fermentation on a household scale. *Food Control*, **8**, 241–258.

Huang, D.Q., Prevost, H. and Divies C. (1995) Principal characteristics of β-galactosidase from *Leuconostoc spp*. *International Dairy Journal*, **5**, 29–43.

Hugenholtz, J. (1993) Citrate metabolism in lactic acid bacteria. *FEMS Microbiology Reviews*, **12**, 165–178.

Jolly, L.L., Vincent, S.J.E, Duboc, P. and Neeser, J.R. (2002) Exploiting exopolysaccharides from lactic acid bacteria. *Antonie van Leeuwenhoek*, **82**, 367–374.

Juodeikiene, G., Šalomskienė, J., Basinskiene, L., Vidmantiene, D., Narbutaite, V. and Kasnauskyte, N. (2009) The influence of novel fermented products on wheat bread spoilage and staling. *Food Chemistry and Technology*, **43**, 36–46.

Kunji, E.R.S., Mierau, I., Hagting, A., Poolman, B. and Konings, W.N. (1996) The proteolytic system of lactic acid bacteria. *Antonie van Leeuwenhoek*, **70**, 187–221.

Ladero, V., Calles-Enríquez, M., Fernández, M. and Alvarez, M.A. (2010) Toxicological effects of dietary biogenic amines. *Current Nutrition and Food Science*, **6**, 145–156.

Lemieux, L. and Simard, R.E. (1992) Bitter flavour in dairy products. II. A review of bitter peptides from caseins: Their formation, isolation and identification, structure masking and inhibition. *Lait*, **72**, 335–382.

Leroy, F. and De Vuyst, L. (2004) Lactic acid bacteria as functional starter cultures for the food fermentation industry. *Trends in Food Science and Technology*, **15**(2), 67–78.

Limsowtin, G.K.Y., Powell, I.B. and Parente, E. (1996) Types of starters. In T.M. Cogan and J.E. Accolas (eds), *Dairy Starter Cultures* (pp. 101–129), New York: VCH.

Liu, M., Bayjanov, J.R., Renckens, B., Nauta, A. and Siezen, R.J. (2010) The proteolytic system of lactic acid bacteria revisited: A genomic comparison. *BMC Genomics*, **11**, 36.

Liu, S.Q., Holland, R. and Crow, V.L. (2001) Purification and properties of intracellular esterases from *Streptococcus thermophilus*. *International Dairy Journal*, **lI**, 27–35.

Lortal, S. and Chapot-Chartier, M.-P. (2005) Role, mechanisms and control of lactic acid bacteria lysis in cheese. *International Dairy Journal*, **15**, 857–871.

Marshall, V.M. (1987) Lactic acid bacteria: Starters for flavour. *FEMS Microbiology Reviews*, **46**, 327–336.

Medina, R.B., Katz, M.B., González, S. and Oliver, G. (2004) Determination of esterolytic and lipolytic activities of lactic acid bacteria. *Methods in Molecular Biology*, **268**, 465–470.

Narbutaite, V., Fernandez, A., Horn, N., Juodeikiene, G. and Narbad, A. (2008) Influence of baking enzymes on antimicrobial activity of five bacteriocin-like inhibitory substances produced by lactic acid bacteria isolated from Lithuanian sourdoughs. *Letters in Applied Microbiology*, **47**, 555–560.

Nes, I.E, Diep, D.B., Havarstein, L.S., Brurberg, M.B., Eijsink, V. and Holo, H. (1996) Biosynthesis of bacteriocins in lactic acid bacteria. *Antonie van Leeuwenhoek*, **70**, 113–128.

Palles, T., Beresford, T., Condon, S. and Cogan, T.M. (1998) Citrate metabolism in *Lactobacillus casei* and *Lactobacillus plantarum*. *Journal of Applied Microbiology*, **85**, 147–154.

Parente, E. and Cogan, T.M. (2004) Starter cultures: General aspects. In P.F. Fox, P.L.H. McSweeney, T.M. Cogan and T.P. Guinee (eds), *Cheese: Chemistry, Physics and Microbiology* (3rd edn, pp. 123–148), London: Elsevier.

Perry, D.B., McMahon, D.J. and Oberg, C.J. (1998) Manufacture of low fat Mozzarella cheese using exopolysaccharide-producing cultures. *Journal of Dairy Science*, **81**, 563–566.

Pina, A.S. and Roque, A.C. (2008) Studies on the molecular recognition between bioactive peptides and angiotensin-converting enzyme. *Journal of Molecular Recognition*, **22**, 162–168.

Poolman, B. (1993) Energy transduction in lactic acid bacteria. *FEMS Microbiology Reviews*, **12**, 125–148.

Powell, I.B., Broome, M.C. and Limsowtin, G.K.Y. (2011) Cheese: Starter cultures: Specific properties. In J.W. Fuquay, P.F. Fox and P.L.H. McSweeney (eds), *Encyclopedia of Dairy Sciences* (pp. 559–566), Amsterdam: Elsevier Academic.

Ramchandran, L. and Shan, N.P. (2008) Proteolytic profiles and angiotensin-I converting enzyme and a-glucosidase inhibitory activities of selected lactic acid bacteria. *Journal of Food Science*, **73**(2), M75–M81.

Riordan, J. (2003) Angiotensin-I-converting enzyme and its relatives. *Genome Biology*, **4**, 225.

Saarela, M., Mogensen, G., Fondén, R., Mättö, J. and Mattila-Sandholm, T. (2000) Probiotic bacteria: Safety, functional and technological properties. *Journal of Biotechnology*, **84**(3), 197–215.

Smit, G., Smit, B.A. and Engels W.J.M. (2005) Flavour formation by lactic acid bacteria and biochemical flavour profiling of cheese products. *FEMS Microbiology Reviews*, **29**, 591–610.

Stadhouders, J. and Leenders, G.J.M. (1984) Spontaneously developed mixed-strain cheese starters: Their behaviour toward phages and their use in the Dutch cheese industry. *Netherland Milk and Dairy Journal*, **38**, 157–181.

Straub, B., Kicherer, M., Schilcher, S. and Hammes, W.P. (1995) The formation of biogenic amines by fermentation organisms. *Zeitschrift für Lebensmittel-Untersuchung und Forschung*, **201**, 79–82.

Tieking, M. and Gänzle, M.G. (2005) Exopolysaccharides from cereal-associated lactobacilli. *Trends in Food Science and Technology*, **16**, 79–84.

Vaughan, E., David, S., Harrington, A., Daly, C., Fitzgerald, G.F. and De Vos, W.M. (1995) Characterization of plasmid-encoded citrate permease (citP) genes from *Leuconostoc* species reveals high sequence conservation with the *Lactococcus lactis* citP gene. *Applied and Environmental Microbiology*, **61**, 3172–3176.

CHAPTER 2
Yeasts as starter cultures

Pietro Buzzini, Simone Di Mauro and Benedetta Turchetti
Department of Agricultural, Food and Environmental Science, Industrial Yeasts Collection DBVPG, University of Perugia, Italy

Together with drying and salting, fermentation is one of the oldest ways to preserve perishable foods and beverages, dating back at least 6000 years (McGovern *et al.* 2004; Sicard and Legras 2011). Nowadays, the importance of fermented products for consumers is underlined by the broad variety of fermented foods and beverages marketed in both developing and industrialized countries, not only for their indisputable benefit of preservation and safety, but also for their highly appreciated sensory attributes. Microorganisms (and their enzymes) contribute to the improvement of some characteristic properties such as taste, aroma, visual appearance, texture, shelf-life and safety (Holzapfel 2002).

The need for inocula for starting the fermentative process was understood early and applied from time immemorial by keeping a sample (sometimes labelled a 'natural culture') from the previous production and using it as a starter. With the discovery of microorganisms, it became possible to improve fermented products by using well-characterized starter cultures. This became routine in the nineteenth century for producing wine, beer, vinegar and bread. In contrast, the dairy and meat industries began to use well-characterized starter cultures only about a century later (Hansen 2002; Holzapfel 2002).

A starter culture may be defined as a preparation containing a large number of (sometimes variable) technological microorganisms, which is inoculated to accelerate and guide a given fermentative process. A typical starter facilitates the control, improvement and predictability of fermentation only if it is well adapted to the substrate (Holzapfel 2002). Food technologists can currently choose either to purchase the starter culture in a ready-to-use and highly concentrated form or to propagate the culture in-house. The preference for one or other of the two methods is currently influenced by the type of fermented product to be obtained; the presence of in-house microbiological expertise and equipment facilities; and the economic impact. Overall, the highest level of safety and flexibility

---

Starter Cultures in Food Production, First Edition. Edited by Barbara Speranza, Antonio Bevilacqua, Maria Rosaria Corbo and Milena Sinigaglia.

is achieved by using commercial starter cultures for direct inoculation. Such starters are usually supplied as dried (or freeze-dried), highly concentrated and active cultures in order to be easily used to inoculate the substrate (Hansen 2002).

Yeasts as starter cultures: General considerations

Although ancient peoples unknowingly used yeasts since antiquity for producing fermented foods and beverages, the awareness of the ability of these microorganisms to convert carbohydrates into ethanol and carbon dioxide (CO_2) dates back to experiments carried out by Louis Pasteur in 1860 (Sicard and Legras 2011). Yeasts are a group of eukaryotic unicellular organisms belonging to the kingdom of fungi and behave in nature as saprotrophs and degraders of organic macromolecules. They are currently used in fermentative processes, mainly because of their ability to utilize a broad variety of feedstock and to produce a number of valuable fermented foodstuffs (Tamang and Fleet 2009; Sicard and Legras 2011).

It has been suggested that the species belonging to the *Saccharomyces sensu stricto* complex (including *Saccharomyces cerevisiae*, commonly labelled 'baker's yeast') were the first example of organisms domesticated by humankind (Sicard and Legras 2011). Accordingly, most people associate yeasts almost exclusively with *Saccharomyces* species. In fact, it is not uncommon in some areas of microbiology, molecular biology and biotechnology to utilize the words 'yeast' and '*Saccharomyces*' as synonyms and to use the species *S. cerevisiae* as the primary model for studying the biology of eukaryotic organisms. This is in spite of evidence that this species represents only an infinitesimal part of the biodiversity existing in the yeast world (Buzzini and Vaughan-Martini 2006). It has been estimated that the number of yeast species so far described (approximately 1500) represents about 1% of the total predictable diversity (Boekhout 2005). Thus, there is enormous potential in studying new yeast species for their possible commercial use. Indeed, an increasing body of academic and industrial research has recently paid attention to several non-*Saccharomyces* species, mainly belonging to the genera *Candida*, *Debaryomyces*, *Kluyveromyces*, *Yarrowia*, *Pichia*, *Zygosaccharomyces* and so on, for possible exploitation as starter cultures for both food and non-food (industrial) technologies (Fleet 2006; Buzzini and Vaughan-Martini 2006; Romano *et al.* 2006).

Yeasts as starter cultures in winemaking

Starter cultures of *S. cerevisiae*

The first evidence of winemaking dates back to 5000 BCE in Mesopotamia and Greece (Bisson *et al.* 2002; Valamoti *et al.* 2007; Legras *et al.* 2007; Sicard and Legras 2011). Grape juice fermentation is a complex biochemical process wherein yeasts play a fundamental role by converting carbohydrates into ethanol, CO_2 and several hundreds of secondary products, sometimes characterized by high volatility (Ciani *et al.* 2010). For many years, wines have been produced by spontaneous fermentation resulting from the competitive activities of a variety of contaminating indigenous yeasts (labelled 'wild yeasts') of the species *Hanseniaspora uvarum* (teleomorph state of *Kloeckera apiculata*), *Torulaspora delbrueckii*, *Pichia* spp., *Candida* spp. and so on. These indigenous yeasts usually dominate the mature grape yeast populations and, despite their inability to achieve complete fermentation, enhance the wine's aroma and flavour during the early stages of the winemaking process. The presence of alcohol-tolerant *S. cerevisiae* strains increases proportionally to the ethanol concentration during the mid to final phases of fermentation at the expense of the indigenous yeasts (Fleet 1999, 2003; Pretorius 2000; Holzapfel 2002; Calabretti *et al.* 2012). The number of indigenous species and their presence during the early phases of fermentation depends on several factors. This consequently determines much of the variation of wine quality from region to region, but also from one year to another (Pretorius 2000).

There is a general assumption that the inoculation of grape must with yeast starter cultures can overwhelm and suppress the growth of indigenous strains and dominate the fermentative process, thus improving the general quality of the wine. This theory has addressed the research of nearly a century into so-called super-selected yeast. Because of their dominance, strains of *S. cerevisiae* have been historically isolated, selected and commercialized for decades as starter cultures for winemaking. Many companies selling yeasts for the food and beverage industries were started in the last 50 years. Some of them conserve an in-house collection of strains, which are regularly subjected to periodic screening surveys for selecting specific starter cultures. Other companies, however, are merely 'sellers' of strains that have been isolated and selected by distinct microbiology laboratories or service culture collections. There is a list of major worldwide companies selling yeast starter cultures in Table 2.1.

The idea that inoculated fermentations can proceed more rapidly and predictably than their spontaneous counterparts is a universally recognized concept. Consequently, yeast starters are regularly utilized by many winemakers

Table 2.1 Major worldwide companies selling yeast starter cultures.

Name	Country	Website	Starters for
Angel Yeast	China	www.angelyeast.com	Beer, bakery products
Lallemand	Canada	www.lallemand.com	Wine, beer, bakery products
Lesaffre/Fermentis	France	www.lesaffre.com www.fermentis.com	Wine, beer, spirits, bakery products
White Labs	USA	www.whitelabs.com	Wine, beer, spirits

worldwide (Calabretti *et al.* 2012). However, molecular ecological studies have now reported that these assumptions are not necessarily correct. Indeed, the indigenous yeasts present in grape and must sometimes continue to contribute to fermentation (Fleet 1999). In order to monitor this phenomenon, a few molecular methods (i.e. mtDNA restriction analysis and comparison of chromosomal DNA profiles) have been proposed to check whether or not fermentation is successfully conducted by the inoculated starter yeasts (Torija *et al.* 2001).

In recent years wine technologists and winemakers have increasingly focused their interest on the use of autochthonous *S. cerevisiae* strains, with the aim of selecting starter cultures better adapted to a specific grape must in order to try to reflect the biodiversity of a given region. This approach is supported by the hypothesis that specific native strains can be associated with a given territory, or even with a particular winery (Torija *et al.* 2001; Lopes *et al.* 2002; Capece *et al.* 2010; Settanni *et al.* 2012). The recent discovery that an overabundance of *S. cerevisiae* living cells is present on the surfaces of wineries has made available a large reservoir of yeast diversity to be used as a source of locally selected starters for winemaking. A few studies have postulated that any winery potentially hosts a local, resident population of *S. cerevisiae* strains, which are technologically optimized for winemaking and adapted to produce a set of peculiar compounds possibly involved in the formation of (sometimes individual) aromas. The logical consequence is that any winery may potentially contain its own 'super-selected' starter producing personalized sensory characteristics (Martini 2003). This approach has also proven to be very effective for selecting commercial 'winery-specific' strains, which are ideal for the production of typical regional wines. Accordingly, a number of researchers have recently characterized *S. cerevisiae* cultures isolated from worldwide wine cellars (Domizio *et al.* 2007; Lopes *et al.* 2007; Valero *et al.* 2007; Capece *et al.* 2010; Settanni *et al.* 2012; Tristezza *et al.* 2012; Mazzei *et al.* 2013; Elmacı *et al.* 2014).

Conventionally, the selection of *S. cerevisiae* starters for winemaking has mainly been approached by using two oenological traits (Martini 2003): primary characteristics, defined as those strictly associated with the formation of ethanol by fermentation; and secondary qualities, related to the production of compounds affecting other parameters, namely the body of a wine (e.g. glycerol), the higher alcohols complex (bouquet) and the appearance of either desirable flavours or undesirable off-flavours. Large-scale screening surveys are still ongoing worldwide particularly aimed at finding the optimal starter for specific wines (often of great value) for both traditional and modern cellars. Wines obtained from different starters have been evaluated for their chemical composition and sensory characteristics (Pretorius 2000; Pretorius *et al.* 2003; Dequin 2001; Bisson 2004; Borneman *et al.* 2007). Advances have been made in yeast fermentation vigour and complete utilization of carbohydrates, and in wine processing (including clarification) and enhanced formation of desired aromas, which is a complex and important aspect of wine quality because the physiology and neurobiology of human olfaction and the assessment of the desired sensory properties have significant impacts on the desirability and economics of wine (Bisson *et al.* 2002). The decrease of possible off-flavours (to enhance the organoleptic qualities of wines) has also been targeted as an additional selection criterion (Pretorius *et al.* 2003; Bisson 2004; Borneman *et al.* 2007).

A number of additional challenges have been addressed in recent years (Moreno-Arribas and Polo 2005). Among them the possible use of starter cultures at low temperatures is worthy of note. It is well known that fermentative processes performed at temperatures below 15 °C lead to more aromatic and paler wines (Bauer and Pretorius 2000; Ribéreau-Gayon 2006). Low temperatures increase the duration of alcoholic fermentation, decrease the rate of yeast growth and modify the ecology of wine fermentation (Torija *et al.* 2003). The pre-adaptation of starter cultures of *S. cerevisiae* to cold conditions could improve fermentation performance, although this improvement is strain dependent. Low-temperature fermentations also determine the reduction of acetic acid and fusel alcohol production and increase the concentrations of glycerol (Llauradó *et al.* 2005). The technological and sensory characteristics of *S. cerevisiae* strains grown at low temperatures have recently been reviewed (Kanellaki *et al.* 2014).

The production of wines with a reduced concentration of ethanol and chemical preservatives represents an additional target for many wine cellars selling their product in developed nations, due to the growing consumer demand for wines containing lower levels of ethanol and chemical additives (labelled 'organic' wines). Both purposes have been pursued by using techniques of DNA mutation or recombination in starter cultures of *S. cerevisiae* (Johnson and Echavarri-Erasun 2011). The first target is related to the increased interest in healthy lifestyles linked to lowering excessive alcohol consumption, as well as concerns related to wine quality, because high alcohol concentrations exert a masking effect on the flavours and aromas of wine (Guth and Sies 2002). In this context, the use of low ethanol–producing yeasts may be considered a cheap opportunity (Rossouw *et al.* 2013). Genetic manipulation of *S. cerevisiae* strains for reducing their ability to accumulate ethanol has been supported by current literature on the regulatory mechanisms of yeast fermentative metabolism (Rossignol *et al.* 2003; Trabalzini *et al.* 2003; Varela *et al.* 2005; Howell *et al.* 2006; Zuzuarregui *et al.* 2006; Marks *et al.* 2008; Rossouw and Bauer 2009). Glycolytic genes are slowly down-regulated as fermentation progresses, with only a few exceptions where isoforms of the same protein are differentially expressed (Varela *et al.* 2005; Marks *et al.* 2008). Under glucose-repressed fermentative conditions, genes encoding the tricarboxylic acid cycle appear to be underexpressed during fermentative metabolism. Additional investigation concerning metabolic C fluxes under simulated fermentation conditions drew attention to discrepancies between these fluxes and the corresponding gene expression patterns (Varela *et al.* 2005). Malherbe *et al.* (2003) expressed the *Aspergillus niger* gene encoding a glucose oxidase in *S. cerevisiae* in order to obtain lower alcohol production and inhibition of spoilage bacteria. *S. cerevisiae* transformants exhibited slightly reduced alcohol production, probably as a consequence of the parallel production of gluconic acid from glucose by glucose oxidase. In contrast, Rossouw *et al.* (2013) screened a set of *S. cerevisiae* mutants exhibiting deletion of genes encoding enzymes involved in central carbohydrate metabolism (i.e. trehalose biosynthesis, central glycolysis, oxidative pentose phosphate pathway and tricarboxylic acid cycle) for their impact on ethanol yields. A TPS1 gene (encoding trehalose-6-phosphate synthase) was selected as a putative candidate to alter flux to ethanol during alcoholic fermentation. The expression of the TPS1 gene was slightly up-regulated, resulting in a decrease in ethanol production and an increase in trehalose biosynthesis. Additional advances in the selection of wine yeasts were realized through DNA technology in *S. cerevisiae* strains to improve a few stress properties, including osmotolerance and ethanol resistance (Johnson and Echavarri-Erasun 2011).

Mixed starters for co-fermentations

It is generally assumed that wine produced by using pure cultures of *S. cerevisiae* can sometimes lack the complexity of taste and sensory characteristics produced by indigenous yeasts in spontaneous fermentative processes. Since the early 2000s this has stimulated the 'rediscovery' of indigenous non-*Saccharomyces* yeasts (as co-starters in association with *S. cerevisiae*) for producing wines characterized by a high aroma content (Romano *et al.* 2003; Cheraiti *et al.* 2005; Calabretti *et al.* 2012). This fascinating topic has attracted the work of a growing number of wine microbiologists in order to study the impact of non-*Saccharomyces* yeasts on the composition, sensory properties and final flavours of wine. Indeed, it is known that the yeast ecology of the fermentative process is more complex than previously thought, and that some non-*Saccharomyces* yeasts can play a relevant role in the fermentation dynamics, metabolic impact and aroma complexity of the final product (Swiegers and Pretorius 2005; Domizio *et al.* 2007; Renouf *et al.* 2007; Fleet 2008; Ciani *et al.* 2010; Calabretti *et al.* 2012). The volatile compounds responsible for varietal aroma in wine are mainly terpenes, wherein free forms of monoterpenes are the most important group because of their high volatility. The glycosylated terpenes can be hydrolysed (by β-glycosidases produced by some non-*Saccharomyces* strains) to the corresponding free forms during the early phases of winemaking (Calabretti *et al.* 2012). Because of their growing importance, most studies recently proposed the use of mixed or sequential inoculation of *S. cerevisiae* and non-*Saccharomyces* strains as a feasible way for improving the complexity and enhancing some specific traits of wines (Romano *et al.* 2003; Clemente-Jimenez *et al.* 2005; Moreira *et al.* 2005, 2008; Rodríguez *et al.* 2010; Ciani *et al.* 2010; Clavijo *et al.* 2011; De Benedictis *et al.* 2011; Viana *et al.* 2011; Calabretti *et al.* 2012; Hong and Park 2013; del Mónaco *et al.* 2014).

Yeasts as starter cultures in brewing

Beer is one of the most widely consumed alcoholic beverages in the world. It was first mentioned in ancient Mesopotamian literature, which dates back to the seventh century BCE (Sicard and Legras 2011). Some studies revealed that the species *S. cerevisiae* (currently used to produce ale beer) includes strains displaying a multiple ploidy and consequently great genome variability (Pedersen 1986). The complete sequencing of the genome of *Saccharomyces pastorianus* (synonym *Saccharomyces carlsbergensis*) W34/70 (a strain largely used for European lager beer) has been deciphered, revealing that it is an allopolyploid interspecies hybrid between *S. cerevisiae* and *Saccharomyces bayanus* (Nakao *et al.* 2009). In contrast, it was recently proposed that other strains of *S. pastorianus* could be originated from an interspecific cross between *S. cerevisiae* and the wild species *Saccharomyces eubayanus*. The draft genome sequence of *S. eubayanus* appears to be 99.5% identical to the non-*S. cerevisiae* portion of the *S. pastorianus* genome sequence. This suggests specific changes in carbohydrate and sulfite metabolism, which were crucial for domestication in the lager brewing environment (Libkind *et al.* 2011).

Many strains of the two closely related species *S. cerevisiae* and *S. pastorianus* have been selected in the last 15 years and proposed for brewing. Targeted properties for starter selection included high-fermentation performances in normal and high-gravity worts; optimal formation of organic acids, volatile compounds, glycerol and other molecules important for beer quality; enhanced flocculation after primary fermentation to favour beer clarification; cell viability as a function of time, temperature and ethanol concentration during storage; and use of a continuous fermentative process utilizing immobilized *S. cerevisiae* cells (Dequin 2001; Verbelen *et al.* 2006; Willaert and Nedovic 2006; Blieck *et al.* 2007; Bleoanca *et al.* 2013).

During the past few decades a number of technological factors have been targeted for the genetic improvement of brewer's yeasts (Bamforth 2000; Dequin 2001). The construction of yeast strains able to secrete heterologous β-glucanases (from *Aspergillus* spp.) to decrease viscosity and to promote more efficient filtration has also been recently proposed (Johnson and Echavarri-Erasun 2011). A strong emphasis has also been devoted to the ability of brewer's yeasts to utilize dextrins, which represent about 25% of malt wort carbohydrates and have a high caloric impact on low-alcohol beers (Johnson and Echavarri-Erasun 2011).

The excess of formation of diacetyl and other vicinal ketones can be considered negative to the savoury properties of beers. Bacterial genes encoding the production of α-acetolactate decarboxylase to enhance the amount of acetoin and to decrease diacetyl have been expressed in brewer's yeasts (Bamforth and Kanauchi 2004). In addition, as altered concentrations of sulfur compounds, as well as other off-flavours (e.g. staling, cardboard flavour attributed to (*E*)-2-nonenal, undesirable aromas derived from lipid oxidation etc.) can be considered nasty by consumers, yeast strains have been developed exhibiting a modified sulfur metabolism and producing superior levels of sulfite with enhanced antioxidant and antibacterial properties (Vanderhaegen *et al.* 2006; Johnson and Echavarri-Erasun 2011).

Yeasts as starter cultures in bakery products

Historically, fermented cereals have played a significant role in human nutrition in all parts of the world where cereals grow (Hammes *et al.* 2005). The first report on bread making dates back to ancient Egypt (Sicard and Legras 2011). Generally, baker's yeast (*S. cerevisiae* is the most common species in bread making) is required to have several technological characteristics, namely a high carbohydrate fermentation rate, sometimes cryotolerance, and a high leavening ability to ensure high-quality baking products (Rollini *et al.* 2007; Wongkhalaung and Boonyaratanakornkit 2007; Giannone *et al.* 2010; Cukier de Aquino *et al.* 2012). Nowadays, the bakery industry offers several commercial starter cultures, the choice of which depends on the type of bread-making technology. Compressed yeast is the form most widely used, but dry yeasts are in successful expansion because they are easy to use, even if their production is time-consuming and they require additional energy costs due to both drying and packaging processes (Papapostolou *et al.* 2012).

Targeted properties for selecting *S. cerevisiae* starters for bakery products include tolerance of high levels of sucrose (doughs can contain up to 30% sucrose, which exerts severe osmotic stress on yeast cells); tolerance of freezing–thawing stress; rapid utilization of maltose; and production of high levels of CO_2 (Verstrepen *et al.* 2006). Like wine and beer yeasts, the use of DNA technology has allowed significant advances in the construction of improved starters for the bakery industry (Johnson and Echavarri-Erasun 2011). *S. cerevisiae* strains exhibiting high sucrose tolerance and rapid utilization of maltose have been proposed for commercial use (Higgins *et al.* 2001). Genes encoding sucrose tolerance and maltose utilization have been studied in depth by using a functional genomics approach. The expression of genes involved in the accumulation and metabolism of glycerol and trehalose, and in resistance to osmotic stress, was demonstrated to be higher in sucrose-tolerant yeasts (Tanaka-Tsuno *et al.* 2007).

The freezing–thawing survival of yeasts is an attribute that would benefit the production and quality of frozen doughs (Rosell and Gomez 2007). Accordingly, cryoresistant *S. cerevisiae* and *T. delbrueckii* strains have been developed (Tanghe *et al.* 2003; Hernández-López *et al.* 2007). The ability of *S. cerevisiae* strains to utilize melibiose is particularly important because raffinose (a prominent component of molasses, an ingredient currently used in some bakery products) is hydrolysed by yeast invertase to fructose and melibiose. The expression of heterologous genes (from *S. pastorianus*) encoding α-galactosidase in

S. cerevisiae increased biomass without alteration of growth rate in model bakery fermentations (Dequin 2001). Another desired property in bread making is a rapid fermentation rate, sometimes related to maltose concentration. Dough amylases release maltose from starch, but many strains of *S. cerevisiae* utilize maltose poorly, primarily due to repression of maltose utilization by other sugars through catabolite repression. The possibility of overcoming this bottleneck could allow improvements in baking productivity (Johnson and Echavarri-Erasun 2011).

Sourdoughs, consisting of a mixed culture of yeasts and lactic acid bacteria, are alternative starter cultures frequently used worldwide in bakery foodstuffs. They have the advantage of improving the nutritional value, sensory qualities and texture of bread, enabling the baking of doughs for rye bread production and increasing the shelf-life of bread (Hammes *et al.* 2005). The yeast diversity of sourdoughs consists of specifically adapted strains, namely *S. cerevisiae* and *Kazachstania exigua* (synonym *Saccharomyces exiguus*; Hammes and Gänzle 1998). Although sourdoughs have been studied in depth in the past few years, research on their microbiological composition underwent a renaissance in the early 2000s, leading to some studies dedicated to characterizing their yeast diversity (Pepe *et al.* 2003; Edema and Sanni 2008; Vogelmann *et al.* 2009; Moroni *et al.* 2010).

Yeasts as starter cultures in dairy products

Cheese

Microbial communities occurring in cheeses are initially dominated by lactic acid bacteria, while yeasts are believed to have a significant role in ripening and flavour development (Viljoen 2001; Hui *et al.* 2004). Fundamental studies on yeast diversity in cheese date back to the early 1990s. Yeasts' wide occurrence in cheese making may be attributed to their ability to tolerate high salt (NaCl) concentrations, low pH and water activity, as well as to their aptitude to grow at low temperatures and to assimilate lactose and lactic acid. In addition, their high proteolysis and lipolysis are considered crucial for releasing soluble amino acids and free fatty acids (Wyder and Puhan 1999).

The use of yeasts of the species *Debaryomyces hansenii* and *Yarrowia lipolytica* as starter cultures for cheese making has been proposed since the 1990s due to their positive impact in cheese ripening, NaCl resistance, ability to grow vigorously in cheese systems, as well as compatibility with lactic acid bacteria in mixed starter cultures (Wyder and Puhan 1999; van den Tempel and Jakobsen 2000; van den Tempel and Nielsen 2000; Guerzoni *et al.* 2001; Suzzi *et al.* 2001). Hence, both *D. hansenii* and *Y. lipolytica* (and occasionally other yeast species) have been proposed as co-starters with lactic acid bacteria, micrococci and/or filamentous fungi in cheese making worldwide (Wyder and Puhan 1999; van den Tempel and Nielsen 2000; Guerzoni *et al.* 2001; Hansen and Jakobsen 2001; Hansen *et al.* 2001; Psomas *et al.* 2001; Suzzi *et al.* 2001; Ferreira and Viljoen 2003; Źarowska *et al.* 2004; Goerges *et al.* 2008; Papapostolou *et al.* 2012; Gkatzionis *et al.* 2014).

Whey

Whey is the pale yellow residual liquid obtained after the flocculation and removal of milk casein during cheese making. This by-product represents approximately 85% of the milk volume and retains 55% of milk nutrients, including lactose, whey proteins, lipids and mineral salts (Dragone *et al.* 2009). Very few yeast species are lactose positive, but most strains are able to utilize the galactose, lactic acid or even citric acid that are present in whey, depending on the cheese-making technology. Whey represents a global environmental problem because of the high volumes produced and high BOD and COD values (Smithers 2008; Guimarães *et al.* 2010). Accordingly, several methods have been proposed for its economic exploitation, among them the production of ethanol (as biofuel) by lactose-positive genetically engineered *S. cerevisiae* strains (Domingues *et al.* 2001, 2010; Guimarães *et al.* 2008).

Fermented milk

Many yeast species have been isolated from commercial fermented milk products, in particular *Kluyveromyces marxianus, D. hansenii, Y. lipolytica* and *Rhodotorula mucilaginosa* (Rohm *et al.* 1990; Jordano *et al.* 1991; McKay 1992). Kefir is yeast-containing fermented milk traditionally produced in-house in Europe and Asia. Although its microbiological composition has been well characterized, considerable variations have been apparently observed among different worldwide cultures. Wyder (1998) found 23 yeast species, in particular *K. marxianus* and *S. cerevisiae*. More recently, some studies have reported the presence of both culturable and non-culturable yeast diversity in kefir grains collected worldwide, predominantly strains belonging to the genera *Candida, Kazachatania, Kluyveromyces, Pichia, Saccharomyces* and *Zygosaccharomyces* (Garbers *et al.* 2004; Witthuhn *et al.* 2004; Jianzhong *et al.* 2009; Magalhães *et al.* 2011; Kök Taş *et al.* 2012; Gao *et al.* 2012, 2013; Leite *et al.* 2012; Miguel *et al.* 2013: Diosma *et al.* 2014).

Yeasts as starter cultures in fermented meat products

The dynamic of microbial communities occurring in fermented meats is similar to that observed in cheeses; these products are primarily colonized by lactic acid bacteria, while yeasts often (together with micrococci and filamentous fungi) play a secondary role in ripening and flavour development (Hui *et al.* 2004). Some studies have shown the impact of the extracellular proteolytic and lipolytic enzymes produced by yeasts on the development of the characteristic tastes and flavours of fermented meats (Durá *et al.* 2004; Flores *et al.* 2004;

Martín *et al.* 2006). Most of the sensory properties are attributed to the hydrolysis of lipids and proteins and to the release of small peptides, amino acids and free fatty acids during the ripening process (Patrignani *et al.* 2007; Andrade *et al.* 2009).

A number of recent studies have characterized yeast communities occurring in fermented meats in order to select strains of *D. hansenii* and *Y. lipolytica* (and occasionally *Candida* spp.) as possible commercial starters, occasionally for mixed fermentation (Coppola *et al.* 2000; Olesen and Stahnke 2000; Bozkurt and Erkmen 2002; Baruzzi *et al.* 2006; Martín *et al.* 2006; Iucci *et al.* 2007; Patrignani *et al.* 2007; Sánchez-Molinero and Arnau 2008; Andrade *et al.* 2009; Purriños *et al.* 2013).

Yeasts as starter cultures in miscellaneous fermented foods and beverages

Fermented olives

It is generally accepted that yeasts can produce compounds exhibiting important organoleptic attributes improving the quality of fermented olives, especially ethanol, glycerol, higher alcohols, organic acids, acetaldehyde, esters and other volatile compounds, which may play an important role in flavour generation during the process (Montaño *et al.* 2003; Sánchez *et al.* 2000; Arroyo-Lopez *et al.* 2008, 2012). The lipolytic activity exhibited by some strains could also improve the volatile profile of these foodstuffs by increasing their free fatty acid content. The biodegradation of polyphenols catalyzed by specific β-glucosidases synthesized by yeasts is another interesting technological feature that could reduce the large quantities of olive wastewater produced during the lye treatment for fruit debittering (Hernández *et al.* 2007; Rodríguez-Gómez *et al.* 2010, 2012).

Against this background, the selection of yeast starters (in particular species of the genera *Candida, Kluyveromyces, Debaryomyces* and *Saccharomyces*) for olive fermentation is considered a key step for improving the process both in the laboratory and at an industrial scale (Hernandez *et al.* 2007; Arroyo-Lopez *et al.* 2012; Bevilacqua *et al.* 2012, 2013; Corsetti *et al.* 2012; Pistarino *et al.* 2013).

Cocoa

The first step of the chocolate-making process involves both pectinolysis and fermentation of cocoa beans. At present, cocoa is almost exclusively transformed by using spontaneous fermentations driven by natural microbial consortia, which include yeasts (in particular *S. cerevisiae*), lactic acid bacteria and acetic acid bacteria (Boekhout and Samson 2005; de Melo Pereira *et al.* 2012). Yeast metabolism in cocoa fermentation (conversion of pulp sugars into ethanol, release of pulp-degrading pectinases and development of chocolate aroma) has recently been elucidated (Nielsen *et al.* 2007; Ho *et al.* 2014). Several studies have investigated yeast diversity in both spontaneous and controlled cocoa fermentations: species of the genera *Candida, Hanseniaspora, Hyphopichia, Kodamaea, Pichia, Meyerozyma, Kluyveromyces, Saccharomyces, Trichosporon* and *Yamadazyma* have been found (Ardhana and Fleet 2003; Boekhout and Samson 2005; Jespersen *et al.* 2005; Nielsen *et al.* 2007; Daniel *et al.* 2009; Papalexandratou and De Vuyst 2011; de Melo Pereira *et al.* 2012; Lefeber *et al.* 2012; Crafack *et al.* 2013; Ho *et al.* 2014).

Coffee

Coffee is one of the most globally appreciated non-alcoholic drinks. Coffee fermentation is a spontaneous process characterized by the presence of different microorganisms, including bacteria, filamentous fungi and yeasts that produce enzymes, namely polygalacturonases and pectin-lyases, which are necessary to depolymerize and hydrolyze the pectins present in the mucilage. Among these, some pectinolytic yeasts of the species *K. marxianus* (syn. *Saccharomyces marxianus*), *Pichia kluyveri, S. bayanus, S. cerevisiae, Schizosaccharomyces* sp. and *Wickerhamomyces anomalus* (syn. *Pichia anomala*) have been proposed as starters for the fermentation of coffee cherries (Kashyap *et al.* 2001; Jayani *et al.* 2005; Masoud and Jespersen 2006; Silva *et al.* 2013).

Fermented fruit and vegetables

The combined effect of lactose intolerance, high cholesterol content, ergenic milk proteins and aspiration for vegetarian alternatives is quickly pushing consumer demand towards the replacement of dairy foodstuffs with products obtained by the fermentation of fruits and vegetables (Heenan *et al.* 2005; Granato *et al.* 2010; Rivera-Espinoza and Gallardo-Navarro 2010). There are a wide variety of traditional non-dairy fermented beverages produced around the world, which represent a huge economic potential for the global food industry. Many of them are non-alcoholic beverages produced using legumes (e.g. soy milk) or cereals as raw materials (Prado *et al.* 2008). Nevertheless, fruit juices and vegetable-based fermented products have also been proposed. There have been studies of the impact of the use of mixed starter cultures (including *S. cerevisiae, Pichia fermentans* and lactic acid bacteria) on their organoleptic, sensory and functional attributes (Rodríguez-Lerma *et al.* 2011; do Amaral Santos *et al.* 2014).

Yeasts as starter cultures in worldwide ethnic fermented foods and beverages

Fermented foods and beverages from Africa and Asia

Some studies on the microbial diversity occurring in some African naturally fermented milk have reported the considerable presence of yeasts as a relevant part of natural microbial consortia (Pedersen *et al.* 2012). Contamination from the environment and the equipment associated with fermentation could be assumed to work as natural inoculum of yeasts in these foods and beverages (Beukes *et al.* 2001; Narvhusa and Gadaga 2003). A number of studies have been carried out to characterize yeast diversity in ethnic fermented foods and beverages in order to select suitable starter cultures (or co-starters, in association with lactic acid bacteria) for improving safety and quality, sensory features and, sometimes, probiotic properties (Annan *et al.* 2003; Nyanga *et al.* 2007; Shetty *et al.* 2007; Vieira-Dalodé *et al.* 2007; Padonou *et al.* 2010; Pedersen *et al.* 2012; Greppi *et al.* 2013). Salient information on the main yeast species found in some African fermented foods and beverages is reported in Table 2.2. In Oriental countries some 'natural starters' (sometimes in the form of dry powders or hard balls made from starchy cereals) are frequently used to inoculate raw materials. There have been studies of yeast diversity in Asian fermented foods and beverages (Tsuyoshi *et al.* 2005; Sridevi *et al.* 2010; Wu *et al.* 2011). Salient information on the main yeast species found in some Asian fermented foods and beverages is listed in Table 2.3.

Fermented foods and beverages from South America

Yeast diversity in South American fermented foods and beverages has been studied recently, in particular in the so-called cachaça, a rum-like spirit produced from sugar cane (Gomes *et al.* 2007; Oliveira *et al.* 2008; Campos *et al.* 2010; Gonçalves de Sousa *et al.* 2012). The spontaneous process of fermenting sugar cane juice usually uses natural microbial starter cultures (containing yeasts that are not well defined; Schwan *et al.* 2001). The isolation of indigenous strains from the local production area and their selection and use as starters could ensure the adequate control of alcoholic fermentation and preserve some positive organoleptic contributions (Gomes *et al.* 2007; Oliveira *et al.* 2008; Campos *et al.* 2010; Gonçalves de Sousa *et al.* 2012). Indeed, some modern industrial processes recommend the addition of starter cultures of *S. cerevisiae* (in the form of active dry yeasts) to speed up the fermentative process, increase the levels of the desired metabolites and prevent the production of deleterious components by microbial contamination (Campos *et al.* 2010).

Yeasts as biocontrol agents in foods and beverages

Spoilage of food and beverages is a serious problem for industry: it can make products unacceptable to consumers and can cause economic losses and potentially severe health hazards. Many spoilage yeasts can grow when good manufacturing practices are not correctly employed (Viljoen *et al.* 2003; Stratford 2006). Food-grade antimicrobial compounds (e.g. sorbic and benzoic acids) are routinely used for prolonging shelf life and the preservation of food quality by inhibiting spoilage microorganisms (Battey *et al.* 2002; Papadimitriou *et al.* 2007).

Table 2.2 Main yeast species found in some African fermented foods and beverages.

Name	Origin	Major ingredients	Yeast species	Other microorganisms	References
Agadagigi	Nigeria, Cameroon	Plantain	*S. cerevisiae*		Sanni and Lonner (1993)
Amasi	Zimbabwe	Cow's milk	*C. kefyr* (anamorph of *K. marxianus*), *C. lipolytica* (anamorph of *Y. lipolytica*), *C. lusitaniae* (anamorph of *C. lusitaniae*), *C. tropicalis*, *Dk. bruxellensis*, *Naumovozyma dairenensis* (syn. *S. dairenensis*), *S. cerevisiae*, *T. Delbrueckii*	LAB	Gadaga *et al.* (2000, 2001)
Burukutu	Nigeria	Sorghum	*S. cerevisiae*		Sanni and Lonner (1993)
Chikokivana	Zimbabwe	Maize, millet	*S.cerevisiae*		Gadaga *et al.* (1999)
Dolo, Pito	Ghana	Sorghum	*S. cerevisiae*, *Candida* spp., *Kl. apiculata, Kluyveromyces* spp, *Sch. pombe*, *T. delbrueckii*, *W. anomalus* (syn. *P. anomala*)	LAB	Kolani *et al.* (1996); Sefa-Dedeh *et al.* (1999); van der Aa Kuhle *et al.* (2001)
Kenkey	Ghana	Maize	*Debaryomyces* spp., *Kluyveromyces* spp., *P. kudriavzevii* (syn. *C. krusei*), *Trichosporon* spp.	LAB	Annan *et al.* (2003); Amoa Awua *et al.* (2007)
Kisra	Sudan	Sorghum	*S. cerevisiae*		Steinkraus (1996); Odunfa and Oyewole (1998)
Masa	Nigeria	Rice	*S. cerevisiae*		E¢uvwevwere and Ezeama (1996)
Mawe	Benin	Maize	*S. cerevisiae*		Hounhouigan *et al.* (1999)
Mbanik	Senegal	Cow's milk	*S. cerevisiae*		Gningue *et al.* (1991)
Munkoyo, Busaa	Zambia, Kenya	Maize	*P. kudriavzevii* (syn. *C. krusei*), *S. cerevisiae*	LAB	Zulu *et al.* (1997)
Nono	Nigeria	Cow's milk	*S. cerevisiae*		Okagbue and Bankole (1992)

(*Continued*)

Table 2.2 (Continued)

Name	Origin	Major ingredients	Yeast species	Other microorganisms	References
Ogi	Nigeria	Maize, sorghum, millet	*C. vini* (anamorph of *Kr. fluxuum*), *S. cerevisiae*, *Z. rouxii*	LAB, AAB	Iwuoha and Eke (1996); Odunfa and Oyewole (1998); Teniola and Odunfa (2001)
Palm wine	Nigeria	Palm sap	*Candida spp.*, *Kl. apiculata*, *S. cerevisiae*, *Sch. pombe*, *Pichia spp.*	LAB, AAB	Owuama and Saunders (1990); Atacador-Ramos (1996); Odunfa and Oyewole (1998)
Rob	Sudan	Cow's, sheep's, goat's milk	*C. kefyr* (anamorph of *K. marxianus*), *S. cerevisiae*	LAB	Abdelgadir *et al.* (2001)

Notes: AAB = acetic acid bacteria; *C.* = *Candida*; *Dk.* = *Dekkera*; *K.* = *Kluyveromyces*; *Kl.* = *Kloeckera*; *Kr.* = *Kregervanrija*; LAB = lactic acid bacteria; *N.* = *Naumovozyma*; *P.* = *Pichia*; *S.* = *Saccharomyces*; *Sch.* = *Schizosaccharomyces*; *T.* = *Torulaspora*; *W.* = *Wickerhamomyces*; *Y.* = *Yarrowia*; *Z.* = *Zygosaccharomyces*

Table 2.3 Main yeast species found in some Asian fermented foods and beverages.

Name	Origin	Major ingredients	Yeast species	Other microorganisms	References
Dhokla	India	Rice and Bengal gram	*W. silvicola* (syn. *P. silvicola*)	LAB	Kanekar and Joshi (1993)
Idli	India, Sri Lanka	Rice and black gram	*S. cerevisiae*	LAB	Koh and Singh (2009)
Jnard, Jaanr, Thumba	Nepal, India, Bhutan	Finger millet, rice, maize, wheat	*S. cerevisiae*, *Sa. fibuligera*, *W. anomalus* (syn. *P. anomala*)	LAB, *Mucor* spp., *Rhizopus* spp.	Tamang *et al.* (1988)
Kombucha, tea fungus	Japan, Indonesia, China, Russia	Tea liquor and sugar	*Brettanomyces spp.*, *Saccharomyces* spp., *Z. Kombuchaensis*	AAB	Mayser *et al.* (1995)
Loog-pang	Thailand	Rice	*Sa. fibuligera*, *W. anomalus* (syn. *P. anomala*)		Limtong *et al.* (2002)
Men	Vietnam	Rice	*S. cerevisiae*	*Amylomyces rouxii*	Dung *et al.* (2005)
Miso	Japan	Rice and soybeans	*Z. rouxii*	*Aspergillus oryzae*, LAB	Ebine (1989)
Murcha/marcha	India, Nepal	Rice	*Hy. burtonii* (syn. *P. burtonii*), *S. cerevisiae*	*Mucor* spp., *Rhizopus* spp.	Shrestha *et al.* (2002); Tsuyoshi *et al.* (2005)
Nan, Kulcha, Bhatura	India, Pakistan, Afghanistan, Iran	Wheat flour	*S. cerevisiae*	LAB	Saono *et al.* (1996)
Palm wines (Toddy, Tari, Tuack, Tuba)	Sri Lanka, Thailand, Malaysia, Indonesia	Sap of coconut, date or Palmyra palm	*Kd. ohmeri*, *S. cerevisiae*, *Sch. Pombe*	LAB, AAB	Joshi *et al.* (1999)
Papad, Papadam	India	Black gram	*P. kudriavzevii* (syn. *C. krusei*), *S. cerevisiae*		Shurpalekar (1986)
Ragi	Indonesia	Rice	*Sa. Fibuligera*	*Amylomyces rouxii*	Hesseltine *et al.* (1988)
Saké	Japan	Rice	*S. cerevisiae* (syn. *S. sake*), *W. anomalus* (syn. *H. anomala*)	*Aspergillus oryzae*	Aidoo *et al.* (2006)

(*Continued*)

Table 2.3 (Continued)

Name	Origin	Major ingredients	Yeast species	Other microorganisms	References
Soy sauce	Japan, China	Soybeans and wheat	*Candida* spp., *Zy. Rouxii*	*Aspergillus oryzae*, *Aspergillus sojae*	Aidoo *et al.* (1994)
Tapé ketan, Tapé ketella, Peujeum	Indonesia	Rice, cassava tubers	*Hy. burtonii* (syn. *P. burtonii*), *Sa. fibuligera*, *W. anomalus* (syn. *P. anomala*)	*Amylomyces rouxii*, *Mucor* spp., *Rhizopus* spp.	Ko (1986)
Tapuy	Philippines	Rice	*C. fennica* (syn. *Tr. fennicum*), *C. parapsilosis*, *D. hansenii*, *R. glutinis*, *Sa. Fibuligera*		Kozaki and Uchimura (1990)
Wadi	India, Pakistan	Black gram	*P. kudriavzevii* (syn. *C. krusei*)	LAB	Sandhu and Soni (1989)
Yakju, Takju	Korea	Rice, wheat, barley, maize, millet	*C. sake*, *S. cerevisiae*, *Sch. polymorphus* var. *polymorphus* (syn. *P. polymorpha*), *Torulaspora delbrueckii* (syn. *T. inconspicua*), *W. anomalus* (syn. *P. anomala*), *W. subpelliculosus* (syn. *H. subpelliculosa*)	*Aspergillus oryzae*, *Aspergillus sojae*, *Rhizopus* spp.	Rhee *et al.* (2003)

Notes: AAB = acetic acid bacteria; *H.* = *Hansenula*; *Hy.* = *Hyphopichia*; *Kd.* = *Kodamaea*; LAB = lactic acid bacteria; *P.* = *Pichia*; *R.* = *Rhodotorula*; *S.* = *Saccharomyces*; *Sa.* = *Saccharomycopsis*; *Sc.* = *Schwanniomyces*; *Sch.* = *Schizosaccharomyces*; *T.* = *Torulaspora*; *Tr.* = *Trichosporon*; *W.* = *Wickerhamomyces*; *Z.* = *Zygosaccharomyces*

Nevertheless, a number of yeasts have been shown to be resistant to many chemical preservatives (Battey *et al.* 2002; Hazan *et al.* 2004). Against this background, an alternative approach could be to select yeast starters on the basis of their aptitude to release specific proteins (also labelled 'killer' proteins) that are able to inhibit the growth of some spoilage yeasts. The use of 'killer' yeasts as biocontrol agents against food and beverage spoilage yeasts is well documented (Lowes *et al.* 2000; Fredlund *et al.* 2002; Fleet 2003; Comitini *et al.* 2004a, b; Golubev 2006; Sangorrin *et al.* 2007; Goretti *et al.* 2009; Liu and Tsao 2009; de Ullivarri *et al.* 2011; Corsetti *et al.* 2012; Oro *et al.* 2014).

Worldwide collections conserving yeast starter cultures

Yeast starters can generally be obtained by companies selling yeast cultures (Table 2.1); by direct isolation from natural or technological environments and subsequent selection by research microbiology laboratories; or from strains conserved in worldwide service culture collections. These institutions are also labelled 'biological resource centres' (BRCs) by the Organisation for Economic Co-operation and Development (OECD 2007; Boundy-Mills 2012). The genetic patrimony preserved in BRCs can be considered as a significant reservoir of gene pools of technological importance and represents a strategic opportunity for industry to select yeast starters (Pretorius 2000). BRCs currently have the facilities and expertise to guarantee proper species identification, minimize the genetic drift that often occurs with repeated transfer, and assure pure and viable cultures for industrial exploitation (Boundy-Mills 2012). There are numerous BRCs around the world, ranging from enormous and well-known collections to smaller, more specialized ones. Many BRCs currently serve as public repositories of microorganisms. Strains deposited include type strains selected by taxonomists to represent a novel species, patented strains, strains cited in publications and genetically modified strains used in research and industry. The accessibility of yeast strains conserved in worldwide BRCs for worldwide utilizers (universities, research centres, industry etc.) is a key task of these centres. The list of the main worldwide BRCs conserving a number of yeast strains >500 is given in Table 2.4.

Conclusion and future outcomes

The use of fermentative processes continues to be considered the best technology for improving the safety and quality of foods and beverages. Modern starter cultures are selected (either as single or mixed strains) for their ability to adapt themselves to specific substrates. Because inoculation with starters (including

Table 2.4 The main worldwide biological resource centres (BRCs) conserving a significant number of yeast strains (>500). Data are obtained from the World Data Centre for Microorganisms' online database (www.wdcm.org) and from recent literature (Boundy-Mills 2012).

Acronym	Name	Country	Approximated number of yeast strains	Website
NRRL	Agricultural Research Service Culture Collection	USA	18,000	http://nrrl.ncaur.usda.gov
ATCC	American Type Culture Collection	USA	10,000	https://www.atcc.org
CBS	Centraalbureau voor Schimmelcultures	The Netherlands	9000	http://www.cbs.knaw.nl
UCD-FST	Phaff Yeast Culture Collection	USA	7000	http://phaffcollection.ucdavis.edu
DBVPG	Industrial Yeasts Collection	Italy	5900	http://www.dbvpg.unipg.it
CGMCC	China General Microbiological Culture Collection Centre	China	4700	http://www.cgmcc.net
BCCM/IHEM	Biomedical Fungi and Yeasts Collection	Belgium	4000	http://bccm.belspo.be
CCY	Culture Collection of Yeasts	Slovakia	3800	http://www.chem.sk/activities/yeast/ccy
NCYC	National Collection of Yeast Cultures	UK	3000	http://www.ncyc.co.uk
VKM	All-Russian Collection of Microorganisms	Russia	3000	http://www.vkm.ru
CECT	Coleccion Española de Cultivos Tipo	Spain	2500	http://www.uv.es/uvweb/spanish-type-culture-collection/en/spanish-type-culture-collection-1285872233521.html
KCTC	Korean Collection for Type Cultures	Korea	2500	http://kctc.kribb.re.kr
VKPM	Russian National Collection of Industrial Microorganisms	Russia	2300	http://eng.genetika.ru/service-offer/vkpm
NCAIM	National Collection of Agricultural and Industrial Microorganisms	Hungary	1700	http://web.uni-corvinus.hu:8089/NCAIM/index.jsp
VTT	VTT Culture Collection	Finland	1500	http://culturecollection.vtt.fi
NBIMCC	National Bank for Industrial Microorganisms and Cell Cultures	Bulgaria	700	http://www.nbimcc.org
IAFB	Collection of Industrial Microorganisms	Poland	500	http://www.ibprs.pl/bazy/cim
DSMZ	Leibniz-Institut DSMZ – Deutsche Sammlung von Mikroorganismen und Zellkulturen GmbH	Germany	500	http://www.dsmz.de

yeasts) does not provide an absolute guarantee against the health hazards associated with foodborne pathogens and/or the failure of fermentative processes, their technological use must be supported by strict compliance with the basic principles of good manufacturing practices (Holzapfel 2002).

In order to select the ideal starter culture, food microbiologists and technologists need to understand the specific function to be attributed to it, and in some cases to have tools to check its technological performance. The research for selecting the best starter culture has until recently been approached by performing screening surveys on a large set of isolates in a laboratory or small-scale food fermentation plant. Therefore, the exploration and *ex situ* conservation of the biodiversity of indigenous yeasts can be an important contribution towards the selection of strains exhibiting specific phenotypes (Capece *et al.* 2010). In this way, some hundreds of potential starter cultures have been isolated in the last few decades based on their ability to give satisfactory technological performance. There is no doubt that this approach will continue to be used in the future to expand the pool of yeasts suitable to be proposed as possible starter cultures. Against this background, BRCs could play an enormously important role. A significant number of yeast strains cited in worldwide research articles are not so far deposited in BRCs. This represents a severe limitation to their possible use for further studies by other researchers or their potential exploitation by third parties. Of course, this limitation reduces enormously the possibility of scaling up scientific results from the laboratory to an industrial scale (Boundy-Mills 2012; Stackebrandt *et al.* 2014). Therefore, the risk that a consistent part of yeast diversity could remain in laboratories in the role of 'Sleeping Beauty' for years, often for decades, and sometimes for ever, is undoubtedly very high. In this context, one possible task of BRCs in the future could be to implement their role as 'connection centres' between research laboratories worldwide and the industry.

In the last two decades the scientific community has greatly increased its efforts around the use of molecular tools allowing for the targeting of individual genes (and related metabolic pathways) responsible for some technologically desired performance (Pretorius 2000; Dequin 2001; Pretorius and Bauer 2002; Bisson 2004; Schuller and Casal 2005; Cebollero *et al.* 2007; Plahutaa and Raspor 2007). This could reinforce in the near future the pressure for the use of mutant selection and genetic engineering to create starters exhibiting superior technological features than those found in nature (Hansen 2002). The advent of functional genomics has undoubtedly created enormous opportunities for improving yeasts for food and beverage fermentation, as well as for other industrial applications (Johnson and Echavarri-Erasun 2011). Because of its long history of safe use, *S. cerevisiae* was among the first organisms to be designated 'generally recognized as safe' (GRAS) and was the first genetically modified organism (GMO) used for recombinant production of food and feed additives. In 1990 a genetically modified strain of *S. cerevisiae* was one of the first GMOs to be approved for food use in the UK and was proposed as a starter for bakery products because of its enhanced production of CO_2 (Aldhous 2000; Johnson and Echavarri-Erasun 2011). A number of genes encoding desirable traits have been cloned in both *Saccharomyces* and non-*Saccharomyces* putative starter cultures (Dequin 2001; Bisson 2004; Pretorius 2003; Schuller and Casal 2005; Coulon *et al.* 2006; Husnik *et al.* 2006; van Rensburg *et al.* 2007). However, although a debate (sometimes very lively) about the scientific and regulatory concerns over GMOs is still ongoing (Pariza and Johnson 2001; Plahutaa and Raspor 2007; Pretorius and Bauer 2002; Cebollero *et al.* 2007), most of these strains are not commercially used, mainly because of public perception hurdles (Fenn 2007). So, despite the wide availability of increasingly sophisticated methods for the genetic manipulation of industrial yeasts, the full approval of genetically engineered yeast strains as starters for the food and beverage industry is far from becoming a universally accepted rule, at least in European countries.

References

Abdelgadir, W.S., Hamad, S.H., Müller, P.L. and Jakobsen, M. (2001) Characterisations of the dominant microbiota of Sudanese fermented milk rob. *International Dairy Journal*, **11**, 63–70.

Aidoo, K.E., Nout, N.J.R. and Sarkar, P.K. (2006) Occurrence and function of yeasts in Asian indigenous fermented foods. *FEMS Yeast Research*, **6**, 30–39.

Aidoo, K.E., Smith, J.E. and Wood, B.J.B. (1994) Industrial aspects of soy sauce fermentations using *Aspergillus*. In K.A. Powell, A. Renwick and J.F. Peberdy (eds), *The Genus Aspergillus: From Taxonomy and Genetics to Industrial Application* (pp. 155–169), New York: Plenum Press.

Aldhous, P. (2000) Modified yeast fine for food. *Nature*, **344**, 186.

Amoa Awua, W.K., Ngunjiri, P., Anlobe, J. *et al.* (2007) The effect of applying GMP and HACCP to traditional food processing at a semi-commercial kenkey production plant in Ghana. *Food Control*, **18**, 1449–1457.

Andrade, M.J., Rodríguez, M., Casado, E.M., Bermúdez, E. and Córdoba, J.J. (2009) Differentiation of yeasts growing on dry-cured Iberian ham by mitochondrial DNA restriction analysis, RAPD-PCR and their volatile compounds production. *Food Microbiology*, **26**, 578–586.

Annan, N.T., Poll, L., Sefa-Dedeh, S., Plahar, W.A. and Jakobsen, M. (2003) Influence of starter culture combinations of *Lactobacillus fermentum*, *Saccharomyces cerevisiae* and *Candida krusei* on aroma in Ghanaian maize dough fermentation. *European Food Research Technology*, **216**, 377–384.

Ardhana, M.M. and Fleet, G.H. (2003) The microbial ecology of cocoa bean fermentations in Indonesia. *International Journal of Food Microbiology*, **86**, 87–99.

Arroyo-López, F.N., Querol, A., Bautista-Gallego, J. and Garrido-Fernández, A. (2008) Role of yeasts in table olive production. *International Journal of Food Microbiology*, **128**, 189–196.

Arroyo-López, F.N., Romero-Gil, V., Bautista-Gallego, J. *et al.* (2012) Potential benefits of the application of yeast starters in table olive processing. *Frontiers in Microbiology*, **3**, 1–4.

Atacador-Ramos, M. (1996) Indigenous fermented foods in which ethanol is a major product. In K.H. Steinkraus (ed.), *Handbook of Indigenous Fermented Foods* (2nd edn, pp. 363–508), New York: Marcel Dekker.

Bamforth, C.W. (2000) Brewing and brewing research: Past, present, future. *Journal of the Science of Food and Agriculture*, **80**, 1371–1378.

Bamforth, C.W. and Kanauchi, M. (2004) Enzymology of vicinal diketone reduction in brewer's yeast. *Journal of the Institute of Brewing*, **110**, 83–93.

Baruzzi, F., Matarante, A., Caputo, L. and Morea, M. (2006) Molecular and physiological characterization of natural microbial communities isolated from a traditional Southern Italian processed sausage. *Meat Science*, **72**, 261–269.

Battey, A.S., Duffy, S. and Schaffner, D.W. (2002) Modeling yeast spoilage in cold-filled ready-to-drink beverages with *Saccharomyces cerevisiae*, *Zygosaccharomyces bailii*, and *Candida lipolytica*. *Applied and Environmental Microbiology*, **68**, 1901–1906.

Bauer, F.F. and Pretorius, I.S. (2000) Yeast stress response and fermentation efficiency: How to survive the making of wine – a review. *South African Journal of Enology and Viticulture*, **21**, 27–51.

Beukes, E.M., Bester, B.H. and Mostert, J.F. (2001) The microbiology of South African traditional fermented milks. *International Journal of Food Microbiology*, **63**, 189–197.

Bevilacqua, A., Beneduce, L., Sinigaglia, M. and Corbo, M.R. (2013) Selection of yeasts as starter cultures for table olives. *Journal of Food Science*, **78**, 742–751.

Bevilacqua, A., Corbo M.R. and Sinigaglia, M. (2012) Selection of yeasts as starter cultures for table olives: A step-by-step procedure. *Frontiers in Microbiology*, **3**, 2–9.

Bisson, L.F. (2004) Biotechnology of wine yeast. *Food Biotechnology*, **18**, 63–96.

Bisson, L.F., Waterhouse, A.L., Ebeler, S.E., Walker, M.A. and Lapsley, J.T. (2002) The present and future of the international wine industry. *Nature*, **418**, 696–699.

Bleoanca, I., Courelas Silva, A.R., Pimentel, C., Rodrigues-Pousada, C. and Menezes Rde, A. (2013) Relationship between ethanol and oxidative stress in laboratory and brewing yeast strains. *Journal of Bioscience and Bioengineering*, **116**, 697–705.

Blieck, L., Toye, G., Dumortier, F. *et al.* (2007) Isolation and characterization of brewer's yeast variants with improved fermentation performance under high-gravity conditions. *Applied and Environmental Microbiology*, **73**, 815–824.

Boekhout, T. (2005) Biodiversity: Gut feeling for yeasts. *Nature*, **434**, 449–451.

Boekhout, T. and Samson, R. (2005) Fungal biodiversity and food. In R.M.J. Nout, W.M. de Vos and M.H. Zwietering (eds), *Food Fermentation* (pp. 29–41), Gelderland: Wageningen Academic.

Borneman, A.R., Chambers, P.J. and Pretorius, I.S. (2007) Yeast systems biology: Modeling the winemaker's art. *Trends in Biotechnology*, **25**, 349–355.

Boundy-Mills, K. (2012) Yeast culture collections of the world: Meeting the needs of industrial researchers. *Journal of Industrial Microbiology and Biotechnology*, **39**, 673–680.

Bozkurt, H. and Erkmen, O. (2002) Effects of starter cultures and additives on the quality of Turkish style sausage (sucuk). *Meat Science*, **61**, 149–156.

Buzzini, P. and Vaughan-Martini, A. (2006) Yeast biodiversity and biotechnology. In C. Rosa and G. Péter (eds), *The Yeast Handbook: Biodiversity and Ecophysiology of Yeasts* (pp. 533–559), Berlin: Springer.

Calabretti, A., La Cara, F., Sorrentino, A. *et al.* (2012) Characterization of volatile fraction of typical Irpinian wines fermented with a new starter yeast. *World Journal of Microbiology and Biotechnology*, **28**, 1433–1442.

Campos, C.R., Silva, C.F., Dias, D.R., Basso, L.C., Amorim, H.V. and Schwan, R.F. (2010) Features of *Saccharomyces cerevisiae* as a culture starter for the production of the distilled sugar cane beverage cachaça in Brazil. *Journal of Applied Microbiology*, **108**, 1871–1879.

Capece, A., Romaniello, R., Siesto, G. *et al.* (2010) Selection of indigenous *Saccharomyces cerevisiae* strains for Nero d'Avola wine and evaluation of selected starter implantation in pilot fermentation. *International Journal of Food Microbiology*, **144**, 187–192.

Cebollero, E., Gonzalez-Ramos, D., Tabera, L. and Gonzalez, R. (2007) Transgenic wine yeast technology: Is it time for transgenic wine? *Biotechnology Letters*, **29**, 191–200.

Cheraiti, N., Guezenec, S. and Salmon, J.M. (2005) Redox interactions between *Saccharomyces cerevisiae* and *Saccharomyces uvarum* in mixed culture under enological conditions. *Applied and Environmental Microbiology*, **71**, 255–260.

Ciani, M., Comitini, F., Mannazzu, I. and Domizio, P. (2010) Controlled mixed culture fermentation: A new perspective on the use of non-*Saccharomyces* yeasts in winemaking. *FEMS Yeast Research*, **10**, 123–133.

Clavijo, A., Calderón, I.L. and Paneque, P. (2011) Effect of the use of commercial *Saccharomyces* strains in a newly established winery in Ronda (Málaga, Spain). *Antonie van Leeuwenhoek*, **99**, 727–731.

Clemente-Jimenez, J.M., Mingorance-Cazorla, L., Martínez-Rodríguez, S., Las Heras-Vázquez, F.J. and Rodríguez-Vico, F. (2005) Influence of sequential yeast mixtures on wine fermentation. *International Journal of Food Microbiology*, **98**, 301–308.

Comitini, F., De Ingeniis, J., Pepe, L., Mannazzu, I. and Ciani, M. (2004a) *Pichia anomala* and *Kluyveromyces wickerhamii* killer toxins as new tools against *Dekkera/Brettanomyces* spoilage yeasts. *FEMS Microbiology Letters*, **238**, 235–240.

Comitini, F., Di Pietro, N., Zacchi, L., Mannazzu, I. and Ciani, M. (2004b) *Kluyveromyces phaffii* killer toxin active against wine spoilage yeasts: Purification and characterization. *Microbiology*, **150**, 2535–2541.

Coppola, S., Mauriello, G., Aponte, M., Moschetti, G. and Villani, F. (2000) Microbial succession during ripening of Naples-type salami, a southern Italian fermented sausage. *Meat Science*, **56**, 321–329.

Corsetti, A., Perpetuini, G., Schirone, M., Tofalo, R. and Suzzi, G. (2012) Application of starter cultures to table olive fermentation: An overview on the experimental studies. *Frontiers in Microbiology*, **3**, 1–6.

Coulon, J., Husnik, J.I., Inglis, D.L. *et al.* (2006) Metabolic engineering of *Saccharomyces cerevisiae* to minimize the production of ethyl carbamate in wine. *American Journal of Enology and Viticulture*, **57**, 113–124.

Crafack, M., Mikkelsen, M.B., Saerens, S. *et al.* (2013) Influencing cocoa flavour using *Pichia kluyveri* and *Kluyveromyces marxianus* in a defined mixed starter culture for cocoa fermentation. *International Journal of Food Microbiology*, **167**, 103–116.

Cukier de Aquino, V., Converti, A., Perego, P. and Caetano da Silva Lannes, S. (2012) Leavening bread dough. *Current Nutrition and Food Science*, **8**, 131–138.

Daniel, H.M., Vrancken, G., Takrama, J.F., Camu, N., De Vos, P. and De Vuyst, L. (2009) Yeast diversity of Ghanaian cocoa bean heap fermentations. *FEMS Yeast Research*, **9**, 774–783.

De Benedictis, M., Bleve, G., Grieco, G., Tristezza, M., Tufariello, M. and Grieco, F. (2011) An optimized procedure for the enological selection of non-*Saccharomyces* starter cultures. *Antonie van Leeuwenhoek*, **99**, 189–200.

de Melo Pereira, G.V., da Cruz Pedrozo, M.G., Ramos, C.L. and Schwan, R.F. (2012) Microbiological and physicochemical characterization of small-scale cocoa fermentations and screening of yeast and bacterial strains to develop a defined starter culture. *Applied and Environmental Microbiology*, **78**, 5395–5405.

de Ullivarri, M.F., Mendoza, L.M., Raya, R.R. and Farías, M.E. (2011) Killer phenotype of indigenous yeasts isolated from Argentinian wine cellars and their potential starter cultures for winemaking. *Biotechnology Letters*, **33**, 2177–2183.

del Mónaco, S.M., Barda, N.B., Rubio, N.C. and Caballero, A.C. (2014) Selection and characterization of a Patagonian *Pichia kudriavzevii* for wine deacidification. *Journal of Applied Microbiology*, **117**, 451–464.

Dequin, S. (2001) The potential of genetic engineering for improving brewing, wine-making, and baking yeasts. *Applied Microbiology and Biotechnology*, **56**, 577–588.

Diosma, G., Romanin, D.E., Rey-Burusco, M.F., Londero, A. and Garrote, G.L. (2014) Yeasts from kefir grains: Isolation, identification, and probiotic characterization. *World Journal of Microbiology and Biotechnology*, **30**, 43–53.

do Amaral Santos, C.C.A., da Silva Libeck, B. and Schwan, R.F. (2014) Co-culture fermentation of peanut-soy milk for the development of a novel functional beverage. *International Journal of Food Microbiology*, **186**, 32–41.

Domingues, L., Guimarães, P.M. and Oliveira, C. (2010) Metabolic engineering of *Saccharomyces cerevisiae* for lactose/whey fermentation. *Bioengineered Bugs*, **1**, 164–171.

Domingues, L., Lima, N. and Teixeira, J.A. (2001) Alcohol production from cheese whey permeate using genetically modified flocculent yeast cells. *Biotechnology and Bioengineering*, **72**, 507–514.

Domizio, P., Lencioni, L., Ciani, M., Di Blasi, S., Pontremolesi, C. and Sabatelli, M.P. (2007) Spontaneous and inoculated yeast populations dynamics and their effect on organoleptic characters of Vinsanto wine under different process conditions. *International Journal of Food Microbiology*, **115**, 281–289.

Dragone, G., Mussatto, S.I., Oliveira, J.M. and Teixeira, J.A. (2009) Characterisation of volatile compounds in an alcoholic beverage produced by whey fermentation. *Food Chemistry*, **112**, 929–935.

Dung, N.T.P., Rombouts, F.M. and Nout, M.J.R. (2005) Developments of defined mixed-culture fungal fermentation starter granulate for controlled production of rice wine. *Innovative Food Science and Emerging Technologies*, **6**, 429–441.

Durá, M.A., Flores, M. and Toldrá, F. (2004) Effect of growth phase and dry-cured sausage processing conditions on *Debaryomyces* spp. generation of volatile compounds from branched-chain amino acids. *Food Chemistry*, **86**, 391–399.

Ebine, H. (1989) Industrialization of Japanese miso fermentation. In K.H. Steinkraus (ed.), *Industrialization of Indigenous Fermented Foods* (pp. 89–126), New York: Marcel Dekker.

Eɛuvwevwere, B.J.O. and Ezeama, C.F. (1996) Influence of fermentation time and an indigenous tenderiser (kanwa) on the microbial profile, chemical attributes and shelf-life of rice masa (a Nigerian fermented product). *Journal of the Science of Food and Agriculture*, **71**, 442–448.

Edema, M.O. and Sanni, A.I. (2008) Functional properties of selected starter cultures for sour maize bread. *Food Microbiology*, **25**, 616–625.

Elmacı, S.B., Özçelik, F., Tokatlı, M. and Çakır İ. (2014) Technological properties of indigenous wine yeast strains isolated from wine production regions of Turkey. *Antonie van Leeuwenhoek*, **105**, 835–847.

Fenn, J.P. (2007) Update of medically important yeasts and a practical approach to their identification. *Lab Medicine*, **38**, 178–183.

Ferreira, A.D. and Viljoen, B.C. (2003) Yeasts as adjunct starters in matured Cheddar cheese. *International Journal of Food Microbiology*, **86**, 131–140.

Fleet, G.H. (1999) Microorganisms in food ecosystems. *International Journal of Food Microbiology*, **50**, 101–117.

Fleet, G.H. (2003) Yeast interactions and wine flavour. *International Journal of Food Microbiology*, **86**, 11–22.

Fleet, G.H. (2006) The commercial and community significance of yeasts in food and beverage production. In A. Querol and G.H. Fleet (eds), *Yeasts in Food and Beverages* (pp. 1–12), Berlin: Springer.

Fleet, G.H. (2008) Wine yeasts for the future. *FEMS Yeast Research*, **8**, 979–995.

Flores, M., Durá, M.A., Marco, A. and Toldrá, F. (2004) Effect of *Debaryomyces* spp. on aroma formation and sensory quality of dry-fermented sausages. *Meat Science*, **68**, 439–446.

Fredlund, E., Druvefors, U., Boysen, M.E., Lingsten, K.J. and Schnürer, J. (2002) Physiological characteristics of the biocontrol yeast *Pichia anomala* J121. *FEMS Yeast Research*, **2**, 395–402.

Gadaga, T.H., Mutukumira, A.N. and Narvhus, J.A. (2000) Enumeration and identification of yeasts isolated from Zimbabwean traditional fermented milk. *International Dairy Journal*, **10**, 459–466.

Gadaga, T.H., Mutukumira, A.N. and Narvhus, J.A. (2001) The growth and interaction of yeasts and lactic acid bacteria isolated from Zimbabwean naturally fermented milk in UHT milk. *International Journal of Food Microbiology*, **68**, 21–32.

Gadaga, T.H., Mutukumira, A.N., Narvhus, J.A. and Feresu, S. (1999) A review of traditional fermented foods and beverages of Zimbabwe. *International Journal of Food Microbiology*, **53**, 1–11.

Gao, J., Gu, F., Abdella, N.H., Ruan, H. and He, G. (2012) Investigation on culturable microflora in Tibetan kefir grains from different areas of China. *Journal of Food Science*, **77**, 425–433.

Gao, J., Gu, F., He, J. *et al.* (2013) Metagenome analysis of bacterial diversity in Tibetan kefir grains. *European Food Research and Technology*, **236**, 549–556.

Garbers, I.M., Britz, T.J. and Witthuhn, R.C. (2004) PCR-based denaturing gradient gel electrophoretic typification and identification of the microbial consortium present in kefir grains. *World Journal of Microbiology and Biotechnology*, **20**, 687–693.

Giannone, V., Longo, C., Damigella, A., Raspagliesi, D., Spina, A. and Palumbo, M. (2010) Technological properties of bakers' yeasts in durum wheat semolina dough. *Journal of Industrial Microbiology and Biotechnology*, **37**, 371–379.

Gkatzionis, K., Yunita, D., Linforth, R.S.T., Dickinson, M. and Dodd, C.E. (2014) Diversity and activities of yeasts from different parts of a Stilton cheese. *International Journal of Food Microbiology*, **177**, 109–116.

Gningue, P.N., Roblain, D. and Thonart, P. (1991) Microbial and biochemical studies of Mbanik, a traditional Senegalese cultures milk. *Cerevisia Biotechnology*, **16**, 32–40.

Goerges, S., Mounier, J., Rea, M.C. *et al.* (2008) Commercial ripening starter microorganisms inoculated into cheese milk do not successfully establish themselves in the resident microbial ripening consortia of a South German red smear cheese. *Applied and Environmental Microbiology*, **74**, 2210–2217.

Golubev, W.I. (2006) Antagonistic interactions among yeasts. In C.A. Rosa and G. Péter (eds), *Biodiversity and Ecophysiology of Yeasts* (pp. 197–219), Berlin: Springer.

Gomes, F.C.O., Silva, C.L.C., Marini, M.M., Oliveira, E.S. and Rosa, C.A. (2007) Use of selected indigenous *Saccharomyces cerevisiae* strains for the production of the traditional cachaça in Brazil. *Journal of Applied Microbiology*, **103**, 2438–2447.

Gonçalves de Souza, A.P., de Araújo Vicente M., Contelli Klein, R. *et al.* (2012) Strategies to select yeast starters cultures for production of flavor compounds in cachaça fermentations. *Antonie van Leeuwenhoek*, **101**, 379–392.

Goretti, M., Turchetti, B., Buratta, M. *et al.* (2009) In vitro antimycotic activity of a Williopsis saturnus killer protein against food spoilage yeasts. *International Journal of Food Microbiology*, **131**, 178–182.

Granato, D., Branco, G.F., Nazzaro, F., Cruz, A.G. and Faria, J.A.F. (2010) Functional foods and non-dairy probiotic food development: Trends, concepts, and products. *Comprehensive Reviews in Food Science and Food Safety*, **9**, 292–302.

Greppi, A., Rantisou, K., Padonou, W. *et al.* (2013) Yeast dynamics during spontaneous fermentation of mawè and tchoukoutou, two traditional products from Benin. *International Journal of Food Microbiology*, **165**, 200–207.

Guerzoni, M.E., Lanciotti, R., Vannini, L. *et al.* (2001) Variability of the lipolytic activity in Yarrowia lipolytica and its dependence on environmental conditions. *International Journal of Food Microbiology*, **69**, 79–89.

Guimarães, P.M., Teixeira, J.A. and Domingues, L. (2008) Fermentation of high concentrations of lactose to ethanol by engineered flocculent *Saccharomyces cerevisiae*. *Biotechnology Letters*, **30**, 1953–1958.

Guimarães, PM., Teixeira, J.A. and Domingues, L. (2010) Fermentation of lactose to bio-ethanol by yeasts as part of integrated solutions for the valorisation of cheese whey. *Biotechnology Advances*, **28**, 375–384.

Guth, H. and Sies, A. (2002) Flavour of wines: Towards an understanding by reconstitution experiments and an analysis of ethanol's effect on odour activity of key compounds. In R.J. Blair, P.J. Williams and P.B. Høj (eds), *Proceedings of the 11th Australian Wine Industry Technical Conference* (pp. 128–139), Adelaide: Australian Wine Industry Technical Conference.

Hammes, W.P. and Gänzle, M.G. (1998). Sourdough breads and related products. In B.J. Wood (ed.), *Microbiology of Fermented Foods* (pp. 199–216), London: Chapman and Hall.

Hammes, W.P., Brandt, M.J., Francis, K.L., Rosenheim, J., Seitter, M.F.H. and Vogelmann, S.A. (2005) Microbial ecology of cereal fermentations. *Trends in Food Science and Technology*, **16**, 4–11.

Hansen, E.B. (2002) Commercial bacterial starter cultures for fermented foods of the future. *International Journal of Food Microbiology*, **78**, 119–131.

Hansen, T.K. and Jakobsen, M. (2001) Taxonomical and technological characteristics of *Saccharomyces* spp. associated with blue veined cheese. *International Journal of Food Microbiology*, **69**, 59–68.

Hansen, T.K., van den Tempel, T., Cantor, M.D. and Jakobsen, M. (2001) *Saccharomyces cerevisiae* as a starter culture in Mycella. *International Journal of Food Microbiology*, **69**, 101–111.

Hazan, R., Levine, A. and Abeliovich, H. (2004) Benzoic acid, a weak organic acid food preservative, exerts specific effects on intracellular membrane trafficking pathways in *Saccharomyces cerevisiae*. *Applied and Environmental Microbiology*, **70**, 4449–4457.

Heenan, C.N., Adams, M.C., Hosken, R.W. and Fleet, G.H. (2005) Survival and sensory acceptability of probiotic microorganisms in a non fermented frozen vegetarian dessert. *Food Science and Technology*, **37**, 461–466.

Hernández, A., Martin, A., Aranda, E., Pérez-Nevado, F. and Córdoba, M.G. (2007) Identification and characterization of yeast isolated from the elaboration of seasoned green table olives. *Food Microbiology*, **24**, 346–351.

Hernández-López, M.J., Pallotti, C., Andreu, P., Aguilera, J., Prieto, J.A. and Randez-Gil, F. (2007) Characterization of a *Torulaspora delbrueckii* diploid strain with optimized performance in sweet and frozen sweet dough. *International Journal of Food Microbiology*, **116**, 103–110.

Hesseltine, C.W., Rogers, R. and Winarno, F.G. (1988) Microbiological studies on amylolytic Oriental fermentation starters. *Mycopathologia*, **101**, 141–155.

Higgins, V.J., Bell, P.J., Dawes, I.W. and Attfield, P.V. (2001) Generation of a novel *Saccharomyces cerevisiae* strain that exhibits strong maltose utilization and hyperosmotic resistance using non recombinant techniques. *Applied and Environmental Microbiology*, **67**, 4346–4348.

Ho, V.T.T., Zhao, J. and Fleet, G.H. (2014) Yeasts are essential for cocoa bean fermentation. *International Journal of Food Microbiology*, **174**, 72–87.

Holzapfel, W.H. (2002) Appropriate starter culture technologies for small-scale fermentation in developing countries. *International Journal of Food Microbiology*, **75**, 197–212.

Hong, Y.A. and Park, H.D. (2013) Role of non-*Saccharomyces* yeasts in Korean wines produced from Campbell Early grapes: Potential use of *Hanseniaspora uvarum* as a starter culture. *Food Microbiology*, **34**, 207–214.

Hounhouigan, D.J., Nout, M.J.R., Nago, C.M., Houben, J.H. and Rombouts, F.M. (1999) Use of starter cultures of lactobacilli and yeast in the fermentation of mawé, an African maize product. *Tropical Science*, **39**, 220–226.

Howell, K., Cozzolino, D., Bartowsky, E., Fleet, G.H. and Henschke, P.A. (2006) Metabolic profiling as a tool for revealing *Saccharomyces* interactions during wine making. *FEMS Yeast Research*, **6**, 91–101.

Hui, Y.H., Meunier-Goddik, L., Hansen, A.S. and Josephsen, J. (eds) (2004) *Handbook of Food Fermentation and Beverage Fermentation Technology*, New York: Marcel-Dekker.

Husnik, J.I., Volschenk, H., Bauer, J., Colavizza, D., Luo, Z. and van Vuuren, H.J. (2006) Metabolic engineering of malolactic wine yeast. *Metabolic Engineering*, **8**, 315–323.

Iucci, L., Patrignani, F., Belletti, N. *et al.* (2007) Role of surface-inoculated *Debaryomyces hansenii* and *Yarrowia lipolytica* strains in dried fermented sausage manufacture. Part 2: Evaluation of their effects on sensory quality and biogenic amine content. *Meat Science*, **75**, 669–675.

Iwuoha, C.I. and Eke, O.S. (1996) Nigerian indigenous fermented foods: Their traditional process operation, inherent problems, improvements and current status. *Food Research International*, **29**, 527–540.

Jayani, R.S., Saxena, S. and Gupta, R. (2005) Microbial pectinolytic enzymes: A review. *Journal of Food Biochemistry*, **40**, 2931–2944.

Jespersen, L., Nielsen, D.S., Honholt, S. and Jakobsen, M. (2005) Occurrence and diversity of yeasts involved in fermentation of West African cocoa beans. *FEMS Yeast Research*, **5**, 441–453.

Jianzhong, Z., Xiaoli, L., Hanhu, J. and Mingsheng, D. (2009) Analysis of the microflora in Tibetan kefir grains using denaturing gradient gel electrophoresis. *Food Microbiology*, **26**, 770–775.

Johnson, E.A. and Echavarri-Erasun, C. (2011) Yeast biotechnology. In C.P. Kurtzman, J.W. Fell and T. Boekhout (eds), *The Yeasts: A Taxonomy Study* (vol. 1, pp. 21–44), New York: Elsevier.

Jordano, R., Medina, L.M. and Salmeson, J. (1991) Contaminating miroflora in fermented milk. *Journal of Food Protection*, **54**, 131–132.

Joshi, V.K., Sandhu, D.K. and Thakur, N.S. (1999) Fruit based alcoholic beverages. In V.K. Joshi and A. Pandey (eds), *Biotechnology: Food Fermentation* (pp. 647–744), Ernakulam: Educational Publishers.

Kanekar, P. and Joshi, N. (1993) *Lactobacillus fermentum*, *Leuconostoc mesenteroides* and *Hansenula silvicola* contributing to acetoin and folic acid during 'dhokla' fermentation. *Indian Journal of Microbiology*, **33**, 111–117.

Kanellaki, M., Bekatorou, A. and Koutinas, A.A. (2014) Low-temperature production of wine, beer, and distillates using cold-adapted yeasts. In P. Buzzini and R. Margesin (eds), *Cold-Adapted Yeasts: Biodiversity, Adaptation Strategies and Biotechnological Significance* (pp. 417–440), Berlin: Springer.

Kashyap, D.R., Vohra, P.K., Chopra, S. and Tewari, R. (2001) Applications of pectinases in the commercial sector: A review. *Bioresource Technology*, **77**, 215–227.

Ko, S.D. (1986) Indonesian fermented foods not based on soybeans. In C.W. Hesseltine and H.L.Wang (eds), *Indigenous Fermented Food of Non-Western Origin* (pp. 67–84), Berlin: J. Cramer.

Koh, B.K. and Singh, V. (2009) Cooking behavior of rice and black gram in the preparation of idli, a traditional fermented product of Indian origin, by viscography. *Journal of Texture Studies*, **40**, 36–50.

Kök Taş, T. Yeşim Ekinci, F. and Guzel-Seydim, Z.B. (2012) Identification of microbial flora in kefir grains produced in Turkey using PCR. *International Journal of Dairy Technology*, **65**, 126–131.

Kolani, S., Delgenes, J.P., Moletta, R., Traore, A. and Doh, A. (1996) Isolation and physiological characterization of yeasts involved in sorghum beer production. *Food Biotechnology*, **10**, 29–40.

Kozaki, M. and Uchimura, T. (1990) Micro-organisms in Chinese starter 'bubod' and rice wine 'tapuy' in the Philippines. *Journal of the Brewing Society of Japan*, **85**, 818–824.

Lefeber, T., Papalexandratou, Z., Gobert, W., Camu, N. and De Vuyst, L. (2012) On-farm implementation of a starter culture for improved cocoa bean fermentation and its influence on the flavour of chocolates produced thereof. *Food Microbiology*, **30**, 379–392.

Legras, J.L., Merdinoglu, D., Cornuet, J.M. and Karst, F. (2007) Bread, beer and wine: *Saccharomyces cerevisiae* diversity reflects human history. *Molecular Ecology*, **16**, 2091–2102.

Leite, A.M., Mayo, B., Rachid, C.T. *et al.* (2012) Assessment of the microbial diversity of Brazilian kefir grains by PCR-DGGE and pyrosequencing analysis. *Food Microbiology*, **31**, 215–221.

Libkind, D., Hittinger, C.T., Valério, E. *et al.* (2011) Microbe domestication and the identification of the wild genetic stock of lager-brewing yeast. *Proceedings of the National Academy of Sciences USA*, **108**, 14539–14544.

Limtong, S., Sintara, S., Suwannarit, P. and Lotong, N. (2002) Yeast diversity in Thai traditional alcoholic starter. *Kasetsart Journal: Natural Science*, **36**, 149–158.

Liu, S.Q. and Tsao, M. (2009) Inhibition of spoilage yeasts in cheese by killer yeast *Williopsis saturnus* var. *saturnus*. *International Journal of Food Microbiology*, **131**, 280–282.

Llauradó, J.M., Rozès, N., Constantí, M. and Mas, A. (2005) Study of some *Saccharomyces cerevisiae* strains for winemaking after preadaptation at low temperatures. *Journal of Agricultural and Food Chemistry*, **53**, 1003–1011.

Lopes, C.A., Rodríguez, M.E., Sangorrín, M., Querol, A. and Caballero, A.C. (2007) Patagonian wines: The selection of an indigenous yeast starter. *Journal of Industrial Microbiology and Biotechnology*, **34**, 539–546.

Lopes, C.A., van Broock, M., Querol, A. and Caballero, A.C. (2002) *Saccharomyces cerevisiae* wine yeast populations in a cold region in Argentinean Patagonia: A study at different fermentation scales. *Journal of Applied Microbiology*, **93**, 608–615.

Lowes, K.F., Shearman, C.A., Payne, J. *et al.* (2000) Prevention of yeast spoilage in feed and food by the yeast mycocin HMK. *Applied and Environmental Microbiology*, **66**, 1066–1076.

Magalhães, K.T., de Melo Pereira, G.V., Campos, C.R., Dragone, G. and Schwan, R.F. (2011) Brazilian kefir: Structure, microbial communities and chemical composition. *Brazilian Journal of Microbiology*, **42**, 693–702.

Malherbe, D.F., Du-Toit, M., Otero, R.R.C., van Rensburg, P. and Pretorius, I.S. (2003) Expression of the *Aspergillus niger* glucose oxidase gene in *Saccharomyces cerevisiae* and its potential applications in wine production. *Applied Microbiology and Biotechnology*, **61**, 502–511.

Marks, V.D., Ho, S.S.J., Erasmus, D. *et al.* (2008) Dynamics of the yeast transcriptome during wine fermentation reveals a novel fermentation stress response. *FEMS Yeast Research*, **8**, 35–52.

Martín, A., Córdoba, J.J., Aranda, E., Córdoba, M.G. and Asensio, M.A. (2006) Contribution of a selected fungal population to the volatile compounds on dry-cured ham. *International Journal of Food Microbiology*, **110**, 8–18.

Martini, A. (2003) Biotechnology of natural and winery-associated strains of *Saccharomyces cerevisiae*. *International Microbiology*, **6**, 207–209.

Masoud, W. and Jespersen, L. (2006) Pectin degrading enzymes in yeasts involved in fermentation of *Coffea arabica* in East Africa. *International Journal of Food Microbiology*, **110**, 291–296.

Mayser, P., Fromme, S., Leitzmann, C. and Gründer, K. (1995) The yeast spectrum of the 'tea fungus kombucha'. *Mycoses*, **38**, 289–295.

Mazzei, P., Spaccini, R., Francesca, N., Moschetti, G. and Piccolo, A. (2013) Metabolomic by 1H NMR spectroscopy differentiates 'Fiano di Avellino' white wines obtained with different yeast strains. *Journal of Agricultural and Food Chemistry*, **61**, 10816–10822.

McGovern, P.E., Zhang, J., Tang, J. *et al.* (2004) Fermented beverages of pre- and proto-historic China. *Proceedings of the National Academy of Sciences USA*, **101**, 17593–17598.

McKay, A.M. (1992) Growth of fermentative and non-fermentative yeasts in natural yoghurt stored in polystyrene cartons. *International Journal of Food Microbiology*, **15**, 383–388.

Miguel, M.G.C.P., Cardoso, P.G., Magalhães-Guedes, K.T. and Schwan, R.F. (2013) Identification and assessment of kefir yeast potential for sugar/ethanol-resistance. *Brazilian Journal of Microbiology*, **44**, 113–118.

Montaño, A., Sánchez, A.H., Casado, F.J., de Castro, A. and Rejano, L. (2003) Chemical profile of industrially fermented green olives of different varieties. *Food Chemistry*, **82**, 297–302.

Moreira, M., Mendes, F., Guedes de Pinho, P., Hogg, T. and Vasconcelos, I. (2008) Heavy sulphur compounds, higher alcohols and esters production profile of *Hanseniaspora uvarum* and *Hanseniaspora guilliermondii* grown as pure and mixed cultures in grape must. *International Journal of Food Microbiology*, **124**, 231–238.

Moreira, M., Mendes, F., Hogg, T. and Vasconcelos, I. (2005) Alcohols, esters and heavy sulphur compounds production by pure and mixed cultures of apiculate wine yeasts. *International Journal of Food Microbiology*, **103**, 285–294.

Moreno-Arribas, M.V. and Polo, M.C. (2005) Winemaking biochemistry and microbiology: Current knowledge and future trends. *Critical Review in Food Science and Nutrition*, **45**, 265–286.

Moroni, A.V., Arendt, E.K., Morrissey, J.P. and Dal Bello, F. (2010) Development of buckwheat and teff sourdoughs with the use of commercial starters. *International Journal of Food Microbiology*, **142**, 142–148.

Nakao, Y., Kanamori, T., Itoh, T. *et al.* (2009) Genome sequence of the lager brewing yeast, an interspecies hybrid. *DNA Research*, **16**, 115–129.

Narvhusa, J.A. and Gadaga, T.H. (2003) The role of interaction between yeasts and lactic acid bacteria in African fermented milks: A review. *International Journal of Food Microbiology*, **86**, 51–60.

Nielsen, D.S., Teniola, O.D., Ban-Koffi, L., Owusu, M., Andersson, T.S. and Holzapfel, W.H. (2007) The microbiology of Ghanaian cocoa fermentations analysed using culture-dependent and culture-independent methods. *International Journal of Food Microbiology*, **114**, 168–186.

Nyanga, L.K., Nout, M.J.R., Gadaga, T.H., Theelen, B., Boekhout, T. and Zwietering, M.H. (2007) Yeasts and lactic acid bacteria microbiota from masau (*Ziziphus mauritiana*) fruits and their fermented fruit pulp in Zimbabwe. *International Journal of Food Microbiology*, **120**, 159–166.

Odunfa, S.A. and Oyewole, O.B. (1998) African fermented foods. In B.J.B. Wood (ed.), *Microbiology of Fermented Foods* (2nd edn, vol. 2, pp. 713–752), London: Blackie Academic and Professional.

Okagbue, R.N. and Bankole, N.O. (1992) Use of starter cultures containing *Streptococcus diacetylactis*, *Lactobacillus brevis* and *Saccharomyces cerevisiae* for fermenting milk for production of Nigerian nono. *World Journal of Microbiology and Biotechnology*, **8**, 251–253.

Olesen, P.T. and Stahnke, L.H. (2000) The influence of *Debaryomyces hansenii* and *Candida utilis* on the aroma formation in garlic spiced fermented sausages and model minces. *Meat Science*, **56**, 357–368.

Oliveira, V.A., Vicente, M.A., Fietto, L.G. *et al.* (2008) Biochemical and molecular characterization of *Saccharomyces cerevisiae* strains obtained from sugar-cane juice fermentations and their impact in cachaça production. *Applied and Environmental Microbiology*, **74**, 693–701.

Organisation for Economic Co-operation and Development (2007) *OECD Best Practice Guidelines for Biological Resource Centres*, https://www.oecd.org/sti/biotech/38777417.pdf (accessed June 2016).

Oro, L., Ciani, M. and Comitini, F. (2014) Antimicrobial activity of *Metschnikowia pulcherrima* on wine yeasts. *Journal of Applied Microbiology*, **116**, 1209–1217.

Owuama, C.I. and Saunders, J.R. (1990) Physiological variants of *Saccharomyces cerevisiae* and *Kloeckera apiculata* from palm wine and cashew juice. *Journal of Applied Bacteriology*, **68**, 491–494.

Padonou, S.W., Nielsen, D.S., Akissoe, N.H., Hounhouigan, J.D., Nago, M.C. and Jakobsen, M. (2010) Development of starter culture for improved processing of Lafun, an African fermented cassava food product. *Journal of Applied Microbiology*, **109**, 1402–1410.

Papadimitriou, M.N.B., Resende, C., Kuchler, K. and Brul, S. (2007) High Pdr12 levels in spoilage yeast (*Saccharomyces cerevisiae*) correlate directly with sorbic acid levels in the culture medium but are not sufficient to provide cells with acquired resistance to the food preservative. *International Journal of Food Microbiology*, **113**, 173–179.

Papalexandratou. Z. and de Vuyst, L. (2011) Assessment of the yeast species composition of cocoa bean fermentations in different cocoa-producing regions using denaturing gradient gel electrophoresis. *FEMS Yeast Research*, **11**, 564–574.

Papapostolou, H., Servetas, Y., Bosnea, L.A., Kanellaki, M. and Koutinas, A. (2012) Novel technology development through thermal drying of encapsulated *Kluyveromyces marxianus* in micro- and nano-tubular cellulose in lactose fermentation and its evaluation for food production. *Applied Biochemistry and Biotechnology*, **168**, 2148–2159.

Pariza, M.W. and Johnson, E.A. (2001) Evaluating the safety of microbial enzyme preparations used in food processing: Update for a new century. *Regulatory Toxicology and Pharmacology*, **33**, 173–186.

Patrignani, F., Lucci, L., Vallicelli, M., Guerzoni, M.E., Gardini, F. and Lanciotti, R. (2007) Role of surface-inoculated *Debaryomyces hansenii* and *Yarrowia lipolytica* strains in dried fermented sausage manufacture. Part 1: Evaluation of their effects on microbial evolution, lipolytic and proteolytic patterns. *Meat Science*, **75**, 676–686.

Pedersen, L.L., Owusu-Kwarteng, J., Thorsen, L. and Jespersen, L. (2012) Biodiversity and probiotic potential of yeasts isolated from Fura, a West African spontaneously fermented cereal. *International Journal of Food Microbiology*, **159**, 144–151.

Pedersen, M.B. (1986) DNA sequence polymorphisms in the genus *Saccharomyces*. III. Restriction endonuclease fragment patterns of chromosomal regions in brewing and other yeast strains. *Carlsberg Research Communications*, **51**, 163–183.

Pepe, O., Villani, F., Oliviero, D., Greco, T. and Coppola, S. (2003) Effect of proteolytic starter cultures as leavening agents of pizza dough. *International Journal of Food Microbiology*, **84**, 319–326.

Pistarino, E., Aliakbarian, B., Casazza, A.A., Paini, M., Cosulich, M.E. and Perego, P. (2013) Combined effect of starter culture and temperature on phenolic compounds during fermentation of Taggiasca black olives. *Food Chemistry*, **138**, 2043–2049.

Plahutaa, P. and Raspor, P. (2007) Comparison of hazards: Current vs. GMO wine. *Food Control*, **18**, 492–502.

Prado, F.C., Parada, J.L., Pandey, A. and Soccol, C.R. (2008) Trends in non-dairy probiotic beverages. *Food Research International*, **41**, 111–123.

Pretorius, I.S. (2000) Tailoring wine yeast for the new millennium: Novel approaches to the ancient art of winemaking. *Yeast*, **16**, 675–729.

Pretorius, I.S. and Bauer, F.F. (2002) Meeting the consumer challenge through genetically customized wine-yeast strains. *Trends in Biotechnology*, **20**, 426–432.

Pretorius, I.S., du Toit, M. and van Rensburg, P. (2003) Designer yeasts for the 21st century. *Food Technology and Biotechnology*, **41**, 3–10.

Psomas, E., Andrighetto, C., Litopoulou-Tzanetaki, E., Lombardi, A. and Tzanetakis, N. (2001) Some probiotic properties of yeast isolates from infant faeces and Feta cheese. *International Journal of Food Microbiology*, **69**, 125–133.

Purriños, L., Carballo, J. and Lorenzo, J.M. (2013) The influence of *Debaryomyces hansenii, Candida deformans* and *Candida zeylanoides* on the aroma formation of dry-cured 'lacón'. *Meat Science*, **93**, 344–350.

Renouf, V., Claisse, O. and Lonvaud-Funel, A. (2007) Inventory and monitoring of wine microbial consortia. *Applied Microbiology and Biotechnology*, **75**. 149–164.

Rhee, S.J., Lee, C.Y.J., Kim, K.K. and Lee, C.H. (2003) Comparison of the traditional (samhaeju) and industrial (chongju) rice wine brewing in Korea. *Food Science and Biotechnology*, **12**, 242–247.

Ribéreau-Gayon, P. (2006) *Handbook of Enology* (2nd edn), Chichester: John Wiley & Sons.

Rivera-Espinoza, Y. and Gallardo-Navarro, Y. (2010) Non-dairy probiotic products. *Food Microbiology*, **27**, 1–11.

Rodríguez, M.E., Infante, J.J., Molina, M., Domínguez, M., Rebordinos, L. and Cantoral, J.M. (2010) Genomic characterization and selection of wine yeast to conduct industrial fermentations of a white wine produced in a SW Spain winery. *Journal of Applied Microbiology*, **108**, 1292–1302.

Rodríguez-Gómez, F., Arroyo-López, F.N., López-López, A., Bautista-Gallego, J. and Garrido-Fernández, A. (2010) Lipolytic activity of the yeast species associated with the fermentation/storage phase of ripe olive processing. *Food Microbiology*, **27**, 604–612.

Rodríguez-Gómez, F., Romero-Gil, V., Bautista-Gallego, J., Fernández, A.G. and Arroyo-López, F.N. (2012) Multivariate analysis to discriminate yeast strains with technological applications in table olive processing. *World Journal of Microbiology and Biotechnology*, **28**, 1761–1770.

Rodríguez-Lerma, G.K., Gutíerrez-Moreno, K., Cárdenas-Manríquez, M. *et al.* (2011) Microbial ecology studies of spontaneous fermentation: Starter culture selection for prickly pear wine production. *Journal of Food Science*, **76**, M346–M352.

Rohm, H., Lechner, F. and Lehner, M. (1990) Microflora of Austrian natural-set yogurt. *Journal of Food Protection*, **53**, 478–480.

Rollini, M., Casiraghi, E., Pagani, M.A. and Manzoni, M. (2007) Technological performances of commercial yeast strains (*Saccharomyces cerevisiae*) in different complex dough formulations. *European Food Research and Technology*, **226**, 19–24.

Romano, P., Capece, A. and Jespersen, L. (2006) Taxonomic and ecological diversity of foods and beverage yeasts. In A. Querol and G.H. Fleet (eds), *Yeasts in Food and Beverages* (pp. 13–54), Berlin: Springer.

Romano, P., Fiore, C., Paraggio, M., Caruso, M. and Capece, A. (2003) Function of yeast species and strains in wine flavour. *International Journal of Food Microbiology*, **86**, 169–180.

Rosell, C.M. and Gomez, M. (2007) Frozen dough and partially baked bread: An update. *Food Reviews International*, **23**, 303–319.

Rossignol, T., Dulau, L., Julien, A. and Blondin, B. (2003) Genome-wide monitoring of wine yeast gene expression during alcoholic fermentation. *Yeast*, **20**, 1369–1385.

Rossouw, D. and Bauer, F.F. (2009) Comparing the transcriptomes of wine yeast strains: Toward understanding the interaction between environment and transcriptome during fermentation. *Applied Microbiology and Biotechnology*, **84**, 937–954.

Rossouw, D., Heyns, E.H., Setati, M.E., Bosch, S. and Bauer, F.F. (2013) Adjustment of trehalose metabolism in wine *Saccharomyces cerevisiae* strains to modify ethanol yields. *Applied and Environmental Microbiology*, **17**, 5197–5207.

Sánchez, A.H., de Castro, A., Rejano, L. and Montaño, A. (2000) Comparative study on chemical changes in olive juice and brine during green olive fermentation. *Journal of Agricultural and Food Chemistry*, **48**, 5975–5980.

Sánchez-Molinero, F. and Arnau, J. (2008) Effect of the inoculation of a starter culture and vacuum packaging during the resting stage on sensory traits of dry-cured ham. *Meat Science*, **80**, 1074–1080.

Sandhu, D.K. and Soni, S.K. (1989) Microflora associated with Indian Punjabi warri fermentation. *Journal of Food Science and Technology*, **26**, 21–25.

Sangorrín, M.P., Lopes, C.A., Giraudo, M.R. and Caballero, A.C. (2007) Diversity and killer behaviour of indigenous yeasts isolated from the fermentation vat surfaces in four Patagonian wineries. *International Journal of Food Microbiology*, **119**, 351–357.

Sanni, A.I. and Lonner, C. (1993) Identification of yeasts isolated from Nigerian traditional alcoholic beverages. *Food Microbiology*, **10**, 517–523.

Saono, S., Gandjar, I. and Basuki, T. (1996) Indigenous fermented foods in which ethanol is a major product. In K.H. Steinkraus (ed), *Handbook of Indigenous Fermented Foods* (pp. 363–508), New York: Marcel Dekker.

Schuller, D. and Casal, M. (2005) The use of genetically modified *Saccharomyces cerevisiae* strains in the wine industry. *Applied Microbiology and Biotechnology*, **68**, 292–304.

Schwan, R.F., Mendonça, A.T., da Silva, Jr J.J., Rodrigues, V. and Wheals, A.E. (2001) Microbiology and physiology of Cachaça (Aguardente) fermentations. *Antonie Van Leeuwenhoek*, **79**, 89–96.

Sefa-Dedeh, S., Sanni, A.I., Tetteh, G. and Sakyi-Dawson, E. (1999) Yeasts in the traditional brewing of 'pito' in Ghana. *World Journal of Microbioogy and Biotechnology*, **15**, 593–597.

Settanni, L., Sannino, C., Francesca, N., Guarcello, R. and Moschetti, G. (2012) Yeast ecology of vineyards within Marsala wine area (western Sicily) in two consecutive vintages and selection of autochthonous *Saccharomyces cerevisiae* strains. *Journal of Bioscience and Bioengineering*, **114**, 606–614.

Shetty, P.H., Hald, B. and Jespersen, L. (2007) Surface binding of aflatoxin B1 by *Saccharomyces cerevisiae* strains with potential decontaminating abilities in indigenous fermented foods. *International Journal of Food Microbiology*, **113**, 41–46.

Shrestha, H., Nand, K. and Rati, E.R. (2002) Microbiological profile of murcha starters and physico-chemical characteristics of poko, a rice based traditional fermented food product of Nepal. *Food Biotechnology*, **16**, 1–15.

Shurpalekar, S.R. (1986) Papads. In N.R.Reddy, M.D. Pierson and D.K.Salunkhe (eds), *Legume-based Fermented Foods* (pp. 191–217), Boca Raton, FL: CRC Press.

Sicard, D. and Legras, J.L. (2011) Bread, beer and wine: Yeast domestication in the *Saccharomyces sensu stricto* complex. *Comptes Rendus Biologies*, **334**, 229–236.

Silva, C.F., Vilela, D.M., de Souza Cordeiro, C., Duarte, W.F., Dias, D.R. and Schwan, R.F. (2013) Evaluation of a potential starter culture for enhance quality of coffee fermentation. *World Journal of Microbiology and Biotechnology*, **29**, 235–247.

Smithers, G.W. (2008) Whey and whey proteins: From 'gutter-to-gold'. *International Dairy Journal*, **18**, 695–704.

Sridevi, J., Halami, P.M. and Vijayendra, S.V.N. (2010) Selection of starter cultures for idli batter fermentation and their effect on quality of idlis. *Journal of Food Science and Technology*, **47**, 557–563.

Stackebrandt, E., Smith, D., Casaregola S. *et al.* (2014) Deposit of microbial strains in public service collections as part of the publication process to underpin good practice in science. *SpringerPlus*, **3**, 208.

Steinkraus, K.H. (1996) *Handbook of Indigenous Fermented Foods* (2nd edn), New York: Marcel Dekker.

Stratford, M. (2006) Food and beverage spoilage yeasts. In A. Querol and G.H. Fleet (eds), *Yeasts in Food and Beverages* (pp. 335–379), Berlin: Springer.

Suzzi, G., Lanorte, M.T., Galgano, F. *et al.* (2001) Proteolytic, lipolytic and molecular characterisation of *Yarrowia lipolytica* isolated from cheese. *International Journal of Food Microbiology*, **69**, 69–77.

Swiegers, J.H. and Pretorius, I.S. (2005) Yeast modulation of wine flavor. *Advances in Applied Microbiology*, **57**, 131–175.

Tamang, J.P. and Fleet, G.H. (2009) Yeasts diversity in fermented foods and beverages. In T. Kunze Satyanarayana (ed.), *Yeast Biotechnology: Diversity and Applications* (pp. 170–193), Berlin: Springer.

Tamang, J.P., Sarkar, P.K. and Hesseltine, C.W. (1988) Traditional fermented foods and beverages of Darjeeling and Sikkim: A review. *Journal of the Science of Food and Agriculture*, **44**, 375–385.

Tanaka-Tsuno, F., Mizukami-Murata, S., Murata, Y. *et al* (2007) Functional genomics of commercial baker's yeasts that have different abilities for sugar utilization and high-sucrose tolerance under different sugar conditions. *Yeast*, **24**, 901–911.

Tanghe, A., Van Dijck, P. and Thevelein, J.M. (2003) Determinants of freeze tolerance in microorganisms, physiological importance, and biotechnological applications. *Advances in Applied Microbiology*, **53**, 129–176.

Teniola, O.D. and Odunfa, S.A. (2001) The effects of processing methods on the level of lysine, methionine and the general acceptability of ogi processed using starter cultures. *International Journal of Food Microbiology*, **63**, 1–9.

Torija, M.J., Rozès, N., Poblet, M., Guillamón, J.M. and Mas, A. (2001) Yeast population dynamics in spontaneous fermentations: Comparison between two different wine-producing areas over a period of three years. *Antonie van Leeuwenhoek*, **79**, 345–352.

Torija, M.J., Rozès, N., Poblet, M., Guillamón, J.M. and Mas, A. (2003) Effects of fermentation temperature on the strain population of *Saccharomyces cerevisiae*. *International Journal of Food Microbiology*, **80**, 47–53.

Trabalzini, L., Paffetti, A., Scaloni, A. *et al.* (2003) Proteomic response to physiological fermentation stresses in a wild-type wine strain of *Saccharomyces cerevisiae*. *Biochemical Journal*, **370**, 35–46.

Tristezza, M., Vetrano, C., Bleve, G. *et al.* (2012) Autochthonous fermentation starters for the industrial production of Negroamaro wines. *Journal of Industrial Microbiology and Biotechnology*, **39**, 81–92.

Tsuyoshi, N., Fudou, R., Yamanaka, S. *et al.* (2005) Identification of yeast strains isolated from marcha in Sikkim, a microbial starter for amylolytic fermentation. *International Journal of Food Microbiology*, **99**, 135–146.

Valamoti, S.M., Mangafa, M., Koukouli-Chrysanthaki, C. and Malamidou, D. (2007) Grape-pressings from northern Greece: The earliest wine in the Aegean? *Antiquity*, **81**, 54–61.

Valero, E., Cambon, B., Schuller, D., Casal, M. and Dequin, S. (2007) Biodiversity of *Saccharomyces* yeast strains from grape berries of wine-producing areas using starter commercial yeasts. *FEMS Yeast Research*, **7**, 317–329.

Van den Tempel, T. and Jakobsen, M. (2000) The technological characteristics of *Debaryomyces hansenii* and *Yarrowia lipolytica* and their potential as starter cultures for production of Danablu. *International Dairy Journal*, **10**, 263–270.

Van den Tempel, T. and Nielsen, M.S. (2000) Effects of atmospheric conditions, NaCl, pH on growth and interactions between moulds and yeasts related to blue cheese production. *International Journal of Food Microbiology*, **57**, 193–199.

van der Aa Kuhle, A., Jespersen, L., Glover, R.L.K., Diawara, B. and Jakobsen, M. (2001) Identification and characterization of *Saccharomyces cerevisiae* strains from West African sorghum beer. *Yeast*, **18**, 1069–1079.

van Rensburg, P., Strauss, M.L.A., Lambrechts, M.G., Cordero Otero, R.R. and Pretorius, I.S. (2007) The heterologous expression of polysaccharidase-encoding genes with oenological relevance in *Saccharomyces cerevisiae*. *Journal of Applied Microbiology*, **103**, 2248–2257.

Vanderhaegen, B., Neven, H., Verachtert, H. and Derdelinckx, G. (2006) The chemistry of beer aging: A critical review. *Food Chemistry*, **95**, 357–381.

Varela, C., Cardenas, J., Melo, F. and Agosin, E. (2005) Quantitative analysis of wine yeast gene expression profiles under wine making conditions. *Yeast*, **22**, 369–383.

Verbelen, P.J., de Schutter, D.P., Delvaux, F., Verstrepen, K.J. and Delvaux, F.R. (2006) Immobilized yeast cell systems for continuous fermentation applications. *Biotechnology Letters*, **28**, 1515–1525.

Verstrepen, K.J., Chambers, P.J. and Pretorius, I.S. (2006) The development of superior yeast strains for the food and beverage industries: Challenges, opportunities, and potential benefits. In A. Querol and G.H. Fleet (eds), *Yeasts in Foods and Beverages* (pp. 399–444), Berlin: Springer.

Viana, F., Belloch, C., Vallés, S. and Manzanares, P. (2011) Monitoring a mixed starter of *Hanseniaspora vineae-Saccharomyces cerevisiae* in natural must: Impact on 2-phenylethyl acetate production. *International Journal of Food Microbiology*, **151**, 235–240.

Vieira-Dalodé, G., Jespersen, L., Hounhouigan, J., Moller, P.L., Nago, C.M. and Jakobsen, M. (2007) Lactic acid bacteria and yeasts associated with gowé production from sorghum in Bénin. *Journal of Applied Microbiology*, **103**, 342–349.

Viljoen, B.C. (2001) The interaction between yeasts and bacteria in dairy environments. *International Journal of Food Microbiology*, **69**, 37–44.

Viljoen, B.C., Lourens-Hattingha, A., Ikalafenga, B. and Peter, G. (2003) Temperature abuse initiating yeast growth in yoghurt. *Food Research International*, **36**, 193–197.

Vogelmann, S.A., Seitter, M., Singer, U., Brandt, M.J. and Hertel, C. (2009) Adaptability of lactic acid bacteria and yeasts to sourdoughs prepared from cereals, pseudocereals and cassava and use of competitive strains as starters. *International Journal of Food Microbiology*, **130**, 205–212.

Willaert, R. and Nedovic, V.A. (2006) Primary beer fermentation by immobilised yeast: A review on flavour formation and control strategies. *Journal of Chemical Technology and Biotechnology*, **81**, 1353–1367.

Witthuhn, R.C., Schoeman, T. and Britz, T.J. (2004). Isolation and characterization of the microbial population of different South African kefir grains. *International Journal of Dairy Technology*, **57**, 33–37.

Wongkhalaung, C. and Boonyaratanakornkit, M. (2007) Characterization of new baker's yeast strains and their leavening ability in bread dough. *Kasetsart Journal of Natural Sciences*, **41**, 751–763.

Wu, P. Xu, X., Xu, Y., Chen, Q. and Pan, S. (2011) *Brettanomyces* as a starter culture in rice-steamed sponge cake: A traditional fermented food in China. *Current Microbiology*, **63**, 458–463.

Wyder, M.T. (1998) Identification and characterization of the yeast flora in kefir and smear ripened cheese: Contribution of selected yeasts to cheese ripening. PhD thesis, Swiss Federal Institute of Technology, Zurich.

Wyder, M.T. and Puhan, Z. (1999) Role of selected yeasts in cheese ripening: An evaluation in aseptic cheese curd slurries. *International Dairy Journal*, **9**, 117–124.

Żarowska, B., Wojtatowicz, M., Połomska, X., Juszczyk, P. and Chrzanowska, J. (2004) Factors affecting killer activity of some yeast species occurring in rokpol cheese. *Folia Microbiologica*, **49**, 713–717.

Zulu, R.M., Dillon, V.M. and Owens, J.D. (1997) Munkoyo beverage, a traditional Zambian fermented maize gruel using Rhynchosia root as amylase source. *International Journal of Food Microbiology*, **34**, 249–258.

Zuzuarregui, A., Monteoliva, L., Gil, C. and del Olmo, M (2006) Transcriptomic and proteomic approach for understanding the molecular basis of adaptation of *Saccharomyces cerevisiae* to wine fermentation. *Applied and Environmental Microbiology*, **72**, 836–847.

CHAPTER 3

Fungal starters: An insight into the factors affecting the germination of conidia

Philippe Dantigny[1] and Antonio Bevilacqua[2]
[1]*Université de Brest, LUBEM, ESIAB, France*
[2]*Department of the Science of Agriculture, Food and Environment, University of Foggia, Italy*

The involvement of moulds (filamentous fungi) in food fermentation goes back to the first records on blue and white moulded cheeses. The first records for the production of well-known cheeses such as Gorgonzola and Roquefort date to 879 and 1070 CE respectively (Jakobsen *et al.* 2002). Nowadays, fungi are used worldwide to produce a variety of indigenous foods (see Table 3.1); they are usually recovered in spontaneous fermentations, although some commercial starter cultures are available on the market (e.g. *Penicillium camemberti*, *Penicillium roqueforti*, *Penicillium nalgiovense*, *Aspergillus oryzae*, *Aspergillus sojae*).

Fungal starters extensively contribute to aroma and texture formation, and in this respect their proteases and lipases play a major role (Geisen and Färber 2002). Some other important traits of fungal starter cultures are the competition with undesired microorganisms, the effect on lactate and the increase of pH, as well as their effect on drying.

Naturally occurring mycobiota represent an important reservoir from which to select promising fungal starters; however, many strains do not fulfil the basic requirements of a possible fungus candidate as a starter (Geisen and Färber 2002); that is, it should not produce mycotoxin, not produce undesired secondary compounds, compete with undesired moulds and foodborne pathogens, be adapted to food products and produce only the desired changes.

Many books and papers have focused on the role of fungi as promising starter cultures. The topic of this chapter is quite different, as it offers an overview of the germination of fungal spores as a key step for the optimal use of fungal starters in foods.

Colonization of food by fungal starters is greatly affected by their germination time. Therefore, it is paramount that the germination time of a fungal starter can be controlled carefully to ensure the domination of inoculated fungi over pathogens and spoilage organisms. Competition may also occur between two

Starter Cultures in Food Production, First Edition. Edited by Barbara Speranza, Antonio Bevilacqua, Maria Rosaria Corbo and Milena Sinigaglia.

Table 3.1 Examples of fungi as starter cultures.

Products	Fungal starters
Dairy products	*Penicillium camemberti* and *Penicillium roqueforti* for the production of blue-veined cheese as secondary starter cultures. They can tolerate medium to high salt concentrations and possess pronounced proteolytic activity.
Meat products	Fungi represent secondary starter cultures for fermented meat products in Italy, Hungary, France, Germany and Spain. *Penicillium nalgiovense* and *Penicillium chrysogenum* represent the most important species for this kind of product. They are involved in many activities, such as an increase of pH, uniform drying and flavour improvement due to their metabolism.
Wine	*Botrytis cinerea* has been labelled a secondary starter culture for the production of sweet white wine.
Indigenous fermented foods	*Aspergillus* spp. and *Penicillium* spp. are used as amilolytic starters for the production of many indigenous foods, such as sake in Japan. In many other products, it is possible to isolate and recovery *Rhizopus* spp. and *Mucor* spp.
Fermented fish products	*Aspergillus penicillioides*, *Aspergillus wentii* and *Eurotium rubrum* can be easily recovered in Indonesian salted fish.

Source: Adapted from Hui and Özgul (2012).

fungal starters. The production of a blue-mould cheese by *P. roqueforti* also involved the use of *Geotrichum candidum* as a ripening agent. Spores of *P. roqueforti* developed germ tubes within 12–15 hours in a rich medium containing carbon and nitrogen sources at pH 6.5 and 26°C (Fan *et al.* 1976). Thereafter the tubes grew at approximately 0.5 mm per day. The milk was already inoculated with *P. roqueforti* spores, whereas spores of *G. candidum* were spread on the cheese just before ripening. Growth of *P. roqueforti* is frequently absent or poor in the outer regions of blue cheese due to localized high salt (NaCl) concentrations (Morris 1964; Godinho and Fox 1981). However, in some cases green spots of *P. roqueforti* appeared on the surface of the cheese before *G. candidum* developed, usually 2–3 days after salting (Rousseau 1984; Chapman and Sharpe 1990), thus preventing these cheeses from being sold. Interspecific interactions also influenced the sporulation of *P. roqueforti*, which was highly affected in the presence of *G. candidum* at 25% carbon dioxide irrespective of the levels of oxygen and NaCl (Van den Tempel and Nielsen 2000). However, in the present example, due to the production of spores by *P. roqueforti*, the green spots were visible on the surface of the cheeses prior to possible inhibition by *G. candidum*. Delayed germination and sporulation of *P. roqueforti* would solve the problem.

Commercially available mould starter cultures are supplied as either concentrated hypertonic liquid suspensions or freeze-dried powder (Sunesen and Stahnke 2003). Frozen cultures generally have a shorter lag phase, although this is somewhat less critical than in dairy fermentations that are to be finished after hours rather than days. It is recommended that freeze-dried powders be resuscitated for several hours before use to allow faster colonization, but solutions should be prepared daily, since spores in suspension rapidly die out (Jessen 1995). Some sausage manufacturers resuscitate lyophilized cultures for several hours before use, but none of them propagates commercial starters *in vitro* any further (Lücke 2000). These practices suggest that controlling the germination time of spores and the growth of moulds should be handled by the producers of starters rather than by cheese and sausage manufacturers.

The viability of spores is also an important factor in controlling the colonization of foods by fungal starters. Conidia of *Botrytis cinerea* stored for 30 months at −80°C, −20°C and 4°C exhibited viability of 79%, 8% and 0.2%, respectively (Gindro and Pezet 2001). The cryopreservation of the fungal starter *G. candidum* was improved by artificial nucleation and temperature downshift control (Missous *et al.* 2007). As a result, a significant decrease of the lethality of *G. candidum* cells subjected to freezing–thawing cycles was observed. Viability and germination time are two key factors that should be assessed for fungal starters. Both can be determined accurately by fitting germination data – that is, the percentage of germinated spores versus time – by means of suitable kinetic models that will be described in this chapter. In order to produce faster or slower commercial fungal starters, it would be highly desirable to control all the factors that may affect the germination time and the viability of fungal spores (Figure 3.1).

For over 15 years, a new field named 'predictive mycology' has been developed to understand and predict the development of fungi and the production of secondary metabolites in foods (Dantigny *et al.* 2005a). This is concerned mainly with foodborne fungi. However, definitions and tools that can be used to describe fungal development should not depend on whether mould is used as a fungal starter or considered as a food spoiler. For example, *Aspergillus niger*, a fungus

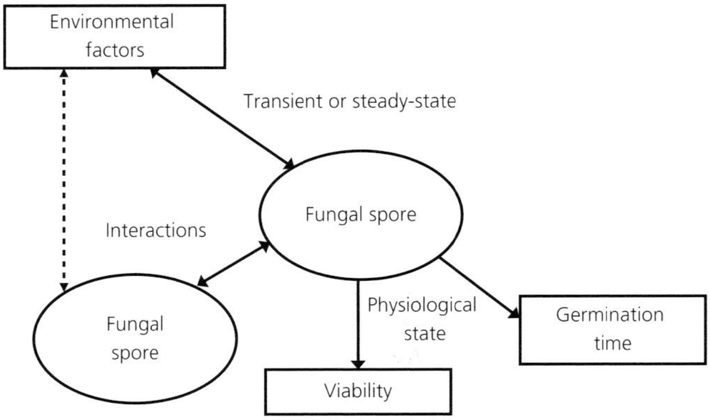

Figure 3.1 Factors that affect the germination and viability of fungal spores. The physiological state can be altered during the production and storage of spores, and during the early phases of germination.

responsible for food spoilage, is also used for citric acid production. It does not seem reasonable that definitions should depend on the use of the fungus (Dantigny *et al.* 2013).

Definitions

The lag time for growth is an important parameter to be estimated for pathogenic bacteria, because after this period microorganisms grow exponentially. However, for fungi the term 'lag time for growth' is not appropriate, because after germination mycelium is already growing and branching, although not visible to the naked eye. It is much better to use the term 'time to visible growth', which is the time required for the colony to reach a diameter in the range of 3–5 mm. Germination of fungal spores can be considered as the main step on which to focus, because visible mycelium appears shortly after germination is completed (Dantigny *et al.* 2005a). Fungal spore germination marks the resumption of vegetative development and the formation of a new individual or colony (Isaac 1998). Germination of a mould spore is a physiological reaction of a resting cell to changes in environmental conditions. Spore germination is a key process common to all fungi. It can be divided into four stages: breaking of spore dormancy; isotropic swelling; establishment of cell polarity; and formation of a germ tube and maintenance of polar growth (d'Enfert 1997; Wendland 2001). Oxygen is required for swelling and the process is energy dependent. During swelling, which represents the major part of the time of the germination process, an isotropic increase in the diameter of the spore is observed. Although the formation of a germ tube does not represent a long period in the germination process, attention is focused on this stage because the definition of a germinated spore is based on a comparison between the length of the germ tube and the length of the swollen spore.

In a need to standardize methods for assessing germination, it was recommended during the first workshop on predictive mycology (Dantigny *et al.* 2006) that the following definition be used: 'a spore had germinated when the length of the longest germ tube was greater than or equal to the greatest dimension of the swollen spore' (Huang *et al.* 2001).

Spores do not germinate at the same time. It is necessary to provide a widely accepted definition of the germination time, t_G, of a population of spores. Those concerned with food spoilage moulds would set t_G at a low percentage (say 10%). The figure of 10% was chosen in preference to larger percentages to obtain a better estimate of the minimum water activity (a_w) allowing germination (Magan and Lacey 1984). In contrast, people concerned with producing starters of fungal metabolites would use a greater percentage (say 90%). It appeared that 50% was a good trade-off between these boundaries (Dantigny *et al.* 2006). However, 50% of germination does not have the same significance whether the maximum percentage of germinated spores equals 60% or 100%.

It is better to base the definition of t_G on a certain percentage of the viable spores, not the inoculated spores. Another practical interest of the definition is that whatever the percentage of viable spores, t_G can be determined (Dantigny *et al.* 2011). For example, if t_G is defined as the time required to have 50% of inoculated spores, this time cannot be determined if the percentage of viable spores is less than 50%. For instance, a maximum percentage of 47.7% was reported for the germination of *Penicillium chrysogenum* conidia submitted to water stress at 0.95 a_w (Judet *et al.* 2008). It would have been impossible to determine t_G based on a 50% germinated conidia. In contrast, 50% of the viable spores was synonymous with 23.9% of the inoculated spores. During the last workshop dedicated to germination, it was recommended that the germination time of a population of spores be defined as the time required to have 50% of the viable spores germinated (Huang *et al.* 2001; Dantigny *et al.* 2006).

Modelling of germination kinetics

The suitability of germination models to describe experimental data is usually evaluated by the goodness of fit. Although this criterion is important, any good model should provide accurate estimations of parameters that are relevant to describe germination kinetics. The parameters that are widely used by mycologists are the percentage of viable spores, which should be less than or equal to 100%, and the germination time, t_G. Primary models for germination should be in accordance with the definitions of these biological parameters. The asymptotic value of the percentage of germination for $t \rightarrow +\infty$ should be equal to the percentage of viable spores; the percentage of germinated spores at $t = t_G$ should be consistent with the definition of the germination time.

The Gompertz equation

The Gompertz equation has been used to fit germination data of *Penicillium digitatum*, *Penicillium italicum* and *G. candidum* (Plaza *et al.* 2003), *Aspergillus ochraceus* (Pardo *et al.* 2005a, b), *P. chrysogenum* (Judet *et al.* 2008), *Penicillium verrucosum* (Pardo *et al.* 2006), *Fusarium moniliforme* and *Fusarium proliferatum* (Marín *et al.* 1996) and some *Aspergillus* and *Penicillium* spp. (Marín *et al.* 1998). The modified Gompertz equation is:

$$P = A.\exp.\left(-\exp\left[\frac{\mu_m.e(1)}{A}(\delta - t) + 1\right]\right) \tag{3.1}$$

where A (%) is the asymptotic P value at $t \rightarrow +\propto$, μ_m (% h^{-1}) is the slope term of the tangent line through the inflection point (t_i) as defined further, δ (h) is the t-axis intercept of the tangent through the inflection point and t is the time (h). The inflection point is $t_i = \delta + A/(\mu_m e(1))$ (Dantigny *et al.* 2003). Parameter A represents the maximum percentage of germinated spores, or the percentage of viable spores. This value should not exceed 100%. Because the model is asymmetric, the estimations of A can exceed this value if germination data in the upper right section of the curve (i.e. close to 100%) are omitted. In the case of an overestimation of A, this parameter can be set to 100%.

Because μ_m represents the growth rate for bacteria, this parameter is sometimes erroneously considered as the growth rate of the germ tubes (Schubert *et al.* 2010). In fact, μ_m could be related to the distribution of the germination time among a population of spores. The greater the μ_m, the lower is the variability of the germination time. Whatever the definition of the germination time, this parameter cannot be determined directly by the Gompertz equation. It has been shown that the accuracy of t_G depends on that of the three parameters of the Gompertz equation, namely A, μ_m and δ (Dantigny *et al.* 2013).

The logistic model

The logistic model has been used to fit germination data of *Mucor racemosus* (Dantigny *et al.* 2002), *P. chrysogenum* (Dantigny *et al.* 2005b; Judet *et al.* 2008) and some *Aspergillus* and *Penicillium* species (Nanguy *et al.* 2010).

The logistic function is:

$$P = \frac{P\max}{1 + \exp\left[k(\tau - t)\right]} \tag{3.2}$$

where P_{max} (%) is the asymptotic P value at $t \rightarrow +\infty$, τ (h) is the inflection point where P equals half of P_{max}, t is the time (h) and $k(h^{-1})$ is related to the slope of the tangent line through the inflection point. The slope of the tangent line at τ is equal to $k.P_{max}/4$ (Dantigny *et al.* 2007). The germination time t_G is equal to τ.

P_{max} represents the maximum percentage of germinated spores. Because the logistic function is symmetric about the inflection point, the upper right section of the germination curve is symmetric to the lower left part for which germination data, in the range 0–50%, are available. Therefore, if data in the range 0–100% are missing, there is less risk than in asymmetric models that overestimation of P_{max} may occur. The germination time t_G, defined as the time at which 50% of the viable spores have germinated, is equal to τ.

The asymmetric model

The main advantage of the logistic model is to provide accurate estimations of germination time. However, in certain cases skewed distributions of individual germination times among spore populations were observed (Judet *et al.* 2008). Attempting to fit a symmetric function to asymmetric data germination sets seems inappropriate. On the other hand, it does not seem realistic to adjust unskewed and skewed distributions with symmetric and asymmetric models, respectively. It is highly desirable that the same model can be used for fitting any data set, regardless of its

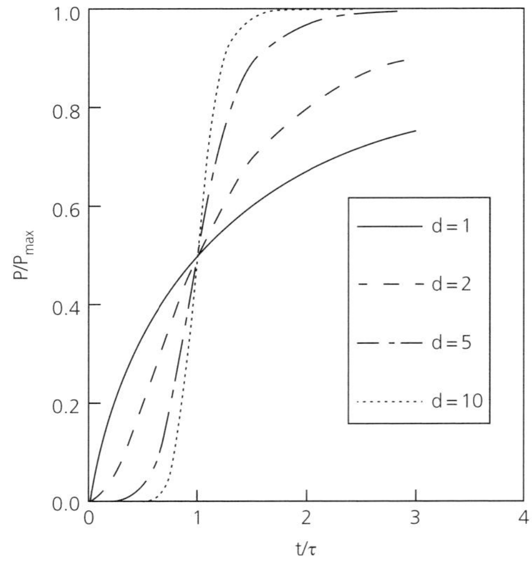

Figure 3.2 Effect of the design parameter d on the shape of the asymmetric model. Source: Dantigny *et al.* 2011.

skewness. For this reason, a versatile model capable of fitting either apparent symmetric and asymmetric germination curves has been proposed (Dantigny *et al.* 2011):

$$P = P_{max}\left[1 - \frac{1}{1+\left(\frac{t}{\tau}\right)^{d}}\right] \tag{3.3}$$

The asymmetric model is derived from the non-competitive inhibition model described by Yano and Koya (1973). P_{max} is the asymptotic P value at $t \rightarrow +\infty$ and τ(h) is the point where P equals half of P_{max}. The germination time t_G is equal to τ. The different shapes of the curves for different values of the design parameter $d>0$ are shown in Figure 3.2. For $d=1$ the shape of the curve differs from the S-shaped germination curve. For $d=2$, the percentage of germinated spores increases shortly after the origin. This is contradictory to the observation of a swelling period before the germinating tube can be formed and the spore germinated. Therefore, in practice the asymmetric model is only a candidate for fitting germination data for d values greater than 2.

Factors that affect germination parameters

Spore density

Intraspecific interactions are especially important during germination. Germination can be inhibited when spores are present in high densities, an effect observed for example in *A. niger* and the zygomycete *Syncephalastrum racemosum* (Hobot and Gull 1980; Barrios-Gonzáles *et al.* 1989). Self-inhibitors inhibit spore germination reversibly. The major function of self-inhibitors is stated as prevention of the premature germination of spores (Chitarra *et al.* 2004). Highly active extracellular siderophores, which are important in conidial germination, were detected in young cultures of *P. chrysogenum* (Charlang *et al.* 1982). When exposed to solutions at low water activity, conidia lost a fraction of their cellular siderophores and subsequent germination failed, or was greatly delayed. However, the influence of the inoculum size on the lag time for growth of *P. chrysogenum* was not significant in the range 10^3–10^5 spores/mL (Sautour *et al.* 2003), thus suggesting that the interaction effect between spores is minor for fungal starters.

Environmental factors

The effects of environmental factors on germination have been described and modelled for *Aspergillus carbonarius* (Mitchell 2006), *A. ochraceus* (Pardo *et al.* 2004), *A. niger, Eurotium amstelodami, Fusarium oxysporum* and *P. chrysogenum* (El Halouat and Debevere 1997), *Aspergillus parasiticus* (Schubert *et al.* 2010), *Fusarium verticillioides* and *F. proliferatum* (Marín *et al.* 1996), *M. racemosus* (Dantigny *et al.* 2002), *P. chrysogenum* (Dantigny *et al.* 2005b), *P. verrucosum* (Pardo *et al.* 2006) and *Ventura nashicola* (Li *et al.* 2003). The effects of temperature and water activity on germination time and viability have usually been assessed, but in a few studies the ethanol concentration and gas composition ($CO_2/O_2/N_2$) were also considered. It has been shown that for conditions close to the optimum, symmetric germination curves were characterized by a sharp increase of the percentage of germination, thus demonstrating that all spores were germinated at about the same moment (Dantigny *et al.* 2007). In contrast, germination curves tended to be asymmetric, in addition to an increase in the germination time, as experimental conditions moved towards reduced water activities and chilled temperatures.

Physiological state

In order to produce spores within the minimum period of time, the environmental conditions were set at optimum growth in terms of water activity and temperature. Additional objectives can be to increase the number of spores produced (Eicher and Ludwig 2002) or to avoid erratic germination (Pitt and Christian 1968). It has been shown that nutritional and environmental conditions prevailing during spore formation may exert a profound influence on their viability (Darby and Mandels 1955), heat resistance (Conner and Beuchat 1987a, b; Beuchat 1988), resistance to preservatives, germination time (Blaszyk *et al.* 1998) and chemistry (Jackson and Schisler 1992). Young conidia of *Aspergillus fumigatus* and *A. niger* showed higher germination rates than old conidia, but age did not affect the germination of *Aspergillus flavus* (Araujo and Gonçalves Rodrigues 2004). A similar study on *Neurospora crassa* exhibited a clear correlation between the germinability of conidia and culture age (Kawanabe 1986). Experiments carried out in our laboratory have also shown that conidia of *P. chrysogenum* harvested from 2-day-old sporulating mycelium exhibited faster germination than other spores harvested from 7-day-old sporulating mycelium. Therefore, there is a trend of an increase in germination time with an increase in the age of the spores. *P. roqueforti* spores produced at 0.88 a_w exhibited shorter germination times than those produced at 0.99 a_w by 5.5 and 4 hours, respectively. Spores of *Penicillium viridicatum* and *Penicillium aurantiogriseum* produced with glycerol exhibited the longest (12 h) and the shortest (6.5 h) germination times, respectively (Blaszyk *et al.* 1998).

Because no clear pattern was evident with regard to germination time, similar experiments were carried out in our laboratory for assessing the influence of the water activities of sporulation, $a_{w\,sp}$, and germination, $a_{w\,ge}$, on the germination time of *P. chrysogenum* (Judet *et al.* 2008). The germination time and its standard deviation were increased at reduced water activity for germination. In contrast, the maximum percentage of germination was less at 0.95 $a_{w\,ge}$ than at 0.99 $a_{w\,ge}$. At 0.99 $a_{w\,ge}$ no significant difference in germination time was shown between conidia produced at 0.95 $a_{w\,sp}$ and 0.99 $a_{w\,sp}$. In contrast, the decrease in germination time observed for spores obtained at 0.95 $a_{w\,sp}$ was clearly exhibited at 0.95 $a_{w\,ge}$.

Incubation of macroconidia at low humidity (0–53% RH) to simulate a drought period suppressed germination and decreased the viability of spores (Beyer *et al.* 2005). The effect of the period of time after discharge of ascospores of *Gibberella zeae* from perithecia on germination was also studied (Beyer and Verreet 2005). It was shown that freshly discharged ascospores germinated within 4 hours at 20°C and 100% RH, but the rate of germination and the percentage of viable ascospores decreased over time. Humidity during storage was a key factor in the germination of *G. zeae*. By incubating ascospores at 53% RH, the percentage of viable spores decreased from 93% to 6% within 10 minutes.

The effects of relative humidity, RH (%), time (day) and temperature (°C) during this period, called 'storage', on the germination time, τ(h), of *A. carbonarius* and *P. chrysogenum* were studied by Lattab *et al.* (2012). A Doehlert design was used in the range 20–100% RH, 2–28 days and 5–25°C. Compared to unstored conidia, the germination time of conidia stored at 60% RH, 15 days, 5 °C was increased by 23% and 28% for *A. carbonarius* and *P. chrysogenum*, respectively. Stored conidia exhibited a minimum τ value at 60% RH and 100% RH for *A. carbonarius* and *P. chrysogenum*, respectively. For these species, τ was minimum for 2 days of storage. The effect of temperature was RH dependent for *A. carbonarius*. The germination time of stored conidia was clearly greater than that of fresh conidia obtained in the laboratory. Minimum germination times for stored conidia were obtained for 2 days, the shortest period of storage. Thereafter there was a trend of an increase in the germination time as the storage duration was increased. It was also shown in that study that the environmental conditions that prevail during storage are paramount. For *P. chrysogenum*, the decrease of germination time observed at 100% RH can be explained by a swelling of the conidia during storage. In fact, it has been observed in the laboratory that some conidia of *P. chrysogenum* are capable of germination at high relative humidity. Conversely, the increase in the germination time observed for the two species at 20% RH can be explained by a decrease in the water content of the conidia during storage. The minimum germination time observed at 60% RH for *A. carbonarius* conidia can be compared to 75% RH, the value at which a loss of viability was reported for one strain of *A. flavus* (Teitell 1958). These observations are not contradictory, as a loss of viability in addition to an increase in the rate of germination was reported earlier for *P. chrysogenum* conidia (Judet *et al.* 2008). The effect of temperature during storage was greatly dependant on the species. This may be because temperatures below 10°C did not allow germination for *Aspergilli*, but did for *Penicillia*. At present, it is difficult to draw conclusions on the effect of environmental conditions on the germination time, since only two species have been studied.

Transients

In the environment, fungal conidia are subject to transient conditions. In particular, temperature varies according to day/night periods. All predictive models for germination assume that fungal spores can adapt instantaneously to changes in temperature. Germination times were determined in steady-state conditions at 10, 15, 20 and 25°C, then temperature shifts (e.g. up-shifts and down-shifts) were applied at 1/4, 1/2 and 3/4 of germination times, with 5, 10 and 15°C magnitudes (Kalai *et al.* 2014). The authors clearly demonstrated that the assumption of instantaneous adaptation of the conidia should be rejected. Temperature shifts during germination led to an induced lag time or an extended germination time compared to the experiments conducted at steady state. In contrast, a decrease in the induced lag time was observed for a 15°C magnitude with a decrease in the instant of the shift. In particular, at 1/4, the induced lag time was negative; that is, the observed germination time was less than the theoretical one. Germination can be defined as a succession of biochemical reactions that occur within the conidia during activation, swelling, polarization and germ tube formation, thus eventually leading to germination. During germination of *A. niger* conidia at optimum temperatures, the number of genes up/down-regulated differed depending on the phase of germination; that is, 917/1986, 856/290, 476/297 and 790/179, for 2 h, 4 h, 6 h and 8 h, respectively (Van Leeuwen *et al.* 2013). This result suggested that germination is probably not a linear process, because many more genes were up/down-regulated in the first hours of germination than in the last ones. The effect of the direction of the shift has already been mentioned for bacteria. Mellefont and Ross (2003) reported that temperature down-shifts induced larger relative lag times than equivalent up-shifts. The same authors also reported the importance of the physiological state of the organisms when shifts are applied. This may explain why the effect of the instant of the temperature shift was significant, although in combination with the other factors.

Conclusion: Applications of these results to fungal starters

Fungal starters are mainly used to preserve food through formation of inhibitory metabolites; to improve food safety through inhibition of pathogens or removal of toxic compounds; to improve nutritional value; and to ensure the organoleptic quality of the food (Bourdichon *et al.* 2012). It would be desirable to control or modify the germination kinetics of fungal starters for some specific applications. Viability is critical, so conservation processes that allow a high viability of starters should be developed and control of the germination time should not be detrimental to this criterion. In this chapter it has been demonstrated that the physiological state of fungal spores is of significant importance to explaining the germination time and viability of fungal spores. The physiological state of fungal spores can be modified by fungal starter producers principally, as food manufacturers resuscitate starters only in some cases. This modification of the physiological state can be obtained by altering the conditions of sporulation, but it cannot be detrimental to productivity. Another direction that could be explored is the production of pregerminated starters to shorten the germination time of fungal starters, but the effect of these modifications on viability needs to be evaluated. In addition, the definitions and the models provided in this chapter would be useful for an accurate determination of the germination time and viability of fungal starters.

References

Araujo, R. and Gonçalves Rodrigues, A. (2004) Variability of germinative potential among pathogenic species of *Aspergillus*. *Journal of Clinical Microbiology*, **42**, 4335–4337.

Barrios-Gonzáles, J., Martinez, C., Aguilera, A. and Raimbault, M. (1989) Germination of concentrated suspensions of spores from *Aspergillus niger*. *Biotechnology Letters*, **11**, 551–554.

Beuchat, L.R. (1988) Thermal tolerance of *Talaromyces flavus* ascospores as affected by growth medium and temperature, age and sugar content in the inactivation medium. *Transactions of the British Mycological Society*, **90**, 359–364.

Beyer, M. and Verreet, J.-A. (2005) Germination of *Gibberella zeae* ascopores as affected by age of spores after discharge and environmental factors. *European Journal of Plant Pathology*, **111**, 381–389.

Beyer, M., Verreet, J.-A. and Ragab, W.S.M. (2005) Effect of relative humidity on germination of ascospores and macroconidia of *Gibberella zeae* and deoxynivalenol production. *International Journal of Food Microbiology*, **98**, 233–240.

Blaszyk, M., Blank, G., Holley, R. and Chong, J. (1998) Reduced water activity during sporogenesis in selected penicillia: Impact on spore quality. *Food Research International*, **31**, 503–509.

Bourdichon, F., Casaregola, S., Farrokh, C. *et al.* (2012) Food fermentations: Microorganisms with technological beneficial use. *International Journal of Food Microbiology*, **154**, 87–97.

Chapman, H.R. and Sharpe, M.E. (1990) Microbiology of cheese. In R.K. Robinson (ed.), *Dairy Microbiology, Vol. 2: The Microbiology of Milk Products* (pp. 203–289), London: Elsevier Applied Science.

Charlang, G., Horowitz, R.M., Lowy, P.H., Ng, B., Poling, S.M. and Horowitz, N.H. (1982) Extracellular siderophores of rapidly growing *Aspergillus nidulans* and *Penicillium chrysogenum*. *Journal of Bacteriology*, **150**, 785–787.

Chitarra, G.S., Abee, T., Rombouts, F.M., Posthumus, M.A. and Dijksterhuis, J. (2004) Germination of *Penicillium paneum* conidia is regulated by 1-octen-3-ol, a volatile self-inhibitor. *Applied and Environmental Microbiology*, **70**, 2823–2829.

Conner, D.E. and Beuchat, L.R. (1987a) Heat resistance of ascospores of *Neosartorya fisheri* as affected by sporulation and heating medium. *International Journal of Food Microbiology*, **4**, 303–312.

Conner, D.E. and Beuchat, L.R. (1987b) Age-related changes in ultrastructure and chemical composition associated with changes in heat resistance of *Neosartorya fisheri* ascospores. *Transactions of the British Mycological Society*, **89**, 539–550.

Dantigny, P., Bensoussan, M., Vasseur, V. *et al.* (2006) Standardisation of methods for assessing mould germination: A workshop report. *International Journal of Food Microbiology*, **108**, 286–291.

Dantigny, P., Guilmart, A. and Bensoussan, M. (2003) Predictive mycology: Some definitions. *Cryptogamie, Mycologie*, **24**, 377–383.

Dantigny, P., Guilmart, A. and Bensoussan, M. (2005a) Basis of predictive mycology. *International Journal of Food Microbiology*, **100**, 187–196.

Dantigny, P., Kalai, S. and Nanguy, S.P.M. (2013) Primary models for germination. In P. Dantigny and E.Z. Panagou (eds), *Predictive Mycology* (pp. 47–62), Boca Raton, FL: Nova Science.

Dantigny, P., Marín, S., Beyer, M. and Magan, N. (2007) Mould germination: Data treatment and modelling. *International Journal of Food Microbiology*, **114**, 17–24.

Dantigny, P., Nanguy, S.P.M., Judet-Correia, D. and Bensoussan, M.A. (2011) A new model for germination of fungi. *International Journal of Food Microbiology*, **146**, 176–181.

Dantigny, P., Soares Mansur, C., Sautour, M., Tchobanov, I. and Bensoussan, M. (2002) Relationship between spore germination kinetics and lag time during growth of *Mucor racemosus*. *Letters in Applied Microbiology*, **35**, 395–398.

Dantigny, P., Tchobanov, I., Bensoussan, M. and Zwietering, M.H. (2005b) Modelling the effect of ethanol vapor on the germination time of *Penicillium chrysogenum*. *Journal of Food Protection*, **68**, 1203–1207.

Darby, R.T. and Mandels, G.R. (1955) Effects of sporulation medium and age on fungus spore physiology. *Plant Physiology*, **30**, 360–366.

d'Enfert, C. (1997) Fungal spore germination: Insights from the molecular genetics of *Aspergillus nidulans* and *Neurospora crassa*. *Fungal Genetics and Biology*, **21**, 163–172.

Eicher, R. and Ludwig, H. (2002) Influence of activation and germination on high pressure inactivation of ascospores of the mould *Eurotium repens*. *Comparative Biochemistry and Physiology Part A*, **131**, 595–604.

El Halouat, A. and Debevere, J.M. (1997) Effect of water activity, modified atmosphere packaging and storage temperature on spore germination of moulds isolated from prunes. *International Journal of Food Microbiology*, **35**, 41–48. DOI: 10.1016/s0168-1605(96)01219-6.

Fan, T.Y., Hwang, D.H. and Kinsella, J.E. (1976) Methyl ketone formation during germination of *Penicillium roqueforti*. *Journal of Agricultural and Food Chemistry*, **24**, 443–448.

Geisen, R. and Färber, P. (2002). New aspects of fungal starter cultures for fermented foods. *Applied Microbiology: Focus on Biotechnology*, **2**, 13–29.

Gindro, K. and Pezet, R. (2001) Effects of long-term storage at different temperatures on conidia of *Botrytis cinerea* Pers.: Fr. *FEMS Microbiology Letters*, **204**, 101–104.

Godinho, M. and Fox, P.F. (1981) Effect of NaCl on the germination and growth of *Penicillium roqueforti*. *Milchwissenschaft*, **36**, 205–208.

Hobot, J.E. and Gull, K. (1980) The identification of a self-inhibitor from *Syncephalastrum racemosum* and its effect upon sporangiospores germination. *Antonie Leeuwenhoek*, **46**, 435–441.

Huang, Y.J., Toscano-Underwood, C., Fitt, B.D.L., Todd, A.D., Koopmann, B. and Balesdent, M.H. (2001) Effects of temperature on germination and hyphal growth from ascospores of A-group and B-group *Leptospaeria maculans* (phoma stem cancer of oilseed rape). *Annals of Applied Biology*, **139**, 193–207.

Hui, Y.H. and Özgul, E. (2012) *Handbook of Food and Beverage Fermentation Technology* (2nd edn), London: CRC Press.

Isaac, S. (1998) What factors influence the germination and outgrowth of fungal spores? *Mycologist*, **12**, 91–92.

Jackson, M.A. and Schisler, D.A. (1992) The composition and attributes of *Colletotrichum truncatum* spores are altered by the nutritional environment. *Applied and Environmental Microbiology*, **58**, 2260–2265.

Jakobsen, M., Cantor, M.D. and Jespersen L. (2002) Production of bread, cheese and meat. In H.D. Osiewacz (ed.), *The Mycota: A Comprehensive Treatise on Fungi as Experimental System and Applied Research. Vol. X: Industrial Applications* (pp. 3–22), Berlin: Springer.

Jessen, B. (1995) Starter cultures for meat fermentation. In G. Campbell-Platt and P.E. Cook (eds), *Meat Science* (pp. 130–159), London: Blakie Academic and Professional.

Judet, D., Bensoussan, M., Perrier-Cornet, J-M. and Dantigny, P. (2008) Distributions of the growth rate of the germ tube and germination time of *Penicillium chrysogenum* conidia depend on water activity. *Food Microbiology*, **25**, 902–907.

Kalai, S., Bensoussan. M. and Dantigny, P. (2014) Lag time for germination of *Penicillium chrysogenum* is induced by temperature shifts. *Food Microbiology*, **42**, 149–153.

Kawanabe, Y. (1986) Responsiveness of *Neurospora crassa* conidia to auxin and gibberellin in relation to culture age. *Agricultural and Biological Chemistry*, **50**, 757–758.

Lattab, N., Kalai, S., Bensoussan, M. and Dantigny P. (2012) Effect of storage conditions (relative humidity, duration, and temperature) on the germination time of *Aspergillus carbonarius* and *Penicillium chrysogenum*. *International Journal of Food Microbiology*, **160**, 80–84.

Li, B., Zhao, H., Li, B. and Xu, X.-M. (2003) Effect of temperature, relative humidity and duration of wetness period on germination and infection by conidia of the pear scab pathogen (*Venturia nashicola*). *Plant Pathology*, **52**, 546–552.

Lücke, F.K. (2000) Utilization of microbes to process and preserve meat. *Meat Science*, **52**, 381–385.

Magan, N. and Lacey, J. (1984) Effect of temperature and pH on water relations of field and storage fungi. *Transactions of the British Mycological Society*, **82**, 71–81.

Marín, S., Sanchis, V., Sáenz, R., Ramos, A.J., Vinas, I. and Magan, N. (1998) Ecological determinants for germination and growth of some *Aspergillus* and *Penicillium* spp. from maize grain. *Journal of Applied Microbiology*, **84**, 25–36.

Marín, S., Sanchis, V., Teixido, A. *et al.* (1996) Water and temperature relations and microconidial germination of *Fusarium moniliforme* and *Fusarium proliferatum* from maize. *Canadian Journal of Microbiology*, **42**, 1045–1050.

Mellefont, L.A. and Ross, T. (2003) The effect of abrupt shifts in temperature on the lag phase duration of *Escherichia coli* and *Klebsiella oxytoca*. *International Journal of Food Microbiology*, **83**, 295–305.

Missous, G., Thammavongs, B., Dieuleveux, V., Guéguen, M. and Panoff, J.M. (2007) Improvement of the cryopreservation of the fungal starter *Geotrichum candidum* by artificial nucleation and temperature downshift control. *Cryobiology*, **55**, 66–71.

Mitchell, D. (2006) Ecological factors affect growth and ochratoxin A production by *Aspergillus* section *Nigri* species on grapes. PhD thesis, Cranfield University.

Morris, T.A. (1964) The manufacture of blue-vein cheese in Queensland. *Australian Journal of Dairy Technology*, **19**, 9–18.

Nanguy, S., P-M., Perrier-Cornet, J-M., Bensoussan, M. and Dantigny, P. (2010) Impact of water activity of diverse media on spore germination of *Aspergillus* and *Penicillium* species. *International Journal of Food Microbiology*, **142**, 273–276.

Pardo, E., Malet, M., Marín, S., Sanchis, V. and Ramos, A.J. (2006) Effects of water activity and temperature on germination and growth profiles of ochratoxigenic *Penicillium verrucosum* isolates on barley meal extract agar. *International Journal of Food Microbiology*, **106**, 25–31.

Pardo, E., Marín, S., Ramos, A.J. and Sanchis, V. (2005a) Effect of water activity and temperature on mycelial growth and ochratoxin A production by isolates of *Aspergillus ochraceus* on irradiated green coffee beans. *Journal of Food Protection*, **68**, 133–138.

Pardo, E., Marín, S., Sanchis, V. and Ramos, A.J. (2004) Prediction of fungal growth and ochratoxin A production by *Aspergillus ochraceus* on irradiated barley grain as influenced by temperature and water activity. *International Journal of Food Microbiology*, **95**, 79–88.

Pardo, E., Ramos, A.J., Sanchis, V. and Marín, S. (2005b) Modelling of effects of water activity and temperature on germination and growth of ochratoxigenic isolates of *Aspergillus ochraceus* on a green coffee-based medium. *International Journal of Food Microbiology*, **98**, 1–9.

Pitt, J.I. and Christian, J.H.B. (1968) Water relations of xerophilic fungi isolated from prunes. *Applied Microbiology*, **16**, 1853–1858.

Plaza, P., Usall, J., Teixidó, N. and Viñas I. (2003) Effect of water activity and temperature on germination and growth of *Penicillium digitatum*, *P. italicum* and *Geotrichum candidum*. *Journal of Applied Microbiology*, **94**, 549–554.

Rousseau, M. (1984) Study of the surface flora of traditional Camembert cheese by scanning electron microscopy. *Milchwissenschaft*, **39**, 129–134.

Sautour, M., Dantigny, P., Guilhem, M.C. and Bensoussan, M. (2003) Influence of inoculum preparation on the growth of *Penicillium chrysogenum*. *Journal of Applied Microbiology*, **95**, 1034–1038.

Schubert, M., Mourad, S. and Schwarze, F.W.M.R. (2010) Statistical approach to determine the effect of combined environmental parameters on conidial development of *Trichoderma viride* (T-15603.1). *Journal of Basic Microbiology*, **50**, 570–580.

Sunesen, L.O. and Stahnke, L.H. (2003) Mould starter cultures for dry sausages-selection, application and effects. *Meat Science*, **65**, 935–948.

Teitell, L. (1958) Effects of relative humidity on viability of conidia of *Aspergilli*. *American Journal of Botany*, **45**, 748–753.

Van den Tempel, T. and Nielsen, M.S. (2000) Effects of atmospheric conditions, NaCl and pH on growth and interactions between moulds and yeasts related to blue cheese. *International Journal of Food Microbiology*, **57**, 193–199.

Van Leeuwen, M.R., Krijgsheld, P., Wyatt, T.T. *et al.* (2013) The effect of natamycin on the transcriptome of conidia of *Aspergillus niger*. *Studies in Mycology*, **74**, 71–85.

Wendland, J. (2001) Comparison of morphogenetic networks of filamentous fungi and yeast. *Fungal Genetics and Biology*, **34**, 63–82.

Yano, T. and Koya, S. (1973) Dynamic behaviour of the chemostat subject to product inhibition. *Journal of General and Applied Microbiology*, **19**, 97–114.

CHAPTER 4

Non-starter bacteria 'functional' cultures

Patricia Ruas-Madiedo and Ana Rodríguez
Instituto de Productos Lácteos de Asturias–Consejo Superior de Investigaciones Científicas (IPLA-CSIC), Spain

Functional cultures

The transformation of raw food materials into fermented products is a natural process that prehistoric farmers began to control empirically. As an example, the earliest evidence of milk processing was dated to the sixth millennium BCE with the discovery of potsherds pierced with small holes found in northern Europe, which have been interpreted as 'cheese strainers' (Salque *et al.* 2013). Nowadays, spontaneous and non-controlled fermentations are the way to obtain traditional products in developing as well as developed countries (Franz *et al.* 2014; Połka *et al.* 2015). Characterization of the microbiota from these wild products helps to maintain biodiversity, since it allows the isolation of novel bacteria to be used in controlled food applications. Indeed, the current production of most fermented foods involves the use of starter cultures, lactic acid bacteria (LAB) being the most extensively used. However, there is a whole 'inventory' of *microbial food cultures* (MFC), composed of bacteria, yeast or moulds, used in food production (Bourdichon *et al.* 2012b). The main reason for using a starter is to begin the fast acidification of diverse raw materials that, in turn, will improve the safety and preservation of food and modify its physicochemical characteristics. Therefore, from the initial activity of LAB, together with the adjunct bacteria and/or the natural secondary microbiota, the raw materials will give a wide variety of fermented products their own sensorial 'identity'. Recently, the definition of starter has evolved to the concept of a *functional culture*, which includes bacteria that 'can contribute to the microbial safety or offer one or more organoleptic, technological, nutritional, or health advantages' (Leroy and De Vuyst 2004; Ravyts *et al.* 2012). Figure 4.1 shows some of the relevant characteristics that functional cultures could have.

The bioprotective potential of specific microorganisms is determined by the production of different metabolites such as organic acids, diacetyl, ethanol, hydrogen peroxidase or bacteriocins, among others (Rouse and van Sinderen 2008). In the

Starter Cultures in Food Production, First Edition. Edited by Barbara Speranza, Antonio Bevilacqua, Maria Rosaria Corbo and Milena Sinigaglia.

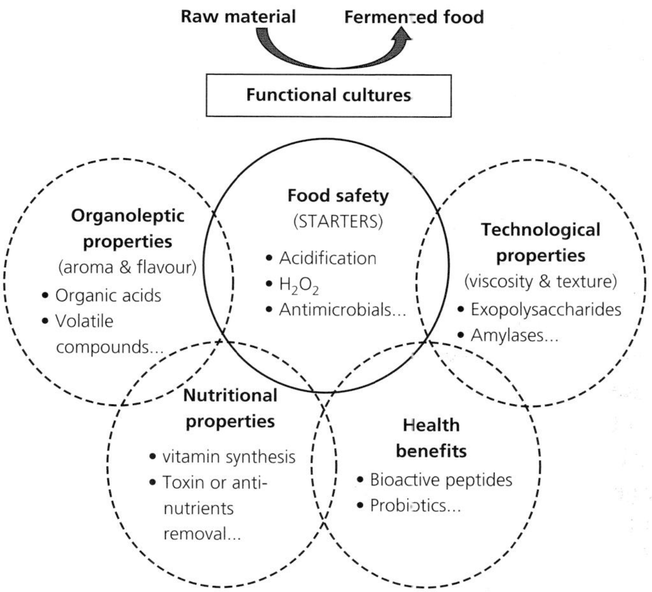

Figure 4.1 Some desirable properties of functional cultures for food fermentation.

case of bacteriocins, these ribosomally synthesized antimicrobial peptides can be effective for the control of foodborne pathogens in the product (Beshkova and Frengova 2012) as well as in the gut environment (Corr *et al.* 2007; Cotter *et al.* 2013). LAB contribute to the aroma and flavour of fermented products through the production of organic acids and volatile compounds from sugar metabolism; the latter can also be obtained from the conversion of amino acids obtained after proteolysis and, to a lesser extent, by lipolysis (Sumby *et al.* 2010; Steele *et al.* 2013). Some LAB are able to synthesize exocellular carbohydrate polymers, known as exopolysaccharides (EPS), which act as natural bio-thickeners or viscosity enhancers for the manufacture of dairy products and sourdoughs; however, EPS-producing LAB are not desirable during the malolactic fermentation of alcoholic beverages since these polymers cause an undesirable viscosity increase (Ruas-Madiedo *et al.* 2002; Gänzle 2009; Badel *et al.* 2011). These polymers could also have beneficial effects for human health, such as antioxidant activity (Xu *et al.* 2011), the capability of counteracting pathogen toxins (Ruas-Madiedo *et al.* 2010) or immune modulating properties, among others (Hidalgo-Cantabrana *et al.* 2012). In addition, it is known that some LAB strains, mainly those belonging to the genus *Lactobacillus*, have health effects (Shah 2007). These are known as probiotics, 'live microorganisms which when administered in adequate amounts confer a health benefit on the host' (FAO-WHO 2006; Hill *et al.* 2014). In some cases the beneficial effects can be achieved through the activity of LAB on food components, for example bioactive peptides, which are encrypted into proteins of plant and animal (including milk) origin and are released by some bacterial peptidases among other proteolytic enzymes (Hernández-Ledesma *et al.* 2014; Singh *et al.* 2014). Another beneficial effect that can be undoubtedly attributed to the activity of LAB on the food matrix is the improvement of lactose intolerance in patients lacking β-galactosidase (WGO 2011). In other cases, specific strains are able to synthesize micronutrients such as vitamins (Capozzi *et al.* 2012) or conjugated linoleic/linolenic acids (Villar-Tajadura *et al.* 2014), thus contributing to the nutritional enhancement of fermented foods. Finally, it is worth noting that the functional analysis of genomes, which is becoming a low-cost technology, can be used to search for specific bacteria of interest, with desired traits allowing improvement of the technological as well as functional properties of fermented foods (Garrigues *et al.* 2013).

As already mentioned, several genera of LAB (belonging to phylum *Firmicutes*) have been used often as starters in food fermentation; however, there are other (non-starter) bacterial groups that actively contribute to the sensorial as well as the beneficial properties of fermented products. The next sections will cover the use of three genera belonging to *Actinobacteria*, a high G + C content Gram-positive phylum, as functional starters (Table 4.1). In some cheese varieties proteolysis is addressed by secondary cultures, among which propionibacteria and corynebacteria are outstanding representatives that grow inside the cheese, producing organic acids and carbon dioxide, or on the cheese surface, respectively. Bifidobacteria are inhabitants of the intestinal tract of animals, but some specific strains have been extensively included in the formulation of probiotic products.

***Propionibacterium* genus**

Propionic acid bacteria (PAB) belong to the *Propionibacteriales* order (Table 4.1) and are mainly isolated from the dairy environment. They are small-rod, non-sporulating, non-motile, anaerobic to aerotolerant, mesophilic bacteria (Stackebrandt *et al.* 1997), with an essential role as secondary microbiota in Swiss-type cheeses such as Emmentaler. The dairy group of propionibacteria comprises four species, *Propionibacterium freudenreichii*, *Propionibacterium acidipropionici*, *Propionibacterium jensenii* and *Propionibacterium thoenii* (Meile *et al.* 2008). The most common species in hard cooked cheeses is *Pr. freudenreichii* subsp. *shermanii*, responsible for the conversion of lactate, produced by LAB from lactose, to propionate, acetate and carbon dioxide through the Wood–Werkman cycle (Figure 4.2). During propionate fermentation, lactate fermenters reach up to 10^8–10^9 colony forming units per gram (cfu/g) of cheese (Langsrud and Reinbold 1973). This secondary fermentation occurs after the temperature of ripening raises to 18–25 °C and a cheese firm mass is developed in such a way that gas is trapped and produces the characteristic holes ('eyes') of Swiss cheeses, while the acidic compounds contribute to their flavour and preservation. Propioniobacteria are strongly peptidolytic but weakly proteolytic bacteria, thus their growth is stimulated in association with *Lactobacillus* species (*Lactobacillus delbrueckii* subsp. *bulgaricus* and *Lactobacillus helveticus*; Piveteau *et al.* 1995). They also produce

Table 4.1 Taxonomic classification of high G+C content bacteria acting as non-starter, 'functional' food cultures.

Phylum	Class	Order	Suborder	Family	Genus
Actinobacteria	Actinobacteria	Actinomycetales	Corynebacterineae	Corynebacteriaceae	*Corynebacterium*
			Micrococcineae	Brevibacteriaceae	*Brevibacterium*
				Micrococcaceae	*Arthrobacter*
				Microbacteriaceae	*Microbacterium*
		Propionibacteriales	–	Propionibacteriaceae	*Propionibacterium*
		Bifidobacteriales	–	Bifidobacteriaceae	*Bifidobacterium*

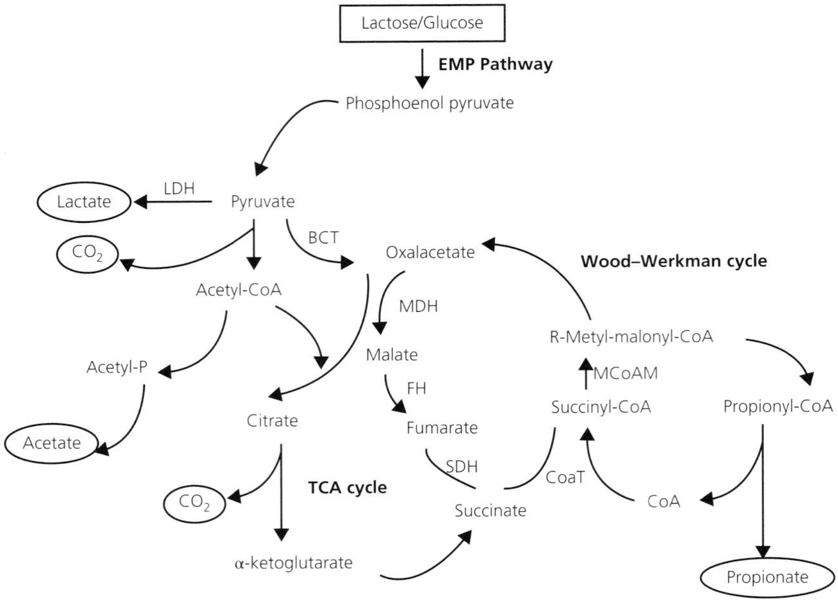

Figure 4.2 Pathway of propionic acid production by *Propionibacterium freuderenchii* (Wood–Werkman cycle) and some of the enzymes involved. Notes: BCT = biotin dependent carboxytransferase; CoAT = CoA transferase; FH = fumarate hydrolase; LDH = lactate dehydrogenase; McoAM = coenzyme B12-dependent methylmalonyl-CoA mutase; MDH = malate dehydrogenase; SDH = succinate dehydrogenase.

several compounds from amino acid and lipid catabolism that account for flavour development. In particular, this species contributes to the formation of free fatty acids derived from lipolysis, branched-chain compounds derived from isoleucine catabolism (2-methylbutanal, 2-methylbutanoland 2-methylbutanoic acid) and from leucine catabolism (3-methylbutanoic acid; Thierry *et al.* 2005; Thierry *et al.* 2011a). Its thermo-tolerance facilitates its survival during the 'cooking' process typical of Swiss-type cheeses. *Pr. freudenreichii* is also able to produce vitamin B12 and its properties as protective cultures for food and feed production have been reported as well (Thierry *et al.* 2011b).

It should be noted that dairy propionibacteria have been assayed as probiotics. Their safety has contributed to this, supported by the widespread consumption of Swiss cheeses. Indeed, *Pr. freudenreichii* is a GRAS (generally regarded as safe) species according to the North American FDA (Food and Drug Administration) and it has been included in the QPS (qualified presumption of safety) list of the EFSA (European Food Safety Authority; Bourdichon *et al.* 2012a; EFSA 2007, 2015). The potential of propionibacteria as probiotics has been studied *in vitro, ex vivo* and *in vivo*, as well as in human clinical trials (Oksaharju *et al.* 2013; Hidalgo-Cantabrana *et al.* 2014). Their ability to modulate the gut microbiota has been suggested, as they are able to decrease the number of pathogens and stimulate the growth of beneficial microorganisms (Cousin *et al.* 2011).

Corynebacterium and related genera

Coryneforms are aerobic, mesophilic, salt-tolerant, non-motile, irregular rod-shaped (i.e. V-shaped) bacteria that belong to the *Actinomycetales* order (Table 4.1). This microbial group has an important role in the ripening processes of several types of cheese, namely smear-ripened cheeses (Munster, Limburger, Tilsit, Marolles, Grana Padano, Parmigiano Reggiano, Gruyère, Appenzeller etc.). Indeed, corynebacteria are components of secondary starters whose main function is the development of the aroma, flavour, texture and appearance of cheese. The source of these microorganisms is most likely the natural microbiota of brine, the environment inside the cheese-processing area and the wooden shelves used for cheese storage during ripening (Bockelmann 2010). Production of orange, pink-red and yellow-brown pigments is a typical characteristic. They show detectable lipolytic and esterolytic activities, in addition to strong proteolytic activity that results in high levels of sulfur-containing volatiles (Wouters *et al.* 2002). Corynebacteria develop in the cheese matrix after lactose is fermented by the primary starter cultures, although the presence of lactate dehydrogenase activity (LDH) and the ability to assimilate lactate have been detected in this microbial group (Mounier *et al.* 2007).

The study of the microbiota of several smear-ripened cheeses has allowed the identification of species belonging to the genus *Corynebacterium* (suborder *Corynebacterineae*; Table 4.1) as well as to genera *Brevibacterium, Arthrobacter* and *Microbacterium* (suborder *Micrococcineae*, Table 4.1; Mounier *et al.* 2007). At the beginning of ripening, most of the lactose has been converted to lactate by LAB, and consequently lactate becomes the main source of carbon on the cheese surface. As ripening progresses, amino acids are also available as a carbon source due to protein hydrolysis, which can be used by corynebacteria. This, along with their tolerance to salt, contributes to corynebacteria's prevalence on the cheese surface at the last stages of ripening. Among these genera, *Brevibacterium linens* is one of the major components of the surface microbiota of cheese and is responsible for the conversion of the amino acid methionine to methanethiol, α-ketobutyrate and ammonia (Bonnarme *et al.* 2001). In fact, in some countries *Br. linens* is used as a commercial adjunct microorganism to smear the young cheeses, but it does not always develop on the cheese surface. Instead, adventitious microbiota from the environment are dominant on the cheese surface (Mounier *et al.* 2006). The presence of *Corynebacterium* species as a component of the surface microbiota has also become important and new species have been described: *Corynebacterium mooreparkense* sp. nov. and *Corynebacterium casei* sp. nov. (Brennan *et al.* 2001). Corynebacteria establish consortia with other microorganisms (yeasts and staphylococci) on the surface of cheeses (Brennan *et al.* 2002). The consortia composition differs among cheese varieties and manufacturers and their complexity has been elucidated by using a combination of biochemical identification methods, 16S rRNA sequencing, amplified ribosomal RNA restriction analysis (ARDRA) and FTIR spectroscopy for species identification, while several genotypic methods

(PFGE, RAPD, rep-PCR etc.) have been used for typing below species level. With this approach, the corynebacteria *Arthrobacter arilaitensis* and *Brevibacterium aurantiacum* were identified in Limburger cheese (Mounier *et al.* 2006; Goerges *et al.* 2008).

The development of a defined smear-surface starter is important to minimize the risk of contamination by pathogenic organisms such as *Listeria monocytogenes* and *Staphylococcus aureus* (Valdés-Stauber *et al.* 1997); thus it can be considered as an alternative to the old–young smearing technique. However, there is strong competition among the components of the natural microbiota on the cheese surface; therefore, only well-adapted corynebacteria strains should be used in commercial starters (Bockelmann 2002).

Bifidobacterium genus

The genus *Bifidobacterium* is included in the family *Bifidobacteriaceae* together with other genera (*Aeriscardovia, Alloscardovia, Bombiscardovia, Gardnerella, Metascardovia, Parascardovia, Pseudoscardovia* and *Scardovia*) according to the NCBI Taxonomy website (http://www.ncbi.nlm.nih.gov/Taxonomy/taxonomyhome.html). Bifidobacteria were identified first by Henry Tissier (Pasteur Institute) at the end of the nineteenth century from faeces of a breastfed baby and named *Bacillus bifidus*. However, it was Sigurd Orla-Jensen in 1924 who proposed its reclassification as the new genus *Bifidobacterium* (Orla-Jensen 1924). Today this genus is constituted by 48 taxa (39 species and 9 subspecies) and the genomes of all type strains have recently been sequenced (Lugli *et al.* 2014). Bifidobacteria are pleomorphic rods than can be slightly bifurcated (Y- or V-shaped), which is the reason for their 'bifid' name, and can present 'spatulated' extremities, being single or organized in chains or clumps (Figure 4.3). They are non-spore forming, non-motile and non-filamentous (Biavati and Mattarelli 2006). Regarding physiology, the optimal growth pH ranges between 6.5 and 7.0 and, in general, bifidobacteria are not able to survive below pH 4.5; thus their tolerance to acid is low, although some species have developed mechanisms to tolerate this stress factor (Sánchez *et al.* 2012). Human-origin bifidobacteria optimally grow at 36–38 °C, while the temperature of those of animal origin is higher (41–43 °C; Cronin *et al.* 2011). One exception to these features is *Bifidobacterium thermacidophilum*, which is able to grow at 49.5 °C and pH 4.0 (Dong *et al.* 2000). Bifidobacteria are considered strict anaerobes, but some species present diverse degrees of aerotolerance when a reducing agent (e.g. L-cysteine) is present (Devries and Stoutham 1969). Indeed, *Bifidobacterium psychraerophilum* is able to grow in a solid medium in the presence of air (Simpson *et al.* 2004) and *Bifidobacterium animalis* subsp. *lactis* displays a notable tolerance to oxygen, which explains its widespread use in probiotic foods (Ruiz *et al.* 2012). Bifidobacteria have a fermentative

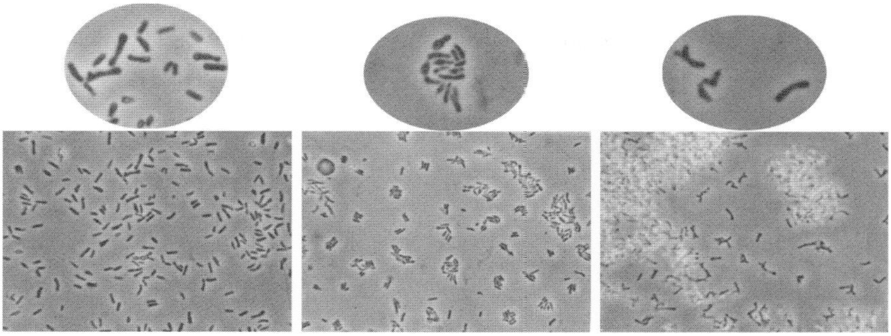

B. animalis subsp. *lactis* *B. longum* subsp. *infantis* *B. bifidum*

Figure 4.3 Optical microscopy photographs showing typical traits of *Bifidobacterium* morphology.

metabolism, although the genome analysis of *Bifidobacterium asteroides* predicted the capability for oxygen-mediated respiration (Bottacini *et al.* 2012).

Genomes of this genus show a high number of genes devoted to the use of complex carbohydrates (Ventura *et al.* 2009; Lee and O'Sullivan 2010). In fact, oligosaccharides are preferentially used over monosaccharides in bifidobacteria (Ruas-Madiedo *et al.* 2005; Amaretti *et al.* 2007). Several studies have demonstrated their ability to use endogenous (human) carbohydrates, such as intestinal mucins, or exogenous (diet) non-digestible oligosaccharides, for example human milk oligosaccharides (Kiyohara *et al.* 2012; Sela *et al.* 2012; Sánchez *et al.* 2013). The metabolism of hexoses is achieved by means of a specific pathway named the 'bifid shunt', or the fructose-6-phosphate phosphoketolase (F6PPK) pathway, which was described in 1965 by Scardovi and Trovatelli and constituted the best test for the identification of bifidobacteria before the development of molecular tools (Sidarenka *et al.* 2008). The central enzyme of this catabolic route is Xfp (xylulose-5-phosphate phosphoketolase), with two activities (X5PPK and F6PPK) described to date (Sánchez *et al.* 2010b). Xfp converts fructose-6-phosphate to erythrose-4-phosphate and acetyl-phosphate (enzyme F6PPK, EC 4.1.2.22) and/or xylulose-5-phosphate to acetyl-phosphate and glyceraldehyde-3-phosphate (enzyme X5PPK, EC 4.1.2.9; Figure 4.4). Pentoses are formed from fructose-6-phosphate and erythrose-4-phosphate by means of a transaldolase (Tal) and a transketolase (Tkt). The glyceraldehyde-3-phosphate generated is metabolized to pyruvate through the last reactions of the Embden–Meyerhof–Parnas (EMP) pathway, which is finally converted to lactate, whereas the acetyl phosphate is converted to acetate. The theoretical yield of this bifid shunt is 2 mol of glucose = 3 mol acetate + 2 mol lactate + 5 molecules of ATP (2 molecules are consumed in the conversion of glucose to glucose-6-P). However, this balance can be altered since pyruvate, in combination with Co-A, can be converted to formate and acetyl-CoA and, finally, the latter to ethanol or acetate (Figure 4.4). Through this alternative means bifidobacteria obtain more energy

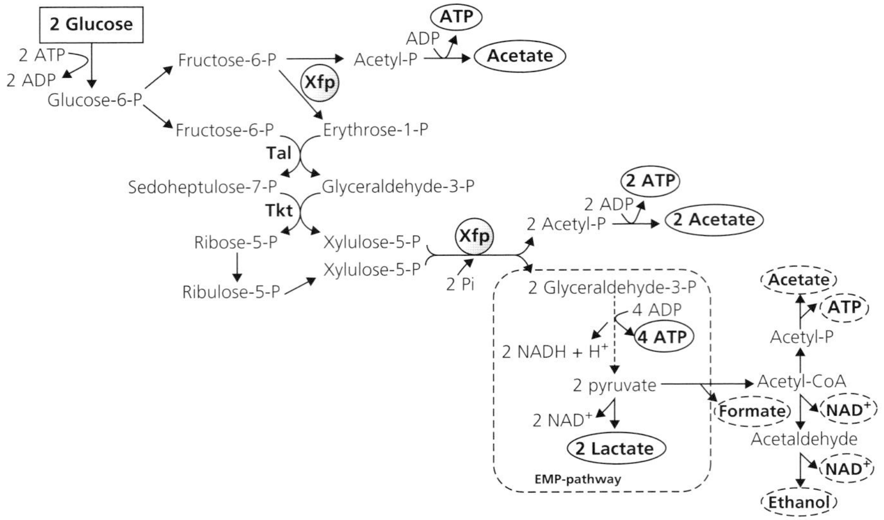

Figure 4.4 Schematized key steps of hexoses metabolism in *Bifidobacterium* through the 'bifid shunt' pathway showing the main final products formed. Notes: EMP = Embden–Meyerhof–Parnas pathway; Tal = transaldolase; Tkt = transketolase; Xfp = xylulose-5-phosphate phosphoketolase/fructose-6-phosphate phosphoketolase.

(ATP) and redox potential (NAD^+); thus, this metabolic reorganization occurs under adverse conditions such as bile, acid and/or oxygen stresses (Ruiz *et al.* 2012; Sánchez *et al.* 2007, 2012).

The ecological niche of the *Bifidobacterium* genus is the gastrointestinal tract of animals, from insects to humans, and only a few of the type strains have been isolated from sewage (probably due to faecal contamination) and from some dairy products (Milani *et al.* 2014). Bifidobacteria are predominant members of the gut microbiota in breast-fed infants, but their number declines with ageing; the most common species in the human gut are *Bifidobacterium adolescentis, Bifidobacterium angulatum, Bifidobacterium bifidum, Bifidobacterium breve, Bifidobacterium catenulatum, Bifidobacterium dentium, Bifidobacterium longum, Bifidobacterium pseudocatenulatum* and *Bifidobacterium pseudolongum,* whereas *Bif. animalis* subsp. *lactis* is the species most often included in probiotic foods and/or food supplements (Masco *et al.* 2005; Turroni *et al.* 2009). Several diseases have been associated with microbiota dysbiosis (an imbalance in the composition, number and/or function of this microbial community) and very often aberrations in bifidobacteria populations have been found (Tojo *et al.* 2014). Thus the oral administration of live bifidobacteria in probiotic foods aims to partially restore its number in the microbiota, thus attenuating or preventing some diseases (Reid *et al.* 2011). Nowadays, there are five species of bifidobacteria included in the EFSA QPS list that could be safely used in the formulation of foods: *Bif. adolescentis, Bif. animalis, Bif. bifidum, Bif. breve* and *Bif. longum* (EFSA 2007). Diverse food matrices have been considered for the oral administration of bifidobacteria (Saarela *et al.* 2011; Rivera-Espinoza and Gallardo-Navarro 2010; Furtado-Martins *et al.* 2013), dairy products being the traditional food used for this purpose (Boylston *et al.* 2004; Prasanna *et al.* 2014). It is worth noting that milk is not a good medium for the growth of *Bifidobacterium* alone due to its weak proteolytic capability; supplementation with additional nitrogen sources or co-culturing with compatible proteolytic bacteria (e.g. a yogurt starter) could favour its growth (Salazar *et al.* 2009). However, milk is a good matrix to protect bifidobacteria and increase their survival in dairy products, as well as during transit through the upper gastrointestinal tract (Sánchez *et al.* 2010a). As already mentioned, *Bif. animalis* subps. *lactis* is the subspecies that was by far the most successfully included in foods, probably due to its robustness, since it is able to deal with and survive several technological challenges (oxygen, low pH etc.). However, specific strains of other species are also able to grow and/or be metabolically active in dairy-based products; this is the case for *Bif. breve* 4L growing in bovine milk (Turroni *et al.* 2011), *Bif. bifidum* PRL2010 cultured in kefir (Serafini *et al.* 2014) and *Bif. bifidum* MF20/5 releasing antihypertensive peptides from milk (Gonzalez-Gonzalez *et al.* 2013).

Conclusion

Fermentation carried out by LAB is an ancient way to preserve and increase the safety of raw food materials, providing additional sensorial, nutritional and health benefits. However, this process can be further optimized if other non-starter 'functional' bacteria are included during the manufacture of fermented products. The use of cultures, such as propionibacteria and corynebacteria, will have a strong impact on the dynamic and function of the food microbiota, thus giving added 'flavour' and textural values to the final product. The addition of non-food-origin bacteria such as bifidobacteria, whose ecological niche is the gut of animals and which do not necessarily achieve fermentation of the food matrix, is a current practice to obtain products with claimed health benefits. Therefore exploring the use of different microorganisms, for the production either of traditional fermented or of novel foods, opens new opportunities for research as well as for increasing the offer of new products to the food market, a sector with a high demand for innovation and development.

Acknowledgements

The authors acknowledge financial support for their research activities and projects from the Spanish Ministry of Economy and Competiveness (MINECO) and from the Regional 'Plan de Ciencia y Tecnología' (PCT, Principado de Asturias), both partially supported by the FEDER funds of the European Union.

References

Amaretti, A., Bernardi, T., Tamburini, E. *et al.* (2007) Kinetics and metabolism of *Bifidobacterium adolescentis* MB 239 growing on glucose, galactose, lactose, and galactooligosaccharides. *Applied and Environmental Microbiology*, **73**, 3637–3644.

Badel, S., Bernardi, T. and Michaud, P. (2011) New perspectives for Lactobacilli exopolysaccharides. *Biotechnology Advances*, **29**, 54–66.

Beshkova, D. and Frengova, G. (2012) Bacteriocins from lactic acid bacteria: Microorganisms of potential biotechnological importance for the dairy industry. *Engineering in Life Sciences*, **12**, 419–432.

Biavati, B. and Mattarelli, P. (2006) The Family *Bifidobacteriaceae*. *Prokaryotes*, **3**, 322–382.

Bockelmann, W. (2002) Development of defined surface starter cultures for the ripening of smear cheeses. *International Dairy Journal*, **12**, 123–131.

Bockelmann, W. (2010) Secondary cheese starter cultures. In B.A. Law and A. Barry (eds), *Technology of Cheesemaking* (pp. 193–230), Chichester: John Wiley & Sons.

Bonnarme, P., Lapadatescu, C., Yvon, E. and Spinnler, H.E. (2001) L-methionine degradation potentialities of cheese-ripening microorganisms. *Journal of Dairy Research*, **68**, 663–674.

Bottacini, F., Milani, C., Turroni, F. *et al.* (2012) *Bifidobacterium asteroides* PRL2011 genome analysis reveals clues for colonization of the insect gut. *PLoS ONE*, **7**, e44229. doi:10.1371/journal.pone.0044229

Bourdichon, F., Berger, B., Casaregola, S. *et al.* (2012a) Safety demonstration of microbial food cultures (MFC) in fermented food products. *Bulletin of International Dairy Federation*, **455**, 1–62.

Bourdichon, F., Casaregola, S., Farrokh, C. *et al.* (2012b) Food fermentations: Microorganisms with technological beneficial use (Review). *International Journal of Food Microbiology*, **154**, 87–97.

Boylston, T.D., Vinderola, C.G., Ghoddusi, H.B. and Reinheimer, J. (2004) Incorporation of bifidobacteria into cheeses: Challenges and rewards. *International Dairy Journal*, **14**, 375–387.

Brennan, N.M., Brown, R., Goodfellow, M. *et al.* (2001) *Corynebacterium mooreparkense* sp. nov. and *Corynebacterium casei* sp. nov., isolated from the surface of a smear-ripened cheese. *International Journal of Systematic and Evolutionary Microbiology*, **51**, 843–852.

Brennan, N.M., Ward, A.C., Beresford, T.P., Fox, P.F., Goodfellow, M. and Cogan, T.M. (2002) Biodiversity of the bacterial flora on the surface of a smear cheese. *Applied and Environmental Microbiology*, **68**, 820–830.

Capozzi, V., Russo, P., Dueñas, M.T., López, P. and Spano, G. (2012) Lactic acid bacteria producing B-group vitamins: A great potential for functional cereals products. *Applied Microbiology and Biotechnology*, **96**, 1383–1394.

Corr, S.C., Li, Y., Riedel, C.U., O'Toole, P.W., Hill, C. and Gahan, C.G. (2007) Bacteriocin production as mechanism for the anti-infective activity of *Lactobacillus salivarius* UCC118. *Proceedings of the National Academy of Sciences of the United States of America*, **104**, 7617–7621.

Cotter, P.D., Ross, R.P. and Hill, C. (2013) Bacteriocins: A viable alternative to antibiotics? *Nature Reviews Microbiology*, **11**, 95–105.

Cousin, F.J., Mater, D.D.G., Foligne, B. and Jan, G. (2011) Dairy propionibacteria as human probiotics: A review of recent evidence. *Dairy Science and Technology*, **91**, 1–26.

Cronin, M., Ventura, M., Fitzgerald, G.F. and van Sinderen, D. (2011) Progress in genomics, metabolism and biotechnology of bifidobacteria. *International Journal of Food Microbiology*, **149**, 4–18.

Devries, W. and Stoutham. A.H. (1969) Factors determining degree of anaerobiosis of *Bifidobacterium* strains. *Archiv für Mikrobiologie*, **65**, 275–287.

Dong, X., Xin, Y., Jian, W., Liu, X. and Ling, D. (2000) *Bifidobacterium thermacidophilum* sp. nov., isolated from an anaerobic digester. *International Journal of Systematic and Evolutionary Microbiology*, **50**, 119–125.

EFSA (2007) Introduction of a Qualified Presumption of Safety (QPS) approach for assessment of selected microorganisms referred to EFSA. *The EFSA Journal*, **587**, 1–16.

EFSA (2015) The 2013 updated list of QPS Status recommended biological agents in support of EFSA risk assessments–2nd revision (new addition). *EFSA Journal*, **13**, 4138.

FAO/WHO (2006) Probiotics in food: Health and nutritional properties and guidelines for evaluation. FAO Food and Nutrition Paper No. 85. Rome: Food and Agriculture Organization of the United Nations/World Health Organization.

Franz, C.M.A.P., Hucha, M., Mathara, J.M. *et al.* (2014). African fermented foods and probiotics. *International Journal of Food Microbiology*, **190**, 84–96.

Furtado-Martins, E.M., Mota-Ramos, A., Lago-Vanzela, E.S., Stringheta, P.C., de Oliveira Pinto, C.L. and Martins, J.M. (2013) Products of vegetable origin: A new alternative for the consumption of probiotic bacteria. *Food Research International*, **51**, 764–770.

Gänzle, M.G. (2009) From gene to function: Metabolic traits of starter cultures for improved quality of cereal foods. *International Journal of Food Microbiology*, **134**, 29–36.

Garrigues, C., Johansen, E. and Crittenden, R. (2013) Pangenomics: An avenue to improved industrial starter cultures and probiotics. *Current Opinion in Biotechnology*, **24**, 187–191.

Goerges, S., Mounier, J., Rea, M.C. *et al.* (2008) Commercial ripening starter microorganisms inoculated into cheese milk do not successfully establish themselves in the resident microbial ripening consortia of a south German red smear cheese. *Applied and Environmental Microbiology*, **74**, 2210–2217.

Gonzalez-Gonzalez, C., Gibson, T. and Jauregi, P. (2013) Novel probiotic-fermented milk with angiotensin I-converting enzyme inhibitory peptides produced by *Bifidobacterium bifidum* MF 20/5. *International Journal of Food Microbiology*, **167**, 131–137.

Hernández-Ledesma, B., García-Nebot, M.J., Fernández-Tomé, S., Amigo, L. and Recio, I. (2014) Dairy protein hydrolysates: Peptides for health benefits. *International Dairy Journal*, **38**, 82–100.

Hidalgo-Cantabrana, C., Kekkonen, R., de los Reyes-Gavilán, C.G. *et al.* (2014) Effect of bacteria used in food industry on the proliferation and cytokine production of epithelial intestinal cellular lines. *Journal of Functional Foods*, **6**, 348–355.

Hidalgo-Cantabrana, C., López, P., Gueimonde, M. *et al.* (2012) Immune modulation capability of exopolysaccharides synthesised by lactic acid bacteria and bifidobacteria. *Probiotics and Antimicrobial Proteins*, **4**, 227–237.

Hill, C., Guarner, F., Reid, G. *et al.* (2014) The International Scientific Association for Probiotics and Prebiotics consensus statement on the scope and appropriate use of the term probiotic. *Natural Review Gastroenterology and Hepatology*, **11**, 506–514.

Kiyohara, M., Nakatomi, T., Kurihara, S. *et al.* (2012) α-N-acetylgalactosaminidase from infant-associated bifidobacteria belonging to novel glycoside hydrolase family 129 is implicated in alternative mucin degradation pathway. *Journal of Biological Chemistry*, **287**, 693–700.

Langsrud, T. and Reinbold, G.W. (1973) Flavor development and microbiology of Swiss cheese: A review. III. Ripening and flavor production. *Journal of Milk and Food Technology*, **36**, 593–609.

Lee, J.H. and O'Sullivan, D.J. (2010) Genomic insights into bifidobacteria. *Microbiology and Molecular Biology Reviews*, **74**, 378–416.

Leroy, F. and De Vuyst, V. (2004) Lactic acid bacteria as functional starter cultures for the food fermentation industry. *Trends in Food Science and Technology*, **15**, 67–78.

Lugli, G.A., Milani, Ch., Turroni, F. *et al.* (2014) Investigation of the Evolutionary Development of the Genus *Bifidobacterium* by Comparative Genomics. *Applied and Environmental Microbiology*, **80**, 6383–6394.

Masco, L., Huys, G., De Brandt, E., Temmerman, R. and Swings, J. (2005) Culture-dependent and culture-independent qualitative analysis of probiotic products claimed to contain bifidobacteria. *International Journal of Food Microbiology*, **102**, 221–230.

Meile, L., Le Blay, G., and Thierry, A. (2008). Safety assessment of dairy microorganisms: *Propionibacterium* and *Bifidobacterium*. *International Journal of Food Microbiology*, **126**, 316–320.

Milani, C., Lugli, G.A., Duranti, S. *et al.* (2014) Genomic encyclopedia of type strains of the genus *Bifidobacterium*. *Applied and Environmental Microbiology*, **80**, 6290–6302.

Mounier, J., Goerges, S., Gelsomino, R. *et al.* (2006) Sources of the adventitious microflora of a smear-ripened cheese. *Journal of Applied Microbiology*, **101**, 668–681.

Mounier, J., Rea, M.C., O'Connor, P.M., Fitzgerald, G.F. and Cogan, T.M. (2007) Growth characteristics of *Brevibacterium*, *Corynebacterium*, *Microbacterium*, and *Staphylococcus* spp. isolated from surface-ripened cheese. *Applied and Environmental Microbiology*, **73**, 7732–7739.

Oksaharju, A., Kooistra, T., Kleemann, R. *et al.* (2013). Effects of probiotic *Lactobacillus rhamnosus* GG and *Propionibacterium freudenreichii* ssp. *shermanii* JS supplementation on intestinal and systemic markers of inflammation in ApoE*3Lieden mice consuming a high-fat diet. *British Journal of Nutrition*, **110**, 77–85.

Orla-Jensen, S. (1924) La classification des bactéries lactiques. *Le Lait*, **4**, 468–474.

Piveteau, PG., Condon, S. and Cogan, T.M. (1995) Interactions between lactic and propionic acid bacteria. *Le Lait*, **75**, 331–343.

Połka, J., Rebecchi, A., Pisacane, V., Morelli, L. and Puglisi, E. (2015) Bacterial diversity in typical Italian salami at different ripening stages as revealed by high-throughput sequencing of 16S rRNA amplicons. *Food Microbiology*, **46**, 342–356.

Prasanna, P.H.P., Grandison, A.S. and Charalampopoulos, D. (2014) Bifidobacteria in milk products: An overview of physiological and biochemical properties, exopolysaccharide production, selection criteria of milk products and health benefits. *Food Research International*, **55**, 247–262.

Ravyts, F., Vuyst, L.D. and Leroy, F. (2012) Bacterial diversity and functionalities in food fermentations. *Engineering in Life Sciences*, **12**, 356–367.

Reid, G., Younes, J.A., Van der Mei, H.C., Gloor, G.B., Knight, R. and Busscher, H.J. (2011) Microbiota restoration: Natural and supplemented recovery of human microbial communities. *Nature Reviews Microbiology*, **9**, 27–38.

Rivera-Espinoza, Y. and Gallardo-Navarro, Y. (2010) Non-dairy probiotic products. *Food Microbiology*, **27**, 1–11.

Rouse, S. and van Sinderen, D. (2008) Bioprotective potential of lactic acid bacteria in malting and brewing. *Journal of Food Protection*, **71**, 1724–1733.

Ruas-Madiedo, P., Hernández-Barranco, A., Margolles, A. and de los Reyes- Gavilán, C.G. (2005) A bile salt-resistant derivative of *Bifidobacterium animalis* has an altered fermentation pattern when grown on glucose and maltose. *Applied and Environmental Microbiology*, **71**, 6564–6570.

Ruas-Madiedo, P., Hugenholtz, J. and Zoon, P. (2002) An overview of the functionality of exopolysaccharides produced by lactic acid bacteria. *International Dairy Journal*, **12**, 163–171.

Ruas-Madiedo, P., Medrano, M., Salazar, N., De Los Reyes-Gavilán, C.G., Pérez, P.F. and Abraham, A.G. (2010) Exopolysaccharides produced by *Lactobacillus* and *Bifidobacterium* strains abrogate in vitro the cytotoxic effect of bacterial toxins on eukaryotic cells. *Journal of Applied Microbiology*, **109**, 2079–2086.

Ruiz, L., Gueimonde, M., Ruas-Madiedo, P. *et al.* (2012) Molecular clues to understand the aerotolerance phenotype of *Bifidobacterium animalis* subsp. *lactis*. *Applied and Environmental Microbiology*, **78**, 644–650.

Saarela, M., Alakomi, H.L., Mättöa, J., Ahonen, A.M., Puhakka, A. and Tynkkynen, S. (2011) Improving the storage stability of *Bifidobacterium breve* in low pH fruit juice. *International Journal of Food Microbiology*, **149**, 106–110.

Salazar, N., Prieto, A., Leal, J.A. *et al.* (2009) Production of exopolysaccharides by *Lactobacillus* and *Bifidobacterium* strains of human origin, and metabolic activity of the producing bacteria in milk. *Journal of Dairy Sciences*, **92**, 4158–4168.

Salque, M., Bogucki, P.I., Pyzel, J. *et al.* (2013) Earliest evidence for cheese making in the sixth millennium BC in northern Europe. *Nature*, **493**, 522–525.

Sánchez, B., Champomier-Vergès, M.C., Stuer-Lauridsen, B. *et al.* (2007) Adaptation and response of *Bifidobacterium animalis* subsp. *lactis* to bile: A proteomic and physiological approach. *Applied and Environmental Microbiology*, **73**, 6757–6767.

Sánchez, B., Fernández-García, M., Margolles, A., de los Reyes-Gavilán, C.G. and Ruas-Madiedo, P. (2010a) Technological and probiotic selection criteria of a bile-adapted *Bifidobacterium animalis* subsp. *lactis* strain. *International Dairy Journal*, **20**, 800–805.

Sánchez, B., Ruiz, L., Gueimonde, M., Ruas-Madiedo, P. and Margolles, A. (2012) Toward improving technological and functional properties of probiotics in foods. *Trends in Food Science and Technology*, **26**, 56–63.

Sánchez, B., Ruiz, L., Gueimonde, M., Ruas-Madiedo, P. and Margolles, A. (2013) Adaptation of bifidobacteria to the gastrointestinal tract and functional consequences. *Pharmacological Research*, **69**, 127–136.

Sánchez, B., Zúñiga, M., González-Candelas, F., de los Reyes-Gavilán, C.G. and Margolles, A. (2010b) Bacterial and eukaryotic phosphoketolases: Phylogeny, distribution and evolution. *Journal of Molecular Microbiology and Biotechnology*, **18**, 37–51.

Scardovi, V. and Trovatelli, L.D. (1965) The fructose-6-phosphate shunt as peculiar pattern of hexose degradation in the genus *Bifidobacterium*. *Annals of Microbiology*, **15**, 19.

Sela, D.A., Garrido, D., Lerno, L. *et al.* (2012) *Bifidobacterium longum* subsp. *infantis* ATCC 15697 α-fucosidases are active on fucosylated human milk oligosaccharides. *Applied and Environment Microbiology*, **78**, 795–803.

Serafini, F., Turroni, F., Ruas-Madiedo, P. *et al.* (2014) Kefir fermented milk and kefiran promote growth of *Bifidobacterium bifidum* PRL2010 and modulate its gene expression. *International Journal of Food Microbiology*, **178**, 50–59.

Shah, N.P. (2007) Functional cultures and health benefits. *International Dairy Journal*, **17**, 1262–1277.

Sidarenka, A.V., Novik, G.I. and Akimov, V.N. (2008) Application of molecular methods to classification and identification of bacteria of the genus *Bifidobacterium*. *Microbiology*, **77**, 251–260.

Simpson, P.J., Ross, R.P., Fitzgerald, G.F. and Stanton, C. (2004) *Bifidobacterium psychraerophilum* sp. nov. and *Aeriscardovia aeriphila* gen. nov., sp. nov., isolated from a porcine caecum. *International Journal of Systematic and Evolutionary Microbiology*, **54**, 401–406.

Singh, B.P., Vij, S. and Hati, S. (2014) Functional significance of bioactive peptides derived from soybean. *Peptides*, **54**, 171–179.

Stackebrandt, E., Rainey, A. and Ward-Rainey, N.L. (1997) Proposal for a new hierarchic classification system, Actinobacteria classis nov. *International Journal of Systematic Bacteriology*, **47**, 479–491.

Steele, J., Broadbent, J. and Kok, J. (2013) Perspectives on the contribution of lactic acid bacteria to cheese flavor development. *Current Opinion in Biotechnology*, **24**, 135–141.

Sumby, K.M., Grbin, P.R. and Jiranek, V. (2010) Microbial modulation of aromatic esters in wine: Current knowledge and future prospects. *Food Chemistry*, **121**, 1–16.

Thierry, A., Deutsch, S.M., Falentin, H., Dalmasso, M., Cousin, F.J. and Jan, G. (2011a) New insights into physiology and metabolism of *Propionibacterium freudenreichii*. *International Journal of Food Microbiology*, **149**, 19–27.

Thierry, A., Falentin, H., Deutsch, S.M. and Jan, G. (2011b) Bacteria, beneficial: *Propionibacterium* spp. In J.W. Fuquay, P.F. Fox and P. McSweeney (eds), *Encyclopedia of Dairy Science* (2nd edn, pp. 403–411), London: Elsevier.

Thierry, A., Maillard, M.B., Richoux, R., Kerjean, J.-R. and Lortal, S. (2005) *Propionibacterium freudenreichii* strains quantitatively affect production of volatile compounds in Swiss cheese. *Le Lait*, **85**, 57–74.

Tojo, R., Suárez, A., Clemente, M.G. *et al.* (2014) Intestinal microbiota in health and disease: Role of bifidobacteria in gut homeostasis. *World Journal of Gastroenterology*, **20**, 15163–15176.

Turroni, F., Foroni, E., Serafini, F. *et al.* (2011) Ability of *Bifidobacterium breve* to grow on different types of milk: Exploring the metabolism of milk through genome analysis. *Applied and Environmental Microbiology*, **77**, 7408–7417.

Turroni, F., Marchesi, J.R., Foroni, E. *et al.* (2009) Microbiomic analysis of the bifidobacterial population in the human distal gut. *ISME Journal*, **3**, 745–751.

Valdés-Stauber, N., Scherer, S. and Seiler, H. (1997) Identification of yeasts and coryneform bacteria from the surface microflora of brick cheeses. *International Journal of Food Microbiology*, **34**, 115–129.

Ventura, M., O'Flaherty, S., Claesson, M.J. *et al.* (2009) Genome-scale analyses of health promoting bacteria: Probiogenomics. *Nature Reviews Microbiology*, **9**, 61–71.

Villar-Tajadura, M.A., Rodríguez-Alcalá, L.M., Martín, V. *et al.* (2014) Production of conjugated linoleic and conjugated α-linolenic acid in a reconstituted skim milk-based medium by bifidobacterial strains isolated from human breast milk. *BioMed Research International*, **2014**, 725406. doi:10.1155/2014/725406

WGO (2011) World Gastroenterology Organisation Global Guidelines: Probiotics and Prebiotics, Milwaukee, WI: World Gastroenterology Organisation. http://www.worldgastroenterology.org/probiotics-prebiotics.html (accessed June 2016).

Wouters, J.T.M., Ayad, E.H.E., Hugenholtz, J. and Smit, G. (2002) Microbes from raw milk for fermented dairy products. *International Dairy Journal*, **12**, 91–109.

Xu, R., Shang, N. and Li, P. (2011) In vitro and in vivo antioxidant activity of exopolysaccharide fractions from *Bifidobacterium animalis* RH. *Anaerobe*, **17**, 226–231.

CHAPTER 5

Industrial production of starter cultures

Sanna Taskila

Chemical Process Engineering, University of Oulu, Finland

The production of a starter culture aims at the reproduction of a selected strain or mixed population of microbes to a density that is likely to survive and be metabolically active in the process. By definition, starter cultures include those strains that initiate fermentation, while other microbes used in food production are commonly referred to as secondary cultures or adjunct cultures (Parente and Cogan 2004). However, secondary cultures are also often discussed under the same heading.

The starter culture may simply be a proportion of the previous process batch inoculated to the following batch. This method, often called backslopping, is still widely used domestically throughout the world, for instance in the production of sourdough, and industrially in the production of, for example, artisanal cheeses. In repeated use backslopping can result in an adaptive selection of strains, enriching those that have desirable features with respect to the manufacturing process (Holzapfel 2002). In low-income regions backslopping is often the sole available approach, and starter cultures may only be an attractive choice if they reduce production costs and fermentation times or increase the shelf-life, sensory quality or safety of the product by a factor that exceeds the costs of adopting new technology (Holzapfel 2002).

Commercial starter culture production takes place in a separate process, including appropriate quality measures. The selection of whether starter cultures are prepared on-site in food production or acquired from commercial producers may influence the safety and flexibility of food manufacture (Hansen 2002). Some food manufacturers also sell their starter cultures to other companies. Application of frozen or freeze-dried starter cultures eliminates the in-house propagation of cultures, and thus reduces the costs associated with bulk culture preparation and lowers the risk of bacteriophage infection (Santivarangkna *et al.* 2007).

Starter Cultures in Food Production, First Edition. Edited by Barbara Speranza, Antonio Bevilacqua, Maria Rosaria Corbo and Milena Sinigaglia.

The development of starter culture production was originally initiated during the early 1800s. Since the early developments by two Danes, Emil Christian Hansen at the Carlsberg Brewery and Christian D.E. Hansen (who later established the Christian Hansen dairy starter manufacturing company), the starter culture industry has grown into a remarkable global business. The first starter cultures were liquid cultures, prepared by cultivation of bacteria in sterilized milk. These cultures suffered from overacidification and thus lost their viability during storage. This loss could be delayed, but not totally avoided, via the addition of calcium carbonate. To overcome the problems of liquid cultures, drying technologies were developed, leading to more stable preparations. Today starter culture development and production follow strict procedures, with quality standards close to those of the pharmaceutical industry.

The main phases of industrial starter culture production are described in this chapter. The emphasis is on the decisive phases where major influences on starter culture functionality and quality can be induced. These are especially related to microbial stress factors during the various phases of production, their effects on starter culture survival during processing and fermentation, and means for their reduction or elimination. Moreover, the current market for commercial starter cultures and some aspects of their growth are discussed.

Production process

Starter cultures are produced either in-house – that is, in the food production process – or by commercial manufacturers. In-house produced cultures, commonly referred to as bulk starter cultures, are used widely in the dairy industry. The main phases in the industrial production of starter cultures are development and construction of a stock culture; preparation of a culture medium; propagation of the stock culture to final cell density in a bioreactor; harvesting and concentration of cells from the medium; and preservation of the culture (Figure 5.1).

The initial phases of production are similar for both starter culture types. The main difference is that bulk starter cultures are used directly, without the need for formulation or preservation. This is an advantage, as these costly process steps are avoided. On the other hand, the production of bulk starter cultures in on-site fermentation increases the risk of bacteriophage contamination in the process. Commercial starter cultures produced by specialized companies, in the dairy industry referred to as direct vat set (DVS) or direct vat inoculation (DVI) cultures, are provided in the form of microbial cells that are usually dehydrated or frozen. Their production thus requires concentration and dehydration of the culture, followed by preservation to avoid spoilage during storage and distribution.

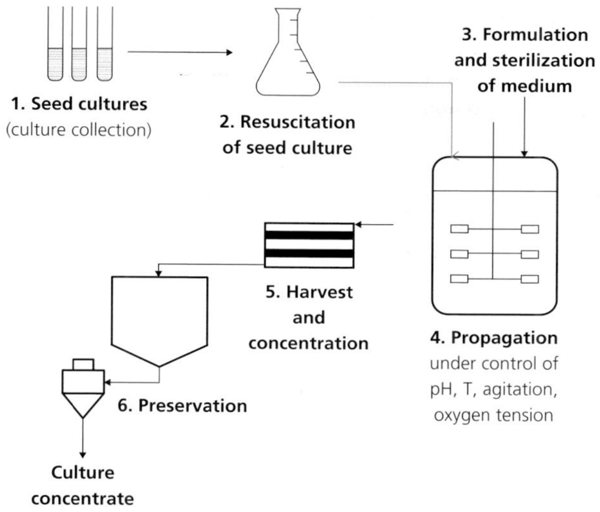

Figure 5.1 The main phases in the industrial production of starter cultures are (1) seed culture development; (2) formulation and sterilization of medium; (3) resuscitation of seed culture; (4) cell propagation; (5) harvest and concentration; and (6) preservation. In bulk starter production the final phases of concentration and preservation may be avoided or reduced.

Quality control

Modern production procedures for commercial starter cultures are close to pharmaceutical standards. Their production requires multidisciplinary knowledge, based on microbiology and microbial physiology combined with process engineering. Furthermore, expertise in cryobiology is required when cryotechnologies are employed for the preservation of cultures. The comprehensive quality control includes several steps, such as testing of raw materials, maintenance and control of plant hygiene, and testing of end-product batches. Personnel are trained regularly in the procedures. It is also essential to maintain inoculum quality and hygiene, as well as to keep aseptic conditions throughout the production chain.

In order to obtain a final product of consistent quality, the conditions for preparing the starter culture should be reproducible. Therefore the quality of batches is monitored via tests of cell viability, and contaminant-detection procedures are employed. Overall, the commercial manufacture of starter cultures is prepared by following the hazard analysis and critical control points system (HACCP), which also contributes to the high quality of the fermentation process (Notermans *et al.* 1995).

To maintain the viability of cells during storage and thus ensure their proper functioning in fermentation, starter cultures should be prepared using conditions that promote cell survival. This applies to each step in manufacturing. Exposure to stress during the different manufacturing stages can result in altered

Table 5.1 Major microbial stress inducers in the production of starter cultures, and possibilities for their reduction or elimination.

Stress	Inducer	Process phase	Reduction options
Starvation	Lack of nutrients	Cultivation	Medium formulation
pH change stress	pH changes	Cultivation	Medium formulation
Oxidative stress (anaerobic microbes)	Reactive oxygen species (such as free radicals)	Cultivation	Anaerobic conditions Medium formulation Strain engineering
		Other phases	Rapid processing
Mechanical stress	Shear forces	Harvesting	Benign processing
Osmotic stress	Ionic strength	Cultivation	Medium formulation
		Harvesting	Rapid processing
		Preservation	Preservation agents
Thermal stress	Too low or high temperatures	Preservation	Low-temperature drying Short heating periods

growth and survival of the starter microbes. In order to ensure sufficient functionality of cultures, it is thus essential to recognize and address stress-inducing factors in each step.Based on the literature, the cultivation phase may expose microbes to starvation, pH changes and metabolite-induced stresses. Oxidative, osmotic, mechanical and thermal stresses are more typical during the later phases of manufacturing, namely harvesting and preservation, and during storage. Typical inducers of microbial stress during starter culture production and respective means for their elimination are summarized in Table 5.1. The description of the process phases gives a more detailed discussion of reported efforts at stress reduction.

Preparation of inoculum culture

Selection of starter strain(s)

Commercial starter cultures usually originate either from raw materials or via the production of traditional foods. For an understanding of the process, it is essential to identify the dominant microorganisms at each process stage, accompanied by an assessment of their influences on the fermentation process and end product. Therefore, process samples from various phases and batches need to be investigated using appropriate techniques to isolate and identify the microbes present and their metabolic products.

Microbes may be enumerated by means of cultivation-based protocols, such as microbial counts prepared by spread-plating on appropriate media. Generally, DeMan, Rogosa and Sharpe agar (MRS-agar) is well suited to lactic

Table 5.2 Typical starter organisms for various substrates based on known indigenous fermentation processes.

Substrate type	Product type	Strains
Cereals and vegetables	Lactic fermented products (Steinkraus 1995; Holzapfel 1997)	*Lb. brevis, Lb. fermentum, Lb. plantarum, Lb. reuteri, Pe. pentosaceus and Pe. acidilactici*
Milk	Dairy products (Hammes and Hertel 1998)	*Lc. lactis*, followed by Lb. *casei (paracasei)* and other *Lactobacillus* spp. during maturation
Meat and fish	Lactic fermented products (Hammes and Hertel 1998)	*Lb. sakei and Lb. curvatus, Ln. mesenteroides*
Sugar-rich plants	Fermented beverages and foods (Holzapfel 2002)	*Saccharomyces, Candida, Torula and Hansenula*

Notes: Lb. = Lactobacillus; Lc. = Lactococcus; Ln. = Leuconostoc; Pe. = Pediococcus.

acid bacteria, while yeasts may be enumerated on, for instance, yeast extract glucose chloramphenicol agar (YGC-agar; Coulin *et al.* 2006). The selection of appropriate cultivation media is based on knowledge of the microbes in the substrates and the process itself. Hints of possible starter organisms for a specific substrate or product can be found in the literature; some typical examples are presented in Table 5.2. Modern molecular identification methods, such as polymerase chain reaction (PCR) or RNA fingerprinting, may offer faster and in some cases also more precise recognition of microbes.

Suitable starter strains are able to compete in a mixed population and retain their viability; produce sufficient quantity of the desired metabolite (usually acid or alcohol); possess antimicrobial properties against pathogens and other contaminating microbes; promote desired organoleptic changes in the substrate; and degrade unwanted antinutritional factors or toxic compounds in the product. A probiotic function would be of further benefit (Holzapfel 1997).

The preparation of pure cultures may be conducted by isolation of strains from mixed populations of traditional fermented foods. It is notable, however, that such isolates may exhibit a wide diversity of metabolic activities, even between strains. Regarding the functionality of starter cultures, critical differences may be seen in growth rate; adaptation to the selected substrate; ability to degrade antinutritive compounds; antimicrobial properties and competitive growth within mixed populations; and flavor and quality attributes (Holzapfel 1997).

Defined or undefined cultures

Inoculum cultures can be prepared either as defined cultures, including a selected strain or multiple strains in consortia, or as undefined cultures (including so-called artisanal cultures). The earlier classification as single-strain, multiple-strain or mixed-population cultures is also still used despite its limitations (Mullan 2014).

The selection of culture types depends on the manufactured product, dictated by the attributes of the substrate and requirements for the production process and the product itself. Single-strain cultures are primarily used in the production of beer, sauerkraut (via backslopping of brine from a previous batch), pickled vegetables and vegetable juices and soy products (Buckenhüskes 1993). Wine and sausage manufacturing may employ both single-strain and multiple-strain cultures. All three starter culture types are used in the dairy industry and sourdough fermentation.

The advantages of single-strain cultures include improved process control and increased predictability of microbial metabolism in the cultures. On the other hand, single-strain cultures are vulnerable to bacteriophage infections, mutations and loss of key physiological properties (especially in the case of plasmid-originating properties), all of which may lead to drastic flaws in the fermentation process. Use of single-strain cultures may also be too demanding for small-scale production with respect to their preparation, handling and application procedures and equipment. However, use of single-strain cultures combined with proper control is the most advisable way to achieve a consistent process and high-quality fermentation products.

Mixed-strain cultures are generally less susceptible to deterioration, a property that, together with less demanding handling procedures, makes them very suitable for small-scale operations. Further benefits of mixed-strain cultures include the possibility of a more complex sensory quality in the product, as well as synergistic advantages related to the degradation of substrate compounds and the generation of flavour. Mixed-strain cultures may increase variation between product batches, which may partially be overcome via proper process control. Undefined cultures are still of industrial significance, despite their poor predictability in fermentation. They are commonly used in Europe for the manufacture of south Italian Mozzarella cheese.

Strain engineering

Strain engineering has been applied for the development of starter strains for some decades (Hassan *et al.* 2001). Emerging modern technologies for this purpose may allow further improvement of the safety, sensory properties and nutritional value of fermented foods, together with an extended shelf-life and further health-promoting functions (Hassan and Frank 2001). High-end tailored starter cultures will thus be available for broader use in the food industry. Moreover, selection may be complemented by genetic tools for the engineering of starter cultures with higher technical and metabolic performance in a specific fermentation. From the production point of view, the major advances of strain engineering relate to the improved stress tolerance of the cells. For instance, the oxidative stress tolerance of *Lactobacillus plantarum* (Noonpakdee *et al.* 2004) as well as the tolerance of *Lactobacillus salivarius* to various stresses (Sheehan *et al.* 2006) have been enhanced via genetic engineering. These improvements are valuable in the manufacture of starter cultures, leading to an increased cell survival rate during concentration and preservation.

Formulation and preparation of media

The components of strain- or culture-specific media are mixed in the medium vessel to give a concentrated solution. This solution is then pumped into the bioreactor, diluted with water to the defined volume and sterilized. Sterilization is commonly achieved by heating in an autoclave at 121 °C and 1 bar positive pressure. More sensitive materials may be treated at lower temperatures for a longer period (e.g. at 90 °C for 45 minutes) or flash sterilized by means of ultra-high temperature (UHT) treatment. After sterilization the medium is cooled down to the cultivation temperature.

Nutrient supplementation

The primary purpose of the medium is to provide the starter microbes with essential nutrients and supplements, respective to the strain and applications. The energy and carbon source for microbes is most commonly either lactose (for lactic acid bacteria, LAB), maltose, glucose or sucrose, provided in concentrations of 10–40 g/L. Suitable nitrogen sources include milk hydrolysates for LAB and peptones for other microbes. Yeast extract is commonly used to provide growth factor. The medium is one major cost factor in starter culture manufacture and it is thus common to utilize agroindustrial by-products and their derivatives for the purpose. Typical examples are whey, corn steep liquor, malt extracts, soy isolates and potato infusions. Sometimes media are supplemented with minerals, amino acids and vitamins to allow the growth of fastidious strains.

Antifoam agents are generally used to prevent excess foaming, and the resulting gas and nutrient transfer problems. The medium recipe might be restricted due to trade, such as specifications of Kosher foods, genetic modification–free products or hypoallergenic products (Vedamuthu 2006). Essentially, all medium components must be food grade.

Role of medium in cell survival and functionality

Another role of the cultivation medium is related to the survival of the starter culture during storage and in food fermentation. Several stress factors may be addressed via proper formulation of the medium.

A variety of medium supplements have been introduced to improve cell survival during preservation (Tamime 2002). In fact, the selection of a carbon source for the microbes might also be important for the functionality of the starter culture. According to Carvalho and co-workers (Carvalho *et al.* 2004), the viability of *Lactobacillus bulgaricus* after freeze drying was clearly lower after cultivation on mannose in comparison to fructose, lactose or glucose. In another study the survival of *Lactobacillus sakei* in spray drying was improved by the addition of sucrose to the medium (Ferreira *et al.* 2005). The addition of trehalose to the growth medium was shown to stabilize the cytoplasmic membrane of *Bradyrhizobium japonicum* during desiccation, which improved the preservation of its cell viability (Streeter 2003). These observations are partly based on the influence of intracellular metabolites, such as mannitol, sorbitol or glutamate, which reportedly improve the survival of cells during dehydration (Wisselink *et al.* 2002). Since the formation of such metabolites depends on the carbon source in the medium, a higher survival rate of cells after preservation may be achieved via the selection of medium components (Kets *et al.* 1996).

A number of other medium components also have effects on the survival of probiotic cultures following dehydration. For instance, excess concentration of sodium chloride (NaCl) has been related to lower cell viability after drying (Linders *et al.* 1997). Moreover, the cultivation of *Lactobacillus reuteri* at pH 5 instead of pH 6 was shown to improve its survival during freeze drying by 50% (Palmfeldt and Hahn-Hägerdal 2000), probably due to the induction of stress-response mechanisms within the cells prior to drying. A similar effect, although with a different cellular mechanism, was induced in bifidobacteria via pH decrease and starvation (Maus and Ingham 2003).

Besides in the production process, the starter culture cells are also exposed to various stress factors during their end use; that is, in the food fermentation process. Stress may originate from temperature fluctuations, acids and other metabolites, pH changes, high osmotic pressure and low availability of nutrients. The tolerance of the culture to these stresses may be enhanced by the induction of a low to intermediate level of microbial stress during the propagation phase. For instance, the cold and acid tolerance of bifidobacteria was significantly increased by the application of sublethal stress to the cells during the final phase of cultivation (Maus and Ingham 2003). However, an understanding of cellular response mechanisms related to each stress factor and interrelations between different factors is a necessity to avoid detrimental damage to the cells.

According to Li and co-workers (Li *et al.* 2003), the uptake of glutathione by *Lactococcus lactis* activates a glutathione–glutathione peroxidase–glutathione reductase system in stationary-phase cells, which catalyses the reduction of hydrogen peroxide (H_2O_2) and thus reduces the oxidative stress of cells. Chelating agents, such as certain citrates, are used to inhibit phage adsorption onto the cells.

Propagation of cells in a bioreactor

The starter culture microbes are preserved in small quantities as stock (or seed) cultures. The reactivation of stock cultures is prepared sequentially, starting from a minor volume culture, such as 0.4 mL, and scaling up to the appropriate quantity. Ideally, each batch is inoculated with approximately 2% of the previous stock culture. The final starter culture fermentations are typically batch processes with a capacity of around 10,000–50,000 L. During sequential cultivations the quantity of cells may be multiplied by a factor of six orders of magnitude or more. To maintain the viability of cells in the starter culture, it is essential not to allow the culture to proceed to the stationary phase. Thus the desired endpoint density is determined for each culture and set of conditions separately during process development.

Although starter cultures are mainly produced in batch processes, continuous operations may become more popular in the future. The advantages of continuous cultures include high cell yield and process volumetric productivity, as well as decreased volumes in downstream processing. Moreover, cells produced during continuous culture are in a controlled physiological state that can be manipulated via environmental parameters, such as medium composition and dilution rate, allowing a more stable specific growth rate of the microorganisms in the system.

The disadvantages of continuous processes are related to a higher contamination risk and the increased possibility of changes in plasmid-based cell characteristics. These issues may be addressed via the immobilization of cells on, for instance, k-carrageenan-based materials (Nedoviç and Willaert 2006). Moreover, immobilization may also protect starter cultures from bacteriophages (Steenson and Swaisgood 1987). A two-reactor system introduced by Doleyres and coworkers (2004) consisted of separate solid-state and submerged cultivation phases, allowing the continuous production of a concentrated mixed culture of *Lactococcus diacetylactis* and *Bifidobacterium longum*.

Process control

The maintenance of starter culture quality requires control of the main process parameters during cultivation. The important parameters in starter culture production are cell density, pH, temperature, dissolved oxygen tension (DOT), agitation and metabolic products (headspace gases) for the desired levels, or compositions in the case of gases; these factors are microbe or even strain specific. Temperature, pH and DOT can be measured on-line in standard equipped reactors, while metabolite analyses may require more complex procedures.

pH

The pH level is one of the major parameters in all microbial cultivations. The fermentation medium usually becomes acidified due to microbial metabolism, which in turn slows down and finally prevents the propagation of cells. The control of pH is thus a prerequisite for cultivation. In the case of LAB, the pH should be kept at 4.5 to maintain growth. To avoid problems associated with pH decrease, culture media buffered to a pH around 6 may be used.

Buffered media can be prepared by the use of cations that are associated with carbonates, hydroxides, phosphates or oxides. Their addition has, however, been criticized for biasing fermentations through the increased risk of phage contamination (Mandel and Higa 1970) or the lower solubility of caseins in cheese production (Duwat *et al.* 2005). The introduction of porphyrin compound in the anaerobic cultivation of LAB allowed pH maintenance without cation addition, and enhanced the viability of cells and their resistance to various stress conditions (Duwat *et al.* 2005). This invention was further upgraded by researchers at Chr. Hansen A/S to produce starter bacteria with porphyrin in order to reduce the oxygen content of foods (Asger *et al.* 2013).

Cell density

The propagation of cells is continued until the culture reaches the early stationary phase. It is thus necessary to monitor the cell density in the reactor. On-line monitoring can be conducted with optical measurements that are typically based on transmittance, absorbance or back scattering. The pH control system may also be used for indirect monitoring of cell growth during fermentation. This is based on the correlation between the pH change in the medium and the speed of carbon metabolism by microbes. Slowing down of the consumption of neutralization chemical thus reflects the stationary phase being reached. This approach allows the elimination of specific cell density measurement.

Temperature

Microbial growth generates heat and the temperature during fermentation is thus maintained at a constant level via cooling, usually via pumping cool water into the jacket of the reactor. Optimal temperatures for the cultivation of psychrophilic, mesophilic and thermophilic starter microbes are 10 °C or less, 20–45 °C and 45 °C or higher, respectively. Traditionally, mesophilic starter strains are used in the production of fermented buttermilk and Dutch and French cheeses, while thermophilic strains are more common in the production of yoghurts, as well as Swiss and Italian cheeses (Parente and Cogan 2004). Nowadays mesophilic and thermophilic strains are often used in mixtures. Psychrophilic – or psychrotrophic – strains may be useful in cases where fermentation typically continues during the storage and distribution of the product or where the growth of psychrophilic pathogens could occur (Lee *et al.* 2006; Eom *et al.* 2007).

Agitation

The control of agitation is required to maintain appropriate DOT in aerobic cultivation. Agitation is complemented by blowing air through a sparger in the bottom of the reactor. Since the solubility of oxygen is relatively low under mild temperatures and pressures, the agitation contributes to the costs of fermentation and induces some level of shear stress to the cells. Nevertheless, providing enough oxygen for the cells is a necessity for the propagation of aerobic microbes, and might even contribute to the formation of desired metabolic products, such as diacetyl (Monnet *et al.* 1994). In anaerobic cultivation, constant agitation is used for mixing to enhance the transfer of medium components in the medium.

Headspace gas analysis

Headspace gas analysis is used for the determination of metabolic activity within a microbial culture. For instance, the formation of diacetyl, which is a volatile aroma compound produced by LAB, can be followed via headspace gas analysis (Monnet *et al.* 1994). Headspace gas composition also reveals shifts from aerobic to anaerobic metabolism, originating from oxygen gradients of too low DOT in the reactor. In some cases the analysis may also be useful for the detection of metabolic by-product accumulation in the reactor, which might reveal a malfunction in the culture, for example related to microbial contamination. Gas chromatography is a commonly utilized technology for the measurement of headspace concentrations of, for instance, acetaldehyde, acetone, di-acetyl, ethanol and methanol.

Harvesting and concentration

After propagation of the cells, the biomass is harvested from the growth medium, often resulting in approximately 10–20 times concentrated slurry. Harvesting is typically carried out by means of centrifugation or ultrafiltration. Another reported technology for starter culture concentration is the diffusion culture method, originally described by Osborne (1977), which is based on pumping cells through a membrane, followed by the diffusion of metabolites through the same membrane towards the fresh medium. This technique requires heat- and chemical-resistant membranes that have a sufficiently large area to permit adequate diffusion of the rapidly produced metabolic by-products. Osborne's method reaches relatively high cell densities.

It is notable that the metabolic activity of the starter concentrate does not necessary correlate to the density of the cell population. However, the concentration of cells prior to preservation is an important factor for cell survival. One likely reason is that the microbial cells shield each other by affecting their proportional area of contact with the environment (Bozoglu *et al.* 1987). The effect of initial cell concentration seems also to be related to the medium used; that is, the performance of protectants such as sucrose may be improved due to the shielding effect.

There is evidence that the timing of cell harvest influences the survival of LAB during freezing (Kim *et al.* 1999; Wouters *et al.* 1999; Fonseca *et al.* 2001; Lee 2004) and spray drying (Corcoran *et al.* 2004). This effect is probably at least partly associated with membrane fatty acid composition (Beal *et al.* 2001). During harvest and concentration, cells are unavoidably submitted to shear stresses. There is some evidence that a certain level of shear stress may be beneficial to the cells, depending on the organism and the intensity of the stress. An intermediate shear stress of 36–54 Pascal seconds (Pa) during cultivation was observed to improve the metabolic activity of *Lb. bulgaricus*, together with causing lengthening of cells (Arnaud *et al.* 1993).

Centrifugation

Centrifugation is a commonly used technique for the harvesting of microbes. It separates particles based on their density under centrifugal force. The main advantages of the technique are a high processing capacity and a high efficiency of separation in industrial applications. The cell mass may be concentrated up to 40% (v/v) (Yavorsky *et al.* 2003), resulting in a starter population of 10^{11}–10^{12} cfu (colony forming units) per mL.

Centrifugation leads to some level of cell injury, although this can be reduced via the design of the centrifuge. Centrifugal damage, induced by high gravitation forces, has been known to alter bacterial cell surface properties and interior structures, including DNA (Peterson *et al.* 2012). Lipopolysaccharides of *Escherichia coli* cell membrane have been observed in the medium after centrifugation, leading to lower stress tolerance in further processing (Gilbert *et al.* 1991; Wyber *et al.* 1994).

Centrifugation is more suitable for the separation of large cells, while particles with a diameter of 1 micrometer or lower are not efficiently sedimented. This may be detrimental in further concentration phases, but does not generally limit the use of centrifugation for harvesting of starter cultures. Further disadvantages of centrifugation are its limited efficiency in the case of viscous media (Foerst and Santivarangkna 2015), low performance for harvesting of certain microbes, indicated by the formation of soft pellets (Reilly and Gilliland 1999), and relatively high investment and maintenance costs.

Filtration

Another method for starter culture concentration is filtration, mainly via membranes. Membrane filtration yields practically all the cells and it is also a cost-efficient technique (Van Reis and Zydney 2007). On the other hand, it is a rather slow separation technique and may also induce changes in cell physiology. The filtration speed may be increased by the use of larger surface area membranes.

The survival of *Lb. bulgaricus* during concentration was lower after microfiltration than after centrifugation. This may relate to lower tolerance to shear stress. On the other hand, cell survival in preservation by freezing and freeze drying was improved when filtration was used instead of centrifugation. Filtration flow velocity and transmembrane pressure have a significant effect on cell survival (Streit *et al.* 2011). The pumping to the membrane may be harmful to the cells, although severe cell damage is not likely to occur (Holst and Mattiasson 1990).

Preservation

Historically, starter cultures were produced by food industries using liquid cultures of their own or of local suppliers. These so-called bulk starters were used, for instance, to inoculate the milk used in the manufacture of dairy products. Over the past few decades, the use of starter cell concentrates has increased, replacing bulk starters in dairy product manufacture. The modern age of starter culture technologies started in the 1960s when suppliers began to employ freeze drying, also known as lyophilization, for the preservation of starter cultures. Common preservation techniques and their attributes are summarized in Table 5.3.

Drying techniques

Drying is nowadays the dominant technique for the preservation of starter cultures. The drying processes developed include freeze drying, vacuum drying, spray drying, drum drying, fluidized bed drying and air drying. Regarding the fixed and operational costs of these processes, freeze drying is by far the most expensive option. In contrast, air drying is estimated to be the cheapest of the drying processes listed, with approximate fixed and manufacturing costs of only 5.3% and 17.9% of those for freeze drying, respectively (Santivarangkna *et al.* 2007).

During dehydration some cell injury will take place regardless of the drying technique employed. Based on various reports, the main site of injury in cells is the cytoplasmic membrane (Lievense and Van't Riet 1994; Gardiner *et al.* 2000), which is indicated by the increased concentration of intracellular components in

Table 5.3 Comparison of preservation techniques for starter cultures.

Technique	Advantages	Disadvantages
Freeze drying	Good stability of products under storage	Cold injury to cells High cost High energy consumption
Spray drying	High speed of drying Possibility of continuous operation Low fixed and operational costs compared to e.g. freeze drying	Thermal damage to cells Low stability of products under storage Difficulties in rehydration of products
Fluidized bed drying	Lower temperature compared to e.g. spray drying, less damage to cells	Shorter drying period compared to e.g. freeze drying
Vacuum drying	Cost-efficient technique	Lack of standard processes Thermal damage to cells
Freezing	Lower requirements for equipment	Risk of thawing Demand for low temperature in storage and transportation leads to high cost

the medium after rehydration of dried cells or by the increased sensitivity of the cells to chemicals (Lievense and Van't Riet 1994). To minimize cell injury in dehydration, drying protectants are commonly employed. The survival of cells during preservation may be improved by the use of polyols, amino acids and amino derivatives, such as betaine, as additives to reduce cellular osmotic stress (Morgan *et al.* 2006). Generally, protectants used in microbial preservation are either amorphous glass or eutectic crystallizing salts (Morgan *et al.* 2006). Protective agents can be added to the cultivation medium or prior to preservation. The appropriate protectant is selected based on the microbe, as described for instance in a review by Morgan and co-workers (2006). Non-fat milk solids, serum, trehalose, glycerol, betaine, adonitol, sucrose, glucose, lactose, dextran and polyethylene glycol function well with many microbial species (Hubalek 2003).

Freeze drying

Despite its high cost compared to other drying techniques, freeze drying is the dominant preservation technique in commercial starter culture production (Santivarangkna *et al.* 2007). Originally, freeze drying was developed for preserving medical supplies, such as blood plasma, during the Second World War. The first food applications were in the preservation of instant coffee. Nowadays, freeze drying is an essential phase of starter culture production in the food industry, maintaining the viability of bacterial cells and increasing product shelf-life.

The freeze-drying process is based on a decrease of the moisture content in the culture down to 5% at a low temperature. The process phases are freezing of material according to optimized time and temperature limits, freeze drying (sublimation) according to defined time, temperature and pressure values for the optimal process, and post-drying of products. The typical batch size in the industrial freeze-drying process is close to 1000 L. The dried material is usually ground to a fine powder, which is then collected in the container and homogenized by mixing. Freeze-dried concentrates can be stored for several months at 4 °C.

The freeze-drying process has some obvious drawbacks. Its main disadvantages relate to high costs and energy consumption, and to the prolonged lag time of the culture in fermentation (Sandine 1996). Cell injury induced by freeze drying leads to changes in the structure, permeability and functionality of the cell membrane, measured as changes in the ratio of unsaturated and saturated fatty acids in the membrane, decreased activity of the membrane-bound enzyme ATPase and a loss of ΔpH (Castro *et al.* 1997).

The exposure to low temperatures during freeze drying also induces other physiological changes in cells, such as lower stability of secondary structures of nucleic acids, inefficient folding of proteins and reduced functionality of ribosome (Phadtare 2004). Furthermore, freeze drying affects the cells' capacity to utilize carbohydrates (Riis *et al.* 1995).

The use of cryoprotectants, such as L-glutaric acid or L-arginine, increases the survival rate of cells during freeze drying (Tamime 2002).

Spray drying

Microencapsulation technologies are emerging in the improvement of cell stability during the storage and application of microbial cultures, especially related to probiotics. Spray drying, regarded as a microencapsulation technology, is considered a very suitable drying method for the long-term preservation of microbial cultures (Riveros *et al.* 2009). The spray-drying process is based on the atomization of wet product at a high velocity, followed by spraying of product droplets into a flow of hot air, at temperatures around 150–200 °C (Peighambardoust *et al.* 2011). Since the atomized droplets have a very large surface area, drying is extremely rapid (Morgan *et al.* 2006; Santivarangkna *et al.* 2007). The technique is thus appropriate for the preservation of bioactive compounds whose functionality must be retained. The commercial applications of spray drying include dehydrated enzymes and isolated proteins, and dehydrated probiotic bacteria (Ortega-Rivas *et al.* 2006; Riveros *et al.* 2009).

The advantages of spray drying include the high speed of drying and the possibility of continuous operation, which enable high processing quantities and low fixed and operational costs compared to freeze drying (Knorr 1998; Santivarangkna *et al.* 2007). However, spray drying has major disadvantages, such as loss of cell activity during drying, low stability of dried products under storage and difficulties in rehydration of products, and it is therefore not commonly used in the commercial preservation of starter cultures (Peighambardoust *et al.* 2011). For instance, a survival rate of 60% at an outlet temperature of 80 °C was reported for *Lactobacillus rhamnosus* GG. The rate could be improved via the use of prebiotic substances in the carrier material, but this was observed to reduce the storage stability of the product. The main reason for cell death was heat-induced damage to the cell membranes, which was directly proportional to the outlet temperature (Ananta *et al.* 2005).

Fluidized bed drying

Fluidized bed drying is also very suitable for the conservation of LAB. In the fluidized bed process a special carrier material is employed to enable application of the substrate via a spray nozzle. The temperature during fluidized bed drying is typically lower than in spray drying to avoid thermal damaging of the cells. Its advantage in comparison to freeze drying is the short drying period. The process starts with the spraying of culture onto the dispersed carrier material to form fine droplets. Dry air is blown in a bottom-up direction through the fluidized bed, leading to fluidization and drying of the particles. The carrier particles coated with cells are covered with a protective coating. After harvesting, the dust and particles are removed in the associated sieving. The typical batch size for fluidized bed drying is around 300–500 kg.

Vacuum drying

Vacuum drying is a promising method for reserving sensitive biological material due to its acceptable cost-effectiveness balance. To avoid cellular damage, in particular the temperature and duration of the vacuum drying process should be optimized (Tymczyszyn *et al.* 2008). Vacuum drying, similar to other heat-associated drying processes, induces thermal stress to the cells. This is likely to cause irreversible damage to the cells and thus the exposure of cells to high temperatures should be minimized. The commercialization of vacuum drying has probably slowed due to the lack of standardized processes, originating from the influence of equipment on drying conditions such as the pressure of the vacuum system (Goderska 2012).

Freezing

Frozen culture concentrates can be prepared by means of conventional freezing at −20 °C, deep freezing at −40–80 °C or ultra-low-temperature freezing in liquid nitrogen at −196 °C (Parente and Cogan 2004). The storage temperature should not be lower than −40 °C (Tamime 2002) and rapid thawing is advisable to minimize cell injury. Frozen cultures typically contain 10^{10}–10^{11} cfu/g cells. This preservation technique is mostly used for dairy cultures. Frozen cultures require low transportation and storage temperatures, which has a major influence on transportation costs (Peighambardoust *et al.* 2011). They are also prone to thawing during storage and distribution. Similar to freeze drying, freezing induces cold shock responses in microbes, which may be reduced via the use of cryoprotectants. Typically employed protecting agents are glycerol, monosodium glutamate, sucrose and lactose. Freezing may be accelerated using a mixture of dry ice and alcohol.

Starter culture market

Current market

The global market for starter cultures has experienced positive growth in the past few years, mainly driven by beverages. The consumption of non-alcoholic beverages in particular is increasing due to emerging health awareness among consumers (MarketsAndMarkets 2014). The global starter culture market is forecast to be worth $1.0 billion by 2018. Yeast cultures, mainly used in the production of alcoholic beverages, dominate the starter culture market. Bacterial starter cultures, applied especially in the dairy industry, represent the second largest market.

Based on reports from 2012, the largest market for starter cultures is Europe, followed by North America. Typical characteristics of European markets are the emerging consumer awareness of the health benefits related to probiotic microbes, and the ability of consumers to pay higher prices for premium products.

Table 5.4 Starter culture products of the largest manufacturers.

Company	Starter products
Angel Yeast	Baker's and brewer's yeast
Chr. Hansen	Dairy cultures, probiotics, meat cultures
Coosheen	Dairy cultures
Csk Food Enrichment	Dairy cultures for milk-based foods and beverages
Danisco	Dairy cultures, probiotics
Lactina	Dairy cultures
Lallemand	Baker's yeast
Lb Bulgaricum Pic	Dairy cultures
Lesaffre Group	Yeast cultures
Wyeast Laboratories	Yeast cultures

The North American market is expected to grow at a compound annual growth rate (CAGR) of 5.9% by 2018. In both Europe and North America the market is dominated by non-alcoholic beverages. Both markets benefit from advances in starter culture technologies.

The fastest growth in the starter culture market is foreseen in the Asia-Pacific region. This market for starter cultures is expected to grow at a CAGR of 6.3% from the mid-2010s. The fast growth in the region is related to favourable market and geographical conditions that promote the production of fermented foods. For cultural reasons, market growth in Asia-Pacific is mainly associated with non-alcoholic beverages.

Currently the leading manufacturers of commercial starter cultures are Angel Yeast Co. Ltd (China), Lallemand Inc. (USA) and Chr. Hansen A/S (Denmark). Among the ten leading starter culture manufacturers, four companies specialize only in yeast cultures and six other products, mostly LAB cultures (Table 5.4). There are also several dairy companies, such as the Finnish company Valio Ltd., which sell their special starter cultures to licensed customers. Additionally, contract manufacturing of starter cultures is provided by a number of companies.

Aspects of future markets

Global development towards the industrialization of food production also promotes the growth of starter culture markets. Emerging business opportunities are related to starter culture markets in developing regions of Africa and Asia, in which fermented foods are a major part of the diet. Fermented foods have an important socioeconomic role in developing countries and contribute to the protein requirements of their populations. Traditional fermentation is usually conducted under primitive conditions that relate to low process performance and poor quality and safety of the products. Moreover, locally produced foods suffer from a lower shelf life and generally less attractive form when compared to imported foods, a fact that has radically changed the food culture in some regions (Achi 2013). Local production is nevertheless a prerequisite to ensure an affordable price for food.

To become a rule instead of an exception, defined starter cultures need to bring clear benefits for the food manufacturer. According to various studies on the use of starters in traditional small-scale fermentation, the prospects for this are promising. As in the case of well-established large-scale manufacturing processes, defined starters have been shown to improve small-scale fermentation processes and their predictability (Sanchez *et al.* 2001; Coulin *et al.* 2006), to enhance the aroma of traditional products (Teniola and Odunfa 2001) and to improve product safety (Valyasevi and Rolle 2002).

In developing countries, especially in rural areas, the use of defined starter cultures is limited due to a lack of appropriate infrastructure and technologies (Holzapfel 2002). Moreover, technologies for small-scale fermentation have been developed over the years based on experience and inherited traditions, and thus the reluctance to accept changes in manufacturing or in the product itself hinders the adoption of defined starter cultures. Close cooperation between the food industry and starter manufacturers and developers has been shown to ease this development by allowing a better information flow between users and developers of starters. For instance the fermentation of traditional Thai soy sauce was significantly improved by such systematic cooperation, leading to a shorter fermentation time, improved bioreactor systems and more efficient waste management (Valyasevi and Rolle 2002).

It is essential to keep the uniqueness and traditional attributes of the product while improving its safety or aroma. For this purpose novel molecular methods, allowing strain engineering for tailored starter cultures, are useful. This has to be carried out in accordance with traditional processes and products.

Conclusion

The main decisive phases in starter culture manufacture are strain selection (including the choice between defined or undefined cultures); medium formulation to support the propagation of cells and to enhance cell survival during further processing; and the selection of harvesting and preservation techniques. All these phases will eventually affect the quality and functionality of produced cultures in food fermentation.

The future production of starter cultures relates to two aspects that have rather different attributes. The markets in developed areas are more and more related to consumers' awareness of health and health-promoting foods and the willingness to pay more for such products. This brings business opportunities for starter developers, perhaps especially for those that are able to employ novel molecular methods to tailor specific starter strains. The geographical development of starter markets, in contrast, relies strongly on manufacturers' ability to overcome mental and technical obstacles to the introduction of novel technologies to traditional food industries. This is eased by raising the consciousness of producers and consumers regarding food safety. The economic benefits of defined starter cultures need to be shown indisputably while the original aroma and other important traditional attributes of the food are maintained. Finally, investment is needed to allow an appropriate level of technology for the purpose.

References

Achi, O.K. (2013) The potential for upgrading traditional fermented foods through biotechnology. *African Journal of Biotechnology*, **4**(5), 375–380.

Ananta, E., Volkert, M. and Knorr, D. (2005) Cellular injuries and storage stability of spray-dried Lactobacillus rhamnosus GG. *International Dairy Journal*, **15**(4), 399–409.

Arnaud, J.P., Lacroix, C., Foussereau, C. and Choplin, L. (1993) Shear stress effects on growth and activity of *Lactobacillus delbrueckii* subsp. *bulgaricus*. *Journal of Biotechnology*, **29**, 157–175.

Asger, G., Kringelum, B.W., Hansen, K.F., Iversen, S.L. and Henriksen, C.M. (2013) Porphyrin containing lactic acid bacterial cells and use thereof. 16-7, Google Patents.

Beal, C., Fonseca, F. and Corrieu, G. (2001) Resistance to freezing and frozen storage of *Streptococcus thermophilus* is related to membrane fatty acid composition. *Journal of Dairy Science*, **84**(11), 2347–2356.

Bozoglu, T.F., Özilgen, M. and Bakir, U. (1987) Survival kinetics of lactic acid starter cultures during and after freeze drying. *Enzyme and Microbial Technology*, **9**(9), 531–537.

Buckenhüskes, H.J. (1993) Selection criteria for lactic acid bacteria to be used as starter cultures for various food commodities. *FEMS Microbiology Reviews*, **12**(1–3), 253–271.

Carvalho, A.S., Silva, J., Ho, P., Teixeira, P., Malcata, F.X. and Gibbs, P. (2004) Effects of various sugars added to growth and drying media upon thermotolerance and survival throughout storage of freeze-dried *Lactobacillus delbrueckii* ssp. *bulgaricus*. *Biotechnology Progress*, **20**(1), 248–254.

Castro, H.P., Teixeira, P.M. and Kirby, R. (1997) Evidence of membrane damage in *Lactobacillus bulgaricus* following freeze drying. *Journal of Applied Microbiology*, **82**(1), 87–94.

Corcoran, B.M., Ross, R.P., Fitzgerald, G.F. and Stanton, C. (2004) Comparative survival of probiotic lactobacilli spray-dried in the presence of prebiotic substances. *Journal of Applied Microbiology*, **96**(5), 1024–1039.

Coulin, P., Farah, Z., Assanvo, J., Spillmann, H. and Puhan, Z. (2006) Characterisation of the microflora of attiéké, a fermented cassava product, during traditional small-scale preparation. *International Journal of Food Microbiology*, **106**(2), 131–136 .

Doleyres, Y., Fliss, I. and Lacroix, C. (2004) Continuous production of mixed lactic starters containing probiotics using immobilized cell technology. *Biotechnology Progress*, **20**(1), 145–150.

Duwat, P., Bravard, A., Sourice, S. and Gruss, A. (2005) Process for preparing starter cultures of lactic acid bacteria, 10-2, Google Patents.

Eom, H.J., Seo, D.M. and Han, N.S. (2007) Selection of psychrotrophic Leuconostoc spp. producing highly active dextransucrase from lactate fermented vegetables. *International Journal of Food Microbiology*, **117**(1), 61–67.

Ferreira, V., Soares, V., Santos, C., Silva, J., Gibbs, P.A. and Teixeira, P. (2005) Survival of *Lactobacillus sakei* during heating, drying and storage in the dried state when growth has occurred in the presence of sucrose or monosodium glutamate. *Biotechnology Letters*, **27**(4), 249–252.

Foerst, P. and Santivarangkna, C. (2015) Advances in starter culture technology: Focus on drying processes. In W. Holzapfel (ed.), *Advances in Fermented Foods and Beverages* (pp. 249–270), Cambridge: Woodhead Publishing.

Fonseca, F., Béal, C. and Corrieu, G. (2001) Operating conditions that affect the resistance of lactic acid bacteria to freezing and frozen storage. *Cryobiology*, **43**(3), 189–198.

Gardiner, G.E., O'Sullivan, E., Kelly, J. *et al.* (2000) Comparative survival rates of human-derived probiotic *Lactobacillus paracasei* and *L. salivarius* strains during heat treatment and spray drying. *Applied and Environmental Microbiology*, **66**(6), 2605–2612.

Gilbert, P., Caplan, F. and Brown, M.R. (1991) Centrifugation injury of Gram-negative bacteria. *Journal of Antimicrobial Chemotherapy*, **27**(4), 550–551.

Goderska, K. (2012) Different methods of probiotics stabilization. In E.C. Rigobelo (ed.), *Probiotics* (pp. 541–550), Rijeka: InTech.

Hammes, W.P. and Hertel, C. (1998) New developments in meat starter cultures. *Meat Science*, **49**, S125–S138.

Hansen, E.B. (2002) Commercial bacterial starter cultures for fermented foods of the future. *International Journal of Food Microbiology*, **78**(1), 119–131.

Hassan, A.N., Frank, J.F. and Shalabi, S.I. (2001) Factors affecting capsule size and production by lactic acid bacteria used as dairy starter cultures. *International Journal of Food Microbiology*, **64**(1), 199–203.

Hassan, A.N. and Frank, J.F. (2001) Starter cultures and their use. In E.H. Marth (ed.), *Food Science and Technology* (pp. 151–206), New York: Marcel Dekker.

Holst, O. and Mattiasson, B. (1990) Cultivation using membrane filtration and cell recycling. *Extractive Bioconversions*, **11**, 11.

Holzapfel, W. (1997) Use of starter cultures in fermentation on a household scale. *Food Control*, **8**(5), 241–258.

Holzapfel, W.H. (2002) Appropriate starter culture technologies for small-scale fermentation in developing countries. *International Journal of Food Microbiology*, **75**(3), 197–212 .

Hubalek, Z. (2003) Protectants used in the cryopreservation of microorganisms. *Cryobiology*, **46**(3), 205–229.

Kets, E.P., Galinski, E.A., De Wit, M., de Bont, J.A. and Heipieper, H.J. (1996) Mannitol, a novel bacterial compatible solute in *Pseudomonas putida* S12. *Journal of Bacteriology*, **178**(23), 6665–6670.

Kim, W.S., Ren, J. and Dunn, N.W. (1999) Differentiation of *Lactococcus lactis* subspecies *lactis* and subspecies *cremoris* strains by their adaptive response to stresses. *FEMS Microbiology Letters*, **171**(1), 57–65.

Knorr, D. (1998) Technology aspects related to microorganisms in functional foods. *Trends in Food Science and Technology*, **9**(8), 295–306.

Lee, J.Y., Kim, C.J. and Kunz, B. (2006) Identification of lactic acid bacteria isolated from kimchi and studies on their suitability for application as starter culture in the production of fermented sausages. *Meat Science*, **72**(3), 437–445.

Lee, K. (2004) Cold shock response in *Lactococcus lactis* ssp. *diacetylactis*: A comparison of the protection generated by brief pre-treatment at less severe temperatures. *Process Biochemistry*, **39**(12), 2233–2239.

Li, Y., Hugenholtz, J., Abee, T., Li, Y. and Chen, J. (2003) Glutathione protects *Lactococcus lactis* against oxidative stress. *Applied and Environmental Microbiology*, **69**(10), 5739–5745.

Lievense, L.C. and Van't Riet, K. (1994) Convective drying of bacteria. In A. Fiechter (ed.), *Biotechnics/Wastewater: Advances in Biochemical Engineering, Biotechnology* (pp. 71–89), Berlin: Springer.

Linders, L.J., Wolkers, W.F., Hoekstra, F.A. and Van't, Riet, K. (1997) Effect of added carbohydrates on membrane phase behavior and survival of dried *Lactobacillus plantarum*. *Cryobiology*, **35**(1), 31–40.

Mandel, M. and Higa, A. (1970) Calcium-dependent bacteriophage DNA infection. *Journal of Molecular Biology*, **53**(1), 159–162.

MarketsAndMarkets (2014) Starter culture market by type (yeast, bacteria, molds), application [alcoholic beverages (beer, wine, whisky), non-alcoholic beverages (dairy-based, cereal-based, kombucha)] and geography: Global trends and forecast to 2018. Pune: MarketsAndMarkets. http://www.marketsandmarkets.com/Market-Reports/starter-culture-market-213083494.html (accessed June 2016).

Maus, J.E. and Ingham, S.C. (2003) Employment of stressful conditions during culture production to enhance subsequent cold- and acid-tolerance of bifidobacteria. *Journal of Applied Microbiology*, **95**(1), 146–154.

Monnet, C., Schmitt, P. and Divies, C. (1994) Method for assaying volatile compounds by headspace gas chromatography and application to growing starter cultures. *Journal of Dairy Science*, **77**(7), 1809–1815.

Morgan, C.A., Herman, N., White, P.A. and Vesey, G. (2006) Preservation of micro-organisms by drying: A review. *Journal of Microbiological Methods*, **66**(2), 183–193.

Mullan, W.M.A. (2014) Starter cultures: Importance of selected genera. In C.A.B. Tortorello (ed.), *Encyclopedia of Food Microbiology* (2nd edn, pp. 515–521), Oxford: Academic Press.

Nedoviç, V. and Willaert, R. (2006) *Applications of Cell Immobilisation Biotechnology*. Dordrecht: Springer.

Noonpakdee, W., Sitthimonchai, S., Panyim, S. and Lertsiri, S. (2004) Expression of the catalase gene katA in starter culture *Lactobacillus plantarum* TISTR850 tolerates oxidative stress and reduces lipid oxidation in fermented meat product. *International Journal of Food Microbiology*, **95**(2), 127–135.

Notermans, S., Gallhoff, G., Zwietering, M.H. and Mead, G.C. (1995) Identification of critical control points in the HACCP system with a quantitative effect on the safety of food products. *Food Microbiology*, **12**, 93–98.

Ortega-Rivas, E., Juliano, P. and Yan, H. (2006) *Food Powders: Physical Properties, Processing, and Functionality*. New York: Springer.

Osborne, R.J.W. (1977) Manufacture of starters: Production of frozen concentrated cheese starters by diffusion culture. *International Journal of Dairy Technology*, **30**(1), 40–44.

Palmfeldt, J. and Hahn-Hägerdal, B. (2000) Influence of culture pH on survival of *Lactobacillus reuteri* subjected to freeze-drying. *International Journal of Food Microbiology*, **55**(1), 235–238.

Parente, E. and Cogan, T.M. (2004) Starter cultures: General aspects. *Cheese: Chemistry, Physics and Microbiology*, **1**, 123–148.

Peighambardoust, S.H., Golshan Tafti, A. and Hesari, J. (2011) Application of spray drying for preservation of lactic acid starter cultures: A review. *Trends in Food Science and Technology*, **22**(5), 215–224.

Peterson, B.W., Sharma, P.K., van der Mei, H.C. and Busscher, H.J. (2012) Bacterial cell surface damage due to centrifugal compaction. *Applied and Environmental Microbiology*, **78**(1), 120–125.

Phadtare, S. (2004) Recent developments in bacterial cold-shock response. *Current Issues in Molecular Biology*, **6**(2), 125–136.

Reilly, S.S. and Gilliland, S.E. (1999) *Bifidobacterium longum* survival during frozen and refrigerated storage as related to pH during growth. *Journal of Food Science*, **64**(4), 714–718.

Riis, S.B., Pedersen, H.M., Sorensen, N.K. and Jakobsen M. (1995) Flow cytometry and acidification power test as rapid techniques for determination of the activity of starter cultures of *Lactobacillus delbrueckii* ssp. *bulgaricus*. *Food Microbiology*, **12**, 245–250.

Riveros, B., Ferrer, J. and Borquez, R. (2009) Spray drying of a vaginal probiotic strain of *Lactobacillus acidophilus*. *Drying Technology*, **27**(1), 123–132.

Sanchez, A.H., Rejano, L., Montano, A. and de Castro, A. (2001) Utilization at high pH of starter cultures of lactobacilli for Spanish-style green olive fermentation. *International Journal of Food Microbiology*, **67**(1–2), 115–122.

Sandine, W.E. (1996) Commercial production of dairy starter cultures. In T.M. Cogan and J.-P. Accolas (eds), *Dairy Starter Cultures* (pp. 191–206), New York: VCH.

Santivarangkna, C., Kulozik, U. and Foerst, P. (2007) Alternative drying processes for the industrial preservation of lactic acid starter cultures. *Biotechnology Progress*, **23**(2), 302–315.

Sheehan, V.M., Sleator, R.D., Fitzgerald, G.F. and Hill, C. (2006) Heterologous expression of BetL, a betaine uptake system, enhances the stress tolerance of *Lactobacillus salivarius* UCC118. *Applied and Environmental Microbiology*, **72**(3), 2170–2177.

Steenson, L.R. and Swaisgood, H.E. (1987) Calcium alginate-immobilized cultures of lactic streptococci are protected from bacteriophages. *Journal of Dairy Science*, **70**(6), 1121–1127.

Steinkraus, K. (1995) *Handbook of Indigenous Fermented Foods*. Boca Raton, FL: CRC Press.

Streeter, J.G. (2003) Effect of trehalose on survival of *Bradyrhizobium japonicum* during desiccation. *Journal of Applied Microbiology*, **95**(3), 484–491.

Streit, F., Athes, V., Bchir, A., Corrieu, G. and Béal, C. (2011) Microfiltration conditions modify *Lactobacillus bulgaricus* cryotolerance in response to physiological changes. *Bioprocess and Biosystems Engineering*, **34**(2), 197–204.

Tamime, A.Y. (2002) Microbiology of starter cultures. In R.K. Robinson (ed.), *Dairy Microbiology Handbook: The Microbiology of Milk and Milk Products* (3rd edn, pp. 261–366), New York: John Wiley & Sons.

Teniola, O.D. and Odunfa, S.A. (2001) The effects of processing methods on the levels of lysine, methionine and the general acceptability of ogi processed using starter cultures. *International Journal of Food Microbiology*, **63**(1–2), 1–9.

Tymczyszyn, E.E., Diaz, R., Pataro, A., Sandonato, N., Gómez-Zavaglia, A. and Disalvo, E.A. (2008) Critical water activity for the preservation of *Lactobacillus bulgaricus* by vacuum drying. *International Journal of Food Microbiology*, **128**(2), 342–347.

Valyasevi, R. and Rolle, R.S. (2002) An overview of small-scale food fermentation technologies in developing countries with special reference to Thailand: Scope for their improvement. *International Journal of Food Microbiology*, **75**(3), 231–239.

Van Reis, R. and Zydney, A. (2007) Bioprocess membrane technology. *Journal of Membrane Science*, **297**(1), 16–50.

Vedamuthu, E.R. (2006) Starter cultures for yogurt and fermented milks. In R.C. Chandam, C. White, A. Kilara and Y.H. Hui (eds), *Manufacturing Yogurt and Fermented Milks* (pp. 89–116), Ames, IA: Blackwell Professional.

Wisselink, H.W., Weusthuis, R.A., Eggink, G., Hugenholtz, J. and Grobben G.J. (2002) Mannitol production by lactic acid bacteria: A review. *International Dairy Journal*, **12**(2), 151–161.

Wouters, J.A., Rombouts, F.M., De Vos, W.M., Kuipers, O.P. and Abee, T. (1999) Cold shock proteins and low-temperature response of *Streptococcus thermophilus* CNRZ302. *Applied and Environmental Microbiology*, **65**(10), 4436–4442.

Wyber, J.A., Andrews, J. and Gilbert, P. (1994) Loss of salt-tolerance and transformation efficiency in *Escherichia coli* associated with sub-lethal injury by centrifugation. *Letters in Applied Microbiology*, **19**(5), 312–316.

Yavorsky, D., Blanck, R., Lambalot, C. and Brunkow, R. (2003) The clarification of bioreactor cell cultures for biopharmaceuticals. *Pharmaceutical Technology*, **27**(3), 62–77.

CHAPTER 6

Safety evaluation of starter cultures

Pasquale Russo, Giuseppe Spano and Vittorio Capozzi
Department of the Science of Agriculture, Food and Environment, University of Foggia, Italy

The use of starter cultures in the management of food fermentations represents a practice that implies a distinct range of benefits for producers and consumers. From a historical point of view, what remains of millennia of unconsciousness microbial management in the food chain is the practice of backslopping or, as recently reported, 'the inoculation of the raw material with a small amount of dough from a previous successful fermentation' (Brandt 2014). In other words, it is a method of using microbes that have been demonstrated as efficient and safe the day before as inoculum for the new product. This approach, which has been declining in the different production chains (following different technological parameters), has been given diverse names in various languages: 'inoculum enrichment', 'sieroinnesto' and 'lattoinnesto' (natural multi-strain starters, which are propagated/prepared in either cheese whey or milk) or 'pied de cuve' (a traditional yeast starter preparation in wineries). The passage from the backslopping regimen to the modern microbial cultures framework paralleled the rise of microbiology as a new science. In fact, the history of microbiology and the history of food microbiology coincide in their initial phases (e.g. Pasteur's 'Études sur le vin' and the development of pure-culturing techniques by Emil Christian Hansen to obtain yeasts for beer or spirit production without contaminants; Jay 2005; Brandt 2014).

To provide a general idea of the present global interest in this industry, it is important to consider that about one-third of food produced for human consumption worldwide is fermented food (Vogel *et al.* 2011). This social attention is well testified to by the famous journalist Michael Pollan, who in a recent book, among rules for eating, even suggested 'eat some foods that have been predigested by bacteria or fungi' (Pollan 2010).

Several fermented products can be considered to be diffused worldwide (yogurt, cheese, beer, salami; Mullan 2014), others to have national/continental diffusion (busa, kimchi, miso, tibi; Mullan 2014), while an extraordinary

---

Starter Cultures in Food Production, First Edition. Edited by Barbara Speranza, Antonio Bevilacqua, Maria Rosaria Corbo and Milena Sinigaglia.

diversity of food matrices and of microbial fermentations characterizes the regional context all over the world. Thus, starter culture technology is crucial to assuring food safety worldwide. In fact, the requirements for innovation regarding the link between starter cultures and food safety are different because of environmental/economic/social dissimilarities among countries. In the so-called Western world, several trends characterize the existing dynamics in the sector: improved and personalized nutrition; new health targets; bio-protection strategies for extended shelf life; bio-protection strategies for ready-to-eat products; 'natural', artisanal, traditional, organic and biodynamic products (Hansen 2002; Capozzi and Spano 2011; Francis *et al.* 2012; Claus and Swann 2013; Oliveira *et al.* 2014). In developing countries, the importance of starter cultures is closely related to the significance of assuring the preservation and safeguarding of food (Holzapfel 2002). A particular role has been reserved for traditional fermented food, where microorganisms may improve the taste of an otherwise bland food, enhance the digestibility of a food that is difficult to assimilate, preserve food from degradation by noxious organisms and increase nutritional value through the synthesis of essential amino acids and vitamins (National Research Council 1992). Traditional fermented foods are sometimes manufactured following primitive know-how, resulting in productive losses. Therefore, the role of microbial starters in increasing the yield of traditional fermented foods is crucial (Achi 2005). All these instances may be inflected as a function of the level of fermented food production: household, traditional and industrial (FAO/WHO 1996). The worldwide net of exigencies and stakeholders, coupled with the global relevance of fermented foods in human health and nutrition, well testifies to the need for continuous updating of strategies for starter culture design and evaluation.

Moreover, it is crucial to underline that starter culture technology has a sort of Janus-faced, dualistic role in contributing to the hygienic qualities of food: it helps to assure food safety by steering fermentation in the desired 'direction', but at the same time it may pose risks for food safety in terms of the safety of the strains used and of the physical, chemical and biological quality of starter cultures.

Food safety, starter cultures and the need for an integrated perspective

The heterogeneous arena described concretizes the typical puzzling environment of food safety issues and solutions. Increasing evidence does in fact indicate that a detailed risk assessment and the design of adequate strategies require a highly multidisciplinary approach, involving scientists and stakeholders from several disciplines (Fischer *et al.* 2005; Havelaar *et al.* 2010; Capozzi *et al.* 2012a). A good example to describe factually the significance of the transdisciplinary evaluation of the safety risks connected with microbial management in food fermentation is represented by the recent tendencies characterizing the sector of traditional, typical and artisanal fermented foods (including geographical indications), but also organic and biodynamic products. In this sector, for different reasons and with different aims, several producers and stakeholders counterpose the use of commercial starter cultures with the exploitation of spontaneous fermentation, preferring this second approach in the management of food fermentation. From this point of view, the return to spontaneous fermentation represents a strategy to restore the tradition, typicality and artisanality that have already been lost. The other leading idea is that the use of commercial starter cultures corresponds to a drift from the 'natural' manufacture of fermented foods, as if it were reasonable that a proportion of 'synthetic fertilizers' are connected to 'plant nutrition' in the same way as 'commercial microbial starter cultures' are connected to 'food fermentations' (Capozzi and Spano, 2011). This point of view is well summarized by Piero Sardo, president of the Slow Food Foundation for Biodiversity Onlus: 'the use of starter cultures implies the disappearance of natural concepts and territory that the cheese should detain' (SlowFood Press Release 2009). Relying on spontaneous fermentation poses serious challenges to the safety and quality of fermented foods. First, during spontaneous fermentation it is possible that there is implantation/domination of microbial strains dangerous for human health. In addition, there is a considerable risk of the development of spoilage microbial communities in food matrices.

Paradoxically, the food microbiology scientific community already possesses at least two possible biotechnological solutions that could combine food safety/quality and the adoption of a microbial regimen in food fermentation compatible with the traditional/typical/artisanal/organic status. These are mainly represented by the design of multi-strain starter cultures based on the selection of ecotypes from spontaneous fermentation; and the application of innovative biotechnologies to monitor spontaneous fermentation. However, currently the approaches proposed do not represent the reference solution for microbial resource management in food fermentation in the sector of traditional, typical and artisanal fermented foods (including geographical indications). What is probably missing is transdisciplinary assessment to close the circle of possible solutions: coherence with existing regulations; coherence with product specifications; the implementation of low-cost production strategies to produce small quantities of microbial biomass; and social research to improve consumers'/stakeholders' understanding and acceptance. In some cases, partial solutions have been proposed (Maqueda *et al.* 2011; Capozzi *et al.* 2012b), but a holistic perspective on the challenges is desirable to assure safe fermentation through a new generation of starter cultures, avoiding the risks connected with spontaneous microbial consortia.

The US regulatory framework: Generally recognized as safe (GRAS) status

According to US regulations, microbial cultures added to a food matrix may be considered either additive or GRAS (generally recognized as safe) substances. With the aim of preventing the employment of food substances potentially harmful to public health, the 1958 Food Additives Amendment of the Food, Drug, and Cosmetic Act states that exogenous ingredients deliberately introduced into the food chain are additives and therefore subject to the approval of the Food and Drug Administration (FDA) prior to marketing. However, based on Section 409 of this amendment, when the use of a substance is GRAS or its employment in food pre-dates 1958, it is exempt from the food additive definition and therefore does not require the pre-market approval of the FDA (FDA 1999).

A GRAS determination is acquired for substances 'generally recognized, among experts qualified by scientific training and experience to evaluate its safety, as having been adequately shown through scientific procedures to be safe under the conditions of its intended use' (FDA 1999, Section 201(s)). In the case of substances used prior to January 1, 1958, GRAS status can be established through either scientific procedures or experience based on a substantial history of use in food by a significant number of consumers. In the latter, GRAS determination does not require the same quantity and quality of scientific evidence as is required to support food additive approval (21 CFR 170.30(c) and 170.3(f)).

Specifically, the assessment requires that evidence of safety must be guaranteed by both a technical point of view and a common knowledge element. The first must be generally available, typically in the form of publications in peer-reviewed scientific journals (21 CFR 170.30(b)), and at the same time a consensus among qualified experts about GRAS status must be achieved. To simplify the achievement of a GRAS determination, in 1997 the FDA issued a proposed rule (62 FR 18938) establishing a notification procedure where any person or company may notify the FDA of a determination that a specific usage of a microorganism is GRAS. Based on the submitted notice, the FDA either resolves to accept the basis for the notifier's GRAS determination or concludes that it does not provide a sufficient basis for a GRAS determination (FDA 2010).

However, from a legal point of view the food company suffers a different degree of liability depending on whether the microbial strain receives GRAS status or the pre-market approval of the FDA. Unlike a food additive, a GRAS determination implies that the company is fully responsible if safety incidents occur related to the consumption of microorganisms. However, according to US law, neither a GRAS determination nor the notification to the FDA of the new use of a microbial strain is mandatory for a food company. Nonetheless, a previous consensus of the FDA on the safety evaluation of a strain for food purposes would reduce the liability of the company, simplifying the resolution of legal controversies (Wessels *et al.* 2004).

In fact, a partial GRAS list of foods for human consumption that may contain or be derived from microorganisms includes milk and cream (21 CFR 131), cheeses and related cheese products (21 CFR 133), bakery products (21 CFR 136) and cereal flours and related products (21 CFR 137). Due to their long safe history in food production, cultures of lactic acid bacteria (LAB) are the most representative microorganisms included in the GRAS list (Stevens *et al.* 2009). In some food formulations, the employment of specific starter cultures is mandatory. Typical examples are the addition of *Lactobacillus bulgaricus* and *Streptococcus thermophilus* for yogurt production (§131.200, §131.203, §131.206) or *Penicillium roquefortii* for blue cheese manufacturing (§133.106).

According to the Code of Federal Regulations (FDA 2015), a substance is safe if there exists reasonable confidence that it is not dangerous under the intended conditions of use. Therefore, it is the specific use of a substance rather than the substance itself that obtains GRAS status. Within a microbiological framework, this means that a starter culture should be considered GRAS not only on the basis of its food application, but also depending on some technological features such as the form in which the microorganisms are added to the matrix (i.e. freeze-dried or liquid cultures) or the microbial concentration achieved in the food.

For example, Danisco USA's *Lactobacillus acidophilus* La-14 has been commercially available in the USA for more than a decade, for the manufacture of fermented milk, yogurt and dietary supplements, with no reports of adverse incidents. Recently, Danisco submitted a notice of a GRAS exemption claim to expand the employment of *Lb. acidophilus* La-14 as an ingredient in several foods, including ready-to-eat breakfast cereals and bars; milk, milk drinks, milk products, fermented milks, yogurt, cheese and ice cream; soy drinks and soy products; bottled water and teas; dry beverages including sports nutrition beverages; fruit juices, fruit nectars, fruit '-ades', fruit drinks, jams and jellies; chewing gum; medical foods; nut and peanut spreads; margarines; snack foods; meal replacements; sauces and condiments; and confections; but excluding infant formula (FDA 2014; Table 6.1). Furthermore, the GRAS notice provided an estimate of the number of viable cells at the end of the shelf life, 10^9 colony forming units (cfu) per 250 g, taking into account a safe upper limit of exposure of the microorganism intake based on a reasonable consumption of the product (FDA 2014). Scientific evidence of the safety of this strain provided by the company was further supported by sequencing the La-14 genome, which has been recently completed and publicly deposited at the National Center for Biotechnology Information under accession number CP 005926 (Stahl and Barrangou 2013). An independent expert panel solidly confirmed the GRAS status of La-14.

A list of GRAS notices on the use of microbial food cultures submitted by food companies to the FDA in recent years is summarized in Table 6.1.

Table 6.1 GRAS notices on the use of microbial food cultures examined by the FDA since 2012.

GRAS Notices	Substance	Notifier	Intended use	Basis	Date of filling	FDA letter
531	*Lb. fermentum* CECT5716	Biosearch Life SA (Spain)	Powdered milk-based infant formula at 10^7 cfu/g of powdered formula	Scientific procedures	14 Aug 2014	Pending
526	*B. coagulans* strain Unique IS2 spores preparation	Unique Biotech Ltd (India)	Probiotic in several foods at a maximum level of 2×10^9 cfu per serving	Scientific procedures	31 Jul 2014	Pending
502	*Lb. acidophilus* La-14	Danisco USA Inc. (WI, USA)	As an ingredient in certain food categories at a level of 10^9 cfu per 250 g serving of food at the time of consumption.	Scientific procedures	27 Feb 2014	19 Aug 2014 FDA has no questions
463	*Lb.acidophilus* NP 28, *Lb. acidophilus* NP51, *Lb.* subsp. *lactis* NP7, *Pe. acidilactici NP3*	Guardian Food Technologies LLC (KS, USA)	As an antimicrobial, up to 1×10^8 cfu/g of food, to control pathogenic bacteria in fresh chopped/ground, whole muscle cuts and carcasses of meat and poultry and ready-to-eat meat products	Scientific procedures	22 Mar 2013	16 Jan 2014 FDA ceased to evaluate the notice
455	*Bif. breve* M-16V	Danone Trading BV (The Netherlands)	As an ingredient in exempt term powdered amino acid-based formulas, at levels providing 10^8 cfu/g of infant formula powder	Scientific procedures	23 Jan 2013	30 Sep 2013 FDA has no questions
454	*Bif. breve* M-16V	Morinaga Milk Industry Co. Ltd (Japan)	As an ingredient in non-exempt powdered term infant formulas (milk or soy based) and exempt powdered term infant formula containing partially hydrolysed milk or soy proteins, at levels up to 10^8 cfu/g of infant formula powder	Scientific procedures	22 Jan 2013	27 Sep 2013 FDA has no questions
453	*Bif. breve* M-16V	Morinaga Milk Industry Co. Ltd (Japan)	As an ingredient in certain food categories at levels up to 5×10^9 cfu per serving	Scientific procedures	22 Jan 2013	27 Sep 2013 FDA has no questions
445	*Bif. animalis* subsp. *lactis* strains HN019, Bi-07, Bl-04 and B420	Danisco USA Inc. (WI, USA)	As ingredients in ready-to-eat breakfast cereals, bars, cheeses, milk drinks and milk products, bottled water and teas, fruit juices, fruit nectars, fruit '-ades' and fruit drinks, chewing gum and confections at a maximum level of 2×10^{11} cfu per serving	Scientific procedures	10 Oct 2012	10 Apr 2013 FDA has no questions

440	*Lb. reuteri* NCIMB 30242	Micropharma Ltd (Canada)	As an ingredient in beverages and beverage bases, breakfast cereals, cheeses, dairy product analogues, fats and oils, frozen dairy desserts, grain products and pastas, milk products, processed fruit and fruit juices, and sugar substitutes at levels ranging from 3.3×10^8 to 10^{10} cfu per serving	Scientific procedures	16 Aug 2012	12 Feb 2013 FDA has no questions
429	*Lb. casei* Shirota	Yakult Honsha Co. Ltd (Japan)	As an ingredient in fermented dairy products at a maximum level of 4×10^8 cfu/mL	Scientific procedures	10 Apr 2012	10 Dec 2012 FDA has no questions
410	*Lb. reuteri* DSM 17938	Nestle Nutrition US (NJ, USA)	As an ingredient in powdered whey-based term infant formula at a minimum level of 10^6 cfu/g, but not higher than 10^8 cfu/g of powdered formula, produced in accordance with current good manufacturing practices	Scientific procedures	16 Nov 2011	26 Mar 2012 FDA has no questions
409	*Lb. reuteri* NCIMB 30242	Micropharma Ltd (Canada)	As an ingredient for use in multiple foods and beverages	Scientific procedures	18 Oct 2011	10 Jan 2012 FDA ceased to evaluate the notice
399	*B. coagulans* GBI-30, 6086 spores	Ganeden Biotech Inc. (OH, USA)	As an ingredient for use in multiple foods and beverages at a maximum level of approximately 2×10^9 cfu per serving	Scientific procedures	23 Aug 2011	31 Jul 2012 FDA has no questions

Notes: *B.* = *Bacillus; Bif.* = *Bifidobacterium;* cfu = colony forming units; *Lb.* = *Lactobacillus; Pe* = *Pediococcus.*

The European regulatory framework: The qualified presumption of safety (QPS) concept

Qualified presumption of safety (QPS) is a generic approach developed by the European Food Safety Authority (EFSA) to evaluate the safety of biological agents deliberately added throughout the food chain, with the objective of harmonizing the risk assessment of microbial cultures for feed and food production across the EFSA Scientific Panels and Units (EFSA 2005). Although QPS status is usually considered as the European correspondent of a GRAS determination for microbial food cultures, some conceptual differences should be highlighted between the two approaches, which arise from the different regulatory frameworks in Europe and the USA. Thus, while GRAS is a voluntary process to notify the safety of a microbial strain under the conditions of intended use by qualified experts external to the FDA, QPS is a pre-assessment at a species level intended to be considered and complemented by a safety assessment of a specific notification. Therefore, QPS never establishes the safety of notified strains; that is within the responsibility of a specific Scientific Panel. From a practical standpoint, QPS is a route whereby EFSA determines with reasonable certainty, via its Panel on Biological Hazards, that a microbial species is harmless. Finally, unlike a GRAS determination, which is based on information about the safety of a specific use of a microorganism, precise modes of use are not detailed in the QPS assessment. For instance, the concentration of a microbial culture in the food is outside QPS approval at any reasonable dose (EFSA 2012a).

QPS is suggested as an operating procedure for risk assessment based on four main pillars: establishing identities; familiarity intended as body of knowledge; end use; and pathogenic potential of QPS candidates (EFSA 2007). The QPS pre-assessment is always carried out at the highest adequately defined taxonomic unit that is appropriate for the purpose for which the evaluation is intended, which is usually the species level (EFSA 2012a; Bourdichon *et al.* 2012). Then, well-defined taxonomic units are evaluated based on the corresponding body of knowledge, a comprehensive approach including, among other factors, a history of apparent safe use, scientific literature and database, ecology, clinical aspects and industrial applications. It follows that familiarity is a critical concept to support evidence of a microbial presumption of safety, in contrast to the notion of novelty, which since 1997 has been covered by a specific EU regulation (Regulation EC no. 258/97).

Coherently, QPS approval cannot be granted to a taxonomic group commonly related to pathological events. However, if pathogenicity is a strain-dependent feature that can be excluded on a case-by-case basis, QPS status for a taxonomic unit might still be approved. On the other hand, even if occasional clinical events attributable to QPS microorganisms can occur, they do not necessarily result in exclusion of the taxonomic unit from the QPS list.

In addition to virulent traits, the pathogenic potential assessment must contemplate the production of undesirable metabolites. Among the toxic compounds, biogenic amines and toxins production are the main concerns related to food poisoning (Spano *et al.* 2010; Hymery *et al.* 2014). Finally, the presence of acquired and transmissible antimicrobial resistance markers is a serious safety concern for the inclusion of bacterial species in the QPS list, unless viable cells are not present in the final product (EFSA 2008). Based on similar considerations, in the 2009 QPS Opinion a qualification regarding absence of antimycotic resistance for yeast was also introduced (EFSA 2009).

If taxonomic units are considered unfit for the QPS list they would remain subject to a full safety assessment, in the form of a notification by the responsible EFSA Scientific Panel (EFSA 2007). This is for example the case of some *Enterococcus* spp., coagulase-negative staphylococci such as *Staphylococcus xylosus* and *Staphylococcus carnosus*, and filamentous fungi usually used in cheese and fermented meat production.

Annually, EFSA's BIOHAZ Panel reviews and updates the list of biological agents recommended for QPS, based on the latest background concerning taxonomic units already assessed through the QPS assessment, and by performing the identification and assessment of taxonomic units not previously considered (Leuschner *et al.* 2010). In 2013, the BIOHAZ Panel expressed an opinion on the safety of Gram-positive bacteria, further divided depending on their sporulating ability, Gram-negative bacteria, yeasts, filamentous fungi and viruses (EFSA 2013).

Gram-positive non-sporulating bacteria play a pivotal role in the production of fermented foods. Microorganisms of this group recommended for the 2013 QPS list include genera belonging to lactic acid bacteria (LAB; namely *Lactobacillus* spp., *Lactococcus* spp., *Leuconostoc* spp., *Pediococcus* spp., *Oenococcus oeni*, *Strep. termophilus*), dairy propionic acid bacteria, *Bifidobacterium* spp. and *Corynebacterium glutamicum*. In addition to their use for food technological applications, LAB and *Bifidobacterium* strains are common inhabitants of the gut of mammals and they often exhibit interesting probiotic properties. Furthermore, due to their QPS status, they have been proposed for other purposes, such as for their protective role against foodborne pathogenic bacteria (Coelho *et al.* 2014) or in order to fortify *in situ* the vitamin content of certain foods (Capozzi *et al.* 2012c). Apart from *Corynebacterium glutamicum*, for which QPS only applies when the species is used for amino acid production, a generic qualification for all QPS bacterial taxonomic units was granted, under the condition that strains should not harbour any acquired antimicrobial resistance genes to clinically relevant antibiotics (EFSA 2013). Nonetheless, *Lactobacillus* species can be sporadically involved in human systemic infections, as reported in Table 6.2.

Although infrequent clinical events do not change the QPS status of the *Lactobacillus* species recommended in the previous EFSA Scientific Opinion, the BIOHAZ Panel suggested that *Lactobacillus rhamnosus* should be closely monitored (EFSA 2013). EFSA has previously expressed a negative opinion on

Table 6.2 Case report of opportunistic infections attributed to QPS Gram-positive non-sporulating bacteria in the years 2012–13, as reported by EFSA (2013).

Presumptive causative agent	Case report	Clinical event	Reference
Bif. longum *Bif. infantis*		Septicaemia	Jenke *et al.* (2012)
Lb. delbrueckii	68-year-old woman	Pyelonephritis and bacteremia	DuPrey *et al.* (2012)
Lb. iners	6-month-old male infant	Bacterial pericarditis	Murata *et al.* (2012)
Lb. acidophilus	28-year-old female with autoimmune hepatitis	Infectious endocarditis	Nishijima *et al.* (2012)
Lactobacillus spp.	65-year-old woman	Anaerobic bacteremia	Hamadah *et al.* (2013)
Lb. rhamnosus GG	17-year-old boy with ulcerative colitis	*Lactobacillus* bacteremia	Vahabnezhad *et al.* (2013)
Lb. casei *Lb. paracasei*	95-year-old woman with underlying chronic diseases	Infection in a prosthetic joint	Orkaby *et al.* (2012)
Lb. acidophilus *Lb. bulgaricus*		Probiotic-related bloodstream infections	Simkins *et al.* (2013)
Lc. lactis cremoris	Middle-aged female patient with diabetes	Necrotic abscess	Hadjisymeou *et al.* (2013)
Lc. lactis	75-year-old man after mitral valve repair	Post-operative infective endocarditis	Rostagno *et al.* (2013)
Lc. cremoris	70-year-old man	Necrotizing pneumonia	Buchelli-Ramirez *et al.* (2013)
Lc. lactis vancomycin-resistant *Enterococcus*	45-year-old Asian male liver transplant recipient	Bacterial infection	Deng *et al.* (2012)
Pediococcus spp.	60-year-old Caucasian male	Necrotizing cellulitis of the abdominal wall	Michalopoulos *et al.* (2013)

Notes: Bif. = Bifidobacterium; Lc. = Lactococcus; Lb. = Lactobacillus.

Enterococcus genus (EFSA 2013) due to a wide susceptibility to antibiotics and the occurrence of virulent factors, resulting in an increasing number of nosocomial infections associated with multidrug-resistant *Enterococcus* strains (Reyes and Zervos 2013). Recently, only *Enterococcus faecium* was considered for inclusion in the QPS list (EFSA 2012a; 2013) because strains of this species are authorized in the European Union (EU) as feed additives. Furthermore, some authors have investigated the probiotic potential of *Ent. faecium* strains (Ahmadova *et al.* 2013; Barbosa *et al.* 2014). However, despite scientific knowledge allowing a differentiation of pathogenic strains from non-pathogenic strains within this species, *Ent. faecium* is not recommended for the QPS list (EFSA 2013).

Gram-positive sporulating bacteria considered for the QPS list belong to the genus *Bacillus* spp., with exclusion of the toxin-producing *Bacillus cereus sensu lato*, which have often been involved in food poisoning (Ceuppens *et al.* 2011; Logan 2012). In 2014 spores of *Bacillus* spp. were investigated for their potential biotechnological applications, including their use as probiotics and in vaccine formulations (De Souza *et al.* 2014; Larsen *et al.* 2014). Furthermore, the genome of a marketed probiotic strain of *Bacillus coagulans* was recently sequenced in order to provide further information on its positive features and safety (Orrù *et al.* 2014).

Gluconobacter oxidans, used in vinegar production as well as industrial biotechnology, was the first Gram-negative bacteria recommended for the QPS list according to the 2013 EFSA Scientific Opinion, although it was subject to a qualification of 'QPS only apply when the species is used for vitamin production' (EFSA 2013).

Yeasts are commonly considered the safest microorganisms and therefore a number of species are included in the QPS list. Among yeasts of food interest, *Saccharomyces cerevisiae* is crucial due to its centuries-long history of application in several food and feed fermentations. Nonetheless, probiotic *S. cerevisiae* subtype *boulardii* strains have been identified as presumptive aetiological agents of opportunistic infections, and therefore this species is contraindicated for infants and adults with underlying disease who have had surgery (Perapoch *et al.* 2000; EFSA 2013; Didari *et al.* 2014). In particular, absence of resistance to antimycotics used for the medical treatment of yeast infections in cases where viable cells are added to the food or feed chain is required to obtain QPS status (EFSA 2009).

In contrast to other microorganisms, filamentous fungi are still ineligible for a QPS recommendation, despite several species (including among others *Aspergillus* spp., *Fusarium* spp., *Paenicillium* spp. and *Trichoderma* spp.) having been notified to EFSA. The negative opinion of the BIOHAZ Panel is mainly due to two critical questions. On the one hand, rapid taxonomic development hinders the establishment of fungi at a species level; on the other hand, a scarce body of knowledge on the toxicological effects of fungal secondary metabolites poses a direct safety concern.

Starter cultures, scientific framework, position papers and regulatory environment

The efficacy of starter cultures in reducing the risks associated with fermentation phases has been demonstrated in several studies. The scientific outcomes justify the worldwide diffusion of the practice to add a suitable (in quantity and quality) microbial biomass to steer food fermentations. This relevance is well testified by several *Codex Alimentarius* standards conceived to contribute to the safety, quality and fairness of this international food trade, in particular those referring to dairy products. In fact, the Codex general standard for cheese reported among permitted ingredients 'starter cultures of harmless lactic acid and/or flavour producing bacteria and cultures of other harmless microorganisms' (CODEX STAN 283-1978). Harmless lactic acid and/or flavour-producing bacteria change each time because of the specific dairy product, for instance, *Strep. thermophilus* and/or *Lactococcus* spp. for Mozzarella (CODEX STAN

262-2006); symbiotic cultures of *Strep. thermophilus* and *Lactobacillus delbrueckii* subsp. *bulgaricus* for yoghurt (CODEX STAN 243-2003); non-gas-forming lactic acid-producing bacteria for cheddar (CODEX STAN 263-1966); *Lactobacillus helveticus, Streptococcus salivarius* subsp. *thermophilus, Lb. delbrueckii* subsp. *bulgaricus* and *Lactobacillus casei*, which are the principal starter culture micro-organisms for Provolone (CODEX STAN 272-1968); while, in the case of Coulommiers, the specification 'other harmless microorganisms' includes *Geotrichum candidum, Brevibacterium linens* and yeast; and 'rind formation and maturation (proteolysis) from the surface to the center is predominantly caused by *Penicillium candidum* and/or *Penicillium camembertii* and *Penicillium caseicolum*' (CODEX STAN 274-1969). The internationally recognized role in assuring food safety is also testified by the importance of starter cultures in the hazard analysis and critical control points (HACCP) analysis of important fermented food products. In general, in HACCP implementation the control of fermentation dynamics through the use of starter cultures or cultures from a previous batch is widely recognized as indispensable for the safe production of fermented products (Bryan 1992). In particular, the technological phase that envisages the addition of starter cultures may belong to the critical control points, at least in the production of cheeses and sausages (Sandrou and Arvanitoyannis 2000; Bover-Cid *et al.* 2000; Silva *et al.* 2003; Mokhtar *et al.* 2012). Other than an indirect commercial relevance in the manufacture of fermented foods, the starter culture industry retains economic significance as a biotechnology market. In 2002, Hansen (2002) estimated the size of the dairy starter culture market as nearly US$250 million. The same author calculated that, considering a possible worldwide diffusion of starter cultures to cheese makers and producers of fermented dairy preparations, the size of the dairy starter culture market would be approximately US$1 billion (Hansen 2002). Giving the importance for food safety and its economic significance, the sector would require the existence of a precise regulatory framework.

However, while starter cultures and their application in foods must comply with official statements and legal requirements regarding food legislation, in a few cases we can find specific regulations for starter cultures. In Danish legislation, starter cultures are categorized as additives, and those introduced after 1973 (the year the law came into force) require notification and approval, on the basis of the documentation presented to assure food safety and efficacy (while cultures present on the market prior to 1973 remain permitted; Wessels *et al.* 2004). The documentation must encompass the following traits: identification ('the micro-organism produced by the producer must be identified by an analytical method'); purity ('there must be conducted studies to ensure that the micro-organism formulation does not contain potentially harmful organisms and/or large amounts of contaminating organisms whose identity is not known'); effects ('the micro-organism concerned must not have any potentially pathogenic properties to humans or animals. If the organism has the ability to produce toxins, it must be shown that these are not formed in harmful quantities in the particular application'); and antibiotic resistance ('It must be proved that the micro-organism does not possess transferable antibiotic resistance'; Danish Veterinary and Food Administration 2013).

While the Danish experience represents a sort of best practice, the legislative situation in the EU is articulate enough to provide an idea of the difficulties encountered in harmonizing this sector, testifying to the wide differences in approaches pursued to production standards and food safety. In the EU legislation, in fact, it is still not clearly defined whether starter cultures belong to additives, processing aids or ingredients (Sundh *et al.* 2012). The use of starter cultures, as an intentional addition of an ingredient into the food chain, must observe the 'safe for consumers' general principle established in Regulation (EC) No 178/2002, satisfying the legal requirements specified. The EU Standing Committee on the Food Chain and Animal Health (SANCO 2006) recommendations, up to now not adopted in EU legislation, suggest that starter cultures should generally be considered as 'processing aids', with the exception of those used in food matrices to achieve a specific technological effect (such as preservation) that can be deemed 'additives' (von Wright 2012). Herody *et al.* (2010), in a paper focused on the legal definition of starter cultures in EU legislation, evaluated the coherence of starter cultures with the status of 'additive' and those with the status of 'processing aid', underlining more generally the possible declaration of starter cultures on the 'list of ingredients' (species with no history of safe use in food are considered 'novel ingredients'). From a safety perspective, this facet is of crucial relevance; in fact, as a function of their recognized status (processing aid, additive, ingredient, novel ingredient), starter cultures have to fulfil different regulatory regimes and to conform to different standards. In other words, as a whole EU harmonization is a Gordian knot that seems to require (as already mentioned) a transdisciplinary approach and a representation of different national/sectoral stakeholders. In fact, the topic not only covers several disciplines (such as microbial ecology, food microbiology, microbial physiology, food science, legislation, risk analysis, epidemiology and medicine), but also synthesizes instances arising from national approaches that are clearly affected by the major fermented food industries of the state and by the prominent national starter culture companies. More generally, to understand the existing differences in legal safety standards for these biotechnological resources, it is relevant to remember the defined contribution of starter cultures to the final fermented products: their biochemisms are indispensable in fermented food manufacture; they assure high standards of hygienic safety; they lead to an increase in shelf life; they may influence the nutritional, sensorial and functional qualities of the final product; and they may degrade toxic or harmful compounds present in the matrix (Vogel *et al.* 2011). If we consider the single aspects, starter cultures may fall into different main categories: ingredients, novel ingredients, additives, processing aids. Effectively, other national food laws solve the puzzle with a clear simplification of the status and/or of the requirements for protechnological microbes added to food to steer food fermentation. This is the situation in US legislation, which reported a very broad definition of food additives also covering starter cultures (which in the absence of GRAS status, as already stated, must be approved before use; Sundh *et al.* 2012); and in Singaporean food law concerning dairy products that requires no pre-market approval of cultures, assigning the responsibility for safe dairy products to the dairy producers. With regard to other countries, some legislation has adopted a positive list of commercial starter cultures/microbial species to regulate the sector, which is the case in Canada for meat starter cultures (Canadian Food Inspection Agency 2014) and in China for probiotics (National Health and Family Planning Commission of the People's Republic of China 2001).

Obviously, a unique scientific and legal name and a definition that includes all microbial cultures used in foods represents a cornerstone and a prerequisite for legislative harmonization aimed at improving the worldwide safety standard for these microbiological resources in the food chain. Astonishingly, as underlined by Vogel *et al.* (2011), a 'shared' definition is currently not available. In Table 6.3, we provide definitions reported in several authoritative scientific publications. Whereas all of these are general in description, some focus on the application in foods (Leroy and De Vuyst 2004; Doyle *et al.* 2013), others emphasize the possible formulations (Herody *et al.* 2010; Bourdichon *et al.* 2012), while some authors cover both aspects (Vogel *et al.* 2011; Hansen 2014). Among the definitions, together with the classic appellation 'starter cultures', appears also the designation 'microbial food cultures (MFC)'. This denomination and the corresponding definition have been proposed by the European Food and Feed Cultures Association (EFFCA) to obviate the highlighted lack in the sector, using the term MFC as a synonym for 'starter cultures, ripening cultures, protective cultures, sourdough starter, dairy starter, sausage starter, wine cultures, meat cultures, malolactic cultures, probiotics, lactic acid, yeast culture, etc.', probiotic cultures and protective cultures (MFC with an antagonistic metabolic activity against undesired microorganisms in food; EFFCA 2013). With the same intent, in a joint project with the International Dairy Federation (IDF), the EFFCA presented an 'inventory of microbial species used in food fermentations' published in the *Bulletin of IDF* (Mogensen *et al.* 2002). Table 6.4 reports the genera 'inventoried' by the EFFCA. It provides a useful map for describing and classifying the frontiers of food safety in term of microbial resources for food fermentation. Even if it can be assimilated to a position paper, the list offers the perspective of the major global/national starter culture firms, through a rigorous scientific lens (Bourdichon *et al.* 2012), providing the basis for a positive comparison and osmosis of the productive stakeholders and the scientific community, to assure safe starter cultures for safe fermented foods.

The distance between species reported in the 'inventory' and species with a recognized QPS status (Table 6.4) offers a measure of 'work in progress' in the safety evaluation of microbial resources in food fermentation. Behind each taxonomic name, we have to consider food transformations, possible biotechnological applications in the food sector, the existing safety assessment, the presence

Table 6.3 A selection of definitions for 'starter cultures' reported in the recent literature.

Definition	Reference
'A *starter culture* can be defined as a microbial preparation of large numbers of cells of at least one microorganism to be added to a raw material to produce a fermented food by accelerating and steering its fermentation process.'	Leroy and De Vuyst (2004)
'*Microbial Food Cultures (MFC)* are live bacteria, yeasts or moulds used in food production. MFC preparations are formulations, consisting of one or more microbial food cultures including unavoidable media components carried over from the fermentation and components, which are necessary for their survival, storage, standardization and to facilitate their application in the food production process. MFC preparations may contain one or several microbial species.'	Herody *et al.* (2010)
'*Starter cultures* are preparations of live microorganisms or their resting forms, whose metabolic activity has desired effects in the fermentation substrate, the food. The preparations may contain unavoidable residues from the culture substrate and additives that support the vitality and technological functionality of the microorganisms (such as antifreeze or antioxidant compounds).'	Vogel *et al.* (2011)
'*Microbial food cultures* are live bacteria, yeasts or molds used in food production. MFC preparations are formulations, consisting of one or more microbial species and/or strains, including media components carried over from the fermentation and components which are necessary for their survival, storage, standardization, and to facilitate their application in the food production process.'	Bourdichon *et al.* (2012)
'*Starter cultures* are food-grade microorganisms of known and stable metabolic activities and other characteristics that are used to produce fermented foods of desirable appearance, body, texture, and flavor.'	Doyle *et al.* (2013)
'*Commercial starter cultures* are standardized inoculum to be used for the production of fermented foods. Starter cultures are produced by specialized manufactures. Rigorous quality assurance and quality control are conducted to ensure performance, composition, and safety of the culture.'	Hansen (2014)

of strain-dependent and/or plasmid-delivered traits dealing with pathogen phenotypes and toxic compound products.

The safety of the starter cultures biomass necessarily follows the assessment of safe use for strains/species. The regulatory environment is just in the middle of these two different facets. When we shift from the safety evaluation of strains constituting starter cultures to the safety evaluation of starter culture formulation, the attention focuses on possible contaminants. The existing high level of attention on the quality of these biotechnological products is testified by the recent joint publication by the International Organization for Standardization (ISO) and the IDF of an international standard on bacterial starter cultures in fermented milk products (ISO 27205: 2010, IDF 149: 2010), including essential composition in terms of contaminants and specifics for safety management.

The 'test trial' suggested by the International Organisation of Vine and Wine (OIV) for the limits of contaminants in starter cultures for alcoholic fermentation (Table 6.5) and in malolactic cultures (Table 6.6) provides us with an idea of the nature of contaminants that might be present in starter cultures.

Table 6.4 List of genera reported in the EFFCA inventory (EFFCA 2012), reporting the number of species for each genus (n.) and in bracket the number of species with a recognized QPS status (according to the 2013 updated list of QPS Status recommended biological agents, revised in 2015*). For each genus the food usage (FU) is reported.

Bacteria (genera)	**n.**	**FU**	*Lactococcus*	3(1)	D,M	*Lachancea*	1(0)	W
Bifidobacterium	8(5)	D,S	*Streptococcus*	3(1)	D,S,V	*Metschnikowia*	1(0)	W
Brevibacterium	3(0)	D	*Acetobacter*	9(0)	Vn,Co, Cf,V	*Pichia*	4(0)	W,D
Corynebacterium	4(0)	D	*Gluconacetobacter*	8(0)	Co,Cf, Vn	*Saccharomyces*	2(2)	D,W, B,V
Brachybacterium	2(0)	D	*Gluconobacter*	1(0)	Vn	*Schwanniomyces*	1(0)	W
Microbacterium	1(0)	D	*Hafnia*	1(0)	D	*Starmerella*	1(0)	W
Arthrobacter	4(0)	D	*Halomonas*	1(0)	M	*Trigonopsis*	1(0)	W
Kocuria	2(0)	D,M	*Zymomonas*	1(0)	B	*Wickerhamomyces*	1(0)	W
Micrococcus	2(0)	D,M	**Yeasts (genera)**	**n.**	**FU**	*Zygosacharomyces*	1(0)	S
Propionibacterium	5(2)	D	*Lecanicillium*	1(0)	D	*Zygotorulaspora*	1(0)	D
Streptomyces	1(0)	M	*Galactomyces*	1(0)	D	*Lachancea*	1(0)	W
Bacillus	3(3)	F,C,S	*Geotrichum*	1(0)	D,M	*Torulaspora*	1(0)	D,W
Carnobacterium	3(1)	D,M,F	*Yarrowia*	1(0)	D	*Schizosaccharomyces*	1(1)	W
Enterococcus	3(0)	D,Sr,M, S,V	*Scopulariopsis*	1(0)	D	*Neurospora*	1(0)	V
Tetragenococcus	2(0)	S,V	*Fusarium*	2(0)	D	**Fungi (genera)**	**n.**	**FU**
Lactobacillus	84(35)	V,Sr,D, F,V,M, W,Co, C,B,S,Fr	*Candida*	10(0)	D,S,V, Sr,W, V,Co	*Aspergillus*	4(0)	T,B,S
Pediococcus	3(3)	M,W	*Cyberlindnera*	2(0)	D,W	*Penicillium*	7(0)	D,M
Leuconostoc	12(4)	M,D,F, V,Cf,W	*Debaryomyces*	1(1)	D,M	*Sporendonema*	1(0)	D
Oenococcus	1(1)	W	*Dekkera*	1(0)	B	*Cystofilobasidium*	1(0)	D
Weissella	9(0)	Sr,V,Co, M,V,F	*Hanseniaspora*	3(1)	W	*Guehomyces*	1(0)	V
Macrococcus	1(0)	D,M	*Kazachstania*	2(0)	D,Sr,	*Mucor*	4(0)	S,D
Staphylococcus	15(0)	M,D, S,F	*Kluyveromyces*	2(2)	D	*Rhizopus*	4(0)	V,S

Notes: * As reported in Table 1 of the official EFSA document (2015); we did not consider that QPS status applied only when the species is used for specific biotechnological functions.

B = beverages; C = cereal; Cf = coffee; Co = cocoa; D = dairy; F = fish; Fr = fruits; M = meat; S = soy; Sr = sourdough; T = tea; V = vegetables; Vn = vinegar; W = wine.

Table 6.5 Test trial suggested by OIV for contaminants and corresponding limits in active dry yeasts (OIV-Oeno 329-2009).

Nature of contamination	Contaminant	Limit
Chemical	Lead	Less than 2 mg/kg of dry matter
Chemical	Mercury	Less than 1 mg/kg of dry matter
Chemical	Arsenic	Less than 3 mg/kg of dry matter
Chemical	Cadmium	Less than 1 mg/kg of dry matter
Microbiological		Less than 10^3 cfu/g for frozen or liquid lactic acid bacteria or 10^4 cfu/g for lyophilisated or driec lactic acid bacteria
Microbiological	Yeasts of species different from the species indicated on the label	Less than 10^5 cfu/g
Microbiological	Mould	Less than 10^3 cfu/g
Microbiological	Lactic acid bacteria	Less than 10^4 cfu/g
Microbiological	Acetic acid bacteria	Less than 10^5 cfu/g
Microbiological	Salmonella	Absence should be checked on a 25 g sample
Microbiological	*Escherichia coli*	Absence should be checked on 1 g sample
Microbiological	Staphylococci	Absence should be checked on 1 g sample
Microbiological	Coliforms	Less than 10^2 cfu/g

Note: cfu = colony forming unit.

Table 6.6 Test trial suggested by OIV for contaminants and corresponding limits in lactic acid bacteria (OIV-Oeno 328-2009, Oeno 494-2012).

Nature of contamination	Contaminant	Limits
Chemical	Lead	Less than 2 mg/kg of dry matter
Chemical	Mercury	Less than 1 mg/kg of dry matter
Chemical	Arsenic	Less than 3 mg/kg of dry matter
Chemical	Cadmium	Less than 1 mg/kg of dry matter
Microbiological	Mould	Less than 10^3 cfu/g
Microbiological	Acetic acid bacteria	Less than 10^3 cfu/g for frozen or liquid lactic acid bacteria or 10^4 cfu/g for lyophilisated or dried lactic acid bacteria
Microbiological	Yeasts	Less than 10^3 cfu/g for lyophilisated or dried lactic acid bacteria or 10^2 cfu/mL for frozen or liquid lactic acid bacteria
Microbiological	Salmonella	Absence should be checked on a 25 g sample
Microbiological	*Escherichia coli*	Absence should be checked on 1 g sample
Microbiological	Staphylococci	Absence should be checked on 1 g sample
Microbiological	Coliforms	Less than 10^2 cfu/g

Note: cfu = colony forming unit.

Starter cultures, microbial contaminants and safety aspects

Starter cultures are ingested in high amounts through the consumption of fermented foods, requiring a more stringent need to demonstrate that their components are totally safe.

From a biological point of view there are three focal theoretical concerns regarding the safety of starter cultures: the occurrence of virulent traits and opportunistic diseases, mainly resulting in bacteremia or endocarditis; the production of toxic compounds; and the transfer of antibiotic resistance determinants. When microorganisms are assessed for their usage as probiotic, these critical points are made more insidious by the threat that hazardous phenomena could occur *in situ* in the gut of the host (Snydman 2008; Sanders 2010).

In general, the risk of opportunistic infections due to the consumption of starter cultures is low, although sporadic events attributable to LAB ingestion have been reported (Table 6.2). According to an eight-year study the incidence of probiotic-related bacteremia due to *Lb. acidophilus* and *Lb. bulgaricus* appeared to be associated with a minimal risk of probiotic-related bloodstream infection (Simkins *et al.* 2013).

A rigorous safety assessment also should exclude the involvement of starter cultures in the production of some toxins of microbial origin, including biogenic amines in fermented foods (Spano *et al.* 2010); ochratoxin A and ethyl carbamate in wine (Pozo-Bayón *et al.* 2012; Petruzzi *et al.* 2014); and a number of toxins in fermented meats and cheeses (Talon and Leroy 2011; Hymery *et al.* 2014). Nonetheless, a comprehensive analysis of toxin production from food-associated coagulase-negative staphylococci revealed that 12% of the meat starter cultures investigated exhibited weak to moderate haemolytic activity with human blood (Zell *et al.* 2008).

Antimicrobial resistance is a serious concern strictly related to the probability of genetic exchange to pathogens and/or other bacteria that can act as reservoirs (Rossi *et al.* 2014). In contrast to intrinsic resistance, which is not a horizontally transferable species-specific trait, the acquired antibiotic resistance of a strain can be due either to the mutation of indigenous genes or via a gain of exogenous DNA, leading to a serious safety risk of lateral transmission (FEEDAP 2008; Sharma *et al.* 2014). Therefore, with the aim of determining the antimicrobial resistance of clinically important microorganisms, the EFSA Panel on Additives and Products or Substances used in Animal Feed (FEEDAP) established rigorous breakpoint values (FEEDAP 2008). The cut-off values defined by FEEDAP within bacterial populations belonging to a single taxonomical unit should be considered as a pragmatic response to identifying strains with acquired resistance from susceptible strains (EFSA 2012b). However, the QPS concept contemplates that the antimicrobial susceptibility of starter cultures should be evaluated by a case-by-case assessment in order to determine whether the antibiotic resistance is an intrinsic trait whose genetic basis has been clearly established (Hummel *et al.* 2007; EFSA 2008; Giraffa 2009; Gueimonde *et al.* 2013).

In an attempt to evaluate the biosafety of probiotic LAB used for human consumption, an extensive screening based on antimicrobial susceptibility was performed covering 473 taxonomically well-characterized isolates of LAB, including 129 isolates used in commercially available probiotic products and 27 isolates used as starter cultures (Klare *et al.* 2007). The results of this study indicated the occurrence of acquired resistance genes in about 10% of the isolates intended for both probiotic use or as starter cultures, with a wide spread of resistance to the clinically important antibiotics streptomycin, erythromycin and tetracycline (Klare *et al.* 2007).

Similarly, the presence of an acquired *tet*(W) gene was responsible for tetracycline resistance in probiotic isolates belonging to the taxa *Bifidobacterium animalis* subsp. *lactis* and *Bifidobacterium bifidum* (Masco *et al.* 2006). The *tet*(W) gene residing on a mobile element has also been reported in the commercial probiotic strains *Bifidobacterium lactis* DSM 10140 and *Lactobacillus reuteri* SD 2112, with the later strain also carrying the lincosamide resistance gene *lnu*(A) (Kastner *et al.* 2005). In the same study, the tetracycline resistance gene *tet*(K) was detected in 5 *Staphylococcus* isolates used as meat starter cultures in Switzerland (Kastner *et al.* 2006). A warning rate of antimicrobial resistance was also found in starters commonly used in dry sausage fermentation, although the incidence was higher among *Pediococcus pentosaceus* than *Staph. carnosus* strains (Cordeiro *et al.* 2010). A comprehensive insight into the incidence of antibiotic susceptibility in food-associated coagulase-negative staphylococci revealed a high percentage (93%) of resistance to the main clinically important antibiotics in *Staph. xylosus* strains obtained from meat starter cultures (Resch *et al.* 2008). In contrast, starter bacteria in Norwegian dairy products do not seem to represent a source for the spread of genes encoding resistance to antimicrobial agents (Katla *et al.* 2001).

More complex is the question of the safety assessment of undefined microbial cultures such as natural whey ('sieroinnesto') and milk ('lattoinnesto') starters, sourdoughs or commercial mixed-strain starters. These matrices are traditionally employed to obtain foods, such as some bread varieties, traditional dried sausages and cheeses (Chamba and Jamet 2008).

To offer insight into this concern, the QPS approach was applied to dominant LAB associated with Grana Padano cheese whey starters, representing the overall genotypic LAB diversity associated with 24 previously collected whey starters (Rossetti *et al.* 2008, 2009). In the opinion of the authors, the proposed approach could be extended to other types of undefined-strain cultures, which are still widely used in food production (Rossetti *et al.* 2009).

A preventive assessment of the safety of microbial cultures cannot fail to consider the risk connected with potential contamination at different levels. Therefore, it is appropriate to propose an HACCP approach to determine the absence of physical, chemical and biological hazards in biomass productions. The eventual occurrence of foreign bodies is definitely related to a very low risk to the health of consumers. In contrast, the presence of some potentially harmful chemicals such as trace metal compounds from the culture media (see Tables 6.5

and 6.6) poses a risk that should be considered. Similarly, the occurrence of allergens in starter growth media could result in contaminated foods. In the last few years, this issue has been the subject of a Food Allergen Labeling Petition that was denied by the FDA (FDA 2006).

Even less obvious, and probably underestimated, is the potential risk related to the presence of microbial contaminants in starter formulation, as already discussed for oenological cultures according to the OIV standards (Tables 6.5 and 6.6). This implies that to some extent, commercial starter cultures can contain a number of microorganisms that have not been subjected to a full safety assessment. However, scientific literature on this topic is scarce and the safety of potential microbial contaminants in starter formulations is still an open question. Among the few studies challenging this matter, Costantini and co-authors (2009) found that some yeast preparations used in winemaking resulted in contamination with biogenic amine-producing bacteria, thus increasing the real risk of these toxic compounds in wine.

It is important to consider that, paraphrasing the concept of probiotics, potential harmful microbial contaminants could be able to colonize the gut environment and exert a detrimental effect on the health of the host. It was reported that the tyramine- and putrescine-producing *Lactobacillus brevis* IOEB 9809 from wine origin was able to produce biogenic amines in co-culture with Caco-2 cell lines, suggesting that exogenous food microorganisms could contribute to increasing the risk of biogenic amines formation at an intestinal level (Russo *et al.* 2012). In a similar way, the role of LAB as a reservoir of antibiotic resistance determinants with transmission potential to pathogens in the gut environment represents a potential health risk that was neglected for a long time (van Reenen and Dicks 2011; Devirgiliis *et al.* 2013; Verraes *et al.* 2013).

Starter culture design for enhanced safety

According to Vogel *et al.* (2011), one of the main advantages of starter cultures in food is connected with the evidence that 'toxic or harmful substances derived from the raw material, such as cyanides, hemagglutinins, goitrogens, proteinase inhibitors, phytic acid, oxalic acid, glucosinolates and indigestible carbohydrates, are partly degraded'. This aptitude of starter cultures opens the way to the tailored selection and design of these microbial resources in order to enhance the safety of the final products via selective degradation in the matrix. It is a property that is found in specific standards in the Codex Alimentarius: the standard for 'Ochratoxin A contamination in wine' (CAC/RCP 63-2007) reported that 'dry active yeasts or inactive yeasts can help reduce the OTA level' and suggested 'for alcoholic or malolactic fermentations, use yeasts or bacteria which have adsorbent properties for OTA', while the 'Code of Practice for the Reduction of Acrylamide in Foods' (CAC/RCP 67-2009) indicated that 'yeast fermentation of wheat bread doughs reduces the free asparagine content'. As highlighted by the

Codex standards, 'food-grade microbes' can reduce chemical contamination (e.g. degradation of organophosphorus pesticides by LAB, Zhang *et al.* 2014; degradation of acrylamide by LAB, Bartkiene *et al.* 2013; nitrite reduction capability by LAB, Paik and Lee 2014; N-nitrosodimethylamine detoxification by LAB, Nowak *et al.* 2014) and/or biological contamination. In the decrease of biological contaminants we have to distinguish direct antagonism (e.g. against pathogens, Jordan *et al.* 2014; against mycotoxigenic fungi, Oliveira *et al.* 2014) and the reduction of microbial toxic compounds (e.g. removal of paralytic shellfish toxins by LAB, Vasama *et al.* 2014; decontamination of ochratoxin A by yeasts, Petruzzi *et al.* 2014; biogenic amines degradation by LAB, Capozzi *et al.* 2012d).

Conclusion

The safe use of microorganisms in food fermentation represents a millenarian tale in human history. Today this 'saga' has different levels of narration, as a function of national, social and economic dimensions. The regulations oscillate between two main approaches: an open process of validation case by case and the definition of a positive list of microbes with a 'safe for use' status. From this point of view, the observation of Hansen (2002) remains valid: 'It is currently difficult to predict in what direction the future regulatory requirements will influence innovation in the food industry.' The big issue remains mediation to ensure both the important applications of starter cultures and the high level of food safety required. The 'open process' model will assure the exploitation of starter cultures as a 'promoting force for the practical use of biotechnology to make better and safer products' (Hansen 2002); the 'positive list' approach will endorse a rigorous application of safety standards. With this concern, it is strongly suggested that the adoption of participatory processes and transdisciplinary approaches would maximize the possible benefits of starter culture technology for human health. From this perspective, the efforts in promoting harmonization and standardization by IDF, EFFCA and OIV represent best practice in the sector. In addition, participation and transdisciplinarity favour an enabling environment for starter culture improvement in developing countries and poor regions (FAO 2010). The infrastructures conceived to create an electronic passport for microbial strains (Verslyppe *et al.* 2014) and a comprehensive dynamic database of microbial resources (Wu *et al.* 2013) are useful tools to enhance coordination in scientific research devoted to strain identification and safety.

A rising issue in the sector concerns the use of live genetically modified microorganisms (GMMs) for food production; wine GM yeasts are commercialized in the USA and Canada (Leòn *et al.* 2011), while no applications have been received by EFSA concerning GMMs for the production of foods in which GMMs or the remains of their cells are still present in the product (Devos *et al.* 2014). Tailored guidelines in this respect have been released by the Codex Alimentarius (Codex Alimentarius Commission 2009). Remaining in the field of advanced biotechnology, next-generation sequencing technologies offer new perspectives on strain safety; in particular, safety assessment based on the complete genome represents an interesting tool (Zhang *et al.* 2012), considering the increasing number of complete genomes of protechnological strains (Labrie *et al.* 2014; Lambie *et al.* 2014), of autochthonous strains belonging to species of protechnological interest (Capozzi *et al.* 2014; Lamontanara *et al.* 2014), of strain producers of biogenic amines (Ladero *et al.* 2013, 2014) and of probiotic strains (Li *et al.* 2014; Treven *et al.* 2014). The recent efforts of Pariza *et al.* (2015) to model previous decision trees (universally applied to assess the safety of microbial enzymes) in order to evaluate the safety of microbial cultures for consumption by humans (and animals) provide a precious concrete basis for scientific debate in the field.

References

Achi, O.K. (2005) The potential for upgrading traditional fermented foods through biotechnology. *African Journal of Biotechnology*, **4**, 375–380.

Ahmadova, A., Todorov, S.D., Choiset, Y. *et al.* (2013) Evaluation of antimicrobial activity, probiotic properties and safety of wild strain *Enterococcus faecium* AQ71 isolated from Azerbaijani Motal cheese. *Food Control*, **30**, 631–641.

Barbosa, J., Borges, S. and Teixeira, P. (2014) Selection of potential probiotic *Enterococcus faecium* isolated from Portuguese fermented food. *International Journal of Food Microbiology*, **191**, 144–148.

Bartkiene, E., Jakobsone, I., Juodeikiene, G., Vidmantiene, D., Pugajeva, I. and Bartkevics V. (2013) Effect of lactic acid fermentation of lupine wholemeal on acrylamide content and quality characteristics of wheat-lupine bread. *International Journal of Food Sciences and Nutrition*, **64**, 890–896.

Bourdichon, F., Casaregola, S., Farrokh, C. *et al.* (2012) Food fermentations: Microorganisms with technological beneficial use. *International Journal of Food Microbiology*, **154**, 87–97.

Bover-Cid, S., Izquierdo-Pulido, M. and Vidal-Carou, M.C. (2000) Mixed starter cultures to control biogenic amine production in dry fermented sausages. *Journal of Food Protection*, **63**, 1556–1562.

Brandt, M.J. (2014) Starter cultures for cereal based foods. *Food Microbiology*, **37**, 41–43.

Bryan, F.L. (1992) *Hazard Analysis Critical Control Point Evaluation: A Guide to Identifying Hazards and Assessing Risks Associated with Preparation and Storage*, Geneva: World Health Organization.

Buchelli-Ramirez, H.L., Alvarez-Alvarez, C., Rojo-Alba, S. *et al.* (2013) Necrotising pneumonia caused by *Lactococcus lactis* cremoris. *International Journal of Tuberculosis and Lung Disease*, **17**, 565–567.

Canadian Food Inspection Agency (2014) List of accepted starter cultures, Ontario: Canadian Food Inspection Agency. http://www.inspection.gc.ca/food/meat-and-poultry-products/manual-of-procedures/chapter-4/annex-g/eng/1370536922521/1370537138760 (accessed November 2014).

Capozzi, V., Russo, P., Dueñas, M.T., López, P. and Spano, G. (2012c) Lactic acid bacteria producing B-group vitamins: A great potential for functional cereals products. *Applied Microbiology and Biotechnology*, **96**, 1383–1394.

Capozzi, V., Russo, P., Ladero, V. *et al.* (2012d) Biogenic amines degradation by *Lactobacillus plantarum*: Toward a potential application in wine. *Frontiers in Microbiology*, **3**, 122.

Capozzi, V., Russo, P., Lamontanara, A., Orrù, L., Cattivelli, L. and Spano, A. (2014) Genome sequences of five *Oenococcus oeni* strains isolated from Nero di Troia wine from the same terroir in apulia, Southern Italy. *Genome Announcements*, **2**(5), e01077-14.

Capozzi, V., Russo, P. and Spano, G. (2012b) Microbial information regimen in EU geographical indications. *World Patent Information*, **34**, 229–231.

Capozzi, V. and Spano, G. (2011) Food microbial biodiversity and 'microbes of protected origin'. *Frontiers in Microbiology*, **2**, 237.

Capozzi, V., Spano, G. and Fiocco, D. (2012a) Transdisciplinarity and microbiology education. *Journal of Microbiology and Biology Education*, **13**, 70–73.

Ceuppens, S., Rajkovic, A., Heyndrickx, M. *et al.* (2011) Regulation of toxin production by *Bacillus cereus* and its food safety implications. *Critical Reviews in Microbiology*, **37**, 188–213.

Chamba, J.F. and Jamet, E. (2008) Contribution to the safety assessment of technological microflora found in fermented dairy products. *International Journal of Food Microbiology*, **126**, 263–266.

Claus, S.P. and Swann, J.R. (2013) Nutrimetabonomics: Applications for nutritional sciences, with specific reference to gut microbial interactions. *Annual Review of Food Science and Technology*, **4**, 381–399.

Codex Alimentarius Commission (2009) *Foods Derived from Modern Biotechnology*, Rome: FAO/WHO.

Coelho, M.C., Silva, C.C.G., Ribeiro, S.C., Dapkevicius, M.L.N.E. and Rosa, H.J.D. (2014) Control of *Listeria monocytogenes* in fresh cheese using protective lactic acid bacteria. *International Journal of Food Microbiology*, **191**, 53–59.

Cordeiro, R.P., Du, T., Mulvey, M.R., Krause, D.O. and Holley, R.A. (2010) Susceptibility of meat starter cultures to antimicrobials used in food animals in Canada. *Journal of Food Protection*, **73**, 916–922.

Costantini, A., Vaudano, E., Del Prete, V., Danei, M. and Garcia-Moruno, E. (2009) Biogenic amine production by contaminating bacteria found in starter preparations used in winemaking. *Journal of Agricultural and Food Chemistry*, **57**, 10664–10669.

Danish Veterinary and Food Administration (2013) List of notified microbial cultures, Glostrup: Danish Veterinary and Food Administration. http://www.foedevarestyrelsen.dk/SiteCollectionDocuments/25_PDF_word_filer%20til%20download/06kontor/Liste%20over%20anmeldte%20kulturer.pdf (accessed November 2014).

Deng, Y., Zhang, Z., Xie, Y., Xiao, Y., Kang, M. and Fan, H. (2012) A mixed infection of *Leuconostoc lactis* and vancomycin-resistant *Enterococcus* in a liver transplant recipient. *Journal of Medical Microbiology*, **61**, 1621–1624.

De Souza, R.D., Batista, M.T., Luiz, W.B. *et al.* (2014) *Bacillus subtilis* spores as vaccine adjuvants: Further insights into the mechanisms of action. *PLoS ONE*, **9**, e87454.

Devirgiliis, C., Zinno, P. and Perozzi, G. (2013) Update on antibiotic resistance in foodborne *Lactobacillus* and *Lactococcus* species. *Frontiers in Microbiology*, **4**, 301.

Devos, Y., Aguilera, J., Diveki, Z. *et al.* (2014) EFSA's scientific activities and achievements on the risk assessment of genetically modified organisms (GMOs) during its first decade of existence: Looking back and ahead. *Transgenic Research*, **23**, 1–25.

Didari, T., Solki, S., Mozaffari, S., Nikfar, S. and Abdollahi, M. (2014) A systematic review of the safety of probiotics. *Expert Opinion on Drug Safety*, **13**, 227–239.

Doyle, D.M.P., Steenson, L.R. and Meng, J. (2013) Bacteria in food and beverage production. In E. Rosenberg, E.F. DeLong, S. Lory, E. Stackebrandt and F. Thompson (eds), *The Prokaryotes* (pp. 241–256), Berlin: Springer.

DuPrey, K.M., McCrea, L., Rabinowitch, B.L. and Azad, K.N. (2012) Pyelonephritis and bacteremia from *Lactobacillus delbrueckii*. *Case Reports in Infectious Diseases*, **2012**, 745743.

EFFCA (2012) Inventory of microorganisms, Brussels: European Food and Feed Cultures Association. http://www.effca.org/content/inventory-microorganisms (accessed November 2014).

EFFCA (2013) Food culture. Brussels: European Food and Feed Cultures Association. http://www.effca.org/content/microbial-food-culture (accessed November 2014).

EFSA (2005) QPS. Qualified presumption of safety of microorganisms in food and feed. EFSA Scientific Colloquium, 13–14 December 2005, Brussels.

EFSA (2007) Opinion of the scientific committee on introduction of a qualified presumption of safety (QPS) approach for assessment of selected microorganisms referred to EFSA. *EFSA Journal*, **587**, 1–16.

EFSA (2008) Scientific opinion of the panel on biological hazards on the maintenance of the list of QPS microorganisms intentionally added to food or feed. *EFSA Journal*, **923**, 1–48.

EFSA (2009) Scientific opinion of the panel on biological hazards on the maintenance of the list of QPS microorganisms intentionally added to food or feed. *EFSA Journal*, **7**, 1–93.

EFSA (2012a) Scientific opinion on the maintenance of the list of QPS biological agents intentionally added to food and feed. *EFSA Journal*, **10**, 3020.

EFSA (2012b) Guidance on the assessment of bacterial susceptibility to antimicrobials of human and veterinary importance. *EFSA Journal*, **10**, 2740–2750.

EFSA (2013) Scientific opinion on the maintenance of the list of QPS biological agents intentionally added to food and feed. *EFSA Journal*, **11**, 3449.

EFSA (2015) Appendix: The 2013 updated list of QPS Status recommended biological agents in support of EFSA risk assessments – 2nd revision (new addition). *EFSA Journal* **13**(6), 4138. http://www.efsa.europa.eu/sites/default/files/scientific_output/files/main_documents/4138ax1.pdf (accessed June 2016).

European Commission (1997) Regulation (EC) n° 258/97 of the European Parliament and of the council of 27 January 1997 concerning novel foods and novel food ingredients. *Official Journal*, L 043, 14/02/1997.

FAO/WHO (1996) Fermentation: assessment and research. Report of a Joint FAO/WHO Workshop on fermentation as household technology to improve food safety. WHO Document WHO/FNU/FOS/96.1, World Health Organization, Geneva.

FAO (2010) Current status and options for biotechnologies in food processing and in food safety in developing countries. In *Biotechnologies for Agricultural Development: Proceedings of the FAO Technical Conference on 'Agricultural Biotechnologies in Developing Countries: Options and Opportunities in Crops, Forestry, Livestock, Fisheries and Agro-Industry to Face the Challenges of Food Insecurity and Climate Change'* (pp. 240–277, Rome: Food and Agriculture Organization of the United Nations.

FDA (1999) *Federal Food, Drug, and Cosmetic Act*. Washington, DC: Food and Drug Administration.

FDA (2006) Agency Response Letter: Objection; Starter Growth Media; FALP No. 001 (Docket No. 2006FL-0238). Silver Spring, MD: US Food and Drug Administration.

FDA (2010) Generally Recognized as Safe (GRAS) Notification Program, Silver Spring, MD: US Food and Drug Administration. http://www.fda.gov/AnimalVeterinary/Products/AnimalFoodFeeds/GenerallyRecognizedasSafeGRASNotifications (accessed November 2014).

FDA (2014) GRAS Notice (GRN) No. 502, Silver Spring, MD: US Food and Drug Administration. http://www.accessdata.fda.gov/scripts/fdcc/?set=GRASNoticesandid=502 (accessed November 2014).

FDA (2015) Code of Federal Regulations (CFR) Title 21, Silver Spring, MD: US Food and Drug Administration. http://www.accessdata.fda.gov/scripts/cdrh/cfdocs/cfcfr/cfrsearch.cfm?fr=101.60 (accessed November 2014).

Federal Register (1997) Department of Health and Human Services, Food and Drug Administration, 21 CFR Parts 170, 184, 186, and 570 [Docket No. 97N–0103], Substances Generally Recognized as Safe. *Federal Register*, **62**(74), 18938–18964. http://www.gpo.gov/fdsys/pkg/FR-1997-04-17/pdf/97-9706.pdf (accessed November 2014).

FEEDAP (2008) Technical guidance: Update of the criteria used in the assessment of bacterial resistance to antibiotics of human or veterinary importance. *EFSA Journal*, **732**, 1–15.

Fischer, A.R.H., De Jong, A.E.I., De Jonge, R., Frewer, L.J. and Nauta, M.J. (2005) Improving food safety in the domestic environment: The need for a transdisciplinary approach. *Risk Analysis*, **25**, 503–517.

Francis, G.A., Gallone, A., Nychas, G.J. *et al.* (2012) Factors affecting quality and safety of fresh-cut produce. *Critical Reviews in Food Science and Nutrition*, **52**, 595–610.

Giraffa, G. (2009) Antibiotic resistance in food-associated lactic acid bacteria. In A.R. Bonilla and K.P. Muniz (eds), *Antibiotic Resistance: Causes and Risk Factors, Mechanisms and Alternatives* (pp. 371–378), New York: Nova Science.

Gueimonde, M., Sánchez, B., de Los Reyes-Gavilán, C.G. and Margolles, A. (2013) Antibiotic resistance in probiotic bacteria. *Frontiers in Microbiology*, **4**, 202.

Hadjisymeou, S., Loizou, P. and Kothari, P. (2013) *Lactococcus lactis cremoris* infection: Not rare anymore? *BMJ Case Reports*, **2013**, bcr2012008479.

Hamadah, A.M., Elrashidi, M.Y. and McDonald, F.S. (2013) 65-year-old woman with confusion. *Mayo Clinic Proceedings*, **88**, e1–e5.

Hansen, E.B. (2002) Commercial bacterial starter cultures for fermented foods of the future. *International Journal of Food Microbiology*, **78**, 119–131.

Hansen, E.B. (2014) Starter cultures: Uses in the food industry. In C.A. Batt and M.L. Tortorello (eds), *Encyclopedia of Food Microbiology* (2nd edn, pp. 529–534), Oxford: Academic Press.

Havelaar, A.H., Brul, S., de Jong, A., de Jonge, R., Zwietering, M.H. and Ter Kuile, B.H. (2010) Future challenges to microbial food safety. *International Journal of Food Microbiology*, **139**, S79–S94.

Herody, C., Soyeux, Y., Hansen, E.B., Gillies, K. (2010) The legal status of microbial food cultures in the European Union: An overview. *European Food and Feed Law Review*, **2010**, 258–269.

Holzapfel, W.H. (2002) Appropriate starter culture technologies for small-scale fermentation in developing countries. *International Journal of Food Microbiology*, **75**, 197–212.

Hummel, A.S., Hertel, C., Holzapfel, W.H. and Franz, C.M.A.P. (2007) Antibiotic resistances of starter and probiotic strains of lactic acid bacteria. *Applied and Environmental Microbiology*, **73**, 730–739.

Hymery, N., Vasseur, V., Coton, M. *et al.* (2014) Filamentous fungi and mycotoxins in cheese: A review. *Comprehensive Reviews in Food Science and Food Safety*, **13**, 437–456.

ISO (2010) Fermented Milk Products – Bacterial Starter Cultures – Standard of Identity. Geneva: International Standardisation Organisation.

Jay, J.M. (2005) History of microorganisms in food. In J.M. Jay, M.J. Loessner and D.A. Golden (eds), *Modern Food Microbiology* (pp. 3–9), New York: Springer, Food Science Text Series.

Jenke, A., Ruf, E.M., Hoppe, T., Heldmann, M. and Wirth, S. (2012) *Bifidobacterium septicaemia* in an extremely low-birthweight infant under probiotic therapy. *Archives of Disease in Childhood: Fetal and Neonatal Edition*, **97**, F217–F218.

Jordan, K., Dalmasso, M., Zentek, J. *et al.* (2014) Microbes versus microbes: Control of pathogens in the food chain. *Journal of the Science of Food and Agriculture*, **94**, 3079–3089.

Kastner, S., Perreten, V., Bleuler, H., Hugenschmidt, G., Lacroix, C. and Meile, L. (2006) Antibiotic susceptibility patterns and resistance genes of starter cultures and probiotic bacteria used in food. *Systematic and Applied Microbiology*, **29**, 145–155.

Katla, A.K., Kruse, H., Johnsen, G. and Herikstad, H. (2001) Antimicrobial susceptibility of starter culture bacteria used in Norwegian dairy products. *International Journal of Food Microbiology*, **67**, 147–152.

Klare, I., Konstabel, C., Werner, G. *et al.* (2007) Antimicrobial susceptibilities of *Lactobacillus, Pediococcus* and *Lactococcus* human isolates and cultures intended for probiotic or nutritional use. *Journal of Antimicrobial Chemotherapy*, **59**, 900–912.

Labrie, S.J., El Haddad, L., Tremblay, D.M. *et al.* (2014) First complete genome sequence of *Staphylococcus xylosus*, a meat starter culture and a host to propagate *Staphylococcus aureus* phages. *Genome Announcements*, **2**, e00671-14.

Ladero, V., Del Rio, B., Linares, D.M. *et al.* (2014) Genome sequence analysis of the biogenic amine-producing strain *Lactococcus lactis* subsp. *cremoris* CECT 8666 (Formerly GE2-14). *Genome Announcements*, **2**, e01088-14.

Ladero, V., Linares, D.M., Del Rio, B., Fernandez, M., Martin, M.C. and Alvarez, M.A. (2013) Draft genome sequence of the tyramine producer *Enterococcus durans* strain IPLA 655. *Genome Announcements*, **1**, e00265-13.

Lambie, S.C., Altermann, E., Leahy, S.C. and Kelly, W.J. (2014) Draft genome sequence of *Lactococcus lactis* subsp. *cremoris* HPT, the first defined-strain dairy starter culture bacterium. *Genome Announcements*, **2**, e00107-14.

Lamontanara, A., Orrù, L., Cattivelli, L., Russo, P., Spano, G. and Capozzi, V. (2014) Genome sequence of *Oenococcus oeni* OM27, the first fully assembled genome of a strain isolated from an Italian wine. *Genome Announcements*, **2**, e00658-14.

Larsen, N., Thorsen, L., Kpikpi, E.N. *et al.* (2014) Characterization of *Bacillus* spp. strains for use as probiotic additives in pig feed. *Applied Microbiology and Biotechnology*, **98**, 1105–1118.

León, C., García-Cañas, V., González, R., Morales, P. and Cifuentes, A. (2011) Fast and sensitive detection of genetically modified yeasts in wine. *Journal of Chromatography A*, **1218**, 7550–7556.

Leroy, F., De Vuyst, L. (2004) Lactic acid bacteria as functional starter cultures for the food fermentation industry. *Trends in Food Science & Technology*, **15**, 67–78.

Leuschner, R.G.K., Robinson, T.P., Hugas, M. *et al.* (2010) Qualified presumption of safety (QPS): A generic risk assessment approach for biological agents notified to the European Food Safety Authority (EFSA). *Trends in Food Science and Technology*, **21**, 425–435.

Li, X.F., Liao, X.Y., Liu, Y.F., Guo, L.-Q., Ye, Z.-W. and Lin, J.F. (2014) Complete genome sequence of probiotic *Lactobacillus plantarum* strain FMNP01, isolated from mango fruit. *Genome Announcements*, **2**, e01207-14.

Logan, N.A. (2012) *Bacillus* and relatives in foodborne illness. *Journal of Applied Microbiology*, **112**, 417–429.

Maqueda, M., Pérez-Nevado, F., Regodón, J.A. *et al.* (2011) A low-cost procedure for production of fresh autochthonous wine yeast. *Journal of Industrial Microbiology and Biotechnology*, **38**, 459–469.

Masco, L., Van Hoorde, K., De Brandt, E., Swings, J. and Huys, G. (2006) Antimicrobial susceptibility of *Bifidobacterium* strains from humans, animals and probiotic products. *Journal of Antimicrobial Chemotherapy*, **58**, 85–94.

Michalopoulos, N., Arampatzi, S., Papavramidis, T.S., Kotidis, E., Laskou, S. and Papavramidisa, S.T. (2013) Necrotizing cellulitis of the abdominal wall, caused by *Pediococcus* sp., due to rupture of a retroperitoneal stromal cell tumor. *International Journal of Surgery Case Reports*, **4**, 286–289.

Mogensen, G., Salminen, S., O'Brien, J. *et al.* (2002) Inventory of microorganisms with a documented history of use in food. *Bulletin of IDF*, **377**, 10–19.

Mokhtar, S., Mostafa, G., Taha, R. and Eldeep, G.S.S. (2012) Effect of different starter cultures on the biogenic amines production as a critical control point in fresh fermented sausages. *European Food Research and Technology*, **235**, 527–535.

Mullan, W.M.A. (2014) Starter cultures: Importance of selected genera. In C.A. Batt and M.L. Tortorello (eds), *Encyclopedia of Food Microbiology* (2nd edn, pp. 515–521), Oxford: Academic Press.

Murata, K., Hoshina, T., Saito, M. *et al.* (2012) Bacterial pericarditis caused by *Lactobacillus iners* in an infant. *Diagnostic Microbiology and Infectious Disease*, **74**, 181–182.

National Health and Family Planning Commission of the People's Republic of China (2001) The Ministry of Health on the issuance of fungal and probiotic health food review provisions (in Chinese), Beijing: National Health and Family Planning Commission of the PRC. http://www.nhfpc.gov.cn/zhuzhan/wsbmgz/201304/a4b531b5586d403183ccbc2068f3fa32.shtml (accessed November 2014).

National Research Council Panel on the Applications of Biotechnology to Traditional Fermented Foods (1992) *Applications of Biotechnology to Fermented Foods: Report of an Ad Hoc Panel of the Board on Science and Technology for International Development*, Washington, DC: National Academies Press.

Nishijima, T., Teruya, K., Yanase, M., Tamori, Y., Mezaki, K. and Oka, S. (2012) Infectious endocarditis caused by *Lactobacillus acidophilus* in a patient with mistreated dental caries. *Internal Medicine*, **51**, 1619–1621.

Nowak, A., Kuberski, S. and Libudzisz, Z. (2014) Probiotic lactic acid bacteria detoxify N-nitrosodimethylamine. *Food Additives and Contaminants. Part A, Chemistry, Analysis, Control, Exposure and Risk Assessment*, **31**, 1678–1687.

Oliveira, P.M., Zannini, E. and Arendt, E.K. (2014) Cereal fungal infection, mycotoxins, and lactic acid bacteria mediated bioprotection: From crop farming to cereal products. *Food Microbiology*, **37**, 78–95.

Orkaby, A.R., Chen, B., Iliaki, E.F., Sulis, C.A. and Oates, D.J. (2012) A curious case of *Lactobacillus casei* in a prosthetic joint: Was it the yogurt? *Journal of the American Geriatrics Society*, **60**, 1177–1178.

Orrù, L., Salvetti, E., Cattivelli, L. *et al.* (2014) Draft genome sequence of *Bacillus coagulans* GBI-30, 6086, a widely used spore-forming probiotic strain. *Genome Announcements*, **2**, e01080-14.

Paik, H.D. and Lee, J.Y. (2014) Investigation of reduction and tolerance capability of lactic acid bacteria isolated from kimchi against nitrate and nitrite in fermented sausage condition. *Meat Science*, **97**, 609–614.

Pariza, M.W., Gillies, K.O., Kraak-Ripple, S.F., Leyer, G. and Smith, A.B. (2015) Determining the safety of microbial cultures for consumption by humans and animals. *Regulatory Toxicology and Pharmacology*, **73**, 164–171.

Perapoch, J., Planes, A., Querol, A. *et al.* (2000) Fungemia with *Saccharomyces cerevisiae* in two newborns, only one of whom had been treated with ultra-levura. *European Journal of Clinical Microbiology and Infectious Diseases*, **19**, 468–470.

Petruzzi, L., Sinigaglia, M., Corbo, M.R., Campaniello, D., Speranza, B. and Bevilacqua, A. (2014) Decontamination of ochratoxin A by yeasts: Possible approaches and factors leading to toxin removal in wine. *Applied Microbiology and Biotechnology*, **98**, 6555–6567.

Pollan, M. (2010) *Food Rules: An Eater's Manual*. Harmondsworth: Penguin.

Pozo-Bayón, M.Á., Monagas, M., Bartolomé, B. and Moreno-Arribas, M.V. (2012) Wine features related to safety and consumer health: An integrated perspective. *Critical Reviews in Food Science and Nutrition*, **52**, 31–54.

Resch, M., Nagel, V. and Hertel, C. (2008) Antibiotic resistance of coagulase-negative staphylococci associated with food and used in starter cultures. *International Journal of Food Microbiology*, **127**, 99–104.

Reyes, K. and Zervos, M. (2013) Endocarditis caused by resistant enterococcus: An overview. *Current Infectious Disease Reports*, **15**, 320–328.

Rossetti, L., Carminati, D., Zago, M. and Giraffa, G. (2009) A qualified presumption of safety approach for the safety assessment of Grana Padano whey starters. *International Journal of Food Microbiology*, **130**, 70–73.

Rossetti, L., Fornasari, M.E., Gatti, M., Lazzi, C., Neviani, E. and Giraffa, G. (2008) Grana Padano cheese whey starters: Microbial composition and strain distribution. *International Journal of Food Microbiology*, **127**, 168–171.

Rossi, F., Rizzotti, L., Felis, G.E. and Torriani, S. (2014) Horizontal gene transfer among microorganisms in food: Current knowledge and future perspectives. *Food Microbiology*, **42**, 232–243.

Rostagno, C., Pecile, P. and Stefàno, P.L. (2013) Early *Lactococcus lactis* endocarditis after mitral valve repair: A case report and literature review. *Infection*, **41**, 897–899.

Russo, P., Fernández De Palencia, P., Romano, A., Fernández, M., Lucas, P., Spano, G. and Lápez, P. (2012) Biogenic amine production by the wine Lactobacillus brevis IOEB 9809 in systems that partially mimic the gastrointestinal tract stress. *BMC Microbiology*, **12**, 247.

SANCO (2006) Summary record of the Standing Committee on the Food Chain and Animal Health held in Brussels on 14 December 2006: Section Toxicological Safety on the Food Chain. Document No. SANCO – D1(06)D/413447, Brussels: Directorate General for Health and Consumers (SANCO). http://ec.europa.eu/food/committees/regulatory/scfcah/toxic/summary23_en.pdf (accessed November 2014).

Sanders, M.E., Akkermans, L.M.A., Haller, D. *et al.* (2010) Safety assessment of probiotics for human use. *Gut Microbes*, **1**, 164–185.

Sandrou, D.K. and Arvanitoyannis, I.S. (2000) Application of hazard analysis critical control point (HACCP) system to the cheese-making industry A review. *Food Reviews International*, **16**, 327–368.

Sharma, P., Tomar, S.K., Goswami, P., Sangwan, V. and Singh, R. (2014) Antibiotic resistance among commercially available probiotics. *Food Research International*, **57**, 176–195.

Silva, I.M.M., Almeida, R.C.C., Alves, M.A.O. and Almeida, P.F. (2003) Occurrence of *Listeria* spp. in critical control points and the environment of Minas Frescal cheese processing. *International Journal of Food Microbiology*, **81**, 241–248.

Simkins, J., Kaltsas, A. and Currie, B.P. (2013) Investigation of inpatient probiotic use at an academic medical center. *International Journal of Infectious Diseases*, **17**, e321–e324.

SlowFood Press Release (2009) A Cheese la tutela per la biodiversità si fa micro. http://www.slowfood.com/press-release/the-joys-of-natural-microflora/ (last accessed 07/09/2016)

Snydman, D.R. (2008) The safety of probiotics. *Clinical Infectious Diseases*, **46**, 2S104–S111.

Spano, G., Russo, P., Lonvaud-Funel, A. *et al.* (2010) Biogenic amines in fermented foods. *European Journal of Clinical Nutrition*, **64**, 3S95–S100.

Stahl, B. and Barrangou, R. (2013) Complete genome sequence of probiotic strain *Lactobacillus acidophilus* La-14. *Genome Announcment*, **1**, e00376-13.

Stevens, H.C. and Nabors, L.O. (2009) Microbial food cultures: A regulatory update. *Journal of Food Technology*, **63**, 36–41.

Sundh, I., Wilcks, A. and Goettel, M.S. (eds) (2012) *Beneficial Microorganisms in Agriculture, Food and the Environment: Safety Assessment and Regulation*, Oxford: CAB International.

Talon, R. and Leroy, S. (2011) Diversity and safety hazards of bacteria involved in meat fermentations. *Meat Science*, **89**, 303–309.

Treven, P., Trmčić, A., Bogovič Matijašić, B. and Rogelj, I. (2014) Improved draft genome sequence of probiotic strain *Lactobacillus gasseri* K7. *Genome Announcements*, **2**, e00725-14.

Vahabnezhad, E., Mochon, A.B., Wozniak, L.J. and Ziring, D.A. (2013) *Lactobacillus* bacteremia associated with probiotic use in a pediatric patient with ulcerative colitis. *Journal of Clinical Gastroenterology*, **47**, 437–439.

van Reenen, C.A. and Dicks, L.M.T. (2011) Horizontal gene transfer amongst probiotic lactic acid bacteria and other intestinal microbiota: what are the possibilities? A review. *Archives of Microbiology*, **193**, 157–168.

Vasama, M., Kumar, H., Salminen, S. and Haskard, C.A. (2014) Removal of paralytic shellfish toxins by probiotic lactic acid bacteria. *Toxins*, **6**, 2127–2136.

Verraes, C., Van Boxstael, S., Van Meervenne, E., *et al.* (2013) Antimicrobial resistance in the food chain: A review. *International Journal of Environmental Research and Public Health*, **10**, 2643–2669.

Verslyppe, B., De Smet, W., De Baets, B., De Vos, P. and Dawyndt, P. (2014) StrainInfo introduces electronic passports for microorganisms. *Systematic and Applied Microbiology*, **37**, 42–50.

Vogel, R.F., Hammes, W.P., Habermeyer, M. *et al.* (2011) Microbial food cultures: Opinion of the Senate Commission on Food Safety (SKLM) of the German Research Foundation (DFG). *Molecular Nutrition and Food Research*, **55**, 654–662.

von Wright, A. (2012) "Microbes for human and animal consumption" In I. Sundh *et al.* (eds.), *Beneficial microorganisms in agriculture, food and environment: safety assessment and regulations* (pp. 27–40), London: CAB International.

Wessels, S., Axelsson, L., Bech Hansen, E. *et al.* (2004) The lactic acid bacteria, the food chain, and their regulation. *Trends in Food Science and Technology*, **15**, 498–505.

Wu, L., Sun, Q., Sugawara, H. *et al.* (2013) Global catalogue of microorganisms (gcm): A comprehensive database and information retrieval, analysis, and visualization system for microbial resources. *BMC Genomics*, **14**, 933.

Zell, C., Resch, M., Rosenstein, R., Albrecht, T., Hertel, C. and Götz, F. (2008) Characterization of toxin production of coagulase-negative staphylococci isolated from food and starter cultures. *International Journal of Food Microbiology*, **127**, 246–251.

Zhang, Y.H., Xu, D., Liu, J.Q. and Zhao, X.H. (2014) Enhanced degradation of five organophosphorus pesticides in skimmed milk by lactic acid bacteria and its potential relationship with phosphatase production. *Food Chemistry*, **164**, 173–178.

Zhang, Z.Y., Liu, C., Zhu, Y.Z. *et al.* (2012) Safety assessment of *Lactobacillus plantarum* JDM1 based on the complete genome. *International Journal of Food Microbiology*, **153**, 166–170.

CHAPTER 7

Management of waste from the food industry: A new focus on the concept of starter cultures

Daniela Campaniello, Salvatore Augello, Fabio de Stefano, Stefano Pignatiello and Maria Rosaria Corbo
Department of the Science of Agriculture, Food and Environment, University of Foggia, Italy

Progress, innovation and industrialization such as the intensification of agriculture are important processes characterizing the economy of a country; they have been achieved through the utilization of natural resources (air, water, forests, minerals etc.). On the other hand, in many cases these resources have been negligently exploited, favouring environmental pollution.

Commercial and industrial practices lead to the accumulation of large quantities of pollutants in the environment and this accumulation can potentially be hazardous for public health or the ecosystem. Soils, groundwater, sediments, surfaces, water and air are contaminated with chemicals that have attained alarming levels; moreover, food wastes contain compounds that could be valorized instead of being thrown out for landfill. Earth and human health are strictly related, thus it is necessary to limit or repair the environmental damage through corrective action.

For this purpose microbiologists suggest the employment of microorganisms as an alternative to the traditional methods used for remediation (i.e. incineration, pyrolysis, landfill etc.).

An effective tool for managing food industry wastes is 'bioremediation', which is defined as the process whereby organic wastes are biologically degraded under controlled conditions to an innocuous state or to a level below concentration limits established by the regulatory authorities (Sharma 2012).

Starter Cultures in Food Production, First Edition. Edited by Barbara Speranza, Antonio Bevilacqua, Maria Rosaria Corbo and Milena Sinigaglia.

This technology involves chemical transformations mediated by organisms (primarily microorganisms) that satisfy their nutritional and energy requirements and act on the environmental contaminants via two different pathways:

- Microorganisms enzymatically attack pollutants, which are transformed to metabolites in less toxic or innocuous forms; their ability to produce enzymes affects their ability to degrade substrates.
- Microorganisms metabolize chemical compounds to produce carbon dioxide or methane, water and biomass.

During the bioremediation process numerous factors must be controlled: microorganisms, energy sources (i.e. the carbon source) and electron acceptor availability, nutrients (i.e. nitrogen, phosphorus and potassium) as well as temperature, absence of toxic compounds and competitors. Food industry wastes are composed of substrates that provide the energy for microbial cell maintenance, favouring the growth of potential microorganisms that are pollutant degrading. For this purpose, some bioremediation processes require the addition of nutrients; carbon, nitrogen and phosphorous (C:N:P) have been suggested, in the ratio 120:10:1.

Bioremediation has numerous applications: clean-up of groundwater, soils, lagoons, sludges and process waste streams. Depending on the place and on the process applied, bioremediation can be realized *in situ* or *ex situ*:

- ***In situ***: bioremediation is applied in place, on the contaminated site. Generally this treatment is attractive because it requires low expenditure and basic equipment, generating a minimal impact on the environment.
- ***Ex situ***: bioremediation requires the excavation or removal of the contaminated soil, which can be manipulated in some way through the use of slurry reactors, composters, biopiles or other technologies.

The main reason for using bioremediation is that it does not require significant equipment or a great deal of work and energy. The main advantages and disadvantages of the bioremediation process are summarized in Table 7.1.

Bioremediation and the starter concept

Bioremediation recalls the concept of a starter culture (a preparation containing live and vital microorganisms with the goal of using their metabolism to start a fermentation process to achieve some specific technological objectives). The employment of useful microorganisms leads the fermentation process to the degradation of complex compounds to simple, less toxic molecules. Microorganisms destroy organic contaminants by using the chemicals for their own growth; in particular, they obtain the energy necessary to survive by the decomposition of organic compounds and inorganic nutrients. A recall on the kinetic microbial growth is reported in Appendix 7.2.

Microorganisms are able to survive in many environmental conditions (in terms of pH, temperature, the presence of oxygen or other electron acceptors and the degree of pollution). Therefore, by adapting their metabolism, they can degrade pollutants.

Table 7.1 Benefits and limits of the bioremediation process.

Benefits	Limits
Low costs	Some pollutants are not amenable to bioremediation
Reduces the risks for personal clean-up	Biological processes are highly specific: they need a high microbial concentration to degrade pollutants, high levels of nutrients and ideal environmental conditions
Complete destruction of pollutants to harmless products	Long treatment
Reduction of volume of organic wastes	Sometimes the products of biodegradation may be more persistent or toxic than the parent compounds
Generally accepted as it involves natural mechanisms through the use of microorganisms	Intensive monitoring activity
Bioremediation is often carried out *in situ*, thus waste transport is not necessary	Difficulty in predicting the potential formation of intermediates throughout the reactions
It can be coupled with other physical or chemical treatments	Difficulty in transferring data resulting from preliminary studies into full-scale bioremediation
It is environmentally compatible and the potential biological hazard of the wastes is controlled	
It kills pathogens	

Source: Bevilacqua *et al*., 2014.

Microbes can grow in a wide temperature range: microorganisms grow from subzero temperatures to extreme heat, with or without oxygen, in the desert or in the presence of excess water. They need an energy resource and a carbon source (Vidali 2001). If a carbon source is present, their population doubles about every 45 minutes.

Sometimes it is necessary to stimulate microbial growth through the addition of nutrients (i.e. carbon, nitrogen, phosphorous and oxygen); this helps the growth of indigenous microorganisms that are able to degrade pollutant compounds.

In a bioremediation process water has a double function: it is used as a vehicle to transport both microorganisms and dissolved substances, including pollutants and their degraded compounds (Pandey and Fulekar 2012). The application of these technologies requires a multidisciplinary approach in order to provide the necessary elements for system implementation on the basis of hydrogeology, microbiological profile and the biochemical mechanisms to apply in place on the site of interest.

Numerous microorganisms including aerobic and anaerobic bacteria and fungi are involved in a bioremediation process. Almost all bioremediation systems are managed under aerobic conditions: nevertheless, a system under anaerobic conditions could allow microbial organisms to degrade otherwise recalcitrant molecules. These microorganisms may be indigenous to the contaminated area or they can be isolated from elsewhere and brought into the contaminated area (Kumar *et al.* 2011).

Different species are responsible for biodegradation processes and can be summarized as follows:

- **Aerobic microorganisms** degrade pollutants in the presence of oxygen. *Pseudomonas, Bacillus, Ochrobacterium, Klebsiella, Alcaligenes, Sphingomonas, Rhodococcus, Acinetobacter* and *Mycobacterium* are the most common. They are able to degrade pesticides and hydrocarbons, both alkanes and polyaromatic compounds. The majority of these bacteria use contaminants as the unique source of carbon and energy (Vidali 2001; Arutchelvan *et al.* 2005).
- **Anaerobic microorganisms** degrade pollutants in the absence of oxygen. There is increasing interest in anaerobic bacteria used for the bioremediation of polychlorinated biphenyls (PCBs) in river sediments, dechlorination of the solvent trichloroethylene (TCE) and chloroform.
- **Ligninolytic fungi** are able to degrade a wide range of persistent or toxic environmental pollutants. *Rhodotorula, Fusarium, Aspergillus, Mucor, Penicillium, Candida* and *Sporobolomyces* are the most common hydrocarbons that fungi degrade (Milić *et al.* 2009).
- **Methylotrophs** are aerobic bacteria able to use methane as a carbon and energy source. Methane monooxygenase (the initial enzyme in the pathway for aerobic degradation) is active against a wide range of compounds (chlorinated aliphatics trichloroethylene and 1,2-dichloroethane).

In the last few years research has focused on the ability of microorganisms to degrade pollutants; volatile organic compounds, BTEX (benzene toluene ethylbenzene and xylene), phenolic compounds, polycyclic aromatic hydrocarbons (PAHs), pesticides, petroleum hydrocarbons and nitroaromatic compounds represent the most important pollutants contained in industrial waste that are potentially suitable for bioremediation. Furthermore, metals and minerals may also be transformed by microbial activity (fungi and bacteria; Gadd 2010). Microorganisms individually are not able to degrade the most hazardous compounds; thus a consortium of microorganisms is necessary to complete the mineralization process through sequential degradation involving synergistic and co-metabolistic activity (Pandey and Fulekar 2012).

Special attention is reserved for heavy metals, which are considered toxic compounds due to their persistence in the ecosystem. The removal of these compounds is quite slow and leads to a tendency to accumulate in the environment. Microorganisms able to bio-accumulate these compounds are not numerous and belong to the following domains:

- **Bacteria**: *Arthrobacter, Bacillus* sp., *Staphylococcus* spp., *Citrobacter, Cupriavidus metallidurans, Cyanobacteria, Enterobacter cloacae, Pseudomonas aeruginosa, Streptomyces* sp., *Zooglea ramigera, Ralstonia pickettii, Sphingomonas* sp.
- **Archea**: Filo *Crenarchaeota, Phanerochaete chrysosporium.*
- **Fungi**: *Aspergillus terreus, Aspergillus niger, Penicillium chrysogenum, Candida utilis, Hansenula anomala, Rhodotorula mucilaginosa, Rhodotorula rubra* GVa5, *Saccharomyces cerevisiae.*

Generally eukaryotes are more sensitive to metal than bacteria.

Finally, the use of genetic engineering has become a popular way to produce pollutant-degrading microorganisms.

Several approaches could be applied for the degradation or detoxification of waste in the food industry. In this chapter we focus on the employment of microorganisms in three kinds of industrial waste treatments (dairy waste, animal-origin wastewater and olive mill wastewater) to understand the mechanism used to degrade pollutants and the microorganisms involved.

Dairy waste management: Potential use for whey beverage production

The dairy industry is considered the largest source of food-processing wastewater in many countries.

Dairy wastewaters are represented by washing waters, composed of water, detergents and milk remaining in pipe machinery, and whey.

Whey is the major by-product of the dairy industry. It is produced in large quantities and presents a high organic load, which affects environmental pollution and makes the treatment cost prohibitive.

Cheese whey is a yellowish liquid resulting from the coagulation of milk characterized by a biological oxygen demand (BOD) of 30–50 g/L and a chemical oxygen demand (COD) of 60–80 g/L (Ghanadzadeh and Ghorbanpour 2012). Generally, wastewater with these characteristics cannot be used for land irrigation purposes and cannot be discharged into public sewers or inland waters; thus, proper treatment of dairy wastewater is necessary before disposal.

Cheese whey has received considerable attention thanks to its biological value. It is defined as the watery part of milk remaining after curd separation. It is the result of milk protein coagulation when acid or proteolytic enzymes are added. Rennet, an industrial coagulant containing chymosin or other coagulant enzymes, induces casein coagulation at a pH value of approximately 6.5; this type of whey is known as 'sweet whey'. It is obtained from rennet-coagulated cheese production (Cheddar, Mozzarella, Swiss and other hard cheeses). There is also another type of whey, named 'acid whey' (pH <5), obtained from the manufacture of acid-coagulated cheeses, using fermentation processes or by adding organic or mineral acids in order to coagulate caseins (Koushki *et al.* 2012).

Since whey proteins have a higher biological value (together with health benefits and therapeutic potential) than other proteins such as those of eggs, soy and even milk caseins (Smithers 2008), it is necessary to valorize these compounds. Whey proteins represent about 20% of the recovered protein fraction and they are concentrated through ultrafiltration and reused for various applications in the food industry. The permeate resulting from those processes is the major pollutant, since it retains lactose, which represents more than 70% of total whey solids and is largely responsible for the whey polluting load.

Whey is readily biodegradable and can be easily treated with conventional techniques like trickling filter or waste stabilization ponds. On the other hand, physical and chemical methods such as coagulation/flocculation, nanofiltration, reverse osmosis membranes (www.lenntech.com) and a coagulation membrane bioreactor (Chen and Liu 2012) have been proved to be successful (The main reactors generally used are briefly discussed in Appendix 7.1). These treatments could lead to safer whey disposal, but they are expensive, thus a big proportion of global whey production is discarded as effluent, causing serious pollution problems since it affects the physical and chemical soil structure, resulting in crop yield decreases. In addition, when it is released into high seas and inland waters, it can reduce aquatic life, depleting water-dissolved oxygen.

Fortunately, the curiosity of researchers has helped to acquire knowledge and a greater understanding of the chemical, biological and nutritional characteristics of this by-product, thus whey could be subjected to specific treatments through which new products can be manufactured. These procedures could eliminate, wholly or partially, the disposal costs. Whey retains many of the solids present in the original whole milk, including functional proteins and peptides, lipids, lactose, water-soluble vitamins and minerals, therefore several applications in the food and pharmaceutical industries could be proposed. In addition, it could be used for biogas production. A different approach to this by-product has been taken thanks to continuous research efforts together with legislation changes concerning effluent disposal. The result is that if in the past whey was considered only as a pollutant, nowadays it is a new resource. For example, it can be considered as a good substrate for microorganism fermentation. Thanks to the microbial activity, this by-product can be remediated and addressed to several uses.

Whey-based fermented beverages

In recent years, many researchers have analysed a large number of potential utilizations of either whole whey or its major components (lactose and proteins). Fermented whey beverage production, using appropriate species of lactic acid bacteria and yeasts, seems to be a good alternative as it brings several advantages. It could help:

- to increase the value of this industrial waste, obtaining a marketable product;
- to reduce environmental pollution;
- to reduce the high costs required for disposal treatments.

Thanks to its high lactose content, whey is a very good material for alcoholic beverage production. Several fermented whey beverages with varying alcohol contents have been successfully produced. Russian researchers have designed a process for producing sparkling whey wine from acid whey. The deproteinized whey is pasteurized, cooled, inoculated with *Lactobacillus acidophilus* or *Lactococcus lactis* and fermented for 2 hours at 41 °C. Then, yeast culture (generally *Kluyveromyces fragilis* and *Streptococcus lactis*) is added along with 8–10% sugar, and the whey is put into champagne-type bottles, sealed and held at 7–10 °C for

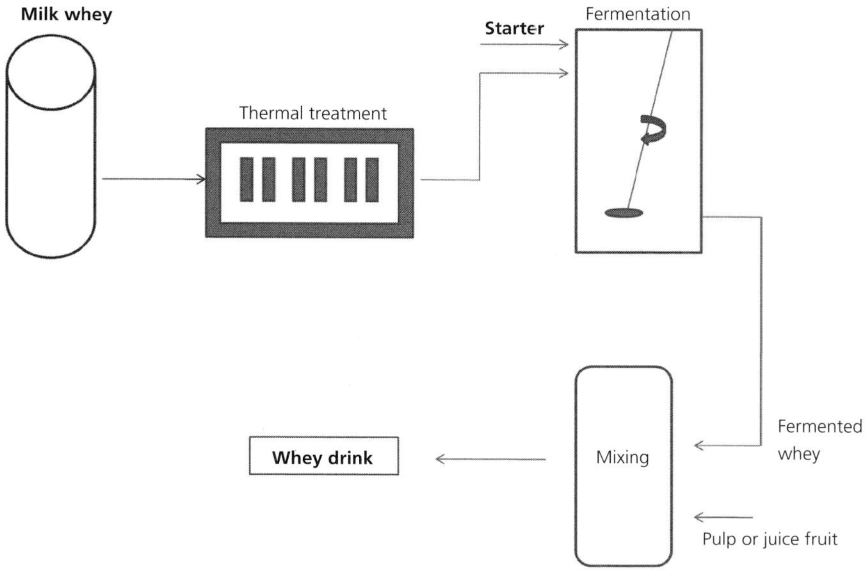

Figure 7.1 Flow chart for fermented whey beverage production.

3–4 days to produce alcohol and carbon dioxide (CO_2). A certain amount of lactose is transformed into lactic acid, which gives a refreshing sour taste to the end product, while the rest of the sugars are fermented to alcohol.

Recently, the emerging field in whey utilization seems to be the production of probiotic drinks. In addition to their role in fermentation processes, some probiotic lactic acid bacteria have been studied as dietary sources of live microorganisms destined to promote a positive impact in the host, improving the properties of the intestinal beneficial microbiota.

A schematic diagram for the production of a fermented whey drink using starter cultures is showed in Figure 7.1.

Naidoo Terroso de Mendonça Brandão *et al.* (2014) studied how to obtain a fermented drink with different quantities of *Lb. acidophilus* and prebiotic fibre in the form of inulin and using total dry whey extract and sucrose. They obtained six samples composed of fruit salad pulp and different percentages of inocula and inulin and evaluated the sensory attributes (colour, flavour, aroma and consistency). Generally, supplementation with fruit pulp or juice covers whey's bitter taste with a potential positive effect on the overall sensory quality of the beverage. These researchers observed that five out of six treatments showed a good acceptability of colour, flavour, aroma and consistency and that although the initial inoculum level decreased, it was above the minimum count of 7 log cfu (colony forming units) per mL recommended by Brazilian legislation.

There are many combinations of microorganisms that can produce beverages with satisfactory sensory characteristics; by combining different strains it is possible to obtain a significant improvement in production performance due to symbiotic interaction between microorganisms. They can also provide the necessary beverage production criteria such as low production cost, functionality and storage stability.

As an example, Bulatović *et al.* (2014) studied the performances of whey-based beverages produced with different strains of *Lactobacillus* bacteria. Individual or mixed cultures containing *Lactobacillus helveticus* ATCC 15009, *Lactobacillus delbrueckii* ssp. *lactis* NRRL B-4525 and *Streptococcus thermophilus* S3 were tested. Beverages obtained by mixed cultures had higher aroma values than those obtained by individual strains. Thus, Bulatović *et al.* (2014) suggested that the improvement in the sensorial characteristics of the products was obtained by consecutive action of the metabolic activity of the strains added: *Str. thermophilus* produced exopolysaccharides and diacetyl; then exopolysaccharides induced *Lb. helveticus* and *Lb. delbrueckii* ssp. *lactis* to produce diacetyl, a major contributor of flavour and aroma. As a consequence, a negligible prolongation of fermentation time was observed, which could be avoided by temperature increases.

These beverages containing proteins with a high nutritional value are an ideal source of energy and nutrients for a large range of consumers (children, the elderly and athletes).

Whey beverages can be also considered as a functional food. Thanks to the presence of lactoferrin (an iron-binding protein), they can be used to improve iron absorption. Furthermore, they improve calcium absorption, thus they are used as a dietary supplement for people with osteoporosis.

Nowadays the consumption of nutraceutical foods is highly encouraged, therefore the production of these beverages has been incremented and it is considered the best proposition to convert this by-product into a value-added product through simple and economical processes.

In conclusion, the main benefits of whey-based fermented drinks are as follows:
- They can be used as dietary supplements since they replace organic and inorganic salts.
- They are suitable drinks for lactose-intolerant people.
- They contribute to reducing environmental pollution.
- They represent added value for the dairy industry, since their reuse allows new marketable products to be obtained, reducing disposal costs.

Animal-origin wastewaters

Animal wastewaters deriving from breeding farms, slaughterhouses and packing houses represent one of the most serious causes of environmental pollution. They are characterized by bad odours and represent a health hazard in almost all developing countries (Karlsson and Ejlertsson 2012).

Breeding farms mainly produce fresh meat – ovine, bovine and poultry – while in the case of dairy cattle breeding they also produce milk.

Slaughterhouses submit fresh meat (whole, half or quarter carcasses or smaller meat cuts) to the following operations: killing, paunch and viscera handling, blood and hair processing.

Packing houses can both slaughter and process fresh meat purchased from outside into smoked, canned and other meat products.

Wastes resulting from the processing of all these products mainly consist of sewage and could contain skin, blood, rumen contents, bones, horns, hoofs, uterus, intestine, ears, meat trimmings, condemned meat, condemned carcasses, hair and poultry offal as well as feathers and heads.

Slaughtering is an activity that requires a large amount of hot water and steam for sterilization and cleaning purposes. During meat processing, various gases are emitted such as CO_2, carbon monoxide (CO), nitrogen oxide (NO_x) and sulfur dioxide (SO_2); in addition, emissions of free ammonia (NH_3) into the air are the result of the evaporation of chilling liquids and freezing machines. Meat processing, meat product smoking and the hair removal process lead to the production of mainly CO_2, CO and NO_x as well as bad odours (EC, No 1069/2009); thus the degree of air pollution is affected by the different levels of emissions generated by these processes.

Treatment of slaughterhouse wastes and the microorganisms involved

Slaughterhouse wastes contain high amounts of protein, other nutrients and fat, therefore they are considered a good substrate for microbial growth; in fact, this type of waste contains several hundred different species of autochthonous microorganisms, including potential pathogens such as *Salmonella* spp., *Staphylococcus* spp. and *Clostridium* spp. (Mead 2004). Only a few animal by-products can be used directly; most of them should be treated to become potential fertilizer (manure) or animal feed.

Wastes can be treated in many different ways. For example, manure could be composted, reducing the pathogen content of the original matrix; this treatment may also reduce the nitrogen content of the compost obtained. Other forms of treatment include incineration and rendering (thermal treatments). Most commonly, animal wastes are treated at rendering plants, which separate meat from fat; the latter could be utilized for manufacturing useful products like soap base. The rest of the scraps have a high potential to produce methane from organic waste and they are usually treated with a digestion method (Rodriguez-Abalde *et al.* 2011).

Anaerobic digestion is a complex biological process, commonly used in Europe as an alternative to other disposal methods because of the strict regulation of landfilling, and is considered the most appropriate way to remove the risks of mad cow disease and other diseases. The process, carried out biologically, destroys a significant portion of the hazardous volatiles and solids present in sludge and reduces harmful and unpleasant odours. Prior to the anaerobic digestion process, slaughterhouse wastes generally pass through screens, settling tanks and filters in order to remove large suspended solids.

Anaerobic sludge digestion is a process consisting of a series of bacterial events that convert organic compounds into methane (CH_4), CO_2 and new bacterial cells. The entire digestion process involves different groups of bacteria and occurs in three different steps: hydrolysis of solids, acidogenesis and methanogenesis (Salminen and Rintala, 2002). These groups work in sequence, and the products of one group become substrates of the next one.

1 **Hydrolysis of solids**: Carbohydrates, proteins and lipids are hydrolysed into sugar, water, amino acids and long-chain fatty acids (Figure 7.2). These compounds are absorbed by bacteria in order to be degraded inside the cells. Hydrolytic bacteria (such as *Clostridium* spp., or *Haliscomenobacter hydrossis* and *Sphaerotilus natans*) and anaerobes (Table 7.2) are the main ones responsible for these reactions.
2 **Acidogenesis**: Acidogenic bacteria ferment amino acids into ammonia, volatile fatty acids (VFAs) and other products including hydrogen, acetic acid and CO_2 (Figure 7.2). The degradation of these compounds results in the production of CO_2, hydrogen gas, alcohols, organic acids, some organic-nitrogen compounds and some other organic-sulfur compounds. Acetate is the most important acid produced because it is used as a substrate by methane-forming bacteria in the third step; it is produced thanks to the activity of acetogenic or acetate-forming bacteria (*Acetobacterium* spp., *Clostridium* spp., *Sporomusa* spp.). The presence of organic-nitrogen compounds and organic sulfur compounds is due to the degradation of amino acids and proteins.

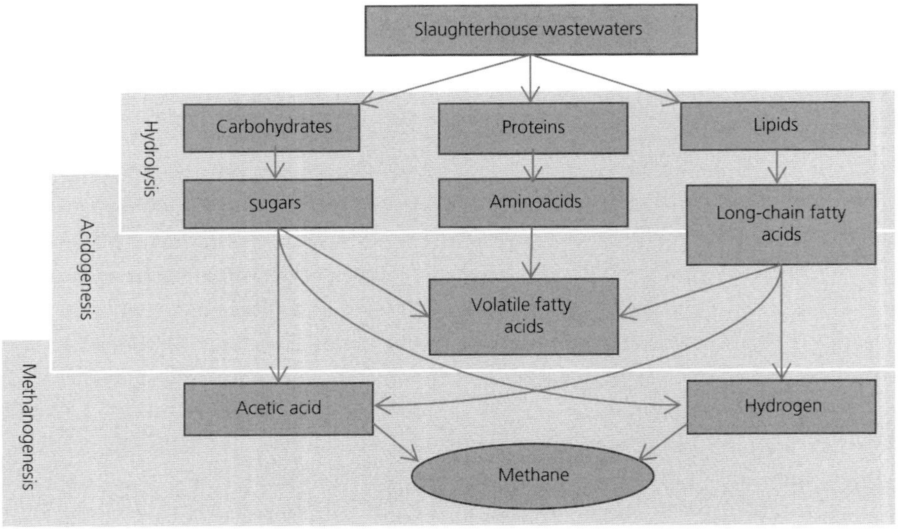

Figure 7.2 Anaerobic digestion bacteria mediated reactions.

Table 7.2 Bacterial species present in anaerobic digesters according to their response to oxygen.

Types of bacteria	Species
Aerobes	*Hb. hydrossis*
	Nitrobacter spp.
	Nitrosomonas spp.
	Escherichia coli
	Bacillus spp.
	Z. ramigera
Anaerobes	*Sph. natans*
	Desulfovibrio spp.
	M. formicium
	Clostridium spp.

Notes: Hb. = Haliscomenobacter; M. = Methanobacterium; Sph. = Sphaerotilus; Z. = Zooglea.

3 **Methanogenesis**: Methanogenic bacteria (*Methanobacterium formicium, Methanobrevibacter arboriphilus*) are involved in the process to complete the conversion of the final products to CH_4 and CO_2 (biogas).

When the first step is inhibited, the substrates for the second and third stages will be limited, thus methane production also decreases.

As reported in Figure 7.3, primary and secondary clarifiers purify primary and secondary sludges, respectively. The primary sludge provides some facultative anaerobes and many anaerobes including methane-forming bacteria and organic particulates, while the secondary sludge is rich in facultative anaerobes. As methane-forming bacteria are strict anaerobes and quickly die in activated sludge (containing oxygen), an anaerobic digester cannot be successfully seeded with secondary sludge alone. Generally, to seed an anaerobic digester with an adequate population of facultative anaerobes and anaerobes including methane-forming bacteria, a ratio of 1:10 secondary:primary sludge may be used.

Sludges are frequently used as inoculum for wastewater treatments; moreover, naturally selected strains are also employed in different forms such as flocs, granules or biofilms. Many studies on the microbial characterization of anaerobic digesters have reported that heterotrophic microorganisms like *Actinomyces, Thermomonospora, Ralstonia* and *Shewanella*, together with *Clostridium* species, are present for the longest time in the first phase of digestion; other microorganisms as *Methanosarcina, Methanobrevibacter* and *Methanobacterium* contribute to methane production (Ike *et al.* 2010). However, it is unusual for a single species to be dominant; a microbial consortium is always responsible for the entire process of digestion.

Chartrain *et al.* (1987) tried to select the microorganisms with the best performance for anaerobic wastewater treatment. In 1989 Chartrain and Zeikus patented EP 0302968 A1, a defined starter culture made with a mixture of live

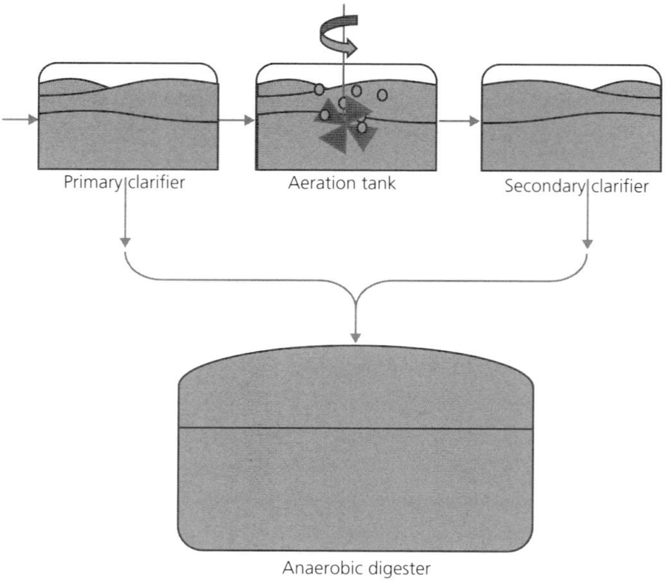

Figure 7.3 Anaerobic digester diagram.

bacterial cells of *Leuconostoc mesenteroides*, *Desulfovibrio vulgaris*, *M. formicicum* and *Methanosarcina barkeri* associated with *Clostridium butyricum*, *Klebisiella oxytoca*, *Clostridium propionicum* and *Methanothrix soehngenii*.

Physical parameters of a digester

A digestion process is generally conducted within three temperature ranges: 0–20 °C (psycrophilic treatment), 20–42 °C (mesophilic treatment) and 42–75 °C (thermophilic treatment; Medina-Herrera *et al.* 2014). The mesophilic conditions are the most commonly used in order to have autochthonous microorganism selection. Acidogenic and methanogenic bacteria present in the second and third steps, respectively, are sensitive to temperature changes. High temperatures could cause the accumulation of a dangerous gas, free ammonia (NH_3), responsible for a possible arrest of the digestion process. In general, lower temperatures result in slower sludge digestion.

The performance of the reactor is also affected by pH. Methanogenic bacteria growth at pH values in a range of 6.8–7.2; in contrast, acid-forming bacteria grow at an acid pH. Sodium bicarbonate is used as a buffer for the maintenance of these pH values, without affecting the microbial population.

Post-treated waste utilization

The main by-products of anaerobic digestion are biogas, acidogenic digestate (or fibre) and methanogenic digestate (or liquor).

The first is composed of CH_4, CO_2 and sulfuric acid (H_2SO_4, in trace amounts). Biogas is normally used in combustion engines or microturbines for the production of electricity, and it can also be used for digester heating. As the gases generated by the digestion process are not released into the atmosphere, it is considered green energy. Moreover, it is also recycled for heat generation.

Acidogenic digestate is rich in organic and mineral compounds together with dead microbial cells. It can be used as compost for domestic and agricultural use.

Methanogenic digestate is called liquor and is abundant in nutrients. If the digest is poor in heavy metals and other toxic substances, concentrated by digestion in the liquid phase, methanogenic digestate can represent an excellent fertilizer.

To evaluate the quality of a digestate, chemical, biological and physical characteristics are considered. Chemical quality is considered in terms of heavy metals and other inorganic contaminants, persistent organic compounds and the content of macro elements (nitrogen, phosphorous and potassium). Biological quality is considered in terms of the total absence of pathogens, which can lead to human, animal or plant diseases if not appropriately managed.

Finally, physical standards include appearance and odour factors as well as the absence of physical impurities such as very large particle sizes, animal identification tags or rubber, glass and wood. Only digestates with high standards of quality should be used.

Bioremediation of olive mill wastewaters

Olive mill wastewaters (OMW) are aqueous residues deriving from olive oil extraction. Table 7.3 reports the main physico-chemical characteristics of OMW.

OMW are a dark, fetid liquid; they represent a longstanding problem because they are produced in large quantities every year in Mediterranean countries and because the disposal of these wastes is very difficult. In particular, ecological problems arise from the presence of phenolic compounds, which include many organic substances characterized by an aromatic ring with one or more hydroxyl groups and a functional side chain (Ramos-Cormenzana *et al.* 1996).

Nowadays the use of green techniques in order to degrade phenolic compounds guides many researchers towards alternative methods to the traditional ones, in an attempt to respect our planet's ecology and compatibility with other environmental systems (Ilyin *et al.* 2004). The use of microorganisms has been proposed to address this question through innovative biotechnologies.

It is important to identify the microorganisms to use as starter; they must be able to adapt themselves to the environmental conditions and to conduct an optimal mineralization process of the phenolic compounds in OMW (Mekki *et al.* 2009). In addition, the absence of pathogenic activity represents an important characteristic to avoid pathogen microbial diffusion.

Table 7.3 Physico-chemical characteristics of untreated olive mill wastewaters.

pH (298 K)	5.10–7.05	Mekki *et al.* (2009); Casacchia *et al.* (2012)
EC (298 K; S/m)	8.54–8.90	Mekki *et al.* (2009); Casacchia *et al.* (2012)
COD (g/L)	72.00–100.00	Mekki *et al.* (2009); Dhouib *et al.* (2006)
BOD (g/L)	13.00–29.40	Dhouib *et al.* (2006); Mekki *et al.* (2009)
BOD/COD	0.24–0.29	
TOC (g/L)	13.60–25.52	Di Serio *et al.* (2008); Casacchia *et al.* (2012)
TN (g/L)	0.60–1.35	Dhouib *et al.* (2006); Mekki *et al.* (2009)
Carbon/nitrogen	18.90–33.00	
Salinity (g/L)	5.04–6.75	Dhouib *et al.* (2006); Mekki *et al.* (2009)
Phenols (g/L)	7.18–9.02	Dhouib *et al.* (2006); Mekki *et al.* (2009)
Total solids (g/L)	52.00–97.40	Dhouib *et al.* (2006); Mekki *et al.* (2009)
Volatile solids (g/L)	5.55–44.00	Dhouib *et al.* (2006); Mekki *et al.* (2009)
Phosphorous (mg/L)	36.00–720.00	Dhouib *et al.* (2006); Mekki *et al.* (2009)
Sodium (mg/L)	118.10–940.00	Mekki *et al.* (2009); Casacchia *et al.* (2012)
Calcium (mg/L)	62.50–1200.00	Mekki *et al.* (2009); Casacchia *et al.* (2012)
Magnesium (mg/L)	14.60–187.00	Mekki *et al.* (2009); Casacchia *et al.* (2012)
Toxicity by LUMIStox (% inhibition)	99.00–100.00	Dhouib *et al.* (2006); Mekki *et al.* (2009)

Notes: BOD = biological oxygen demand; COD = chemical oxygen demand; EC = electrical conductivity; TOC = total organic carbon; TN = total nitrogen.

Bevilacqua *et al.* (2013) reported that specific bacteria are involved in OMW phenolic compound degradation; that is, *Bacillus pumilus, Pseudomonas putida, Arthrobacter, Azotobacter vinelandii, Azotobacter chroococcum* and *Ralstonia* spp. In addition, several fungi, including white-rot fungi (WRF), showed the ability to degrade OMW phenols. These fungi include *Ph. chrysosporium, Trametes versicolor, Pleurotus* spp., *Funalia trogii* and *Lentinus edodes, A. niger* and *A. terreus*. Also yeasts can be used for waste treatment; *Candida, Geotrichum, Pichia, Saccharomyces, Trichosporon* and *Yarrowia* have been proposed and tested successfully in detoxifying OMW.

Use of microorganisms for OMW detoxification

OMW biodegradation is very hard due to the high polyphenol and organic content, thus the use of microorganisms (as an alternative technique) and their metabolic capacity for phenolic degradation were the subject of numerous studies. Bacteria have two different metabolic pathways involved in phenolic degradation: meta and ortho. Through the meta pathway phenol is transformed into acetaldehyde and pyruvate, while the ortho pathway transforms phenols into trichloroacetic acid. This latter way is the most productive pathway for organisms as it involves less energy expenditure (Indu Nair *et al.* 2008). In particular, phenolic compounds are used as a source of energy and carbon, thus microbial strains open the aromatic ring and through the action of different enzymes produce acetaldehyde and pyruvate or trichloroacetic acid as the final compound. These enzymes include oxygenases, hydroxylases, peroxidases, tyrosinases and oxidases.

Some extracellular enzymes are also involved in phenol degradation; these enzymes are produced by some strains (*Streptomyces psammoticus*) and act to remove pollutants in OMW just beyond the adsorption of phenols (Indu Nair *et al.* 2008).

El Asli *et al.* (2005) reported that the toxicity of phenols could be reduced when microorganisms are grown in immobilized forms. They isolated some acetic acid– and formic acid–degrading bacteria and characterized *Kl. oxytoca* as the main phenol-degrading strain; in particular, its ability to reduce the toxicity of phenol was higher when it is grown in immobilized form. The authors supposed that this behaviour was probably due to physiological changes in the cells.

WRF are also able to degrade pollutants; for the degradation of polyphenolic compounds they are more effective than bacteria (Bevilacqua *et al.* 2013). Martinez (2002) proposed *Ph. chrysosporium* as an active strain in OMW treatment. Other researchers reported that *Basidiomycetes*, mainly of the WRF group, showed an ability to mineralize lignin (Trovaslet *et al.* 2007) and phenolic compounds (Pointing 2001).

In recent years Alaoui *et al.* (2008) studied the ability of WRF to detoxify OMW. They proposed different cultivation modes for four WRF strains (*Ph. chrysosporium, T. versicolor, Coriolopsis polyzona* and *Pycnoporus coccineus*) able to mineralize lignin; the targets were cultivated in the form of free mycelium, mycelium immobilized in alginate beads and solid state cultivation on Petri dishes. WRF showed different efficiency for OMW treatment on the basis of the cultivation mode: *Co. polyzona* and *Py. coccineus* were slightly affected by the cell cultivation procedure; in contrast, *Ph. chrysosporium* and *T. versicolor* preferred, respectively, solid state and alginate immobilization procedures.

In addition, other authors showed that the use of WRF requires pretreatment (physico-chemical and aerobic) before their application on soil to limit the negative impact on the biological activities of soil (Paraskeva and Diamadopoulus 2006). According to these results, Dias *et al.* (2004) and Ahmadi *et al.* (2006) reported that, generally, heat treatment is necessary to facilitate the growth of introduced fungi.

Chtourou *et al.* (2004) studied the use of yeasts to reduce or eliminate the pollutant substances in OMW. *Trichosporon cutaneum* was able to remove low molecular weight phenolic compound. The results of this biotransformation were a decrease in phenolic content and hence a reduction in the phytotoxic effects of the effluent after yeast treatment. *Tr. cutaneum* used only phenolic substances as a unique source of carbon and energy such as 3,4-di-OH-benzoic acid, *p*-OH-phenol acetic acid, caffeic acid and other aromatic compounds.

Bevilacqua *et al.* (2013) proposed the use of indigenous yeast strains to detoxify OMW. For this purpose the strains were isolated from OMW, characterized (through a technological and functional protocol) and identified. Successively, only the strains with the best performance in relation to their ability to remove phenolic compounds and to reduce COD/BOD values were studied. These authors concluded that the isolated yeasts could be used in the biotechnological treatment of OMW because many strains were able to grow in this stressful environment. If compared to moulds, yeasts showed a lower yield in phenol removal; nevertheless, as biomass could absorb phenols and release them throughout prolonged storage, the use of yeast strains is advisable.

Appendix 7.1 Typology of digesters

Different types of bioreactors are available, but they can be summarized in three types:

- **Fixed film**: This consists of a stainless steel tank with an inner biofilm support structure made of hard rock particles, activated carbon, polyvinyl chloride (PVC) or ceramic rings used for bacteria immobilization. The support structure is fully submerged and the wastewater flow can pass through the support structure in direct contact with microorganisms (Weeks 2003).
- **Up-flow anaerobic sludge blanket (UASB)**: This consists of a gas/solids separator to retain the sludge in the reactor and a sludge distribution system. In this digester the microorganisms stick themselves to other bacteria or to small sludge particles in order to form agglomerates named 'sludge granules'. These granules form the active sludge blanket at the bottom of the tank (Ren *et al.* 2008).
- **Fluidized bed**: Here the bacteria are attached to very small sand particles or activated carbon ones, and they are kept distant from each other by drag forces. In this case the contact between bacteria and substrate is at maximum (Xing *et al.* 2010).

Appendix 7.2 Kinetic microbial growth

Microbial growth is the key to the bioremediation process, as microorganisms are able to decrease the concentration of polluting organic compounds. Thus, knowledge of microbial kinetics is fundamental.

Numerous models have been used to describe how microorganisms grow, such as the Monod and Haldane-Andrews equations:

$$Monod\left(1949\right): r_S = \frac{r_{Smax} S}{K_s + S}$$

$$Haldane\left(\text{Andrews } 1968\right): r_S = \frac{r_{Smax} S}{K_s + S + \frac{S^2}{K_i}}$$

where K_s, K_i, r_S, r_{Smax} and S are the half-saturation constant (mg/L), inhibition constant (mg/L), specific phenol (substrate) consumption rate (mg/mg/h), maximum specific phenol (substrate) and substrate concentration (mg/L), respectively.

The Monod equation is applied when any inhibition due to substrate is observed. In contrast, when substrate inhibits microbial growth the Andrews equation is used.

The specific phenol (substrate) consumption rate is an important parameter to evaluate the specific growth rate and, consequently, the biomass generation.

These equations allow the researcher to measure microbial growth in the presence of a constant concentration of phenolic compounds and, successively, to compare these data to those obtained in the presence of different phenolic substances.

References

Ahmadi, M., Vahabzadeh, F., Bonakdarpour, M. and Mofarrah, E. (2006) Phenolic removal in olive oil mill wastewater using loofah-immobilized *Phanerochaete chrysosporium*. *World Journal of Microbiology and Biotechnology*, **22**(2), 119–127.

Alaoui S.M., Merzouki, M., Penninckx, M.J. and Benlemlih, M. (2008) Relationship between cultivation mode of white rot fungi and their efficiency for olive oil mill wastewaters treatment. *Electronic Journal of Biotechnology*, **11**(4). doi:10.2225/vol11-issue4-fulltext-15.

Andrews, J. F. (1968) A mathematical model for the continuous culture of microorganisms utilizing inhibitory substance. *Biotechnology and Bioengineering*, **10**(6), 707–723.

Arutchelvan, V., Kanakasabai, V., Nagarajan, S. and Muralikrishnan, V. (2005) Isolation and identification of novel high strength phenol degrading bacterial strains from phenol-formaldehyde resin manufacturing industrial wastewater. *Journal of Hazardous Materials*, **127**, 238–243.

Bevilacqua, A., Petruzzi, L., Corbo, M.R. and Sinigaglia, M. (2013) Bioremediation of olive mill wastewater by yeasts: A review of the criteria for the selection of promising strains. In Y.B. Patil and P. Rao (eds), *Applied Bioremediation: Active and Passive Approaches*, Rijeke: InTech. doi:10.5772/56916

Bulatović, M.L.J., Rakin, M.B., Mojović, L.V., Nikolić, S.B., Vukašinović-Sekulić, M.S. and Đukić-Vuković, A.P. (2014) Improvement of production performance of functional fermented whey-based beverage. *Chemical Industry and Chemical Engineering Quarterly*, **20**(1), 1–8.

Casacchia, T., Sofo, A., Zelasco, S., Perri, E. and Toscano, P. (2012) In situ olive mill residual co-composting for soil organic fertility restoration and by-product sustainable reuse. *Italian Journal of Agronomy*, **7**, 167–170.

Chartrain, M., Bhatnagar, L. and Zeikus, J.G. (1987) Microbial ecophysiology of whey biomethanation: Comparison of carbon transformation parameters, species composition, and starter culture performance in continuous culture. *Applied and Environmental Microbiology*, **53**, 1147–1156.

Chartrain, M. and Zeikus, J.G. (1989) Starter cultures and method for biomethanation. EP 0 302 968 A.

Chen, W. and Liu, J. (2012) The possibility and applicability of coagulation-MBR hybrid system in reclamation of dairy wastewater. *Desalination*, **285**, 226–231.

Chtourou, M., Ammar, E., Nasri, M. and Medhioub, K. (2004) Isolation of a yeast, *Trichosporon cutaneum*, able to use low molecular weight phenolic compounds: Application to olive mill waste water treatment. *Journal of Chemical Technology and Biotechnology*, **79**, 869–878.

Dhouib, A., Ellouz, M., Aloui, F., Sayadi, S. (2006) Effect of bioaugmentation of activated sludge with white-rot fungi on olive mill wastewater detoxification. *Letters in Applied Microbiology*, **42**, 405–411.

Dias, A., Bezerra, R.M. and Periera, A.N. (2004) Activity and elution profile of laccase during biological decolorization and dephenolization of olive mill wastewater. *Bioresource Technology*, **92**(1), 7–13.

Di Serio, M.G., Lanza, B., Mucciarella, M.R., Russi, F., Iannucci, E., Marfisi, P. and Madeo, A. (2008) Effects of olive mill wastewater spreading on the physic-chemical and microbiological characteristics of soil. *International Biodeterioration & Biodegradation*, **62**, 403–407.

El Asli, A., Errachidi, F., Bennisse, R., Qatibi, A.-I. and Errami, M. (2005) Effect of cell immobilization on the treatment of olive mill wastewater by a total phenols, acetic acid and formic acid degrading bacterium strain. *Grasas y Aceites*, **56**(2), 116–120.

European Community (2002) Regulation (EC) No 1069/2009 of the European Parliament and of the Council of 21 October 2009 laying down health rules as regards animal by-products and derived products not intended for human consumption and repealing Regulation (EC) No 1774/2002 (Animal by-products Regulation). *Official Journal*, No. L 300.

Gadd, G.M. (2010) Metals, minerals and microbes: Geomicrobiology and bioremediation. *Microbiology*, **156**, 609–643.

Ghanadzadeh, H. and Ghorbanpour, M. (2012) Optimization of ethanol production from cheese whey fermentation in a batch-airlift bioreactor. *Journal of Bioengineering & Biomedical Sciences*, **2**(111). doi:10.4172/2155-9538.1000111.

Ike, M., Inove, D., Miyano, T. *et al.* (2010) Microbial population dynamics during start up of a full-scale anaerobic digester treating industrial food waste in Kyoto Ecoenergy Project. *Bioresource Technology*, **10**, 3952–3957.

Ilyin, V.K., Smirnov, I.A., Soldatov, P.E. *et al.* (2004) Microbial utilisation of natural organic wastes. *Acta Astronautica*, **54**, 357–361.

Indu Nair, C., Jayachandran, K. and Shashidhar, S. (2008) Biodegradation of phenol. *African Journal of Biotechnology*, **7**(25), 4951–4958.

Karlsson, A. and Ejlertsson, J. (2012) Addition of HCl as a means to improve biogas production from protein-rich food industry waste. *Biochemical Engineering Journal*, **61**, 43–48.

Koushki, M., Jafari, M. and Azizi, M. (2012) Comparison of ethanol production from cheese whey permeate by two yeast strains. *Journal of Food Science and Technology*, **49**(5), 614–619.

Kumar, A., Bisht, B.S., Joshi, V.D. and Dhewa, T. (2011) Review on bioremediation of polluted environment: A management tool. *International Journal of Environmental Sciences*, **1**(6), 1079–1093.

Martínez, A.T. (2002) Review: Molecular biology and structure-function of lignin-degrading heme peroxidases. *Enzyme and Microbial Technology*, **30**(4), 425–444.

Mead, G.C. (2004) Microbiological quality of poultry meat: A review. *Brazilian Journal of Poultry Science*, **6**(3), 135–142.

Medina-Herrera, M., Rodríguez-García, A., Montoya-Herrera, L. *et al.* (2014) Anaerobic digestion of slaughterhouse solid waste for the optimization of biogas production. *International Journal of Environmental Research*, **8**, 483–492.

Mekki, A., Dhouib, A. and Sayadi, S. (2009) Evolution of several soil properties following amendment with olive mill wastewater. *Progress in Natural Science*, **19**, 1515–1521.

Milić, J.S., Beškoski, V.P., Ilić, M.V., Ali, S.A.M., Gojgić-Cvijović, G.D. and Vrvić, M.M. (2009) Bioremediation of soil heavily contaminated with crude oil and its products: Composition of the microbial consortium. *Journal of Serbian Chemical Society*, **74**(4), 455–460.

Monod, J. (1949) The growth of bacterial cultures. *Annual Review of Microbiology*, **3**, 371–394.

Naidoo Terroso de Mendonça Brandão, W.A.P.L., Naidoo Terroso de Mendonça Brandão, S., Naidoo Terroso de Mendonça Brandão, H.C.A.D., *et al.* (2014) Acceptability of prebiotic fiber-treated whey drink fermented with *Lactobacillus acidophilus*. *African Journal of Biotechnology*, **13**(25), 2565–2574.

Pandey, B. and Fulekar, M.H. (2012) Bioremediation technology: A new horizon for environmental clean-up. *Biology and Medicine*, **4**(1), 51–59.

Paraskeva, P. and Diamadopoulos, E. (2006) Technologies for olive mill wastewater (OMW) treatment: A review. *Journal of Chemical Technology and Biotechnology*, **81**, 1475–1485.

Pointing, S.B. (2001) Feasibility of bioremediation by whiterot fungi: Mini-review. *Applied Microbiology and Biotechnology*, **57**(7), 20–33.

Ramos-Cormenzana, A., Juárez-Jiménez, B. and Garcia-Pareja, M.P. (1996) Antimicrobial activity of olive mill waters (alpechin) and biotransformed mill wastewater waste-olive oil. *International Biodeterioration & Biodegradation*, **1996**, 283–290.

Ren, T.T., Mu, Y., Yu, H.Q., Harada, H. and Li, Y.Y. (2008) Dispersion analysis of an acidogenic UASB reactor. *Chemical Engineering Journal*, **142**, 182–189.

Rodriguez-Abalde, A., Fernandez, B., Silvestre, G. and Flotats, X. (2011) Effects of thermal pre-treatments on solid slaughterhouse waste methane potential. *Waste Management*, **31**(7), 1488–1493.

Salminen, E. and Rintala, J. (2002) Anaerobic digestion of organic solid poultry slaughterhouse waste: A review. *Bioresource Technology*, **83**, 13–26.

Sharma, S. (2012) Bioremediation: Features, strategies and applications. *Asian Journal of Pharmacy and Life Science*, **2**, 202–213.

Smithers, G.W. (2008). Whey and whey proteins: From 'gutter-to-gold'. *International Dairy Journal*, **18**, 695–704.

Trovaslet, M., Enaud, E., Guivarc'h, Y., Corbisier, A.-M. and Vanhulle, S. (2007) Potential of a *Pycnoporus sanguineus* laccase in bioremediation of wastewater and kinetic activation in the presence of an anthraquinonic acid dye. *Enzyme and Microbial Technology*, **41**(3), 368–376.

Vidali, M. (2001) Bioremediation: An overview. *Pure and Applied Chemistry*, **73**(7), 1163–1172.

Weeks, S.A. (2003) Anaerobic fixed-film digester system for dairy manure. Northeast Agricultural and Biological Engineering Conference, Storrs, CT, August.

Xing, W., Ngo, H.H., Guo, W. *et al.* (2010) Enhancement of the performance of anaerobic fluidized bed bioreactors (AFBBRs) by a new starch based flocculant. *Separation and Purification Technology*, **72**, 140–146.

CHAPTER 8

A new frontier for starter cultures: Attenuation and modulation of metabolic and technological performance

Antonio Bevilacqua, Barbara Speranza, Mariangela Gallo and Maria Rosaria Corbo
Department of the Science of Agriculture, Food and Environment, University of Foggia, Italy

A starter culture is any active microbial preparation intentionally added during product manufacture to initiate desirable changes. Starters can be attenuated to create populations of both lysed and dormant cells with the technological potential to control and/or accelerate the process (Yarlagadda *et al.* 2014). Thus, attenuation can be defined as a technological method for enhancing the total pool of intracellular enzymes relased into the matrix, positively influencing the flavour and quality of the final product.

Principally, attenuated cultures are lactic acid bacteria (LAB) that do not have the ability to synthesize lactic acid during cheese making, but contain enzymes that may influence cheese quality during ripening (Klein and Lortal 1999). Producing an excess of lactic acid might be considered one of the major drawbacks that negatively influence flavour and texture in cheese making, especially when high cell densities of mesophilic lactobacilli are added to milk. The use of attenuated starters was first proposed by Petterson and Sjöström (1975) to accelerate the ripening of Svecia, a Swedish semi-hard cheese, by thermal treatment (e.g. 69 °C for 15 s). At this time, their main purpose was to accelerate proteolysis and shorten ripening time. Today, in most developed countries the hygienic quality of milk is high, thus the main issue is to improve the flavour. In this case, an attenuated starter can improve both the rate and the quality of cheese ripening for light cheeses, which commonly exhibit poor flavour development and a rubbery texture (Klein and Lortal 1999; Upadhyay *et al.* 2007).

In the case of cheeses, it is known that modification of the bacterial enzymes contained in these products has a direct influence on the rate of cheese ripening and the final flavour, because ripening is a time-consuming process involving

Starter Cultures in Food Production, First Edition. Edited by Barbara Speranza, Antonio Bevilacqua, Maria Rosaria Corbo and Milena Sinigaglia.

complex biochemical reactions balanced between glycolysis, proteolysis and lipolysis of milk components and the further degradation of amino acids required for the development of taste, flavour and texture (Klein and Lortal 1999; Madkor *et al.* 2000). The degradation products of milk fats and proteins (peptides, free amino acids, mono- and diglycerides, free fatty acids and other degradation compounds such as aldehydes, ketones, esters, alpha-ketoacids etc.) contribute to the specific taste, flavour and texture particular to the cheese in question.

Klein and Lortal (1999) define an attenuated starter as a strain (or mixture of strains) that is easy to produce and to treat in a repeatable and cost-effective way, able to positively and significantly influence the ripening process (reduce time and/or increase flavour) at the lowest inoculation level in milk. This implies good retention of the starter in cheese curd, a significant enzyme content and adequate lysis of the attenuated starter. El-Soda *et al.* (2000) define attenuated starters as strains added to milk for the purpose of increasing the sensorial attributes and improving the textural properties of cheese.

Attenuated starters can be prepared by various treatments. Klein and Lortal (1999) report attenuating treatments such as heating, freezing, spray drying, freeze drying, fragilization using lysozyme or solvents, or the selection of lactose-negative mutants treatment. A number of chemical and physical techniques intended for such use have been extensively reviewed (Klein and Lortal 1999; El-Soda *et al.* 2000; Geciova *et al.* 2002).

According to Yarlagadda *et al.* (2014), these techniques can be generally divided into two types of treatments:

- Chemical treatments using hexadecyltrimethylammonium bromide (CTAB), ethylenediaminetetraacetic acid (EDTA), isopropyl alcohol (IPA), sodium dodecyl sulphate (SDS) or n-butanol (Exterkate 2006; Doolan and Wilkinson 2009).
- Physical treatments including heat or freeze shocking and/or mechanical treatments such as sonication (Exterkate 2006), bead mill, high-pressure homogenization and microfluidization (Geciova *et al.* 2002).

Specific methods show benefits and limits. Chemical treatments using chelating agents, such as EDTA, have strain-specific effects affected by buffers and are more effective towards Gram-negative bacteria. SDS is used mainly for Gram-negative bacteria; however, it can also cause denaturation of proteins. The use of alkanols such as n-butanol increases the permeabilization process of lactococcal cells; however, the enzymes are sensitive to irreversible inactivation (Exterkate 2006). On the other hand, use of mechanical treatments like bead milling is more efficient in yeasts or moulds compared with bacteria, but the effectiveness is dependent on the size of the beads. Sonication is a laboratory-scale method that not only increases cell lysis, but also significantly increases the degradation of enzymes due to heat denaturation. Mechanical treatments, as described by Geciova *et al.* (2002), use high-pressure homogenization, which is widely employed in the pharmaceutical or biotech industries to disrupt bacteria and yeasts. The main limit of homogenizers is the need for external cooling that can only be applied after cell disruption.

Industrial use of attenuated starters is not widespread, due to some limits. This may be due to many reasons:

- The extent of attenuation is often strain dependent and the selection of a given strain or species is frequently arbitrary.
- Each attenuation method shows benefits and limits.
- Approximate costs for the use of attenuated starters are hard to ascertain.

Attenuation by heat treatment

The most studied form of starter attenuation is heat treatment. This method is based on the sub-lethal heat treatment of a cell suspension. *Lactobacillus helveticus* is a bacterial species that is often heat treated, because of its high peptidase activity compared to other LAB (Klein and Lortal 1999). This method has one drawback; that is, it is important to correctly define the combination temperature/time, so as to delay acidification without denaturation of enzymes. Petterson and Sjöström (1975) recommend a temperature of 59 °C for 15 seconds for mesophilic starters (lactococci) and 69 °C for 15 seconds for thermophilic starters (lactobacilli such as *Lb. helveticus*); thus, they found a delay in acid production between 5 and 10 hours.

Bie and Sjöström (1975) performed experiments with a mixed-starter strain culture for cheese production and reported that a temperature exposure of 54 °C for 20 seconds was adequate to attenuate the cells, obtaining, in this case, a delay in acid production by 2–3 hours.

Abdel Baky *et al.* (1986) reported that inoculation of cheese curd with a heat-shocked culture of either *Lactobacillus casei* or *Lb. helveticus* did not affect cheese composition, but did influence flavour intensity. When comparing the influence of heat shock for *Lb. helveticus, Lactobacillus bulgaricus* or *Streptococcus thermophilus* on the ripening and quality of Gouda cheese, Bartels *et al.* (1987a) found that *Lb. helveticus* gave the best results (Klein and Lortal 1999).

Exterkate *et al.* (1987) added a maximum of 5% mixed-strain mesophilic heat-treated starter cultures with the primary starter to cheese milk. They found that cooling cells to 8 °C and then maintaining them at 56.5 °C for 20 seconds (included a warm-up period) inhibited acid production for few hours. Gouda cheese produced with the addition of treated culture and primary starter was characterized by enhanced flavour and reduced ripening time (Briggs 2003).

The critical factor in this technique was reported by Castaneda *et al.* (1990); that is, the narrow range between adequate and excessive heat treatments. For example, heat treatment at 64 °C for 18 seconds reduced 90% of the acidification activity in *Lb. helveticus*, but only 60% of the cell wall proteinase and 10% of the aminopeptidase activities. An increase of 2 °C from 64 °C to 66 °C dramatically inactivated proteinase and peptidase activities. Asensio *et al.* (1995) found a similar reduction for *Lb. casei* or *Lactobacillus plantarum* when heated at 54 °C for 15 seconds compared to 50 °C.

Frey *et al.* (1986) suggested that attenuation by heat may occur at different levels, such as damage to the enzymes involved in the transport of substrates through cell walls and membrane, physical damage of the membrane itself or denaturation of β-galactosidase. The major contribution of heat-attenuated starters to cheese making was the hydrolysis of medium-sized peptides (Ardö and Pettersson 1988; Vafopoulo *et al.* 1989). Peptidases were shown to be intracellular in LAB (Kunji *et al.* 1996), therefore their presence in cheese would indicate that cell lysis occurred (Lortal *et al.* 1997). Castaneda *et al.* (1990) found that heat-treated cells of *Lb. helveticus* (64 °C for 18 seconds), observed by electron microscopy, showed no apparent damage to membrane and cell wall, but they exhibited fewer ribosomes and DNA aggregation compared to the control. Lysed cells were not evident in micrographs of heat-treated bacteria.

Attenuation by freezing–thawing

The first cheese study involving freeze-shocked cells was performed by Bartels *et al.* (1987b). This treatment involved washing a tenfold concentration of cells, freezing at −20 °C and thawing at 40 °C prior to the addition of milk. A similar study was conducted by El-Tanboly *et al.* (2010a) to determine the effects of *Lactobacillus acidophilus* on the sensory attributes, ripening time and composition of Gouda cheese, as well as the survival of the best microorganism subjected to freeze shocking at −10/−20 °C for 24/96 hours. In this experiment the mixed strains of mesophilic starter LAB and adjunct lactobacilli were harvested by centrifugation and the resultant pellet was washed with saline solution, resuspended in 10% sterile skim, attenuated by freeze shocking, thawed at 40 °C and added to milk.

Freezing cells in suboptimal conditions reduces the viability of LAB. Stressed cells do not contribute significantly to lactic acid production during cheese making, but they may retain protease and peptidase activity (El-Tanboly *et al.* 2010a). Frozen cells may lyse to a greater extent than non-frozen cells, and thus release intracellular enzymes (Bartels *et al.* 1987b).

El-Soda *et al.* (1991) found that freeze-shocked cultures did not significantly affect the moisture level and the acidity of cheeses. However, significantly enhanced proteolysis occurred and the flavour of low-fat cheese was greatly improved. Furthermore, there was an increase in the content of volatile fatty acids (Klein and Lortal 1999). El-Tanboly *et al.* (2010a) used attenuated lactobacilli to produce Gouda cheese and suggested that the combination of attenuated lactobacilli and rennet exerted a positive effect on maturation.

Moreover, attenuated starters reduced bitterness due to their enhanced proteolysis (El-Tanboly *et al.* 2010b).

Freeze-shocked cells of *Lb. helveticus*, when compared to *Lb. casei* (Aly 1990), *Lb. bulgaricus* and *Strep. thermophilus* (Aly 1996), showed significantly higher proteolysis and flavour development. However, attenuated *Lb. helveticus* cells sometimes gave a sweet, nutty flavour, which was not normally associated with cheese (Johnson *et al.* 1995). Assays carried out in cheese slurries have confirmed that *Lb. helveticus* increased proteolysis and lipolysis and led to a sweet or light sulfuric character depending on the strain (Madkor *et al.* 1999).

The physico-chemical-biochemical effects of freezing are multiple and complex. Bacterial cells can freeze by intracellular ice formation or by dehydration; freezing removes water from both the external and internal environments and increases the concentration of solutes both inside and outside the cells (Ray and Speck 1973). Ice crystals can cause damage to the cell membrane. After freezing and thawing treatment, bacterial cells can lose many micro- and macromolecular cellular components (Klein and Lortal 1999). Several factors, including lowered temperature, both extra- and intracellular ice formation and both extra- and intracellular solute concentration, either singly or in combination, can damage frozen bacterial cells. Cell wall injury was also observed in *Lb. acidophilus*.

Spray and freeze drying

Johnson *et al.* (1995) compared the attenuation of *Lb. helveticus* CNRZ-32 by spray drying and freeze drying to freezing. According to these authors, spray-dried (SD) adjuncts might be more economical than frozen or freeze-dried (FD) adjuncts, because the frozen adjuncts occupy a large volume, are heavy and must be stored at subzero temperatures. On the other hand, the FD adjuncts are dry, stable and occupy small volumes. In addition, FD adjuncts might have delayed acid production because cells are subjected to attenuation by dehydration effects and thermal effects when heat is applied to speed drying.

Johnson *et al.* (1995) concluded that frozen and FD cells had the highest viability, compared to SDHT (spray-dried at a high outlet air temperature of 120 °C) cells. In addition, in SDHT samples, aminopeptidase and β-galactosidase activities were almost completely deleted, lactic acid production was delayed by 5 hours and permeability was increased more than in the other samples. Instead, for frozen, FD and SDLT (spray-dried at a low outlet air temperature of 82 °C) cells, lactic acid production was not delayed and the rates of lactic acid production were similar, even though SDLT cells contained less than one-third of the viable cells in the other two samples. Aminopeptidase and β-galactosidase activities of the SDLT cells were several times higher than those of the frozen and FD cells.

Attenuation by high pressure and high-pressure homogenization

High-pressure (HP) treatment dates back to the end of the 19th century (Hite 1899), as a suitable mean for reducing food contamination by pathogens and spoiling microorganisms. It is defined as non-thermal treatment that uses pressure (300–700 megapascals [MPa], in some cases up to 1000 MPa) as the main preservation method. Due to the fact that the pressure increase is achieved through a fluid (for example water), this process has also been referred to as high hydrostatic pressure (HHP) as opposite to high-pressure homogenization (HPH), where the increase in pressure is obtained by forcing the product through a small valve (homogenizing valve; Bevilacqua *et al.* 2010).

During HP treatment, a sample receives instantaneous and uniform pressure. Some studies have suggested the use of this treatment to accelerate cheese ripening (O'Reilly *et al.* 2000a, 2003; Saldo *et al.* 2002). In addition, HP treatment has also been used to control spoilage organisms in cheese (O'Reilly *et al.* 2000b), as HP can cause perturbation of the bacterial cell wall and membranes (Cheftel 1995). LAB have shown resistance to lysis at high pressures (Casal and Gomez 1999). Moreover, activities of the enzymes of LAB have been found to be unaffected when exposed to pressures up to 200 MPa (Malone *et al.* 2003).

Upadhyay *et al.* (2007) studied attenuation of starter bacteria using HP treatment. They treated *Lactococcus lactis* spp. *cremoris* HP and *Lc. lactis* spp. *cremoris* 303 through pressure at 100, 200 and 300 MPa, in vacuum bags and vacuum packed, for 1 s, 10 min and 20 min at 20 °C; thus, they found that HP treatment could impair acid production without causing cell lysis. Moreover, these high pressure–treated bacteria, when used in combination with primary strains, did not produce acid during cheese making. Use of high pressure–treated starters accelerated secondary proteolysis in Cheddar cheese during ripening, as shown by the higher levels of most free amino acids.

From 1990, a new generation of homogenizers, referred to as high-pressure homogenizers, has been developed and used by pharmaceutical, cosmetic, chemical and food industries for the preparation or stabilization of emulsions and suspensions or for creating physical changes, such as viscosity changes, in products. Other applications are cell disruption of yeasts or bacteria (in order to release intracellular products such as recombinant proteins) and/or reduction of the microbial cell load of liquid products (Bevilacqua *et al.* 2008). A homogenizer consists of a positive displacement pump and a homogenizing valve. In the homogenizing valve (often referred to as a radial diffuser), the fluid is forced under pressure by a pump, through a small orifice between the valve and the valve seat (Diels and Michiels 2006).

HPH is a technology that has been proposed for non-thermal fluid food microbial decontamination (Lanciotti *et al.* 2007; Bevilacqua *et al.* 2010).

Cavitation and viscous shear have been identified as the primary mechanisms of microbial cell disruption during HPH treatment (Middelberg 1995; Kleinig and Middelberg 1998). Furthermore, it appears that even this technology is active on food constituents, especially proteins, leading to changes in their functional properties and activities (Kheadr *et al.* 2002; Vannini *et al.* 2004). HPH treatment of skim and whole milk has been reported to modify the ratio of the nitrogen fractions and the soluble forms of calcium and phosphorous, and to improve the coagulation characteristics of milk as well as increase cheese yields (Lanciotti *et al.* 2007). HPH treatment of milk was associated with an acceleration of lipolysis activities found in Crescenza (a soft cheese), when produced using HPH-treated milk at 100 MPa (Lanciotti *et al.* 2004). Temperature increase during the process is a key parameter for enzyme activity modification and microbial inactivation (Hayes *et al.* 2005). According to Grandi *et al.* (2005), temperature increase derives from pressure energy transformation into thermal energy and corresponds to about 12 °C per 50 MPa.

Activation of endogenous and microbial proteolytic enzymes was observed in cheeses obtained from cow's milk and goat's milk treated at 100 MPa (Lanciotti *et al.* 2006), suggesting that the treatment increased the activity of milk and extracellular enzymes and of the enzyme located on the cell envelope (Lanciotti *et al.* 2007).

Lanciotti *et al.* (2007) studied the effect of HPH treatment on the proteolytic and metabolic activities of different strains belonging to the *Lactobacillus* species, involved in dairy product fermentation and ripening, in particular:

1 Fermentation kinetics of HPH-treated cells inoculated in milk.
2 Metabolic profiles.
3 Release of intracellular proteolytic enzymes.
4 Activity of extracellular or cellular wall proteolytic enzymes.

The suspensions were treated for one cycle at 50, 100 and 150 MPa or for two cycles at 50 and 100 MPa, with the inlet temperature of samples at 10 °C and the increase rate of temperature of 3 °C/10 MPa. The researchers concluded that the response to HPH varied according to the species and the characteristics of the individual strains. In general, HPH treatment did not have a significant effect on cell viability (the viability loss did not exceed 1.3 log colony forming units [cfu] per mL after the higher treatments applied), but displayed an important influence for points 1, 2 and 3.

Concerning the last point, HPH treatment had a positive effect on the proteolytic activity of some strains tested. This result confirmed the electrophoretic profiles of α- and/or β-casein incubated with different cell-free filtrates. Moreover, HPH treatment affected the acidification rate and the primary metabolism of some strains; in fact, the levels of the principal fermentation products like lactic or acetic acid and ethanol, and flavour molecules such as acetoin and 2-methyl butyric acid, were significantly increased.

Attenuation by microfluidization

'Microfluidization' could have interesting potential (Geciova *et al.* 2002). A microfluidizer (Microfluidics, Newton, MA, USA) operates on a principle different from that of the high-pressure valve homogenizer. A stream of cell suspension is impacted at high velocity with adjustable pressure against a stationary surface in an interaction chamber that disrupts cell integrity. The operation pressure is a function of flow rate and thus the rate of cell disruption increases with increasing pressure and the number of passes through the chamber. The reduction of enzyme activity by thermal degradation can be minimized by controlling the temperature within the chamber. This technique therefore can be used to create specific populations of live, permeabilized or lysed cells for use as adjuncts in cheese ripening (Yarlagadda *et al.* 2014). Yarlagadda *et al.* (2014) found that the addition to Gouda of strains attenuated by microfluidization affected proteolysis and volatile and sensory attributes after 12 weeks of ripening.

Attenuation by sonication

Sonication is a laboratory-scale method that not only increases cell lysis, but also significantly increases the degradation of enzymes due to heat denaturation (Geciova *et al.* 2002), as ultrasonic waves have the potential to exert a significant effect on microorganisms and living cells (Tabatabaie and Mortazavi 2010).

Ultrasound is defined as a pressure wave with a frequency of 20 kHz or more. Generally, it uses frequencies from 20 kHz to 10 MHz, and for industrial purposes it has two main requirements: a liquid medium, even if the liquid element forms only 5% of the overall medium; and a source of high-energy vibrations (the ultrasound; Patist and Bates 2008). The effects of destructive ultrasound rely on acoustic cavitation (Leighton 1998; Soria and Villamiel 2010; Cárcel *et al.* 2012). The antimicrobial effect depends on the amplitude of the waves, temperature and duration of the treatment, the volume and composition of the food (Chemat *et al.* 2011) and microorganism shape and dimensions (Heinz *et al.* 2001). Gram-positive bacteria are more resistant than Gram-negative ones (Drakopoulou *et al.* 2009) and cocci are more resistant than rods (Chemat *et al.* 2011).

During the process, the sonic wave encounters a liquid medium, creating longitudinal waves that generate regions of high pressure alternating with areas of low pressure. These regions of different pressure cause cavitation and the formation of gas bubbles that gradually increase in volume until they implode, creating regions of high temperature and pressure. Pressure resulting from these implosions causes the main bactericidal effect of ultrasound.

Thus, ultrasonication may break down the cell wall and cause autolysis (Tabatabaie and Mortazavi 2010).

Tabatabaie and Mortazavi (2010) tested ultrasound on cheese starters (*Strep. thermophillus, Lactobacillus delbrueckii* ssp *bulgaricus, Lb. helveticus*), with the aim of investigating the viability of starter strains during ripening of cheese and to determine whether ultrasound treatment led to an enhanced rate of autolysis of the starter culture during a four-week ripening period. These starters were treated at 20 kHz frequency from 0–20 min at 20 °C, with an amplitude of 80%. The inactivation of total viable counts and cell lysis release of lactate dehydrogenase enzyme (LDH) due to autolysis of LAB were examined during cheese ripening. This study showed that lactococci were more sensitive to ultrasound than lactobacilli up to 20 min. In conclusion, the degree of inactivation was found to be affected by exposure time.

Lysozyme treatment

Attenuation using lysozyme consists of treating a cell suspension with this enzyme (Ristagno *et al.* 2012). The cells treated with lysozyme break down in the curd at the salting step and thus release their intracellular enzymes into the cheese matrix (Briggs 2003).

Lysozyme is active throughout a wide pH range (4–10); however, high ionic strength (>0.2 M salt) was shown to have an inhibitory effect on its activity (Chung and Hancock 2000). Lysozyme belongs to a class of enzymes that lyses the cell walls of Gram-positive bacteria, as it specifically splits the bond between N-acetylglucosamine and N-acetylmuramic acid of the peptidoglycan in the bacterial cell walls. Extensive hydrolysis of peptidoglycan by exogenous lysozymes results in cell lysis and death in a hypo-osmotic environment, but some exogenous lysozymes can also cause lysis of bacteria by stimulating autolysin activity on interaction with the cell surface (Nakimbugwe *et al.* 2006).

Law *et al.* (1976) treated a suspension of streptococci cells in deionized water for 15 min at 37 °C with lysozyme (0.1–2 mg/mL; optimum at 0.5 mg/mL). The results showed that the cells were not lysed by the treatment and these lysozyme-sensitive cells (LSC) did not produce acid in skim milk over a 6-hour period. Proteolytic activity was reduced by 36%, which can be explained by a partial loss of cell wall material, to which proteases were associated. In Cheddar cheese experiments, the addition of 3×10^9 LSC/g of cheese led to a 1.8 increase in free amino acids, compared to the control without LSC. A sevenfold increase (2×10^{10} LSC/g) of cells led to a 2.5 times increase in free amino acids, which is significant but not proportional. However, the final product did not exhibit a more pronounced flavour. A limitation of this approach is the high cost of the enzyme (Klein and Lortal 1999; Ristagno *et al.* 2012).

Attenuation by treatment with various solvents

Chemical treatment using chelating agents, such as EDTA, has strain-specific effects that are influenced by buffers and are most effective towards Gram-negative bacteria. SDS is used mainly for Gram-negative bacteria; however, it can also cause denaturation of proteins (Middelberg 1995; Geciova *et al.* 2002).

Exterkate (2006) discovered that the use of alkanols like n-butanol increased the permeabilization process of lactococcal cells; however, the resulting enzymes were sensitive to irreversible inactivation.

Starter attenuation by treatment with various solvents changed the lipid structure of the cell membrane (Jain *et al.* 1978), in particular n-butanol, making the cell unable to produce lactic acid (Klein and Lortal 1999; Ristagno *et al.* 2012). Exterkate (1984) reported increases in peptidase activity. The addition of butanol-treated lactococci to Gouda cheese reduced bitterness, compared to control cheeses (Stadhouders *et al.* 1983). Even if solvents are found to give the desired attenuation effect, this approach may be impractical because of cost, health hazards and legal barriers (Klein and Lortal 1999; Ristagno *et al.* 2012).

Lactose-negative mutants as attenuated starters

Lortal and Klein (1999) considered the use of lactose-negative mutant strains as attenuated starters.

Briggs (2003) reported that a lactose-negative mutant can be defined as an organism that is unable to ferment lactose; therefore, it is unable to produce lactic acid, but is still able to provide the necessary abundance of enzymes that will slowly be released as the cells lyse, which will be available to enhance proteolysis, lipolysis and glycolysis, leading to a final product that will demonstrate an overall reduced bitterness and a reduction of ripening time.

Dulley *et al.* (1978) isolated variants of *Streptococcus lactis* C2 that spontaneously lost their ability to ferment lactose (Lac^-). Klein and Lortal (1999) reported that these variants could be classified as attenuated starters, even though they were not derived by physical treatment.

Many authors found a positive effect by the addition of mutant starter, such as an increase in flavour score, reduction in bitterness, a quick ripening period and an enhanced texture. However, these natural cultures, which lose their ability to convert lactose to lactic acid, are difficult to isolate and therefore most of them are genetically engineered (Briggs 2003).

References

Abdel Baky, A., El-Neshawy, A., Rabie, A. and Ashour, M. (1986) Heat-shocked lactobacilli for accelerating flavor development of Ras cheese. *Food Chemistry*, **21**, 301–313.

Aly, M.E. (1990) Utilization of freeze-shocked lactobacilli for enhancing flavour developement of Ras cheese. *Egyptian Journal of Dairy Science*, **18**, 143–156.

Aly, M.E. (1996) Evaluation of some freeze-shocked lactic starters for accelerated ripening of Domiati cheese made at various salt levels. *Egyptian Journal of Dairy Science*, **24**, 47–60.

Ardö, Y. and Pettersson, H.E. (1988) Accelerated cheese ripening with heat treated cells of *Lactobacillus helveticus* and a commercial proteolytic enzyme. *Journal of Dairy Research*, **55**, 239–245.

Asensio, C., Gomez, R. and Pelaez, C. (1995) Effect of heat treatment on the proteolytic activity of mesophilic bacteria isolated from goats' milk cheese. *Letters in Applied Microbiology*, **21**, 25–30.

Bartels, H.J., Johnson M.E. and Olson, N.F. (1987a) Accelerated ripening of Gouda cheese: Effect of freeze-shocked *Lactobacillus helveticus* on proteolysis and flavor development. *Milchwissenschaft*, **42**, 139–144.

Bartels, H.J., Johnson, M.E. and Olson, N.F. (1987b) Accelerated ripening of Gouda cheese: Effect of heat-shocked thermophilic lactobacilli and streptococci on proteolysis and flavor development. *Milchwissenschaft*, **42**, 83–88.

Bevilacqua, A., Campaniello, D. and Sinigaglia, M. (2010) Use of high pressure processing for food preservation. In A. Bevilacqua, M.R. Corbo and M. Sinigaglia (eds), *Application of Alternative Food-Preservation Technologies to Enhance Food Safety and Stability* (pp. 114–142), Sharjah: Bentham Science.

Bevilacqua, A., Corbo, M.R. and Sinigaglia, M. (2008) *Alicyclobacillus acidoterrestris*: New method for inhibiting spore germination. *International Journal of Food Microbiology*, **125**, 103–110.

Bie, R. and Sjöström, G. (1975) Autolytic properties of some lactic acid bacteria used in cheese production. Part II: Experiments with fluid substrates and cheese. *Milchwissenschaft*, **30**, 739–747.

Briggs, S. (2003) Evaluation of lactic acid bacteria for the acceleration of cheese ripening using pulsed electric fields. Department of Agricultural and Biosystem Engineering, Macdonald Campus of McGill University, Montreal.

Cárcel, J.A., García-Peréz, J.V., Benedito, J. and Mulet, A. (2012) Food process innovation through new technologies: Use of ultrasound. *Journal of Food Engineering*, **110**, 200–207.

Casal, V. and Gomez, R. (1999) Effect of high pressure on the viability and enzymatic activity of mesophilic lactic acid bacteria isolated from caprine cheese. *Journal of Dairy Science*, **82**, 1092–1098.

Castaneda, R., Vassal, L., Gripon, J.C. and Rousseau, M. (1990) Accelerated ripening of a Saint-Paulin cheese variant by addition of a heat-shocked *Lactobacillus* suspension. *Netherland Milk and Dairy Journal*, **44**, 49–62.

Cheftel, J.C. (1995) High-pressure, microbial inactivation and food preservation. *Food Science and Technology International*, **1**, 75–90.

Chemat, F., Zill-e-Huma and Khan, M.K. (2011) Applications of ultrasound in food technology: Processing, preservation and extraction. *Ultrasonics Sonochemistry*, **18**, 813–835.

Chung, W. and Hancock, R.E.W. (2000) Action of lysozyme and nisin mixtures against lactic acid bacteria. *International Journal of Food Microbiology*, **60**, 25–32.

Diels, A.M.J. and Michiels, C.W. (2006) High pressure homogenization as a non-thermal technique for the inactivation of microorganisms. *Critical Reviews in Microbiology*, **32**, 201–216.

Doolan, I.A. and Wilkinson, M.G. (2009) Comparison of the effects of various attenuation methods on cell permeability and accessibility of intracellular enzymes in *Lactococcus lactis* strains. *International Dairy Journal*, **19**, 215–221.

Drakopoulou, S., Terzakis, S., Fountoulakis, M.S., Mantzavinos, D. and Manios, T. (2009) Ultrasound-induced inactivation of gram-negative and gram-positive bacteria in secondary treated municipal wastewater. *Ultrasonics Sonochemistry*, **16**, 629–634.

Dulley, J.R., Brooks, D.E.J. and Grieve, P.A. (1978) The possible use of Lac$^-$ starter strains to accelerate cheese ripening and a method for their detection in cheese. *XX International Dairy Congress*, Paris, p. 485.

El-Soda, M., Chen, B., Riesterer, B. and Olson, N. (1991) Acceleration of low-fat cheese ripening using lyophilised 15 extracts or freeze shocked cells of some cheese related micro-organism. *Milchwissenschalft*, **46**, 358–360.

El-Soda, M., Madkor, S.A. and Tong, P.S. (2000) Adjunct cultures: Recent developments and potential significance to the cheese industry. *Journal of Dairy Science*, **83**, 609–619.

El-Tanboly, E., El-Hofi, M., Abd-Rabou, N.S. and El-Desoki, W. (2010a) Contribution of mesophilic starter and adjunct lactobacilli to proteolysis and sensory properties of semi hard cheese. *Journal of American Science*, **6**, 697–703.

El-Tanboly, E., El-Hofi, M., Youssef, Y.B., El-Desoki, W. and Jalil, R.A. (2010b) Influence of freeze-shocked mesophilic lactic starter bacteria and adjunct lactobacilli on the rate of ripening Gouda cheese and flavor development. *Journal of American Science*, **6**, 465–471.

Exterkate, F.A. (1984) Location of peptidases outside and inside the membrane of *Streptococcus cremoris*. *Applied and Environmental Microbiology*, **47**, 177–183.

Exterkate, F.A. (2006) Controlled permeabilization of *Lactococcus lactis* cells as a means to study and influence cheese ripening processes. *International Dairy Journal*, **16**, 788–796.

Exterkate, F.A., de Veer, G.J.C.M. and Stadhouders, J. (1987) Acceleration of the ripening process of Gouda cheese by using heat-treated mixed-strain starter cells. *Netherlands Milk and Dairy Journal*, **41**, 307–320.

Frey, J.P., Marth, E.H., Johnson, M.E. and Olson, N.F. (1986) Heat- and freeze-shocking cause changes in peptidase and protease activity of *Lactobacillus helveticus*. *Milchwissenschaft*, **41**, 681–685.

Geciova, J., Bury, D. and Jelen, P. (2002) Methods for disruption of microbial cells for potential use in the dairy industry: A review. *International Dairy Journal*, **12**, 541–553.

Grandi, S., Rainieri, S., Guerzoni, M.E. and Pagliarini, G. (2005) Performance of an ultra high pressure homogeniser for a suspension of *Saccharomyces cerevisiae*. In G. Pagliarini and S. Rainieri (eds), *Heat and Mass Transfer in Food Processing: Eurotherm Seminar* 77 (pp 307–310), Parma: Università degli Studi di Parma.

Hayes, M.G., Fox, P.F. and Kelly, A.L. (2005) Potential applications of high pressure homogenisation in processing of liquid milk. *Journal of Dairy Research*, **72**, 25–33.

Heinz, V., Alvarez, I., Angersbach, A. and Knorr, D. (2001) Preservation of liquid foods by high intensity pulsed electric fields: Basic concepts for process design. *Trends in Food Science & Technology*, **12**, 103–111.

Hite, B.H. (1899) The effect of pressure in the preservation of milk. *Bulletin of West Virginia University Agricultural Experimental Station*, **58**, 15–35.

Jain, M.K., Gleeson, J., Upreti, A. and Upreti, G.C. (1978) Intrinsic perturbing ability of alkanols in lipid bilayers. *Biochimica et Biophysica Acta (BBA) – Biomembranes*, **509**, 1–8.

Johnson, J.A.C., Etzel, M.R., Chen, C.M. and Johnson, M.E. (1995) Accelerated ripening of reduced-fat Cheddar cheese using four attenuated *Lactobacillus helveticus* CNRZ-32 adjuncts. *Journal of Dairy Science*, **78**, 769–776.

Kheadr, E., Vachon, J.F., Paquin, P. and Fliss, I. (2002) Effect of dynamic pressure on microbiological, rheological and microstructural quality of Cheddar cheese. *International Dairy Journal*, **12**, 435–446.

Klein, N. and Lortal, S. (1999) Attenuated starters: An efficient means to influence cheese ripening –a review. *International Dairy Journal*, **9**, 751–762.

Kleinig, A.R. and Middelberg, A.P.J. (1998) On the mechanisms of microbial cell disruption in high-pressure homogenisation. *Chemistry and Engineering Science*, **53**, 891–898.

Kunji, E.R.S., Mierau, I., Hagting, A., Poolman, B. and Konings, W.N. (1996) The proteolytic systems of lactic acid bacteria. *Antonie Van Leeuwenhoek*, **70**, 187–221.

Lanciotti, R., Chaves-Lòpez, C., Patrignani, F. *et al.* (2004) Effects of milk treatment with HPH on microbial population as well as on the lipolytic and proteolytic profiles of Crescenza cheese. *International Journal of Dairy Technology*, **57**, 19–25.

Lanciotti, R., Patrignani, F., Iucci, L., Saracino, P. and Guerzoni, M.E. (2007) Potential of high pressure homogenization in the control and enhancement of proteolytic and fermentative activities of some *Lactobacillus* species. *Food Chemistry*, **102**, 542–550.

Lanciotti, R., Vannini, L., Patrignani, F. *et al.* (2006) Effect of high pressure homogenisation of milk on cheese yield and microbiology, lipolysis and proteolysis during ripening of Caciotta cheese. *Journal of Dairy Research*, **73**, 216–226.

Law, B.A., Castanon, M.J. and Sharpe, M.E. (1976) The contribution of starter streptococci to flavour development in Cheddar cheese. *Journal of Dairy Research*, **43**, 301–311.

Leighton, T.G. (1998) The principles of cavitation. In M.J.W. Povey and T.J. Mason (eds), *Ultrasound in Food Processing* (pp. 151–182), London: Chapman & Hall.

Lortal, S., Valence, F., Bizet, C. and Maubois, J.L. (1997) Electrophoretic pattern of peptidoglycan hydrolases, a new tool for bacterial species identification: Application to ten *Lactobacillus* species. *Research in Microbiology*, **148**, 461–474.

Madkor, S.A., El-Soda, M. and Tong, P.S. (1999) Evaluation of commercial adjuncts for use in cheese ripening: Ripening aspects and flavor development in cheese curd slurries prepared with adjunct lactobacilli. *Milchwissenschaft*, **54**, 133–137.

Madkor, S.A., Tong, P.S. and El-Soda, M. (2000) Ripening of Cheddar cheese with added attenuated adjunct cultures of lactobacilli. *Journal of Dairy Science*, **83**, 1684–1691.

Malone, A.S., Wick, C., Shellhammer, T.H. and Courtney, P.D. (2003) High pressure effects on proteolytic and glycolytic enzymes involved in cheese manufacturing. *Journal of Dairy Science*, **86**, 1139–1146.

Middelberg, A.P.J. (1995) Process-scale disruption of microorganisms. *Biotechnology Advances*, **13**, 491–551.

Nakimbugwe, D., Masschalck, B., Anim, G. and Michiels, C.W. (2006) Inactivation of Gram negative bacteria in milk and banana juice by hen egg white and lambda lysozyme under high hydrostatic pressure. *International Journal of Food Microbiology*, **112**, 19–25.

O'Reilly, C.E., Kelly, A.L., Oliveira, J.C., Murphy, P.M., Auty, M.A.E. and Beresford, T.P. (2003) Effect of varying high-pressure treatment conditions on acceleration of ripening of Cheddar cheese. *Innovative Food Science and Emerging Technologies*, **4**, 277–284.

O'Reilly, C.E., O'Connor, P.M., Kelly, A.L., Beresford, T.P. and Murphy, P.M. (2000a) Use of hydrostatic pressure for inactivation of microbial contaminations in cheese. *Applied and Environmental Microbiology*, **66**, 4890–4896.

O'Reilly, C.E., O'Connor, P.M., Murphy, P.M., Kelly, A.L. and Beresford, T.P. (2000b) The effect of exposure to pressure of 50 MPa on Cheddar cheese ripening. *Innovative Food Science and Emerging Technologies*, **2**, 109–117.

Patist, A. and Bates, D. (2008) Ultrasonic innovations in the food industry: From the laboratory to commercial production. *Innovative Food Science and Emerging Technologies*, **9**, 147–154.

Petterson, H.E. and Sjöström, G. (1975) Accelerated cheese ripening: A method for increasing the number of lactic starter bacteria in cheese without detrimental effect to the cheese-making process, and its effect on the cheese ripening. *Journal of Dairy Research*, **42**, 313–326.

Ray, B. and Speck, M.L. (1973) Freeze injury in bacteria. *Critical Reviews in Clinical Laboratory Sciences*, **4**, 161–213.

Ristagno, D., Hannon, J.A., Beresford, T.P. and McSweeney, P.L.H. (2012) Effect of a bacteriocin-producing strain of *Lactobacillus paracasei* on the nonstarter microflora of Cheddar cheese. *International Journal of Dairy Technology*, **65**, 523–530.

Saldo, J., McSweeney, P.L.H., Sendra, E., Kelly, A.L. and Guamis, B. (2002) Proteolysis in caprine milk cheese treated by high pressure to accelerate cheese ripening. *International Dairy Journal*, **12**, 35–44.

Soria, A.C. and Villamiel, M. (2010). Effect of ultrasound on the technological properties and bioactivity of food: A review. *Trends in Food Science & Technology*, **21**, 323–331.

Stadhouders, J., Hup, G., Exterkate, F.A. and Visser, S. (1983) Bitter formation in cheese. 1. Mechanism of the formation of the bitter flavour defect in cheese. *Netherland Milk and Dairy Journal*, **37**, 157–167.

Tabatabaie, F. and Mortazavi, S.A. (2010) Effects of ultrasound treatment on viability and autolysis of starter bacteria in hard cheese. *American Eurasian Journal of Agricultural and Environmental Science*, **8**, 301–304.

Upadhyay, V.K., Huppertz, T., Kelly, A.L. and McSweeney, P.L.H. (2007) Use of high pressure treatment to attenuate starter bacteria for use as adjuncts for Cheddar cheese manufacture. *Innovative Food Science and Emerging Technologies*, **8**, 485–492.

Vafopoulo, A., Alichanidis, E. and Zerfiridis, G. (1989) Accelerated ripening of Feta cheese, with heat-shocked cultures or microbial proteinases. *Journal of Dairy Research*, **56**, 285–296.

Vannini, L., Lanciotti, R., Baldi, D. and Guerzoni, M.E. (2004) Interactions between high pressure homogenization and antimicrobial activity of lysozyme and lactoperoxidase. *International Journal of Food Microbiology*, **94**, 123–135.

Yarlagadda, A.B., Wilkinson, M.G., O'Sullivan, M.G. and Kilcawley, K.N. (2014). Utilisation of microfluidisation to enhance enzymatic and metabolic potential of lactococcal strains as adjuncts in Gouda type cheese. *International Dairy Journal*, **38**, 124–132.

CHAPTER 9

The role of the pangenome concept in selecting new starter cultures

Antonio Bevilacqua, Francesca Fuccio, Maria Clara Iorio, Martina Loi and Milena Sinigaglia
Department of the Science of Agriculture, Food and Environment, University of Foggia, Italy

Identification is the process by which an isolate is attributed to a specific *taxon*, or a taxonomic group consisting of microorganisms that share a high degree of similarity. The species is the fundamental taxonomic unit; for bacteria it can be defined as a set of strains having no less than 70% of homology between their molecular DNA/DNA and a similarity >98% in the 16S rDNA gene.

The choice of using the 16S rRNA sequence is due to the origin of the bacteria from a common ancestor and their gene complement is the result of a combination of extensive gene loss and horizontal gene transfer during evolution (Makarova *et al.* 2006). 16S rRNA is considered a *gold standard*, as it presents a base sequence identical for all strains belonging to a given species. It is also mainly preserved in the same species, like all *housekeeping genes*, thus its sequencing is reliable for identifying the microorganisms belonging to the same species.

Although the value of 16S rRNA in assessing evolutionary links is universally accepted, it is a matter of debate whether the 16S sequence is enough for defining bacteria species (Fraser *et al.* 2009). Bacteria are able to perform horizontal gene transfer (HGT), thus the recombination of genes varies from clonal lineages to highly recombinant, known as panmictical strains (Smith *et al.* 1993).

Some sources of error in genome studies are the lack of replicates, experiments across time and use of different primers, which in turn result in different sequences of the same genomes (Klappenbach *et al.* 2001; Engelbrektson *et al.* 2010).

Therefore, a new approach is emerging, which is based on the comparison of complete genome sequences of a number of members of the same species and is referred to as pangenomics, or a 'study of the whole genome'.

The pangenome is the global gene repertoire of a bacterial species; it assesses both the 'core genome' – that is, the pool of genes shared by all the strains of the

Starter Cultures in Food Production, First Edition. Edited by Barbara Speranza, Antonio Bevilacqua, Maria Rosaria Corbo and Milena Sinigaglia.

same bacterial species – and the 'dispensable genome' – the pool of genes present in some strains of the same bacterial species (Medini *et al.* 2005).

The core genome generally contains all genes responsible for the basic aspects of the biology of a species and its major phenotypical traits. It encodes for such basic and housekeeping functions as ribosomal proteins, H^{+}ATPase, host signaling, D-alanylation of lipothecoic acid, responses to the environment, prolidase and carbon catabolite control and fibronectin-binding protein (De Vos 2005; Schick *et al.* 1999; Courtney *et al.* 1994). On the other hand, dispensable genes contribute to species diversity and might encode supplementary metabolic pathways and functions that are not essential for bacterial growth. However, they confer selective advantages, such as antibiotic resistance, colonization of new hosts or niches and adaptation to a hostile environment, or virulence. These genes are usually clustered in 'genomic islands' flanked by SRS (short reported sequences), characterized by a high G+C content (Daubin and Ochman 2004). Hence, core and dispensable genes represent the essence and the diversity of the species, respectively.

Bacterial species cannot be fully described, as new genes are always added to the genome of the species with each new genomic sequence. As more genome sequences are determined, the size of the core genome is likely to decrease while the pangenome will increase (Medini *et al.* 2005). There are some genes that are usually recovered only in a small number of genomes; these genes are referred to as ORFans.

The acquisition of these new genes may happen through:

- duplication and diversification of existing genes, which represent a slow route to adaptation (Brussow *et al.* 2004);
- HGT from related or unrelated organisms; this process represents a fast route, which enables an organism to adapt quickly to a changing environment (Alcaraz 2014);
- bacteriophage infection;
- plasmid exchange;
- accumulation of mutations during clonal expansion (Ochman *et al.* 2000).

There are both open and close pangenomes. An open pangenome is typical of those species that colonize multiple environments and have multiple ways of exchanging genetic material, like Streptococci, Meningococci, *Helicobacter pylori*, *Salmonella* and *Escherichia coli*. On the other hand, *Bacillus anthracis*, *Mycobacterium tuberculosis* and *Chlamydia trachomatis* are characterized by closed pangenomes, because the genome sequences available on international databases can describe the species (Tettelin *et al.* 2005; Mira *et al.* 2010).

The pangenome concept arose when comparing *Streptococcus agalactiae* strains, a major cause of disease in newborns, infants and the elderly (Doran and Nizet 2004; Schuchat and Wenger 1994); they produce the same symptomatic illnesses and host similar 16S rRNA sequences. However, when comparing the genome sequences of the *Strep. agalactiae* isolated from different patients against the reference genome an unexpected result was observed: each strain shared 20% of the genes. Thus, even if they were supposed to belong to same species, there were differences within them.

Building a pangenome is helpful for having a full inventory of the metabolic capabilities of a given group of organisms. Differences in the unique genes of closely related bacteria could be a partial answer to local adaptation to particular niches (i.e. free living, host associate and virulence). Thus, it might be useful to have pangenome catalogues for all the known groups of bacteria with sequenced genomes, and to build confidence intervals based on gene presence/absence within particular bacteria. In fact, genome sequencing is useful for ruling out the presence of undesirable content such as antibiotic resistance genes and virulence factors in individual strains or in a species pangenome (Bennedsen *et al.* 2011).

Pangenome analysis is a relatively new concept for industrial starter cultures, with only a limited number of studies reported. Early publications were based on comparative genome hybridization and used microarrays with few strains represented owing to the low number of genome sequences available (Garrigues *et al.* 2005).

The pangenomic approach offers the opportunity to compare the genetic repertoire of different species and genera, pathogens and non-pathogens, as well as organisms from different ecological niches, including foods; it could be the background to formulating new hypotheses in commensal, probiotic and pathogenic relationships with the host.

Use of the pangenome approach to select starter cultures

HGT, infection of lysogenic phages, mutations of the DNA sequence and conjugation are able to create genetic diversity and produce unique phenotypes without affecting the housekeeping genes; these phenotypical traits could be retained, acquired or maintained as active only in a limited number of strains.

Functional genomics have recently revealed the presence of ORFans, annotated genes with no known homologue even in related genomes (Daubin and Ochman 2004). These genes are exclusive to a particular genome and may encode for novel uncovered functions of a species. Thus, analysis of the microbial dispensable genome is essential to detect strain specificity and to identify and optimize a potential new strain to be used as a new starter culture in the food industry.

Through the pangenomic approach, strain-specific characteristics that are set in the dispensable genome can be identified and elucidated; gene variation among different strains can be correlated to the presence (or absence) of phenotypical traits; and molecular biomarkers can be identified and used to operate genome-assisted selection for the detection of interesting or undesired traits (Smokvina *et al.* 2013).

A pangenomic approach has been proposed for a new selection of probiotics. When different strains elicit a positive physiological effect in a host, it is difficult to identify the responsible molecule through the classic biochemical or genetic approaches. On the other hand, postgenomic studies allow for identification of the common genes that are shared and, possibly, encode for that particular phenotypical feature, thus elucidating the molecular mechanisms of action by which probiotics exert their health-promoting functions. The application of a pangenomic approach to probiotics leads to the development of so-called probiogenomics (Ventura *et al.* 2009).

The most important factor leading to genetic differences within a species is assumed to be HGT, which probably occurs in matrices with a high microbial concentration, like foods.

Although many LAB genomes have already been sequenced (i.e. *Lactobacillus plantarum, Lactobacillus delbrueckii* subsp. *bulgaricus, Lactobacillus rhamnosus*), we know the sequence of different strains only for the species *Oenococcus oeni* (Borneman *et al.* 2012). There are several ongoing sequencing projects concerning *Bifidobacterium* and *Lactobacillus* genomes, but to date less than the 10% of pangenomes have been annotated (Sanchez *et al.* 2013).

Some studies have focused on the strain specificity of three traits (exopolysaccharide production, the protelytic system and tyramine production) through a pangenomic approach.

Exopolysaccharide (EPS) production is an important technological characteristic for the production of fermented products, such as wine and yoghurt. EPSs play an important role in modulating the components of wine and therefore its organoleptic quality, while in yoghurt they avoid syneresis, especially in skimmed milk variants. Furthermore, EPS can act as stabilizing, gel-forming and/or water-binding agents in various foods, thus improving their rheological properties (De Vuyst *et al.* 2001).

However, composition, structure, molecular mass (MM), yield and functionalities rely on a very specific glucidic structure, which is genetically determined. More than 10 clusters of EPS have been identified and sequenced in the *Staphilococcus thermophilus* pangenome, while there are several others in the pangenome of *O. oeni, Lb. delbrueckii* subsp. *bulgaricus* and *Lactobacillus paracasei*.

EPS synthesis requires the action of several genes, glucosyltransferase (gtf), the phosphotransferase system (PTS) and lipopolysaccharide N-acetylglucosaminyltransferase being the most important. These genes are arranged in clusters; some of them may be either absent or present, while unique genes may appear or be active in a single strain. EPS clusters of different strains seem to be an almost 'random' mix of different EPS genes, thus a single cluster is responsible for synthesizing a hexasaccharide-repeating unit composed of galactose, glucose, rhamnose, N-acetylglucosamine and N-acetylgalactosamine in a unique ratio (Hao *et al.* 2011).

The proteolytic system of lactic acid bacteria converts proteins to peptides and peptides to free amino acids. Amino acids may be also converted to flavour compounds such as aldehydes, alcohols and esters.

The ability to break down and use proteins is a selective advantage for those strains that live in protein-rich niches (i.e. milk) or may confer a particular flavour on an end product. Thus, the selection of a strain can be done based on the presence of a strong proteolytic system, consisting of proteinases, peptidases, aminotransferases, enzymes for the biosynthesis of amino acids, and transport systems for peptides and amino acids.

Strains have different rates of proteolysis and efficiency, according to the cleavage specificity of the peptidase. Apart from cysteine proteases, serine proteases and metalloproteinases that show cleavage specificity for dipeptides containing cysteine, serine and glycine, respectively, other enzymes, belonging to the C69 family, act on a wide variety of dipeptides, with the exception of those containing proline (Zheng *et al.* 2012). In milk and related products, casein is the most important source of nitrogen; since it has many proline residues in its primary structure, the presence of proline peptidases may be a critical feature to be screened.

The cell-wall-bound proteinase (CEP) responsible for the initial protein breakdown was found by Liu and collaborators (2010) in 14 of 39 strains of *Lactococcus lactis*. The absence of such an enzyme in strains of dairy origin may be explained by a repression or loss mechanism due to adaptation to a nitrogen-rich environment (casein hydrolizate, peptons, oligopeptides).

Another tool to discriminate strains is to screen the presence and type of transporters through which the peptides enter the cell. There are two types of transporters, oligopeptide-binding proteins (Opp) and di/tripeptide-binding proteins (Dpp/Tpp). Even though in the majority of cases both transporters are present, some strains differ in the number of existing genes coding for a specific type of transporter. This is the case for three strains of *Strep. thermophilus* (CNRZ1066, LMG18311 and LMD9), which have four, three and two genes, respectively, coding for oligopeptide-binding protein A (OppA; Liu *et al.* 2010).

Similarly, strains may differ for the type and number of genes coding for internal peptidase that catalyse the hydrolysis of peptides into free amino acids.

The genomic-assisted selection of a starter culture can be useful to detect desired traits, as well as undesired ones, such as the production of biogenic amines (BA). Tyramine derives from the decarboxylation of the amino acid tyrosine and can be found in numerous foods, especially in fermented ones, where the decarboxylation is catalysed by tyrosine decarboxylase. It has been established that the production of BA is not a species trait but is strain dependent. In particular, a recent publication demonstrated that two out of three *Lactobacillus brevis* strains, which are not distinguishable by RAPD and 16S-23S rRNA gene sequencing, are tyramine producers. Tyramine has no impact on the organoleptic attribute of a product, hence its presence has to be monitored because it exerts a negative physical effect on sensitive consumers. Therefore, it is indispensable to choose strains that do not produce this compound (Coton and Coton 2009).

Tyrosine decarboxylase is one of the genes set in a metabolic island together with a tyrosine permease, a Na^+/H^+ antiporter and a tyrosil-tRNA synthetase, which may have been acquired through HGT.

This strain-dependent aptitude for producing biogenic amines has been found in some strains of *O. Oeni* and *Lactobacillus hilgardii* able to produce histamine and putrescine (Coton *et al.* 1998; Marcobal *et al.* 2006).

Tools and software to study the microbial pangenome

The main way to study the pangenome is by sequencing multiple strains from a given species. Despite the rapid development in DNA sequencing technologies, whole-genome sequencing is too expensive for routine studies; thus, strain typing and the assessment of genetic similarity based on core gene sequences are not yet practical. An alternative approach is comparative genome hybridization (CGH) by microarrays of distributed gene probes, both to type strains and to estimate relationships among them. If the genome of a representative species is sequenced, all open reading frames (ORFs) are typically spotted on a microarray slide that serves as a probe against which a strain of unknown gene content is tested. Polymorphisms for gene deletions and insertions can be detected as a change in the ratio of fluorescence emitted when the two labelled genomic DNA samples are hybridized to the microarray (Gibson 2002). DNA/DNA hybridization techniques allow the identification of genomic differences even between bacterial species (Lindroos *et al.* 2005).

This technology is an effective tool for studying intraspecific differences, but one of the most important limitations is that the presence or absence of the sequences is established in relation to the reference genome and therefore unique genes on the tested strains cannot be identified. This problem can be circumvented by suppression subtractive hybridization (SSH), in which DNA from a test strain is depleted by hybridization to sequences from a reference strain. The remaining DNA is enriched in test-strain-specific genes, which are then cloned and/or sequenced (Zhang *et al.* 2005).

As reported in Table 9.1, a microarray based on a single strain of *Bifidobacterium animalis* subsp. *lactis* was used to compare a selection of commercially available *Bifidobacterium* strains (Garrigues *et al.* 2005). A pangenome microarray of *Strep. thermophilus* based on three genome sequences was used to characterize 47 industrial *Strep. thermophilus* strains (Rasmussen *et al.* 2008). A pangenome microarray of 2 *Lb. casei* genome sequences was used to compare 21 *Lb. casei* isolates from various environmental niches (Cai *et al.* 2009). Microarrays based on 4 *Ln. lactis* genome sequences were used to compare 39 *Lb. lactis* isolates of plant or dairy

Table 9.1 Application of pangenomic tools and software packages.

Software/tool	Species	Origin	Application	Reference
Comparative genome hybridization by microarrays	*Bif. animalis* subsp. *lactis*	Commercial dairy starters	Characterization, comparison and selection of commercial starters	Garrigues *et al.* (2005)
	S. thermophilus	Commercial dairy starters		Rasmussen *et al.* (2008)
	Lb. casei	Environmental origin		Cai *et al.* (2009)
	Lc. lactis	Plant and dairy origin		Meijerink *et al.* (2010)
	Lb. plantarum	Plant and human origin		Siezen *et al.* (2011)
PANSEQ	*Lc. lactis* subsp. *lactis*	Wheat sourdough	Determination of strain-specific characteristics	Passerini *et al.* (2013)
PGAP	*Bif. animalis* subsp. *lactis*	Fermented milk and human origin	Pangenome calculation, determination of ORF content	Milani *et al.* (2013)
	Bif. breve	Human milk	Identification of new genes	Bottaccini *et al.* (2014)
PANCGH	*Lc. lactis* subsp. *cremoris*	Environmental origin	Identification of strain	Siezen *et al.* (2011)

Notes: *Bif.* = *Bifidobacterium*; *Lb.* = *Lactobacillus*; *Lc.* = Lactococcus; ORF = open reading frame; *Strep.* = *Streptococcus*.

origin, and microarrays based on a single *Lb. plantarum* strain were used to characterize 60 *Lb. plantarum* isolates (Meijerink *et al.* 2010; Siezen *et al.* 2011).

Another approach for pangenome analysis is the use of metagenomic sequences, which is already applicable by data mining from the growing metagenomic databases. However, the ideal situation for metagenomic analysis of a species pangenome is the availability of a sequenced genome together with an environmental population of the same species obtained from the same site. Such data would enable evaluation of how representative an individual is within the population and would reveal the form and distribution of genetic variability in the core and the accessory gene pool (Mira *et al.* 2010).

In order to make a pangenome analysis more useful for one bacterial population, several software tools have been designed, like Panseq (Laing *et al.* 2010) and PGAT (Brittnacher *et al.* 2011). Panseq could identify a single nucleotide polymorphism (SNP) on a core genome and a strain's specific region, while PGAP was designed to perform five analytical functions with only one command (Zhao *et al.* 2012).

In particular, Panseq is a freely available online program for quickly finding and extracting strain- or group-specific novel accessory genomic information as well as the complete pangenome for a group of genomic sequences based on user-defined parameters. Panseq produces alignments of the core genome of each sequence and determines the distribution of accessory regions among all sequences analysed. This tool uses the MUMmer alignment algorithm for whole-genome comparisons and the BLASTn algorithm for local sequence comparisons, and can efficiently compute values for large numbers of sequences. In addition, it is able quickly to identify the most variable and discriminatory *loci* set in an iterative manner from single-character tabular data.

However, Panseq is unable to present the pangenome profiles of given strains, to trace the evolutionary history with multiple materials or to point out the variation and function enrichment of functional genes.

PGAT allows for comparison of gene content and sequences across multiple microbial genomes, thus it allows the discovery of genetic differences that may explain observed phenotypes. This application supports database queries to identify genes that are present or absent in user-selected genomes, comparison of sequence polymorphisms in sets of orthologous genes, multigenome display of regions surrounding a query gene, comparison of the distribution of genes in metabolic pathways and manual community annotation. PGAT integrates many features of current online resources such as Integrated Microbial Genomes (IMG; Markowitz *et al.* 2010), the *Burkholderia* genome database (Winsor *et al.* 2008) and the *Neisseria* base (Kislyuk *et al.* 2010). Its main strong point is the homogenization of gene features across the genomes and the integrated functionality to compare gene content, single nucleotide polymorphisms (SNPs) in orthologous genes, and the resulting impact of SNPs and indels on the encoded proteins.

However, PGAT only provides analytical results for limited species in the database and cannot analyse genome data from users. In 2012, Zhao and colleagues developed a new stand-alone program called the pangenome analysis pipeline (PGAP), which has integrated multiple-function models and could be used to study the evolutionary history of bacteria, discover pathogenic mechanisms and prevent and control epidemics (Zhao *et al.* 2012).

PGAP is a revolution in the pipeline of genome analysis because it has integrated five analysis modules that are commonly used in genome research. Users can perform five analytical tasks for their research with just one command, including cluster analysis of functional genes, pangenome profile analysis, genetic variation analysis of functional genes, species evolution analysis and function enrichment analysis of gene clusters.

Therefore, PGAP could cluster all genes into different clusters, detect genetic variation in each gene cluster and construct phylogenetic trees with different methods and data. These data could be used for studying species evolution and microbial typing in epidemics, and they are also helpful to discover pathogenic mechanisms.

Nevertheless, PGAP could only deal with small-scale genomes in the pangenome profile analysis module; thus, Zhao and colleagues have developed a tool, named PanGP, that is a highly efficient tool for large-scale bacterial pangenome profile analysis with sampling algorithms (Zhao *et al.* 2014). Infact, PanGP has integrated two sampling algorithms, totally random (TR) and distance guide (DG). The DG algorithm draws sample strain combinations on the basis of the genomic diversity of bacterial populations and exhibits an overwhelming advantage in accuracy and stability over the TR algorithm. PanGP requires ortholog information as the input data, which can be generated by a series of software packages, such as PGAP (Zhao *et al.* 2012) and PanOCT (Fouts *et al.* 2012).

A pangenome ortholog clustering tool, called PanOCT, is a tool for the pangenomic analysis of closely related prokaryotic species or strains and uses conserved gene neighbourhood (CGN) information to better separate very recently diverged paralogs into orthologous clusters where homology-only clustering methods cannot.

In pangenomes, orthologous genes can be defined as homologous genes that diverge from a single ancestral gene after a speciation event and are more likely to conserve their functions across organisms. These orthologous genes (strain orthologs) share different levels of nucleotide sequence identity with paralogous genes, which are homologous genes derived by a duplication event from a single sequence. Effective genotyping can be achieved by grouping genes into ortholog groups (OGs) and subsequently genotyping at the level of OGs.

Bayjanov and colleagues have published PanCGH, an algorithm that assigns OG presence/absence to each strain analysed by pangenome microarrays (Bayjanov *et al.* 2009). Pangenome microarrays contain probes that target all known genes within related strains of the same species (Tettelin *et al.* 2005) and allow the genomic content of bacterial strains to be determined more accurately than conventional comparative genome hybridization (CGH) by microarrays (Castellanos *et al.* 2009).

Bayjanov and colleagues have also developed a web tool – PanCGHweb – that uses this algorithm to effectively genotype strains based on pangenome microarray data (Bayjanov *et al.* 2010). The main steps of PanCGHweb are orthology prediction among genes of the selected reference genomes; alignment of microarray probes to the individual gene members of each OG; and genotype calling using the PanCGH algorithm, which enables researchers to analyse the complex hybridization data in an easy and transparent way to understand genomic diversity among related strains.

In Table 9.1, some food applications of pangenomic tools are reported.

There has also been the publication of GET_HOMOLOGUES, an open-source software package released under a GNU General Public License, specifically designed and tested for the pangenomic and comparative genomic analysis of bacterial strains at different phylogenetic distances on Linux/Mac OSX computer systems. This software implements a fully automatic and highly customizable analysis pipeline, including genome data download, extraction of user-selected sequence features, running of BLAST and HMMER jobs, and indexing, clustering and parsing of results (Contreras-Moreira and Vinuesac 2013).

References

Alcaraz, L.D. (2014) Pan-genomics: Unmasking the gene diversity hidden in the bacteria species. *PeerJ PrePrints*, https://peerj.com/preprints/137

Bayjanov, J.R., Siezen, R.J. and van Hijum, S.A.F.T. (2010) PanCGHweb: A web tool for genotype calling in pangenome CGH data. *Bioinformatics*, **26**(9), 1256–1257.

Bayjanov, J.R., Wels, M., Starrenburg, M., van Hylckama Vlieg, J.E.T., Siezen, R.J. and Molenaar, D. (2009) PanCGH: A genotype-calling algorithm for pangenome CGH data. *Bioinformatics*, **25**, 309–314.

Bennedsen, M., Stuer-Lauridsen, B., Danielsen, M. and Johansen, J. (2011) Screening for antimicrobial resistance genes and virulence factors via genome sequencing. *Applied and Environmental Microbiology*, **77**, 2785–2787.

Borneman, A.R., McCarthy, J.M., Chambers, P.J. and Bartowsky, E.J. (2012) Comparative analysis of the *Oenococcus oeni* pan genome reveals genetic diversity in industrially-relevant pathways. *BMC Genomics*, **13**, 373.

Bottacini, F., O'Connell Motherway, M., Kuczynski, J. *et al.* (2014) Comparative genomics of the Bifidobacterium breve taxon. *BMC Genomics*, **15**, 170.

Brittnacher, M.J., Fong, C., Hayden, H.S., Jacobs, M.A., Radey, M. and Rohmer, L. (2011) PGAT: A multi-strain analysis resource for microbial genomes. *Bioinformatics*, **27**, 2429–2430.

Brussow, H., Canchaya, C. and Hardt, W.D. (2004) Phages and the evolution of bacterial pathogens: From genomic rearrangements to lysogenic conversion. *Microbiology and Molecular Biology Reviews*, **68**, 560–602.

Cai, H., Thompson, R., Budinich, M.F., Broadbent, J.R. and Steele, J.L. (2009) Genome sequence and comparative genome analysis of *Lactobacillus casei*: Insights into their niche-associated evolution. *Genome Biology and Evolution*, **1**, 239–257.

Castellanos, E., Aranaz, A., Gould, K.A. *et al.* (2009) Discovery of stable and variable differences in the *Mycobacterium avium* subsp. *paratuberculosis* type I, II, and III genomes by pan-genome microarray analysis. *Applied and Environmental Microbiology*, **75**, 676–686.

Contreras-Moreira, B. and Vinuesac, P. (2013) GET_HOMOLOGUES, a versatile software package for scalable and robust microbial pangenome analysis. *Applied and Environmental Microbiology*, **79**, 7696–7701.

Coton, E. and Coton, M. (2009) Evidence of horizontal gene transfer as origin of strain to strain variation of the thiramine production trait in *Lactobacillus brevis*. *Food Microbiology*, **26**(1), 1051–1071.

Coton, E., Rollan, G., Bertrand, A. and Lonvaud-Funel, A. (1998) Histamine producing lactic acid bacteria in wines: Early detection, frequency and distribution. *American Journal of Enology and Viticolture*, **49**(2), 199–204.

Courtney, H.S., Li, Y., Dale, J.B. and Hasty, D.L. (1994) Cloning, sequencing, and expression of a fibronectin/fibrinogen-binding protein from group A streptococci. *Infection and Immunity*, **62**(9), 3937–3946.

Daubin, V. and Ochman, H. (2004) Bacterial genomes as new genes homes: The genealogy of ORFans in *E.coli*. *Genome Research*, **14**(6), 1036–1042.

De Vos, W.M. (2005) Lipotechoic acid in lactobacilli: d-Alanine makes the difference. *Proceedings of National Academy of Sciences of the United States of America*, **102**(31), 10763–10764.

De Vuyst, L., De Vin, F., Vaningelgem F. and Degeest, B. (2001) Recent developments in the biosynthesis and applications of heteropolysaccharides from lactic acid bacteria. *International Dairy Journal*, **11**(9), 687–707.

Doran, K.S. and Nizet, V. (2004) Molecular pathogenesis of neonatal group B streptococcal infection: No longer in its infancy. *Molecular Microbiology*, **54**(1), 23–31.

Engelbrektson, A., Kunin, V., Wrighton, K.C. *et al.* (2010) Experimental factors affecting PCR-based estimates of microbial species richness and evenness. *ISME Journal*, **4**(5), 642–647.

Fouts, D.E., Brinkac, L., Beck, E., Inman, J. and Sutton, G. (2012) PanOCT: Automated clustering of orthologs using conserved gene neighborhood for pan-genomic analysis of bacterial strains and closely related species. *Nucleic Acids Research*, **40**(22), e172.

Fraser, C., Alm, E.J., Polz, M.F., Spratt, B.G. and Hanage, W.P. (2009) The bacterial species challenge: Making sense of genetic and ecological diversity. *Science*, **323**, 741–746.

Garrigues, C., Stuer-Lauridsen, B. and Johansen, E. (2005) Characterisation of *Bifidobacterium animalis* subsp. *lactis* BB-12 and other probiotic bacteria using genomics, transcriptomics and proteomics. *Australian Journal of Dairy Technology*, **60**, 84–92.

Gibson, G. (2002) Microarrays in ecology and evolution: A preview. *Molecular Ecology*, **11**, 17–24.

Hao, P., Zheng, H., Yu, Y. *et al.* (2011) Complete sequencing and pan-genomic analysis of *Lactobacillus delbrueckii* subsp. *bulgaricus* reveal its genetic basis for industrial yogurt production. *PLoS ONE*, **6**(1), e15964.

Kislyuk, A.O., Katz, L.S., Agrawal, S. *et al.* (2010) A computational genomics pipeline for prokaryotic sequencing projects. *Bioinformatics*, **26**, 1819–1826.

Klappenbach, J.A., Saxman, P.R., Cole, J.R. and Schmidt, T.M. (2001) rrndb: The Ribosomal RNA Operon Copy Number Database. *Nucleic Acids Research*, **29**(1), 181–184.

Laing, C., Laing, C., Buchanan, C. *et al.* (2010) Pan-genome sequence analysis using Panseq: An online tool for the rapid analysis of core and accessory genomic regions. *BMC Bioinformatics*, **11**, 461.

Lindroos, H.L., Mira, A., Repsilber, D. *et al.* (2005) Characterization of the genome composition of *Bartonella koehlerae* by microarray comparative genomic hybridization profiling. *Journal of Bacteriology*, **187**(17), 6155–6165.

Liu, M., Bayjanov, J.R., Renckens, B., Nauta, A. and Siezen, R.J. (2010) The proteolytic system of lactic acid bacteria revisited: A genomic comparison. *BMC Genomics*, **11**, 36.

Makarova, K., Slesarev, A., Wolf, Y. *et al.* (2006). Comparative genomics of the lactic acid bacteria. *Proceedings of National Academy of Sciences of the United States of America*, **103**(42), 15611–15616.

Marcobal, A., Belos Rivas, B., Moreno-Arribas, M.V. and Muñoz, R. (2006) Evidence for horizontal gene transfer as origin of putrescine production in *Oenococcus oeni* RM83. *Applied and Environmental Microbiology*, **72**(12), 7954–7958.

Markowitz, V.M., Chen, I-M.A., Palaniappan, K. *et al.* (2010) The integrated microbial genomes system: An expanding comparative analysis resource. *Nucleic Acids Research*, **38**, D382–D390.

Medini, D., Donati, C., Tettelin, H., Masignani, V. and Rappuoli, R. (2005) The microbial pangenome. *Current Opinion in Genetics & Development*, **15**, 589–594.

Meijerink, M., van Hemert, S., Taverne, N. *et al.* (2010) Identification of genetic loci in *Lactobacillus plantarum* that modulate the immune response of dendritic cells using comparative genome hybridization. *PLoS ONE*, **5**, e10632.

Milani, C., Duranti, S., Lugli, G.A. *et al.* (2013) Comparative genomics of *Bifidobacterium animalis* subsp. *Lactis* reveals a strict monophyletic bifidobacterial taxon. *Applied and Environmental Microbiology*, **79**(14), 4304–4315.

Mira, A., Martin-Cuadrado, A.B., D'Auria, G. and Rodríguez-Valera, F. (2010) The bacterial pan-genome: A new paradigm in microbiology. *International Microbiology*, **13**, 45–57.

Ochman, H., Lawrence, J.G. and Groisman, E.A. (2000) Lateral gene transfer and the nature of bacterial innovation. *Nature*, **405**, 299–304.

Passerini, D., Coddeville, M., Le Bourgeois, P. *et al.* (2013) The carbohydrate metabolism signature of *Lactococcus lactis* strain A12 reveals its sourdough ecosystem origin. *Applied and Environmental Microbiology*, **79**(13), 5844–5852.

Rasmussen, T.B., Danielsen, M., Valina, O., Garrigues, C., Johansen, E. and Pedersen, M.P. (2008) *Streptococcus thermophilus* core genome: Comparative genome hybridization study of 47 strains. *Applied and Environmental Microbiology*, **74**(15), 4703–4710.

Sanchez, B., Ruiz, L., Gueimonde, M. and Margolles, A. (2013) Omics for the study of probiotic microorganisms. *Food Research International*, **54**, 1061–1071.

Schick, J., Weber, B., Klein, J.R. and Henrich, B. (1999) PepR1, a CcpA-like transcription regulator of *Lactobacillus delbrueckii* subsp. *lactis*. *Microbiology*, **145**, 3147–3154.

Schuchat, A. and Wenger, J.D. (1994) Epidemiology of group B streptococcal disease: Risk factors, prevention strategies, and vaccine development. *Epidemiologic Reviews*, **16**, 374–402.

Siezen, R.J., Bayjanov, J.R., Felis, G.E. *et al.* (2011) Genome-scale diversity and niche adaptation analysis of *Lactococcus lactis* by comparative genome hybridization using multi-strain arrays. *Microbial Biotechnology*, **4**(3), 383–402.

Smith, J.M., Smith, N.H., Rourke, M.O. and Spratt, B.G. (1993) How clonal are bacteria? *PNAS*, **90**(10), 4384–4388.

Smokvina, T., Wels, M., Polka, J. *et al.* (2013) *Lactobacillus paracasei* comparative genomics: Towards species pan-genome definition and exploitation of diversity. *PLoS ONE*, **8**(7), e68731.

Tettelin, H., Masignani, V., Cieslewicz, M.J. *et al.* (2005) Genome analysis of multiple pathogenic isolates of *Streptococcus agalactiae*: Implications for the microbial 'pan-genome'. *Proceedings of National Academy of Sciences of the United States of America*, **102**(39), 13950–13955.

Ventura, M., O'Flaherty, S., Claesson, M.J. *et al.* (2009) Genome-scale analysis of health-promoting bacteria: Probiogenomics. *Nature Reviews Microbiology*, **7**, 61–71.

Winsor, G.L., Khaira, B., Van Rossum, T., Lo, R., Whiteside, M.D. and Brinkman, F.S.L. (2008) The Burkholderia Genome Database: Facilitating flexible queries and comparative analyses. *Bioinformatics*, **24**, 2803–2804.

Zhang, Q., Melche, U., Zhou, L., Najar, F.Z., Roe, B.A. and Fletcher, J. (2005) Genomic comparison of plant pathogenic and nonpathogenic *Serratia marcescens* strains by suppressive subtractive hybridization. *Applied and Environmental Microbiology*, **71**, 7716–7723.

Zhao, Y., Jia, X., Yang, J. *et al.* (2014) PanGP: A tool for quickly analyzing bacterial pan-genome profile. *Bioinformatics*, **30**(9), 1297–1299.

Zhao, Y., Wu, J., Yang, J., Sun, S., Xiao, J. and Yu, J. (2012) PGAP: Pan-genomes analysis pipeline. *Bioinformatics*, **28**, 416–418.

Zheng, H., Lui, E., Hao, P., Konno, T., Oda, M. and Ji, Z.S. (2012) In silico analysis of aminoacid biosynthesis and proteolysis of *Lactobacillus delbrueckii* subsp. *bulgaricus* 2038 and the implications for bovine milk fermentation. *Biotechnology Letters*, **34**(8), 1545–1551.

CHAPTER 10

Commercial starters or autochtonous strains? That is the question

Maria Rosaria Corbo, Angela Racioppo, Noemi Monacis and Barbara Speranza
Department of the Science of Agriculture, Food and Environment, University of Foggia, Italy

The high and stable quality of traditional fermented foods is one of the consumer's main demands and a leading factor in developing product uniformity is starter activity. In fact, the role of starters is crucial in fermented foods, because without correct fermentation by yeasts and bacteria, some essential foods like meats, cheeses, beer, olives, wine, bread and vinegar do not exist.

It is generally recognized that there are two main ways to pilot a fermentation process: some producers prefer to rest the quality of their products on the performance of 'virtuous' natural microflora, whereas others prefer to use commercial starter cultures (biomass of microbial origin) to ensure a correct and predictable process and avoid fermentation arrests or the production of undesired metabolites. However, in the first case the risk of incurring health problems and a non-standardized product and process is very high, thus the use of commercial starters is actually the most common practice.

A strain could be considered as a potential starter if it responds better than others to predetermined selection criteria, depending on the type of action and the product to be obtained. The list of selection criteria is continually growing and includes not only the absence of unwanted characteristics, but also the presence and expression of the desired characteristics. By definition, starters must ensure the success of the transformation, make its trajectory predictable and guarantee the quality of the final product. If their use is properly performed, starter cultures overcome indigenous microflora, which can remain viable (at least for a certain period of time), but it is able to grow massively and compete with the inoculated strains. Modern techniques of molecular biology have permitted confirmation that, when the operations of the starter inoculum are well conducted, there is absolute dominance of the

Starter Cultures in Food Production, First Edition. Edited by Barbara Speranza, Antonio Bevilacqua, Maria Rosaria Corbo and Milena Sinigaglia.

inoculated strain at all stages of the fermentation process: one strain and only one strain.

The starters available on the market are apparently not few in number and each year that number increases. Each strain is presented by the manufacturers as 'really capable' of operating as described in the technical information (provided with the commercial preparation) and at least one strain exists for every type of production. On the other hand, this is not true, because the number of strains available on the market is absolutely meagre when compared to the number and types of fermented foods produced, from which it follows that, in many cases, the strain used as starter in a specific product has nothing to do with the raw material and the geographical area where the typical product is realized. In general, the main factor that drives food manufacturers towards the use of commercial starters is the fear of failure, but this choice often hampers the possibility of obtaining a 'unique' product through the exploitation of microbial biodiversity. In the dairy and wine industries, for example, the use of the same commercial starters for the production of different product types is causing a 'flattening' of the sensory quality of the products obtained, which are no longer distinguishable by production technology and geographical origin. This is a great loss of typical products representing priceless treasures of human communities able to symbolize their heritage and the sociocultural aspects of their ethnicity. Food prepared by different peoples should remain unique and distinguishable on the basis of geographical location, environmental factors, food preference and the availability of plant or animal sources.

Fortunately, to guarantee the production of a 'unique' fermented product, a third way has recently appeared as a synthesis of the two previous methods (natural microflora or commercial starter); that is, the use of 'typical' microbial starters formulated from autochthonous microbial strains appropriately selected as representatives of 'virtuous' microbial biodiversity. The dominant autochthonous strains should be isolated from the specific fermented foods (or similar ones) and evaluated as potential starter cultures for the same product; these strains are, in fact, most adapted to the specific intrinsic ecology of a specified fermented matrix and possess favourable technological and hygienic properties for food applications. To be used as potential starter cultures, the strains must be phenotypically and genotypically characterized, including technological and safety features (Ammor and Mayo 2007). However, even if results *in vitro* (at given laboratory conditions) could be promising, the performance of the isolated cultures in real food fermentation must be demonstrated. Thus, in the last few years several authors have focused their studies on the identification, selection and use of wild strains to pilot a standardized fermentation process, in order to achieve the desired fermentation parameters specific to the product type (Casquete *et al.* 2012a, b).

Application of selected wild strains in food: Some case studies

An overview of studies conducted on the selection, characterization, identification and use of autochthonous strains as starters in fermented foods during the last decade is shown in Table 10.1. As can be inferred, studies have been performed on different food matrices, such as meat, fish, dairy products, vegetables, fermented beverages, sourdoughs and other fermented products from all over the world.

In general, the starters used in these studies were previously isolated from the same matrix or from a different one and inoculated at different levels ranging from 10^4 to 10^9 colony forming units (cfu) per gram (or per mL), with the general aim of improving process control and preserving the typical characteristics of the final products.

Meat and fish products

In meat fermentation, many products still base their success on the growth of lactic acid bacteria (LAB) within the natural microflora of the ingredient mix, and the selective effect of the environment and processing conditions is vital in ensuring satisfactory fermentation. Sometimes this can be stimulated by a procedure known as *backslopping*, where a mix from a previous successful batch is retained and added to the new batch to ensure the introduction of a substantial inoculum. However, even where commercial starter cultures are used, the selectivity of the environment plays an important role since, unlike in many fermented dairy products, it is not possible to pasteurize the raw materials to eliminate competitors. Numerous studies have identified the particular species of LAB associated with different fermented meats: *Lactobacillus* species are dominant at the end of fermentation, with the most common species encountered being *Lactobacillus sake*, *Lactobacillus curvatus* and *Lactobacillus plantarum* (Leroy *et al.* 2006), whereas Gram-positive, catalase-positive cocci such as *Staphylococcus xylosus*, *Staphylococcus carnosus*, *Staphylococcus saprophyticus* and *Kocuria varians* (formerly known as *Micrococcus varians*) would normally be most active at the start of fermentation and tend to be inhibited as the LAB begin to dominate and the pH declines.

Commercial starter cultures have been used in fermented meats for about 50 years (Adams 1986; Jessen 1995) and nowadays different combinations of strains are available on the market to simplify handling and give a more rapid initiation of fermentation.

The majority of the studies listed in Table 10.1 focused just on meat products, and particularly on dry-fermented sausage, which is the fermented meat product *par excellence* and consequently the most used matrix to isolate wild strains, characterize them phenotypically and technologically and reuse them as starters to pilot fermentation. Only in one study (Cenci-Goga *et al.* 2008)

Table 10.1 Overview of recent studies conducted on the selection, characterization, identification and use of autochthonous strains as starters in some fermented foods.

Product	Autochthonous starter and its origin	Aim	Results obtained	Reference
MEAT				
Chorizo	*Pe. acidilactici* MC184, MS198, MS200 and *Staph. vitulus* RS34, isolated from traditional Iberian dry-fermented sausages	Effect of the different ripening processes at an industrial level on the growth and development of starter cultures	Autochthonous starter cultures were able to compete and colonize the sausages, improving their homogeneity and safety	Casquete *et al.* (2012a)
Chouriço	*Lb. sakei* ST153, isolated from Chouriço, a cured/smoked pork product	Antilisterial effect while preserving typical sensory characteristics	A significant inhibition of pathogen growth was observed and good sensory characteristics were recovered	Jácome *et al.* (2014)
Galician chorizo	*Lb. sakei* LS131, *Staph. equorum* SA25, *Staph. epidermidis* SA49, *Staph. saprophyticus* SB12, isolated from Androlla and Botillo, two traditional Galician fermented sausages	Effect of different autochthonous starter cultures on volatile compounds profile and sensory properties	Autochthonous strains were able to guarantee homogeneous production with desired typical sensory characteristics obtained during artisanal elaboration	Fonseca *et al.* (2013)
Greek sausage	*Lb. sakei* (8416 and 4413), isolated from naturally fermented sausages	Inactivation of *L. monocytogenes* and *E. coli* O157:H7	A significant inhibition of pathogen growth was observed	Pragalaki *et al.* (2013)
Greek sausage	*Lb. sakei* (8416, 4413, 8426), *Lb. plantarum* 7423 and *Lb. curvatus* 8427, isolated from naturally fermented sausages	Production of low-acid fermented sausages	*Lb. sakei* 4413 was identified as the best autochthonous starter in contributing to sensory attributes, reducing biogenic amines and preventing lipid oxidation	Baka *et al.* (2011)
Salame nostrano	*Lc. lactis* ssp. *lactis* (16, 340), *Lb. casei* ssp. *casei* (208), isolated from traditional cheeses	Effect of a formulation of dairy-origin lactic acid bacteria on the microbiological, chemical and sensory characteristics of Salame nostrano	Selected starter cultures prevented the growth of safety indicators and increased the acceptability of full-ripened salami	Cenci-Goga *et al.* (2008)
Salchichòn	*Pe. acidilactici* MS200 and *Staph. vitulus* RS34, isolated from Iberian dry-fermented sausage	Improvement of sensory characteristics and prevention of biogenic amine formation	The use of the proposed starters determined a higher amount of amino nitrogen and volatile compounds, and a reduction of biogenic amine accumulation	Casquete *et al.* (2011a)

(Continued)

Table 10.1 (Continued)

Product	Autochthonous starter and its origin	Aim	Results obtained	Reference
Salchichòn	*Pe. acidilactici* (MS198, MS200) and *Staph. vitulus* RS34, isolated from Iberian dry-fermented sausage	Improvement of process control and of homogeneity/healthiness of the product	Autochthonous starter cultures produced a positive effect on the sensory characteristics and improved homogeneity and healthiness of the product	Casquete *et al.* (2011b)
Salchichòn	*Pe. acidilactici* (MS198, MS200) and *Staph. vitulus* RS34, isolated from Iberian dry-fermented sausage	Ability of selected autochthonous starter cultures to become implanted during product manufacture and effect on flavour quality	The implantation of autochthonous starter cultures was adequate and did not significantly modify the flavour of the traditional product	Casquete *et al.* (2011c)
Salchichòn	*Pe. acidilactici* (MS198, MS200) and *Staph. vitulus* RS34, isolated from Iberian dry-fermented sausage	Development of sensory characteristics in dry-fermented sausages	Autochthonous starters determined higher amounts of non-protein nitrogen and volatile compounds and reduced biogenic amine accumulation in comparison with commercial starters	Casquete *et al.* (2012b)
Sardinian sausage	*Staph. xylosus*, *Lb. curvatus* and *Lb. plantarum*, isolated from Sardinian traditional fermented sausages	Effect on microbiological, biochemical and sensory properties	Autochthonous starters produced a rapid acidification of substrates, reduced the number of spoilage microorganisms and the content of total biogenic amines and obtained the best organoleptic properties	Mangia *et al.* (2013)
Sausage	*Lb. sakei* F08F202, *Staph. equorum* F08bF15 and *Staph. succinus* F08bF19, isolated from traditional dry-fermented sausages	Improvement of safety while preserving typical sensory characteristics	A significant inhibition of pathogen growth was observed and good sensory characteristics were recovered	Talon *et al.* (2008)
Sausage	*Lb. sakei* (DBPZ0062, DBPZ0098, DBPZ0338, DBPZ0329, DBPZ0416), *Staph. equorum* (DBPZ0241, DBPZ0248, DBPZ0044), *Staph. succinus* (DBPZ0251) and *Staph. xylosus* (DBPZ0224), isolated from traditional Italian fermented sausages	Effects of addition of starter cultures on ripening	Products with good and stable organoleptic characteristics were obtained	Bonomo *et al.* (2011)

Soppressata of Vallo di Diano	*Staph. xylosus* (AVS3, AVS4, AVS5, BVS10, IVS37, IVS38, IVS39, LVS48, LVS49), *Lb. sakei* (BVL6, BVL7), *Lb. curvatus* (CVL12), *Lb. sakei* (DVL17, FVL26, HVL36) and *Lb. curvatus* (IVL41, IVL42) isolated from soppressata	Isolation, selection and use of suitable autochthonous strains of lactic acid bacteria and staphylococci in manufacturing the sausages	*Staph. xylosus* AVS5 and *Lb. sakei* DVL17 were the most suitable strains to manufacture this product	Villani *et al.* (2007)
FISH				
Whole carp	*Lb. plantarum* 120, *Lb. plantarum* 145 and *Pe. pentosaceus* 220, isolated from Chinese traditional low-salt fermented carp	Determination of the effects of autochthonous lactic acid bacteria on the characterization of whole carp during fermentation	The strains tested were able to reduce fermentation time and improve the quality of samples	Zeng *et al.* (2014)
Surimi	*Lb. plantarum* (LP-2, LP-7, LP-15, LP-21), isolated from Chinese traditional low-salt fermented whole-fish product 'Suan yu'	Characterization and selection of autochthonous strains to be used as starters	Lp-15 and Lp-21 stand up as good starter strains with optimal technological properties	Zeng *et al.* (2014)
Fish sauce	*Pe. acidilactici* (17 and 51) and *Pe. pentosaceus* (41), isolated from salted anchovies	Characterization and selection of autochthonous strains to be used as starters	The use of the proposed strains reduced the fermentation time (2 days), but also improved the microbiological quality of the final product	Speranza *et al.* (2014)
DAIRY PRODUCTS				
Caciocavallo Pugliese cheese	*Lb. paracasei* subsp. *paracasei* (B44f3t, B25f3) and *Lb. parabuchneri* (B51f5, B23f3), isolated from Caciocavallo Pugliese cheese	Standardization of Caciocavallo Pugliese cheese processing, safeguarding the peculiar traits of the product	The inoculated strains guaranteed higher levels of free amino acids and typical sensory characteristics	Morea *et al.* (2007)
Fior di Latte	*Lb. casei*, *Lb. plantarum*, *Pe. acidilactici* and *Staph. thermophilus* isolated from whey and Fior di Latte	Selection of autochthonous strains to standardize *Fior di Latte* production	Three isolates were identified as good acidifiers of milk	Speranza *et al.* (2014)

(Continued)

Table 10.1 (Continued)

Product	Autochthonous starter and its origin	Aim	Results obtained	Reference
Fiore Sardo cheese	*Lc. lactis* subsp. *lactis* CFM 7, *Lb. casei* subsp. *casei* Lc101 and *Lb. plantarum* Lp 17, isolated from raw ewes' milk and Fiore Sardo cheese	Improvement of the microbiological and physicochemical characteristics of the cheese	Selected starters contributed to a balanced ratio of chemical constituents, a reduced number of spoilage microorganisms and the absence of production waste	Mangia *et al.* (2008)
Fresh cheese	*Lb. fermentum* A8 and *Ent. faecium* A7 isolated from artisanal fresh cheese	Identification, characterization and implementation of autochthonous starter cultures in the production of traditional fresh cheese	Enrichment of the flavour of industrially produced fresh-type cheese under controlled conditions	Leboš Pavunc *et al.* (2012)
Goats' milk	*Lb. rhamnosus* and *Lb. plantarum*, isolated from raw goats' milk in western Algeria	Isolation of lactic acid bacteria–producing bacteriocines, able to inhibit poisoning bacteria	Two lactobacilli species were able to strongly inhibit *Staph. aureus* growth	Anas *et al.* (2012)
Manchego cheese	*Lc. lactis* subsp. *lactis* CECT 7883 (B8W3), *Lc. lactis* subsp. *lactis* CECT 7884 (A0W2) and *Lb. paracasei* subsp. *paracasei* CECT 7882 (PBL226), isolated from Manchego cheese	Improvement of volatile composition and odour sensory characteristics	Intensification and improvement of industrial Manchego cheese aroma greatly similar to the traditional one	Poveda *et al.* (2014)
Mozzarella cheese	*Lb. helveticus* 2B, *Lb. delbrueckii* subsp. *lactis* 20F, *Lb. plantarum* 18A and *Strep. thermophilus* 22C, isolated from natural whey starter cultures for Italian Mozzarella cheese	Improvement of texture, sensory properties and shelf life of high-moisture traditional cows' Mozzarella cheese	The shelf life was extended to 12–15 days instead of the 5–7 days of traditional high-moisture Mozzarella cheese	De Angelis *et al.* (2008)
Pecorino Sardo cheese	*Lc. lactis* subsp. *lactis* 3PS103, *Lb. casei* subsp. *casei* SPS1, *Strep. thermophilus* LbPS2 and *Lb. helveticus* LPS31, isolated from raw ewes' milk and traditional Pecorino Sardo	Improvement of physico-chemical parameters	Production of experimental cheese with a significantly higher level of essential free amino acids	Madrau *et al.* (2006)
Tenerife cheese	*Lc. lactis* subsp. *lactis* TF53, *Lb. plantarum* TF191 and *Ln. mesenteroides* subsp. *mesenteroides* TF756, isolated from artisanal Tenerife cheese	Comparison of microbiological, physico-chemical, proteolytic and sensory characteristics of Tenerife cheese made with commercial starter or an autochthonous one	Uniformity of the product with constant quality and improved sensory characteristics	Gonzàlez and Rate (2012)

Yogurt	*Strep. thermophilus* and *Lb. delbrueckii* ssp. *bulgaricus*, isolated from artisanal yogurt	Improvement of microbial, chemical and organoleptic characteristics	Increase in acidity and acetaldehyde content, significant decrease in pH and microbial population during cold storage	Pourahmad and Assadi (2007)
VEGETABLES				
Carrots, beans and marrows	*Lb. plantarum* M1, *Pe. pentosaceus* C4 and *Ln. mesenteroides* C1 isolated from carrots, french beans or marrows	Identification and selection of a mixed autochthonous starter to be used for fermented vegetables	Autochthonous starters were preferred to allochthonous starters for the rapid decrease of pH, inhibition of *Enterobacteriaceae* and yeasts, higher total concentration of vitamin C, improved colour and fragrance	Di Cagno *et al.* (2008)
Chinese sauerkraut	*Ln. mesenteroides* NCU1426, *Lc. lactis* NCU1315, *Lb. plantarum* NCU1121 and *Lb. casei* NCU 1222, isolated from Chinese sauerkraut	Selection of autochthonous strains to be used as starters	The four lactic acid bacteria strains showed optimal fermentative properties	Xiong *et al.* (2014)
Grains	*Pe. pentosaceus* F16A and *Lb. curvatus* F18A, isolated from Acha flour; *Pe. pentosaceus* 16I and *Lb. plantarum* 13I, isolated from Iburu flour	Evaluation of technological and nutritional properties of African cereals Acha and Iburu	Improvement of nutritional and sensorial qualities	Coda *et al.* (2011a)
Green olives (cv. *Arbequina*)	*Lb. pentosus*, *Candida diddensiae* and *Lb. plantarum*, isolated from olive brines	Evaluation of the potential of *Lb. pentosus* to be used as a starter for table olive fermentation	Reduction of the survival time of *Enterobacteriaceae* and decrease of pH to desirable levels	Hurtado *et al.* (2010)
Kedong sufu	*Kc. rosea* KDF3, isolated from traditional Kedong sufu	Investigation of the effects of *Kc. rosea* KDS3 and protease KP3 as adjuncts for the acceleration of Kedong sufu ripening	*Kc. rosea* KDF3 or its protease KP3 can hasten sufu maturation and bring desired characteristics	Feng *et al.* (2014)

(Continued)

Table 10.1 (Continued)

Product	Autochthonous starter and its origin	Aim	Results obtained	Reference
Leek	*Lb. plantarum* IMDO 788, *Ln. mesenteroides* IMDO 1347 and *Lb. sakei* IMDO 1358, isolated from spontaneous leek fermentation	Application and validation of selected strains of LAB to perform controlled leek fermentation	The mixed-starter culture of *Lb. plantarum* IMDO 788 and *Ln. mesenteroides* IMDO 1347 resulted in fermented leeks of good microbiological quality and in more extensive carbohydrate consumption	Wouters *et al.* (2013)
Millet grains	*Lb. fermentum* ULAG2, *Lb. plantarum* ULAG11, *Lb. plantarum* ULAG 24, *Pe. pentosaceus* ULAG23, *Strep. gallolyticus* subsp. *macedonicus* ULAG45 and *B. cereus* ULAG84, isolated from ogi (fermented white, yellow maize and red sorghum) and kunu-zaki (fermented millet)	Evaluation of the effects of selected strains on safety quality and nutritional enhancement of cereal fermentation	*Lb. plantarum* ULAG11 demonstrated amylase production in both *in vitro* and *in situ* laboratory-scale fermentations, indicating that this strain can be used during small- and industrial-scale cereal fermentation	Oguntoyinbo and Narbad (2012)
Red and yellow peppers	*Lb. plantarum* PE21 and *Lb. curvatus* PE4, isolated from raw red and yellow peppers	Setting up the protocol for fermentation of red and yellow peppers that aimed at ensuring the shelf life of peppers at room temperature	All autochthonous strains persisted during processing and storage, showing rapid decrease of pH, inhibition of total *Enterobacteriaceae* and yeasts, higher firmness and colour indexes	Di Cagno *et al.* (2009a)
Red sweet peppers, pattypan squash and tomatoes	Strains of *Lb. plantarum*, *Lb. brevis* and *Pe. pentosaceus* isolated from peppers, squash and tomatoes	Selection of autochthonous strains from spontaneously fermented pattypan squash, red sweet peppers and tomatoes to be used as potential starters	The use of isolated strains of *Lb. plantarum*, *Lb. brevis* and *Pe. pentosaceus* gave fermented products with sensorial qualities equal to those obtained through spontaneous fermentation	Piasecka-Jóźwiak *et al.* (2013)
Sweet cherry puree	*Pe. pentosaceus* SWE5 and *Lb. plantarum* FP3, isolated from sweet cherry (*Prunus avium L.*)	Use of mixed autochthonous starter to improve safety, antioxidant, texture, colour and sensory properties of cherry puree	Fermentation by autochthonous lactic acid bacteria positively interferes with antioxidant activity, viscosity, colour and sensory attributes	Di Cagno *et al.* (2011)

Table olives	*Lb. pentosus* OM13 and *Lb. coryniformis* OM68, isolated from Green olives from the cultivar 'Nocellara del Belice'	Evaluation of the effects of selected strains on the microbiological and sensory profile of table olives	Autochthonous starter showed optimal adaptation to the environment, rapid acidification and low presence of off-odours	Aponte *et al.* (2012)
Tomato juice	*Lb. plantarum* POM1, POM8, POM27, POM35 and POM43, *Lb. brevis* POM2, *Ent. faecium/ faecalis* POM3, *We. cibaria* POM11, *Pe. pentosaceus* POM10 and *Lactobacillus* sp. POM44, isolated from tomatoes	Use of selected strains for fermentation to obtain optimal viscosity, colour, antioxidant activity and volatile compounds	Juice fermented with autochthonous strains had the highest viscosity, elevated values of ascorbic acid, glutathione and total antioxidant activity and improved colour indexes	Di Cagno *et al.* (2009b)
FERMENTED BEVERAGES AND WINE				
Emmer beverage	*Lb. plantarum* 6E, isolated from emmer or spelt flours, *Lb. plantarum* PL9, *We. cibaria* WC4, *We. cibaria* WC3, *We. cibaria* WC9 and *Pe. pentosaceus* PP1 isolated from wheat sourdoughs	Manufacturing and characterizing the physical, chemical, functional and sensory properties of non-alcoholic emmer beverages	Autochthonous starter showed high suitability to produce functional beverages from emmer	Coda *et al.* (2011b)
Pineapple juice	*Lb. plantarum* 1LE12 and *Lb. rossiae* 2LC10, isolated from pineapples	Assurance of microbiological, antioxidant, texture, colour and sensory properties of the juice	Autochthonous starters were preferred for the highest antioxidant activity and firmness, besides better preservation of the natural colours and odour	Di Cagno *et al.* (2010)
Shalgam (black carrot beverage)	*Lb. plantarum*, *Lb. brevis*, *Lb. paracasei* subsp. *paracasei*, *Lb. buchneri*, *Lb. pentosus*, *Lb. delbrueckii* subsp. *delbrueckii*, *Lb. fermentum*, *Lc. lactis* subsp. *lactis*, *Ln. mesenteroides* subsp. *mesenteroides*, *Ln. mesenteroides* subsp. *mesenteroides/ dextranicum*, *Ln. mesenteroides* subsp. *cremoris* and *Pe. pentosaceus* isolated from shalgam	Selection of the most suitable strains as starter culture for shalgam production	*Lb. plantarum* bx, *Lb. fermentum* and *Lb. paracasei* subsp. *paracasei* showed the best potential for use as lactic starter cultures to standardize shalgam production	Tangüler and Erten (2013)

(Continued)

Table 10.1 (Continued)

Product	Autochthonous starter and its origin	Aim	Results obtained	Reference
Wine	*S. cerevisiae*, isolated from natural fermentation of must obtained from Susumaniello grapes	Characterization of indigenous strains denoted by excellent oenological properties, to be used for the preparation of a native fermentation starter	Five yeast strains dominated the fermentation process and produced wines characterized by peculiar oenological and organoleptic features	Tristezza *et al.* (2014)
Wine	*Kl. apiculata* mc1, *S. cerevisiae* mc2 and *O. Oeni* X2L, isolated from Malbec grape musts and red wine	Comparative analysis of chemical and sensory characteristics of Argentinean typical red wines fermented by indigenous starter	Inclusion of autochthonous *Kl. apiculata* mc1 as an adjunct culture to *S. cerevisiae* mc2 during Malbec must fermentation improved the organoleptic properties of red wines	Mendoza *et al.* (2011)
Wine	*S. cerevisiae* JP88, E7AR1 and SMR16-5A, isolated from Spanish wineries	Analysis of the microbial quality and the oenological properties of the fresh yeast culture concentrate for elaboration of good-quality wine	The performance of these yeasts was excellent, by dominating fermentation and improving physico-chemical parameters and organoleptic quality	Maqueda *et al.* (2011)
Wine	*S. cerevisiae* isolated from grape samples in a single vineyard of Uva di Troia variety (harvest 2008) from Apulia (southern Italy)	Selection of autochthonous *S. cerevisiae* strains as wine starter using a polyphasic approach and ochratoxin A removal	The novelty of this study relies on the use of the ability to remove ochratoxin A as a primary trait for the selection of potential wine starters	Petruzzi *et al.* (2014b)
Wine	*S. cerevisiae* W13 isolated from Uva di Troia grape (harvest 2008), a typical cultivar from Apulia (southern Italy).	Selection of a wild strain of *S. cerevisiae* able to remove ochratoxin A as a potential starter	*S. cerevisiae* W13 showed the ability to remove ochratoxin A	Petruzzi *et al.* (2014a)
Wine	*S. cerevisiae*, isolated from a Spanish cellar	Selection of the best yeast strains among autochthonous and commercial to obtain organic wines with high organoleptic qualities	Autochthonous yeasts produced wines with higher volatile compound content and organoleptic quality	Callejon *et al.* (2010)
Yogurt-like beverages (cereal, soy and grape must)	*Lb plantarum* 6E, isolated from emmer flour, and *Lb. plantarum* M6, isolated from blackberries	Effects on microbiological, textural, nutritional and sensory properties of the final products	The tested strains increased total free amino acids and concentration of polyphenolic compounds and ascorbic acid	Coda *et al.* (2012)

SOURDOUGH AND FERMENTED CEREALS				
Wheat flour sourdough	*Lb. plantarum* DB200, 3DM, CF1, 2MF8, G10C3, 12H1 and LP20, isolated from Italian sourdough	Exploration of the robustness of selected *Lb. plantarum* during propagation of wheat flour sourdough	Autochthonous strains of *Lb. plantarum* dominated the flour microbiota	Minervini *et al.* (2010)
Wheat flour sourdough	*Lb. plantarum* AELLI12 and EMRS4, *Ln. citreum* BELLI7 and *Ln. mesenteroides cremoris* AMSE2	Evaluation of the behaviour of starters during a complete wheat sourdough breadmaking process	Each starter ensured the production of breads with overall acceptance	Robert *et al.* (2006)
Barley flour sourdough	*Lb. plantarum* (PB12, PB98, PB161), *We. kimchii* (PB32, PB151, PB162), *Lb. sanfranciscensis* (PB211, PB221, PB276), *Lb. paralimentarius* (PB230, PB264), *Lb. brevis* (PB55), *Lb. curvatus* (PB1, PB115), *Ln. pseudomesenteroides* (PB172, PB285), *Ln. citreum* (PB220), *Lb. helveticus* (PB189), *We. confusa* (PB150), *S. cerevisiae* (PL15, PL33, PL80) and *C. humilis* (PL11, PL12, PL13)	Microbiological and technological characterization of laboratory-made sourdoughs for use in barley flour–based breadmaking	A few species among lactobacilli and yeasts (*Lb. brevis*, *Lb. plantarum* and *S. cerevisiae*) showed robustness to environmental conditions and microbial competitors	Zannini *et al.* (2009)
Sourdough	*Lb. plantarum* 6, 7, 12, 13, 38 and 54, *S. cerevisiae* 17 and *C. humilis* 40 isolated from sourdough of Altamura bread	Selection of promising starter cultures for sourdough	Six strains also possessed good performance in real systems	Corbo *et al.* (2013)

Notes: B. = Bacillus; C. = Candida; Ent. = Enterococcus; E. = Escherichia; Kc. = Kocuria; Kl. = Kloeckera; Ln. = Leuconostoc; Lb. = Lactobacillus; Lc. = Lactococcus; L. = Listeria; O. = Oenoccocus; Pe. = Pediococcus; Sacch. = Saccharomyces; Staph. = Staphylococcus; Strep. = Streptococcus; We. = Weissella.

were the starters used isolated from traditional cheeses and then used to perform fermentation of a *salame nostrano*, a typical Italian product.

It is well recognized that starters used for meat product fermentation have to play some essential roles: acidification of the substrate, inhibition of undesirable organisms, promotion of bioconservation of the final product with contribution to its ripening, and improvement of organoleptic characteristics through proteolysis and lipolysis processes (Casquete *et al.* 2011a, b, c). In fact, as shown in Table 10.1, most of the studies performed aimed to develop an autochthonous starter culture able to improve safety and inhibit pathogen growth, while preserving the typical sensory characteristics of the traditional fermented product.

On this issue, the study conducted in 2012 by Casquete *et al.* (2012b) from the University of Extremadura (Spain) stands out as particularly interesting. These authors investigated the effects of selected autochthonous starter cultures on the development of sensory characteristics in *salchichon* (a traditional Iberian dry-fermented sausage), using a commercial starter as a comparison. More specifically, lyophilized strains of *Pediococcus acidilactici* MC184 and *Staphylococcus vitulus* RS34, previously isolated from indigenous populations of Iberian dry-fermented sausage, were used to start fermentation of a mixture of Iberian pork, salt (NaCl), white and black pepper and a commercial compound of spices and additives. The inoculum was performed at a concentration of 5×10^7 cfu/g. Another batch, with the same mixture, was added with a commercial starter culture, according to the manufacturer's instructions. The sausages were ripened for 86 days in different conditions of temperature and relative humidity and then subjected to different analyses: moisture, water activity, pH determination, microbiological analysis, determination of the parameters related to lipid and protein fraction, sensory analysis, instrumental determinations of colour, volatile compounds and texture.

The results showed that the autochthonous starter cultures were able to compete well and control the process, comparable to commercial starter, but the autochthonous starters showed a higher inhibitory effect on *Enterobacteriaceae* and coliform flora than the commercial one, guaranteeing better microbiological and hygienic quality. Another very important result was found for biogenic amines, since the autochthonous starter was better at reducing their accumulation than the commercial one, ensuring the healthiness and homogeneity of the products and without producing a negative effect on the sensory characteristics of the traditional fermented sausage. In fact, products realized using autochthonous starter cultures showed higher amounts of both non-protein nitrogen (NPN) and volatile compounds, derived from amino acid catabolism, promoting a flavour associated with traditional dry-fermented meat products. In this area the results obtained were very promising, as the tested autochthonous starters were able to guarantee the unique qualities of this fermented meat product, maintaining strong attention on the industrial exigencies of product standardization. Similar results were also obtained in the other reported studies, so opening up a new, successful way to pilot a fermentation process.

Compared to fermented meat products, the use of starters in fish products is still limited, but more recently the use of pure bacterial cultures to produce fish-type products is attracting increasing interest. Fish fermentation remains one of the most common methods of seafood preservation, as it has many benefits and could be used as a low-cost, convenient technique for the preservation of fish muscle, improving its organoleptic qualities and increasing the nutritional value and/or digestibility of the raw material. Nowadays, fermented fish products are largely confined to east and southeast Asia, although some products are being produced elsewhere and exported from Oriental countries to Europe and North America (Adams 2009). LAB are found to be the dominant microorganisms in many fermented fish products, where their primary role is to perform carbohydrate fermentation, causing a decrease in pH. The combination of low pH (below 4.5) and organic acids (mainly lactic acid) is the main preservation factor in this kind of product (Kose and Hall 2011). However, most of the fermented fish products are still produced on a cottage industry or domestic scale; to avoid spontaneous fermentation that is not easily controllable, in this case it would also be appropriate to select appropriate starters, to optimize and control the evolution of biochemical processes that occur during the production cycle of the products concerned. A new approach to isolate, characterize and select autochthonous starters for fermented fish products was proposed by Speranza *et al.* (2014); these authors aimed to individuate promising strains as potential starter cultures for fermented fish products using a step-by-step procedure. Their study focused on 59 isolates of bacteria recovered from salted anchovies (*Engraulis encrasicholus*). The isolates were phenotypically characterized through Gram staining, catalase activity, glucose metabolism, hydrogen sulfide (H_2S) and indole production, nitrate reduction, citrate utilization and hydrolysis of arginine, esculin, casein, gelatine, starch, Tween 80 and urea. Then Gram-positive isolates (44 out of 59) were studied for their growth at different temperatures, with salt added and at various pH values, for acidification of a laboratory medium. Thus, three promising strains were selected and identified as members of the genus *Pediococcus*. After selection, the research finished with evaluation of the acidification performance of the promising isolates throughout the fermentation of a fish sauce: the results were very promising, since the use of the proposed strains not only reduced fermentation time (2 days), but also improved the microbiological quality of the final product by standing as effective substitutes for commercial starters.

Dairy products

As for the other products mentioned, the natural microbiota of milk is inefficient, uncontrollable and unreliable; moreover, it is generally destroyed by the heat treatments applied in some processes. Consequently, modern technology makes use of carefully selected microorganisms that are intentionally added to pasteurized or sterilized milk for controlling the fermentation process in a more predictable way. Depending on the principal function, added microorganisms are referred to as starters or primary cultures (if they participate in acidification) and adjunct, maturing or secondary cultures (for flavour, aroma and maturing activities; Topisirovic *et al.* 2006). Primary cultures include species such as *Lactococcus lactis*, *Leuconostoc* spp., *Streptococcus thermophilus*, *Lactobacillus delbrueckii* and *Lactobacillus helveticus* (Parente and Cogan 2004). Secondary cultures include *Propionibacterium freudenreichii*, *Brevibacterium linens*, *Debaryomyces hansenii*, *Geotrichum candidum*, *Penicillium roqueforti* and *Penicillium camemberti*.

Strains available today as starters are derived from 'natural starters' (NS) of undefined composition reproduced daily in fermentation facilities by backslopping, especially by artisan and industrial manufacturers. The best NS propagated under controlled conditions result in 'mixed-strain starters' (MSS), an undefined mixture of LAB species. As opposed to MSS, 'defined-strain starters' (DSS) are composed of one or more strains (up to 13–15; Parente and Cogan 2004).

Specific starters isolated for dairy products should preserve the sensorial characteristics of traditional products and allow reproduction of the organoleptic properties of fermented products made of raw milk from pasteurized (or sterilized) milk. In fact, a large part of the studies listed in Table 10.1 aimed at the identification and selection of candidate strains for designing specific 'virtuous' starters able to satisfy these needs. In these studies, the strains were generally isolated from raw milk, natural whey and cheese and inoculated in this same matrix to 10^6–10^8 cfu/mL, to obtain yogurt and/or different kinds of fresh and ripened cheese.

As previously mentioned, the primary function of dairy product starters is the production of lactic acid from lactose (milk acidification), but there are other important aspects, including flavour, aroma and alcohol production, proteolytic and lipolytic activities, inhibition of undesirable organisms and improvement of the cheese's keeping quality (Parente and Cogan 2004). Most of the studies presented in Table 10.1 only aimed at selecting autochthonous starter cultures able to perform the functions mentioned, also having a good effect on physicochemical and organoleptic characteristics and safeguarding the peculiar traits of the product.

The use of autochthonous cultures has been studied in several dairy products, such as Tetilla cheese (Menéndez *et al.* 2004), Proosdij-type cheese (Ayad *et al.* 2003), Reggianito Argentino cheese (Candioti *et al.* 2002) and New Zealand Cheddar cheese (Crow *et al.* 2001). Among these products, Caciocavallo is a typical *pasta filata* (spun paste) cheese obtained from natural microflora and made from cows' milk. Most Caciocavallo cheeses produced in Italy use a natural whey starter, corresponding to backslopping from a previously successful cheese batch. As in other food fermentations, natural whey starters are preferred as they contribute to the typical flavour and aroma of the final cheese, qualities that are attributed to the complex microflora (Parente and Cogan 2004). On the other hand, an interesting study by Morea *et al.* (2007) proposed some autochthonous strains to standardize Caciocavallo Pugliese cheese processing, safeguarding the peculiar traits of this typical southern Italian dairy product. Four strains previously isolated from Caciocavallo Pugliese cheese (Gobbetti *et al.* 2002) and belonging to *Lactobacillus paracasei* subsp. *paracasei* and *Lactobacillus parabuchneri* species were added to natural whey and their contribution to the proteolysis of cheese was evaluated.

Two Caciocavallo batches were undertaken: one was performed traditionally (traditional cheese), with milk and curd fermentation being carried out by the microflora present in non-selected whey culture, whereas a second batch (experimental cheese) was conducted by adding known amounts of four lactobacilli as adjuncts to the natural whey inoculum. The results demonstrated that the *Lb. paracasei* subsp. *paracasei* strains used were particularly suitable for Caciocavallo cheese applications, surviving the stretching and brining steps and remaining at high levels even after two months of ripening. In addition, all the peptidase activities assayed, as well as the content of total free amino acids, were consistently higher in experimental cheese than in traditional cheese.

Even in this case, the use of autochthonous strains as specific starters for dairy products appears to be the ideal solution to preserve the sensorial characteristics of traditional products while allowing both the reproduction of their organoleptic properties and the standardization of the process.

Fermented vegetables

Fermented vegetables are another food product in which the use of starter cultures is suggested. Table 10.1 also lists some work carried out on these matrices (green olives, sweet peppers, carrots, leeks) in the last decade, with the common aim being to individuate and select autochthonous strains able to control fermentation, improve safety and provide good organoleptic quality for the final products. As a general trend, the microorganisms isolated were mostly LAB (e.g. *Lb. plantarum*, *Lactobacillus brevis*, *Leuconostoc mesenteroides*, *Pediococcus* spp.) and their use guaranteed good acidification, while preserving the original and unique flavour of traditional spontaneously fermented vegetables.

Among the most important European fermented vegetables are table olives, and recently interest in developing effective starter cultures to be used during their production has been increasing, since industrial experience suggests that appropriate inoculation reduces the probability of spoilage contamination and guarantees an improved and more predictable fermentation process (Buckenheuskes 2001). On this issue, an interesting study was performed by Perricone *et al.* (2010), who proposed the use of a strain of *Lb. plantarum* to control the fermentation of 'Bella di Cerignola' table olives, a traditional variety from the Apulia region (southern Italy); the strain was previously isolated from table olives and proposed as a starter for this kind of food (Bevilacqua *et al.* 2010). The research focused on the interaction of the proposed starter with the naturally occurring microflora, the quantitative/qualitative composition of the yeast population, the decrease of pH and the content of organic acids. After a preliminary characterization, three strains of *Lb. plantarum*, selected for their probiotic and technological performances, were used as a multiple-strain starter and inoculated (approximately 2%) in olives, processed according to the Spanish style, brined at 8% and 10% of NaCl and with 0.5% of glucose added. The results obtained highlighted that the use of the starter in combination with glucose assured a correct fermentation course, decreasing the pH up to a safe value (4.3–4.5) and controlling yeast growth. The concentrations of both L- and D-lactic acids increased throughout the fermentation, while citric and malic acids (both isomers D and L) remained at low levels (0.2–0.4 g/L). Concerning yeast species, *Candida guilliermondii* was mainly isolated at the beginning (7–14 days), while *Candida famata* prevailed at the end of fermentation. The innovative aspect of this study was that the starter strains used in this research possessed both 'good technological' properties and probiotic characteristics, thus highlighting the suitability of Bella di Cerignola table olives as a new kind of functional food.

Besides table olives, different studies were also performed on other fermented vegetables, such as sauerkraut, peppers, carrots, leeks, soybeans, grams and, recently, some ethnic fermented vegetables, as well as bamboo shoot products from the Himalayas (Tamang 2010). For this kind of product, the results for the use of autochthonous strains as starters were also very promising.

Fermented beverages

Fermented beverages represent a vast diversity of products ranging from ethnic, alcoholic drinks and distilled alcoholic products to wine and beer. These products are generally prepared by starch hydrolysis and fermentation is accomplished by amylolytic moulds and yeasts, followed by alcohol-producing yeasts and also flavour-enhancing LAB. Yeasts associated with fermented beverages are species of *Saccharomyces*, *Saccharomycopsis*, *Schizosaccharomyces*, *Pichia*, *Hansenula*, *Candida*, *Kluyveromyces*, *Debaryomyces*, *Torulopsis* and *Zygosaccharomyces*, whereas species of *Pediococcus* and *Lactobacillus* are frequently found as LAB (Tamang and Fleet 2009). Yeasts are involved with the production of beer and wine, which depends on their ability to ferment sugar rapidly and efficiently into ethanol (Dung *et al.* 2005, 2006).

As could be expected, most of the research on fermented beverages has been performed on wine, one of the oldest alcoholic drinks that has been commercialized, mass produced and studied; moreover, this product plays a very important role in economic terms, thus stimulating several studies about the selection of wine starters. The technological and qualitative key traits for a good wine starter are high ethanol production and tolerance, sugar exhaustion, high sugar concentration and high temperature growth, good glycerol production, low hydrogen sulfide and volatile acidity production, sulfur dioxide resistance and a good enzymatic profile (Nikolaou *et al.* 2006). More recently, a particular kind of functional trait for wine yeast selection was individuated into the ability to remove ochratoxin A (OTA); this trait is receiving special focus because this mycotoxin is considered the principal safety hazard in the winemaking process (Delage *et al.* 2003).

In 2014, Petruzzi *et al.* studied an autochthonous strain of *Saccharomyces cerevisiae* (W13) as a potential starter for wine production, also able to remove OTA. Three different strains of *S. cerevisiae* were used throughout this study: two commercial strains (*S. cerevisiae* EC1118 and *S. cerevisiae* DBVPG 6500) were compared to *S. cerevisiae* W13, a wild strain isolated from the Nero di Troia grape (harvest 2008), a typical cultivar of the Apulia region (southern Italy). These strains were studied for their OTA-removal ability, growth pattern and alcoholic fermentation profile at two different temperatures (25 and 30 °C) and two different sugar levels (200 and 250 g/L), with or without supplementation of the medium with diammonium phosphate (DAP). The results obtained showed that all the strains were able to conclude fermentation, but *S. cerevisiae* W13 was able both to remove OTA (6–57.21% of the initial amount) under different fermentation conditions (with the highest removal effect observed at 30 °C with 250 g/L sugar) and to produce more ethanol and glycerol than the commercial strains. In addition, the proposed strain showed a high tolerance to single and combined stress conditions, β-D-glucosidase, pectolytic and xylanase activities, a low level of hydrogen sulfide production, a low-to-medium parietal interaction with phenolic compounds and no biogenic amine formation, so suggesting the potential of using *S. cerevisiae* W13 for wine production with improved qualitative and food safety characteristics.

Moreover, in the case of winemaking, works selecting ecotypes from spontaneous fermentation by reason of their pro-technological properties are very numerous (Capozzi *et al.* 2011; Tristezza *et al.* 2014) and results really highlighted the possibility of enriching the 'wild' starter culture approach (Capozzi and Spano 2011).

Sourdough and fermented cereals

It is known that sourdoughs are a mixture of water and flour that is fermented by a heterogeneous microbial community, including LAB and yeasts, which clearly improves the dough properties, texture, flavour and nutritional value of the end products (Zannini *et al.* 2009). This is due to the LAB metabolism, responsible for the production of organic acids and, along with yeasts, the production of aromatic compounds. In addition to this, starters delay the staling process and prevent bakery products from mould and bacterial spoilage (Robert *et al.* 2006).

Although fewer, some studies have also been conducted to evaluate the possibility of selecting autochthonous starter cultures, both LAB and yeasts, for use in the breadmaking process. Of particular importance is the study conducted by Corbo *et al.* (2013), which proposed a simple approach to select some LAB as promising starter cultures for sourdoughs. In particular, 54 strains of LAB were isolated from a single factory in Altamura (Apulia region, southern Italy), identified by molecular tools and studied for their growth at different NaCl concentrations, temperatures, pH and acidification in MRS broth. Through a statistical analysis, six strains (6, 7, 12, 13, 38 and 54), identified as *Lb. plantarum*, able to grow in a wide range of conditions and/or able to decrease the pH by 1.77–2.0 units, were selected and tested in a model system. They were inoculated to 10^7 cfu/mL in a mixture of commercial flour and tap water, to study acidification after 24 hours and their viability after 14 days. After 8 hours, lactobacilli caused a decrease of pH ranging from 1 (strains 12 and 13) to 1.5 (strains 6 and 7), while after 24 hours the reduction in pH was circa 2.6–2.7, without significant differences among the strains. After the assessment of acidification, the strain viability in dough was evaluated for a prolonged running time (14 days); during the storage period, neither water nor flour was added, in order to assess cell viability and resistance under unfavourable conditions. Some strains underwent a drastic decrease of cell numbers, below the detection limit after 8 days, whereas others retained cell viability for a longer period and at the end of the running time (14 days) the cell level was circa 6 log cfu/g. Robustness is an important trait for starter cultures to be selected for sourdough of type I and it could be referred to as the ability to prevail over the autochthonous microbiota and persist (i.e. maintain an active metabolism and high cell number) for a long time. The results of LAB viability in sourdoughs for 14 days showed that at least four strains (12, 13, 38 and 54) were robust.

As a final step for validation, dough was inoculated with both lactobacilli and yeasts isolated from the same source: lactobacilli were not affected by yeasts, while LAB influenced yeast growth in a positive way. In conclusion, the six strains selected through this research were very promising microorganisms, and laboratory validation showed that they could also demonstrate good performance in real systems. The novelty of this paper does not rely on starter selection and the evaluation of technological performance (performed similarly to other studies), but on the methodological approach followed throughout the research. In fact, starter selection was addressed using a simple methodology (evaluation of growth through absorbance reading and pH determination) as well as a simple index (Growth Index).

Besides wheat, lactic acid cereal fermented foods and beverages are made from a great variety of cereals, such as rye, millet, maize, rice, barley, oat and sorghum; sometimes the same cereals can be germinated to produce malt for their use in brewing cereal slurries to make beers through alcoholic fermentation due to yeasts (mainly *S. cerevisiae*) associated with LAB (Jespersen 2003; Maoura *et al.* 2005). However, this kind of production remains mainly on a domestic scale, relying for its success on the performance of natural microflora, thus stimulating further studies on the use of autochthonous strains as starters.

Conclusion

We started by wondering whether it is better to use commercial starters or autochthonous strains. What is the answer? If we summarize what has been highlighted by the various studies cited (see Figure 10.1), we can strongly affirm that the winner is the selected autochthonous microflora. Even if further investigations are required, the awareness that autochthonous bacteria are most adapted and competitive in food systems allows us to state that the diversity of microbial communities undoubtedly represents great potential for remarkable scientific, social and economic impact. Starting from the selected strains, fermented food industries might formulate their own multistrain starter culture in order to guarantee preservation of the sensorial characteristics of their traditional products while allowing standardization of the process. In addition, identification of characteristic autochthonous strains typical of a specific environment and/or of production of a

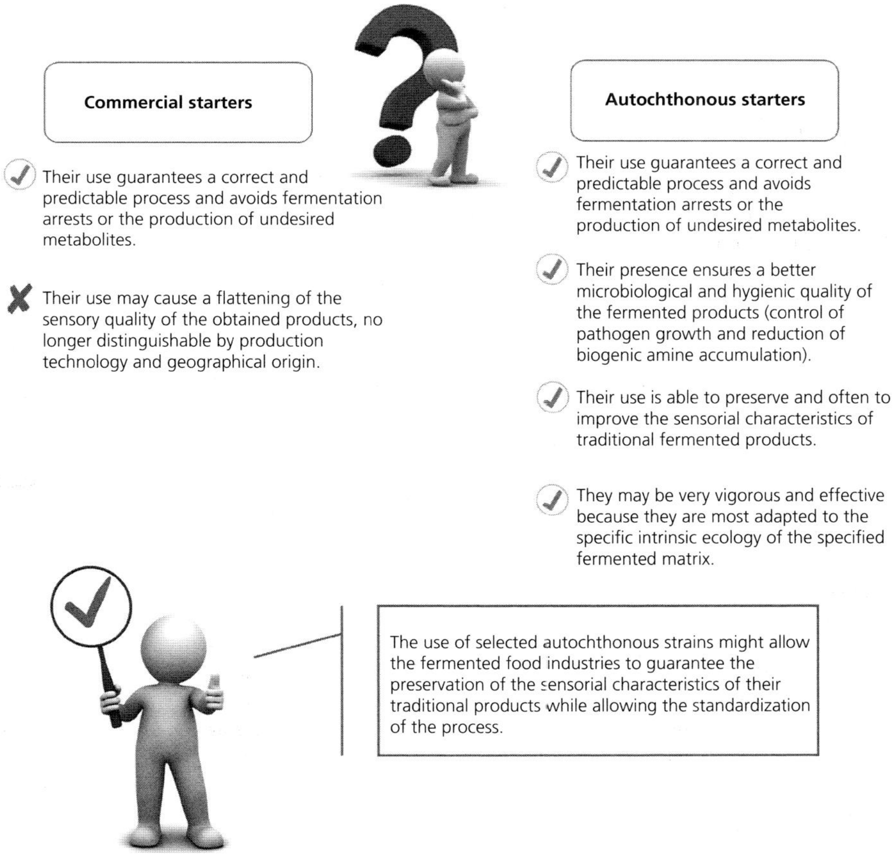

Figure 10.1 Commercial starters or autochthonous strains? This is the answer.

particular food should be a key factor in the context of fermented food Geographical Indications (GIs), to correlate each microbial attribute (including the list of autochthonous strains used as starters) to a unique territorial origin.

References

Adams, M.R. (1986) Fermented flesh foods. *Progress in Industrial Microbiology*, **23**, 159–198.

Adams, M. (2009) Fermented fish. In F. Rhea (ed.), *Microbiology Handbook Fish and Seafood* (pp. 123–140), Leatherhead: Leatherhead Publishing.

Ammor, M.S. and Mayo, B. (2007) Selection criteria for lactic acid bacteria to be used as functional starter cultures in dry sausage production: An update. *Meat Science*, **76**, 138–146.

Anas, M., Zinedine, B.A., Rizk, H.A., Eddine, H.J. and Mebrouk, K. (2012) Screening of autochthonous *Lactobacillus* species from Algerian raw goats' milk for the production of bacteriocin-like compounds against *Staphylococcus aureus*. *African Journal of Biotechnology*, **11**, 4595–4607.

Aponte, M., Blaiotta, G., La Croce, F. *et al.* (2012) Use of selected autochthonous lactic acid bacteria for Spanish-style table olive fermentation. *Food Microbiology*, **30**(1), 8–16.

Ayad, E.H.E., Verheul, A., Bruinenberg, P., Wouters, J.T.M. and Smit, G. (2003) Starter culture development for improving the flavour of Proosdij-type cheese. *International Dairy Journal*, **13**, 159–168.

Baka, A.M., Papavergou, E.J., Pragalaki, T., Bloukas, J.G. and Kotzekidou, P. (2011) Effect of selected autochthonous starter cultures on processing and quality characteristics of Greek fermented sausages. *LWT – Food Science and Technology*, **44**, 54–61.

Bevilacqua, A., Altieri C., Corbo, M.R., Sinigaglia, M. and Ouoba, L.I. (2010) Characterization of lactic acid bacteria isolated from Italian Bella di Cerignola table olives: Selection of potential multifunctional starter cultures. *Journal of Food Science*, **75**, M536–M544.

Bonomo, M., Ricciardi, A. and Salzano, G. (2011) Influence of autochthonous starter cultures on microbial dynamics and chemical-physical features of traditional fermented sausages of Basilicata region. *World Journal of Microbiology and Biotechnology*, **27**, 137–146.

Buckenheuskes, H.J. (2001) Fermented vegetables. In M.P. Doyle, L.R. Beuchat and T.J. Montville (eds), *Food Microbiology, Fundamentals and Frontiers* (2nd edn, pp. 665–679), Washington, DC: American Society for Microbiology.

Callejon, R.M., Clavijo, A., Ortigueirac, P., Troncoso, A.M., Paneque, P. and Morales, M.L. (2010) Volatile and sensory profile of organic red wines produced by different selected autochthonous and commercial *Saccharomyces cerevisiae* strains. *Analytica Chimica Acta*, **660**, 68–75.

Candioti, M.C., Hynes, E., Quiberoni, A., Palma, S., Sabbag, N. and Zalazar, C. (2002) Reggianito Argentino cheese: Influence of *Lactobacillus helveticus* strains isolated from natural whey cultures on cheese making and ripening processes. *International Dairy Journal*, **12**, 923–931.

Capozzi, V. and Spano, G. (2011) Food microbial biodiversity and 'microbes of protected origin'. *Frontiers in Microbiology*, **2**, 1–3.

Capozzi, V., Ladero, V., Beneduce, L. *et al.* (2011) Isolation and characterization of tyramine-producing *Enterococcus faecium* strains from red wine. *Food Microbiology*, **28**(3), 434–439.

Casquete, R., Benito, M.J., Martín, A., Ruiz-Moyano, S., Aranda, E. and Córdoba, M.G. (2012a) Use of autochthonous *Pediococcus acidilactici* and *Staphylococcus vitulus* starter cultures in the production of 'chorizo' in 2 different traditional industries. *Journal of Food Science*, **71**, M70–M79.

Casquete, R., Benito, M.J., Martín, A., Ruiz-Moyano, S., Córdoba, J.J. and Córdoba, M.G. (2011a) Role of an autochthonous starter culture and the protease EPg222 on the sensory and safety properties of a traditional Iberian dry-fermented sausage 'salchichon'. *Food Microbiology*, **28**, 1432–1440.

Casquete, R., Benito, M.J., Martín, A., Ruiz-Moyano, S., Hernández, A. and Córdoba, M.G. (2011b) Effect of autochthonous starter cultures in the production of 'salchichon', a traditional Iberian dry-fermented sausage, with different ripening processes. *LWT – Food Science and Technology*, **44**, 1562–1571.

Casquete, R., Benito, M.J., Martín, A., Ruiz-Moyano, S., Pérez-Nevado, F. and Córdoba, M.G. (2012b) Comparison of the effects of a commercial and an autochthonous *Pediococcus acidilactici* and *Staphylococcus vitulus* starter culture on the sensory and safety properties of a traditional Iberian dry-fermented sausage 'salchichon'. *International Journal of Food Science and Technology*, **47**, 1011–1019.

Casquete, R., Martín, A., Benito, M.J., Ruiz-Moyano, S., Nevado, F.P. and de Guía Córdoba, M. (2011c) Impact of pre-selected autochthonous starter cultures on the flavor quality of Iberian dry-fermented 'salchichon' sausage with different ripening processes. *Journal of Food Science*, **76**(9), S535–S544.

Cenci-Goga, B.T., Ranucci, D., Miraglia, D. and Cioffi, A. (2008) Use of starter cultures of dairy origin in the production of Salame nostrano, an Italian dry-cured sausage. *Meat Science*, **78**, 381–390.

Coda, R., Di Cagno, R., Rizzello, C.G., Nionelli, L., Edema, M.O. and Gobbetti, M. (2011a) Utilization of African grains for sourdough bread making. *Journal of Food Science*, **76**(6), M329–M335.

Coda, R., Lanera, A., Trani, A., Gobbetti, M. and Di Cagno, R. (2012) Yogurt-like beverages made of a mixture of cereals, soy and grape must: Microbiology, texture, nutritional and sensory properties. *International Journal of Food Microbiology*, **155**, 120–127.

Coda, R., Rizzello, C.G., Trani, A. and Gobbetti, M. (2011b) Manufacture and characterization of functional emmer beverages fermented by selected lactic acid bacteria. *Food Microbiology*, **28**, 526–536.

Corbo, M.R., Bevilacqua, A., Campaniello, D., Speranza, B. and Sinigaglia, M. (2013) Selection of promising lactic acid bacteria as starter cultures for sourdough: Using a step-by-step approach through quantitative analyses and statistics. *Journal of the Science of Food and Agriculture*, **94**, 1772–1780.

Crow, V., Curry, B. and Hayes, M. (2001) The ecology of non-starter lactic acid bacteria (NSLAB) and their use as adjuncts in New Zealand Cheddar. *International Dairy Journal*, **11**, 275–283.

De Angelis, M., de Candia, S., Calasso, M.P. *et al.* (2008) Selection and use of autochthonous multiple strain cultures for the manufacture of high-moisture traditional Mozzarella cheese. *International Journal of Food Microbiology*, **125**, 123–132.

Delage, N., d'Harlingue, A., Colonna, B., Ceccaldi, B.C. and Bompeix, G. (2003) Occurrence of mycotoxins in fruit juices and wine. *Food Control*, **14**, 225–227.

Di Cagno, R., Cardinali, G., Minervini, G. *et al.* (2010) Taxonomic structure of the yeasts and lactic acid bacteria microbiota of pineapple (*Ananas comosus* L. *Merr.*) and use of autochthonous starters for minimally processing. *Food Microbiology*, **27**, 381–389.

Di Cagno, R., Surico, R.F., Minervini, G., De Angelis, M., Rizzello, C.G. and Gobbetti, M. (2009a) Use of autochthonous starters to ferment red and yellow peppers (*Capsicum annum* L.) to be stored at room temperature. *Food Microbiology*, **130**(2), 108–116.

Di Cagno, R., Surico, R.F., Minervini, G. *et al.* (2011) Exploitation of sweet cherry (*Prunus avium* L.) puree added of stem infusion through fermentation by selected autochthonous lactic acid bacteria. *Food Microbiology*, **28**, 900–909.

Di Cagno, R., Surico, R.F., Paradiso, A. *et al.* (2009b) Effect of autochthonous lactic acid bacteria starters on health-promoting and sensory properties of tomato juices. *International Journal of Food Microbiology*, **128**(3), 473–483.

Di Cagno, R., Surico, R.F., Siragusa, S. *et al.* (2008) Selection and use of autochthonous mixed starter for lactic acid fermentation of carrots, French beans or marrows. *International Journal of Food Microbiology*, **127**, 220–228.

Dung, N.T., Rombouts, F.M. and Nout, M.J.R. (2005) Development of defined mixed-culture fungal fermentation starter granulate for controlled production of rice wine. *Innovative Food Science and Emerging Technologies*, **6**, 429–441.

Dung, N.T., Rombouts, F.M. and Nout, M.J. (2006) Functionality of selected strains of moulds and yeasts from Vietnamese rice wine starters. *Food Microbiology*, **23**(4), 331–340.

Feng, Z., Chen, H., Lv, X.T. *et al.* (2014) Accelerated ripening of Kedong sufu with autochthonous starter cultures *Kocuria rosea* KDF3 and its protease KP3 as adjuncts. *Journal of Applied Microbiology*, **116**, 877–889.

Fonseca, S., Cachaldora, A., Gómez, M., Franco, I. and Carballo, J. (2013) Effect of different autochthonous starter cultures on the volatile compounds profile and sensory properties of Galician chorizo, a traditional Spanish dry fermented sausage. *Food Control*, **33**, 6–14.

Gobbetti, M., Morea, A., Baruzzi, F. *et al.* (2002) Microbiological, compositional, biochemical and textural characterization of Caciocavallo Pugliese cheese during ripening. *International Dairy Journal*, **12**, 511–523.

González, L. and Rate, V. (2012) Influence of an autochthonous starter culture and a commercial starter on the characteristics of Tenerife pasteurised goats' milk cheese. *International Journal of Dairy Technology*, **65**(4), 542–547.

Hurtado, A., Reguant, C., Bordons, A. and Rozès, N. (2010) Evaluation of a single and combined inoculation of a *Lactobacillus pentosus* starter for processing cv. Arbequina natural green olives. *Food Microbiology*, **27**, 731–740.

Jácome, S.L., Fonseca, S., Pinheiroa, R. *et al.* (2014) Effect of lactic acid bacteria on quality and safety of ready-to-eat sliced cured/smoked meat products. *Chemical Engineering Transactions*, **38**, 403–408.

Jespersen, L. (2003) Occurrence and taxonomic characteristics of strains of *Saccharomyces cerevisiae* predominant in African indigenous fermented foods and beverages. *Fems Yeast Research*, **2**, 191–200.

Jessen, B. (1995) Starter cultures for meat fermentations. In G. Campbell-Platt and P.E. Cook (eds), *Fermented Meats* (pp. 130–159), Glasgow: Blackie Academic and Professional.

Kose, S. and Hall, G.M. (2011) Sustainability of fermented fish-products. In G.M. Hall (ed.), *Fish Processing: Sustainability and New Opportunities* (pp. 138–166), Leatherhead: Leatherhead Publishing.

Leboš Pavunc, A., Beganović, J., Kos, B., Uroić, K., Blažić, M. and Šušković, J. (2012) Characterization and application of autochthonous starter cultures for fresh cheese production. *Food Technology and Biotechnology*, **50**(2), 141–151.

Leroy, F., Vertuyten, J. and De Vuyst, L. (2006) Functional meat cultures for improved sausage fermentation. *International Journal of Food Microbiology*, **106**, 270–285.

Madrau, M.A., Mangia, N.P., Murgia, M.A. *et al.* (2006) Employment of autochthonous microflora in Pecorino Sardo cheese manufacturing and evolution of physicochemical parameters during ripening. *International Dairy Journal*, **16**, 876–885.

Mangia, N.P., Murgia, M.A., Garau, G., Sanna, M.G. and Deiana, P. (2008) Influence of selected lab cultures on the evolution of free amino acids, free fatty acids and Fiore Sardo cheese microflora during the ripening. *Food Microbiology*, **25**, 366–377.

Mangia, N.P., Trani, A., Di Luccia, A. *et al.* (2013) Effect of the use of autochthonous *Lactobacillus curvatus*, *Lactobacillus plantarum* and *Staphylococcus xylosus* strains on microbiological and biochemical properties of the Sardinian fermented sausage. *European Food Research Technology*, **236**, 557–566.

Maoura, N., Mbaiguinam, M., Nguyen, H.V., Gaillardin, C. and Pourquie, J. (2005) Identification and typing of the yeast strains isolated from *bili bili*, a traditional sorghum beer of Chad. *African Journal of Biotechnology*, **4**, 646–656.

Maqueda, M., Pérez-Nevado, F., Regodón, J.A. *et al.* (2011) A low-cost procedure for production of fresh autochthonous wine yeast. *Journal of Industrial Microbiology and Biotechnology*, **38**, 459–469.

Mendoza, L.M., Merín, M.G., Morata, V.I. and Farías, M.E. (2011) Characterization of wines produced by mixed culture of autochthonous yeasts and *Oenococcus oeni* from the northwest region of Argentina. *Journal Industrial of Microbiology and Biotechnology*, **38**, 1777–1785.

Menéndez, S., Godìnez, R., Hermida, M., Centeno, J.A. and Rodríguez-Otero, J.L. (2004) Characteristics of 'Tetilla' pasteurized milk cheese manufactured with the addition of autochthonous cultures. *Food Microbiology*, **21**, 97–104.

Minervini, F., De Angelis, M., Di Cagno, R. *et al.* (2010) Robustness of *Lactobacillus plantarum* starters during daily propagation of wheat flour sourdough type I. *Food Microbiology*, **27**, 897–908.

Morea, M., Matarantea, A., Di Cagno, R., Baruzzi, F. and Minervini, F. (2007) Contribution of autochthonous non-starter lactobacilli to proteolysis in Caciocavallo Pugliese cheese. *International Dairy Journal*, **17**, 525–534.

Nikolaou, E., Soufleros, E.H., Bouloumpasi, E. and Tzanetakis, N. (2006) Selection of indigenous *Saccharomyces cerevisiae* strains according to their oenological characteristics and vinification results. *Food Microbiology*, **23**, 205–211.

Oguntoyinbo, F.A. and Narbad, A. (2012) Molecular characterization of lactic acid bacteria and *in situ* amylase expression during traditional fermentation of cereal foods. *Food Microbiology*, **31**(2), 254–262.

Parente, E. and Cogan T.M. (2004) Starter cultures: General aspects. In P.O. Fox (ed.), *Cheese: Chemistry, Physics and Microbiology* (3rd edn, pp. 123–147), Oxford: Elsevier.

Perricone, M., Bevilacqua, A., Corbo, M.R. and Sinigaglia, M. (2010) Use of *Lactobacillus plantarum* and glucose to control the fermentation of 'Bella di Cerignola' table olives, a traditional variety of Apulian region (southern Italy). *Journal of Food Science*, **75**(7), M430–M436.

Petruzzi, L., Bevilacqua, A., Baiano, A., Beneduce, L., Corbo, M.R. and Sinigaglia, M. (2014a) Study of *Saccharomyces cerevisiae* W13 as a functional starter for the removal of ochratoxin A. *Food Control*, **35**, 373–377.

Petruzzi, L., Bevilacqua, A., Corbo, M.R., Garofalo, C., Baiano, A. and Sinigaglia, M. (2014b) Selection of autochthonous *Saccharomyces cerevisiae* strains as wine starters using a polyphasic approach and Ochratoxin A removal. *Journal of Food Protection*, **77**(7), 1168–1177.

Piasecka-Jóźwiak, P., Rozmierska, J., Chabłowska, B. *et al.* (2013) Starter cultures for lactic acid fermentation of sweet pepper, pattypan squash and tomatoes. *Polish Journal of Food and Nutrition Science*, **63**(2), 95–102.

Pourahmad, R. and Assadi, M.M. (2007) Use of isolated autochthonous starter cultures in yogurt production. *International Journal of Dairy Technology*, **60**(4), 259–262.

Poveda, J.M., Nieto-Arribas, P., Seseña, S. *et al.* (2014) Volatile composition and improvement of the aroma of industrial Manchego cheese by using *Lactobacillus paracasei* subsp. *paracasei* as adjunct and other autochthonous strains as starters. *European Food Research and Technology*, **238**, 485–494.

Pragalaki, T., Bloukas, J.G. and Kotzekidou, P. (2013) Inhibition of *Listeria monocytogenes* and *Escherichia coli* O157:H7 in liquid broth medium and during processing of fermented sausage using autochthonous starter cultures. *Meat Science*, **95**, 458–464.

Robert, H., Gabriel, V., Lefebvre, D., Rabier, P., Vayssier, Y. and Fontagné-Faucher, C. (2006) Study of the behaviour of *Lactobacillus plantarum* and *Leuconostoc* starters during a complete wheat sourdough breadmaking process. *LWT – Food Science and Technology*, **39**, 256–265.

Speranza, B., Bevilacqua, A., Corbo, M.R., Altieri C. and Sinigaglia, M. (2014) Selection of autochthonous strains as promising starter cultures for *Fior di Latte*, a traditional cheese of southern Italy. *Journal of the Science of Food and Agriculture*, **95**, 88–97.

Talon, R., Leroy, S., Lebert, I. *et al.* (2008) Safety improvement and preservation of typical sensory qualities of traditional dry fermented sausages using autochthonous starter cultures. *International Journal of Food Microbiology*, **126**, 227–234.

Tamang, J.P. (2010) *Himalayan Fermented foods: Microbiology, Nutrition and Ethnic Values*, New York: CRC Press/Taylor & Francis.

Tamang, J.P. and Fleet, G.H. (2009) Yeasts diversity in fermented foods and beverages. In T. Satyanarayana and G. Kunze (eds), *Yeasts Biotechnology: Diversity and Applications* (pp. 170–193), Berlin: Springer.

Tangüler, H. and Erten, H. (2013) Selection of potential autochthonous starter cultures from shalgam, a traditional Turkish lactic acid-fermented beverage. *Turkish Journal of Agriculture and Forestry*, **37**, 212–220.

Topisirovic, L., Kojic, M., Fira, D., Golic, N., Strahinic, I. and Lozo, J. (2006) Potential of lactic acid bacteria isolated from specific natural niches in food production and preservation. *International Journal of Food Microbiology*, **114**, 36–42.

Tristezza, M., Fantastico, L., Vetrano, C. *et al.* (2014) Molecular and technological characterization of *Saccharomyces cerevisiae* strains isolated from natural fermentation of Susumaniello grape must in Apulia, southern Italy. *International Journal of Microbiology*, **2014**, 897428. doi:10.1155/2014/897428.

Villani, F., Casaburi, A., Pennacchia, C., Filosa, L., Russo, F. and Ercolini, D. (2007) Microbial ecology of the soppressata of Vallo di Diano, a traditional dry fermented sausage from southern Italy, and in vitro and in situ selection of autochthonous starter cultures. *Applied and Environmental Microbiology*, **73**(17), 5453–5463.

Wouters, D., Bernaert, N., Anno, N. *et al.* (2013) Application and validation of autochthonous lactic acid bacteria starter cultures for controlled leek fermentations and their influence on the antioxidant properties of leek. *International Journal of Food Microbiology*, **165**, 121–133.

Xiong, T., Li, X., Guan, Q., Peng, F. and Xie, M. (2014) Starter culture fermentation of Chinese sauerkraut: Growth, acidification and metabolic analyses. *Food Control*, **41**, 122–127.

Zannini, E., Garofalo, C., Aquilanti, L., Santarelli, S., Silvestri, G. and Clementi, F. (2009) Microbiological and technological characterization of sourdoughs destined for bread-making with barley flour. *Food Microbiology*, **26**, 744–753.

Zeng, X., Xia, W., Wang, J. *et al.* (2014) Technological properties of *Lactobacillus plantarum* strains isolated from Chinese traditional low salt fermented whole fish. *Food Control*, **40**, 351–358.

CHAPTER 11

Sourdough and cereal-based foods: Traditional and innovative products

Luca Settanni

Department of Agricultural and Forest Sciences, University of Palermo, Italy

Cereal-based foods have been key components of human diets for thousands of years (Alfonzo *et al.* 2013); several historical sources evidence that baking of leavened doughs was a daily practice in several cultures (Corsetti and Settanni 2007). Furthermore, as revealed by the discovery of fossil kernels, humans' utilization of cereals commenced in the Neolithic era (Settanni and Moschetti 2010). Cereals remain a major source of nutrition, particularly in developing and overpopulated countries (Blandino *et al.* 2003). Indeed, the history of several cultures is directly defined by cereals and, consequently, many human populations are identified by the cereals they eat: Chinese are 'rice people', South and Central Americans are 'maize people', North Americans and Mediterranean people are 'wheat people', North Europeans are 'oats and rye people' and Africans are 'millet and sorghum people' (Gifford and Baer-Sinnot 2007).

Millennia BCE, wheat was already one of the most important crops grown in the Mediterranean basin. Following mass migrations, its cultivation underwent a huge expansion, resulting in its production worldwide (Toderi 1989). Among cereals, wheat is critical in the Mediterranean diet: it provides approximately one-third of the daily protein and energy requirements (~2400 kcal) for an adult (Cannella and Piredda 2006). For this reason, wheat surpasses other cereals in terms of the number of hectares dedicated to its cultivation worldwide (Gifford and Baer-Sinnot 2007). However, the world's major sources of energy for humans are rice, wheat and maize (Spiertz and Ewert 2009).

Due to their potential for nutritional enhancement and the fact that their consumption substantially lowers the risk of significant diet-related diseases (Topping 2007), cereals also assume a basic role in the diet of industrialized countries. They are generally consumed after boiling or after fermentation.

Starter Cultures in Food Production, First Edition. Edited by Barbara Speranza, Antonio Bevilacqua, Maria Rosaria Corbo and Milena Sinigaglia.

In the latter case, cereals are first ground and mixed with water (Salovaara 1998). In this matrix, different microbial groups, including mainly bacteria and yeasts, transform raw materials into final products (Galati *et al.* 2014).

As already mentioned, the process of cereal fermentation has been important for humans since ancient times (Spicher 1999; De Vuyst and Neysens 2005). The acidic product, depending on the proportion of water, may represent a food or a beverage; sometimes it undergoes baking. Either way, fermentation contributes to the microbial stability of the final products (Mensah 1997). Hence, fermented cereal-based foods are complex microbial ecosystems, whose activities confer on the resulting products characteristic features such as palatability, high sensory quality, structure and texture, stability, nutritional and healthful qualities (Corsetti and Settanni 2007) and, when they are in a living form at the moment of consumption (unbaked products), potential probiotic properties (Perricone *et al.* 2014).

Traditional fermented foods prepared from most common types of cereals (such as rice, wheat, maize or sorghum) are well known in many parts of the world. In general, most of these foods are typical to restricted geographical areas, where the cereal substrates (alone or mixed with other cereals or legumes or tubers), in the form of flour, are processed into different products (Blandino *et al.* 2003). Among these foods, bread is common to many societies; it is produced almost everywhere, even in southeast Asian countries that have not been traditional bread consumers (Jenson 1998).

Products derived from cereal flours (e.g. bread, cereal snacks and breakfast cereals) are useful food vehicles to provide micronutrients, but sometimes the amount needed for correct alimentation is quite limited. For this reason, policy and programme responses of several countries, including those located in developing areas, promote food-based strategies, such as food fortification, to prevent micronutrient malnutrition. Food fortification allows delivery of the required nutrients to many populations without requiring radical changes in food consumption (Allen *et al.* 2006).

Bread is mainly produced from wheat, but rye and barley are often used for this purpose. Due to their gluten content, these cereals are toxic to people affected by coeliac sprue (CS), also known as gluten-sensitive enteropathy, an autoimmune disease of the small intestinal mucosa (Silano and De Vincenzi 1999). Although several attempts are being made to decrease the CS-inducing effects of gluten by enzymatic treatment (Caputo *et al.* 2010), to date a strict, life-long gluten-free diet is the only safe and efficient treatment available for this disease (Tack *et al.* 2010). For this reason, the development or enhancement of gluten-free products continues to grow (O'Shea *et al.* 2014). Due to the lower-quality characteristics of these products compared to those made with wheat flour, several studies are in progress to determine the best flour combinations and, for fermented foods, to select the starter strains able to enhance the quality aspects of gluten-free products. Another strategy to produce products compatible with a CS diet is to hydrolyse the toxic components of flours that contain gluten. From this standpoint, many attempts have been made to apply different microorganisms, mainly lactic acid bacteria (LAB) and microbial proteases (Rizzello *et al.* 2014b), but much more work still needs to be done before the efficacy and effectiveness of the microbial activities are proven to obtain safe foods for CS-affected people.

Cereals used in fermented food production

To date, more than half of the arable land in the world has been planted with cereals, mainly wheat, rice, maize, barley, sorghum and millet (Toderi 1993). Wheat clearly dominates in terms of hectares dedicated to its cultivation. However, thanks to improved plant breeding, rice production per hectare is higher than that for wheat. Other cereal crops are also relevant in several areas throughout the world (Dahlberg 2007).

Thanks to the great adaptability of its several varieties to different climatic conditions, including extreme values of temperature and/or humidity, wheat is cultivated almost everywhere for food and feed production. The species most commonly involved in food production are *Triticum durum* and *Triticum aestivum* (Toderi 1993). *T. durum*, which requires high temperatures, is cultivated in temperate regions, while *T. aestivum*, well suited to lower temperatures, is mainly cultivated in colder areas. In Italy, where cereal-based foods constitute the major part of the daily diet, *T. durum* and *T. aestivum* are traditionally used for pasta and bakery products (bread and leavened baked goods, such us breakfast or recurrent products), respectively. However, in some southern regions, durum wheat flour is used alone or in combination with soft wheat flour for bread production (Settanni *et al.* 2008).

The reason for the great expansion of wheat cultivation is that the other cereals are not as suitable for the production of foods characterized by nutritional value, shelf life and taste as those obtained from wheat flour (Macrae *et al.* 1993; Francis 2000; Cannella *et al.* 2010). In addition to wheat, rye is particularly common for bread production in Scandinavian countries, Germany, Poland and Russia (Bushuk 2001), maize is used in Portugal (Rocha and Malcata 1999) and sorghum in Sudan (Hamad *et al.* 1997), while rice is commonly used for gluten-free bread production (Neumann and Bruemmer 1997; Meroth *et al.* 2004).

The most cultivated cereal in developing countries is maize (Giardini and Vecchiettini 2000). It has been used for human consumption for centuries. The different varieties belonging to the species *Zea mays* are mainly distinguished by colour, basically white or yellow. The latter cultivars are particularly rich in carotenoids (pro-vitamin A; Schober and Bean 2008). The current maize production could be considered sufficient to meet the caloric needs of nearly 2 billion people (Giardini and Vecchiettini 2000). The peculiar element of maize is starch, which is used in the confectionery industry.

In Asia and Africa, sorghum constitutes an integral part of the basic diet for millions of people (Schober and Bean 2008). Millet is considered to be the oldest cereal, widely cultivated in Asia and Africa and in some areas of eastern Europe (Taylor and Emmambux 2008). Barley is the main food source for a large number of people living in cold or semiarid areas where wheat does not fit (Mosca and Toniolo 2000).

The cereals mentioned are used to produce fermented products throughout the world (Blandino *et al.* 2003; De Vuyst and Neysens 2005; Galati *et al.* 2014). In particular, millet, maize, sorghum, rice and wheat are mainly used in Africa; maize, rice and wheat in America and Australia; wheat, maize, rye, barley and millet in Europe; rice, millet and wheat in Asia.

Technological properties of cereal flours

The choice of raw materials is crucial to obtain bakery products able to satisfy "consumers' needs", and this is particularly important in breadmaking. For this purpose, wheat cultivars have been selected by bread producers since ancient times (Bottega *et al.* 2010). The technological aptitudes of wheat flour depend on its ability to form gluten, defining of its versatility. Gluten is generated by the interaction between two groups of proteins, gliadins and glutenins, when flour is mixed with water and the mixture is allowed to stand for a while (Cannella *et al.* 2010). The resulting dough is characterized by a viscoelastic behaviour that is responsible for its extensibility during kneading, leavening and the early stages of cooking, but also confers toughness and elasticity, which allow it to maintain its shape and to develop mass with regularity (Bottega *et al.* 2010). Although all wheat varieties are able to form gluten, the presence of particular protein subunits (different in structure and relationship gliadin/glutenin, composition and molecular weight) ensures the formation of tenacious nets (MacRitchie 1992; Shewry 2003; Bottega *et al.* 2010).

Other cereals, such as rye, contain protein subunits similar to those of wheat, but their ability to expand to give a well-developed leavened product is limited (He and Hoseney 1991; Flander *et al.* 2007). Even more critical is the workability of wheat dough enriched with gluten-free flours (Mariotti *et al.* 2006, 2008; Schoenlechner *et al.* 2006). These flours negatively influence the leavening of the dough. Doughs made from durum wheat semolina are characterized by high strength and limited extensibility (Pogna *et al.* 1996). During leavening, therefore, dough tends to have a development by volume lower than that of a good soft wheat flour (Raffo *et al.* 2003).

Cereal microflora

Cereal grains are naturally contaminated by eukaryotic (moulds and yeasts) and procaryotic (bacteria) organisms. The total microbial population and the relative species proportion on wheat grains can be affected by many factors, mainly climatic conditions such as temperature and rainfall, physical damage due to insects or mould attacks and application of insecticides and fungicides. Microorganisms of grains might follow different phases of flour preparation and, since flour does not undergo thermal treatment, it is a source of living and active microorganisms that can be found in the resulting fermented foods (Corsetti and Settanni 2007). The microorganisms that contaminate cereals are generally concentrated in the outer layers of kernels, and they tend to stay in fractions rich in bran during milling. Consequently, flour obtained from milling should theoretically contain a lower bacterial load than caryopses, but the subsequent conditioning phase can increase its microbial content (Berghofer *et al.* 2003).

The levels of living microorganisms present on cereals might range between 10^4 and 10^7 colony forming units (cfu) per gram, while they reach cell densities up to 10^6 cfu/g in the corresponding flours (Stolz 1999). The bacteria, mainly mesophilic, include Gram-negative aerobes (e.g. *Pseudomonas*) and facultative anaerobes (*Enterobacteriaceae*) and Gram-positive species (De Vuyst and Neysens 2005; Minervini *et al.* 2014). Among the latter bacterial groups, LAB, which together with yeasts are relevant for the process of food fermentation, might be found in several spontaneously fermented cereal-based products (Galati *et al.* 2014).

Although several studies have focused on the identification and characterization of LAB in the final fermented products, only a few works have investigated the microbial ecology of raw materials used for cereal-based food production. Some studies are available on the cereals, and the corresponding flours, used in breadmaking. The first document dates to 1987 (Galli and Franzetti 1987), when several samples of Italian wheat flours were analysed for the presence of different microbial groups. Subsequently, Corsetti *et al.* (1996) isolated and identified LAB and yeasts from common wheat and organic flours. Both studies were performed with a phenotypical/biochemical approach that revealed several lactobacilli (*Lactobacillus alimentarius, Lactobacillus plantarum, Lactobacillus rhamnosus, Lactobacillus confusus* and *Lactobacillus viridescens*), even though some species were then reclassified as weissellas (*Weissella confusa* and *Weissella viridescens*).

The cultivable LAB populations associated with durum wheat kernels, cultivated in several Italian regions, as well as bran and non-conventional flours (amaranth, chickpea, maize, rice, quinoa and potato) used to produce gluten-free baked goods, were found at levels ranging between 1.00 and 2.16 log cfu/g. The isolates were genetically investigated by applying a polyphasic strategy consisting of randomly amplified polymorphic DNA-polymerase chain reaction (RAPD-PCR) analysis, partial 16S rRNA gene sequencing, species-specific and multiplex PCRs (Corsetti *et al.* 2007a). Besides *Lactobacillus* (*Lactobacillus coryniformis, Lactobacillus curvatus* and *Lactobacillus graminis*), the work revealed the presence of other LAB genera: *Aerococcus* (*Aerococcus viridans*), *Lactococcus* (*Lactococcus garvieae*), *Enterococcus* (*Enterococcus durans, Enterococcus casseliflavus, Enterococcus faecalis, Enterococcus faecium* and *Enterococcus mundtii*) and *Pediococcus* (*Pediococcus pentosaceus*). De Vuyst and Neysens (2005) reviewed the presence of several LAB species (*Lactobacillus casei, Lb. coryniformis, Lb. curvatus, Lb. plantarum, Lactobacillus salivarius, Lactobacillus brevis, Lactobacillus fermentum, Ent. faecalis, Lactococcus lactis, Pediococcus acidilactici, Pediococcus parvulus, Pe. pentosaceus, Leuconostoc* and *Weissella*) on cereal grains.

More recently, Alfonzo *et al.* (2013) investigated different wheat (*T. durum* and *T. aestivum*) flours used for the production of traditional sourdough breads in Sicily (southern Italy) using culture-dependent phenotypical and genotypical tools, as well as a culture-independent method based on the denaturing gradient gel electrophoresis (DGGE) technique. The last approach was useful to test the technological performance of the dominant strains (which reached concentrations up to 4.75 log cfu/g) and also to detect the species present at undetectable (subdominant) levels and/or as dormant (non-cultivable) flora. *Ent. mundtii, Lactobacillus sanfranciscensis, Lb. plantarum, Lactobacillus sakei, Lc. lactis, Leuconostoc mesenteroides, Leuconostoc pseudomesenteroides, Leuconostoc citreum, Pe. pentosaceus, Weissella cibaria* and *We. confusa* were identified. In particular, the most prevalent species detected were *We. cibaria, Lb. plantarum, Ln. pseudomesenteroides* and *Ln. citreum*. DGGE analysis confirmed the detection of the genera to which most isolates belonged (*Lactobacillus, Enterococcus, Leuconostoc* and *Weissella*), but only two species, *Lb. plantarum* and *Ln. citreum*, were clearly identified. The apparent lack of correspondence between culture-dependent and culture-independent methods was explained by the fact that when bacteria isolated by plating are not detected with DGGE analysis based on 16S rRNA gene amplification, they are not a major component of the microbial community being investigated (Shinohara *et al.* 2011). For this reason, a combined approach consisting of both methodologies provides the best strategy for detection of microbial communities within complex food matrices (Carraro *et al.* 2011).

Regarding eukaryotic organisms, yeasts are also detected both on the cereal surface and in flour samples ranging from a few cells to 10^4 and 10^3 cfu/g, respectively. The species most commonly found are *Candida, Cryptococcus, Pichia, Rhodotorula, Saccharomyces, Sporobolomyces, Torulaspora* and *Trichosporon*. Among fungi (circa 10^4 cfu/g), *Alternaria, Cladosporium, Drechslera, Fusarium, Helminthosporium* and *Ulocladium* (from the field), and *Aspergillus* and *Penicillium* (from storage), are found (De Vuyst and Neysens 2005).

Furthermore, it is also important to consider that the microorganisms performing the fermentation of cereal flours may originate from the equipment used in the milling and/or production process (Berghofer *et al.* 2003).

Cereal fermentation

Since the beginning of human civilization there has been an intimate relationship between the human being and the fermentative activities of microorganisms. These activities have been utilized in the production of fermented foods and beverages, which are defined as those products that have been subordinated to the effect of microorganisms or enzymes determining desirable biochemical changes (Settanni and Moschetti 2014). Fermentation represents the oldest and most economical method of producing and preserving food (Chavan and Kadam 1989; Billings 1998). In fact, fermentation helps in the production of safe and stable foods with a longer shelf life than their raw materials. These foods are more digestible and appealing than unprocessed substrates because they acquire new desired organoleptic characteristics (Settanni and Moschetti 2014). In addition, fermentation provides a natural way to reduce the volume of the material to be transported, to destroy undesirable components, to enhance the nutritive value and to reduce the energy required for cooking (Simango 1997). During cereal fermentation several volatile compounds are formed, which contribute to a complex blend of flavours in the processed products (Chavan and Kadam 1989).

The microorganisms responsible for fermentation may be the microflora indigenously present on the substrate (and this occurred unknowingly for millennia) or they may be added as starter cultures (Harlander 1992). The latter strategy commenced when the microorganisms had been isolated and their activities discovered and studied.

By one biochemical definition, fermentation is an anaerobic process for deriving energy from the oxidation of organic compounds using an endogenous electron acceptor, which is usually an organic compound (Prescott *et al.* 2005). Following this process, the carbohydrates are partially oxidized and several microorganisms produce energy by means of this metabolic pathway. Among the several fermentations employed to produce different foods (Soni and Sandhu 1990), the two main processes that are defining for the transformation of cereal flours are alcoholic and lactic acid fermentations. Alcohol fermentation results in the production of ethanol and yeasts are the predominant organisms involved; lactic acid fermentation is carried out by LAB (Corsetti and Settanni 2007).

In general, natural fermentation of cereals leads to a decrease in the level of carbohydrates as well as some non-digestible poly- and oligosaccharides. This process determines the saccharification of starch and increases the availability of proteins (Blandino *et al.* 2003). Certain amino acids may be synthesized and the availability of B-group vitamins might be improved. Fermentation also provides optimum pH conditions for the enzymatic degradation of phytate, which is present in cereals in the form of complexes with polyvalent cations such as iron, zinc, calcium, magnesium and proteins. Such a reduction in phytate may increase the amount of soluble iron, zinc and calcium severalfold (Haard *et al.* 1999; De Angelis *et al.* 2003).

Cereal-based fermented products

Based on the type of flour (alone or in combination), the fermenting agent(s) and the technological process(es) applied, different products can be obtained. The fermenting microorganisms for almost all cereal-based fermented products are mainly LAB and a few yeast species. In several cases, their combined action allows the production of foods and beverages with the desired quality characteristics (Galati *et al.* 2014). However, the microbial populations responsible for the fermentation of several niche cereal-based products, especially those produced in countries where this process is not driven by selected starter cultures, remain unknown.

Cereal-based fermented foods are spread all over the world and undoubtedly bread is the main product. Bread is the typical and oldest food of leavened products and it is a symbol of religion. The earliest records date back to the second millennium BCE, when, after a flood of the Nile that covered the grain reserves, the Egyptians realized that grain flour when mixed with water increased in volume over time (Di Giandomenico 2010). During the Second World War, due to the scarcity of cereals, a large number of cereals (maize, rye, oat, barley, rice, sorghum) or other vegetable sources (bean, cassava, soy, potato, chestnut flour etc.) have been used in breadmaking in place of wheat (INSOR 2012).

To date, unlike bread, where biotechnologies are under control during the transformation process carried out at an industrial level, the preparation of many other cereal-based fermented foods is restricted to limited areas and they are mostly produced on a domestic scale (Owczarek *et al.* 2004). However, some of them are particularly important in different countries and have been the object of study for many research groups (Table 11.1).

Bread

The term 'bread' refers to a food of any shape and dimension obtained from a dough prepared with flour and water, with or without salt, fermented naturally or with the addition of yeasts and subsequently subjected to cooking. Breadmaking technology (Figure 11.1) is quite simple (Pagani *et al.* 2006).

The production of leavened bakery products can be summarized as the semi-solid mass transformation of the dough, a particular 'emulsion', characterized by

Table 11.1 Traditional fermented cereal-based foods and beverages that have been the object of scientific investigation.

Products	Cereals	Countries
Amgba	Sorghum, millet	Cameroon, Chad
Ang-kak	Rice	China, Indonesia, Thailand
Atole	Maize	Guatemala, Mexico
Ben saalga	Millet	Burkina Faso
Boza	Wheat, millet, maize	Albania, Bulgaria, Romania, Turkey
Bread	Wheat, rye, barley, sorghum, rice	Five continents
Bushera	Sorghum, millet	Uganda
Chicha	Maize	Peru
Chikokivana	Maize, millet	Zimbabwe
Chongju	Rice	Korea
Dolo	Sorghum	Burkina Faso
Ikikage	Sorghum	Rwanda
Jaanr	Millet	Northeastern Himalayas
Kachasu	Maize	Zimbabwe
Kaffir beer	Maize	South Africa
Koko	Maize	Ghana
Kunu-Zaki	Millet	Nigeria
Lao-chao	Rice	China, Indonesia
Mantou	Wheat	China
Mirin	Rice	Japan
Muramba	Sorghum	Uganda
Mutwiwa	Maize	Zimbabwe
Ogi	Maize, sorghum, millet	Nigeria, West Africa
Pito	Maize, sorghum	Ghana, Nigeria
Pozol	Maize	Mexico
Puto	Rice	Philippines
Sake	Rice	Japan
Sourdough bread	Wheat, rye, barley, sorghum, rice	Five continents
Takju	Rice, wheat	Korea
Tape ketan	Rice	Indonesia
Tapuy	Rice	Philippines
Tchoukoutou	Sorghum, millet, maize	Benin, Togo
Tesguino	Maize	Mexico
Togwa	Maize, sorghum, millet	Tanzania

a continuous phase represented by hydrated gluten that surrounds the starch granules in which are dispersed microbubbles of air, in a 'foam', that is to say in a product in which the continuous phase retains significant volumes of gas (Bottega *et al.* 2010).

The leavening process is of paramount importance during breadmaking. It determines the increase of dough volume, the development of precursors of the aroma compounds and the improvement of the nutritional characteristics of flour.

Figure 11.1 General process of bread production.

The volume expansion can be achieved in different ways: the biological approach (thanks to carbon dioxide [CO_2] resulting from yeasts and/or LAB metabolism); the use of chemicals added to the formulation (which exert their action especially during cooking); and the physical approach, with the inclusion of air as a result of intensive mechanical action and typical leavening of some formulations rich in fat (Bottega *et al.* 2010).

Biological leavening assumes a basic importance for the sensory properties of the final product. This process can be carried out by baker's yeast, mainly represented by *Saccharomyces cerevisiae*, and a mixture of yeasts and LAB. In the latter case, the technology applied is that of sourdough, referred to as a mixture of flour and water in which the development of LAB results in the production of lactic acid and acetic acid. LAB developing in the dough may originate as contaminants of flour and/or the bakery environment or might derive from a starter culture containing one or more known species of LAB (De Vuyst and Neysens 2005).

Sourdough

Sourdough is a mixture of cereal flour and water in which a heterogeneous population composed of LAB and yeasts is metabolically active, either by spontaneous fermentation or by fermentation initiated through the addition of a sourdough starter culture, whether or not it involves backslopping (De Vuyst *et al.* 2009).

Based on the technology applied, sourdoughs have been grouped into type I, type II and type III (Böcker *et al.* 1995). Type I sourdoughs are traditionally characterized by continuous, daily refreshment to keep the microorganisms in an active state; fermentation is carried out at room temperature until the pH reaches a final value of around 4.0 (Corsetti and Settanni 2007). Type II sourdoughs are semifluid silo preparations fermented at temperatures higher than 30 °C for long periods (at least 2 days); these sourdoughs are generally added during bread preparation as dough-souring supplements (Böcker *et al.* 1995; Hammes and Gänzle 1998). Type III sourdoughs are dried preparations containing LAB resistant to the drying process (Hammes and Gänzle 1998). Types II and III sourdoughs require the addition of *S. cerevisiae* as a leavening agent. Some authors

(De Vuyst and Neysens 2005) also reported on type 0 dough. This product, although prepared exclusively from baker's yeast and not made with sourdough technology, also contains LAB species as contaminants of the yeast inoculums. These are especially lactobacilli rather than *Pediococcus*, *Lactococcus* and *Leuconostoc* spp. (Jenson 1998), and contribute only to a small degree to the acidification and aroma development of dough because of the short processing time.

The addition of sourdough improves the texture, flavour, nutritional aspects and shelf life (Gänzle *et al.* 2007) of wheat bread produced from baker's yeast due to the synthesis of aroma compounds (Czerny and Schieberle 2002; Hansen and Schieberle 2005), enzymes and antibacterial (Settanni *et al.* 2005) and antifungal (Ryan *et al.* 2008; Poutanen *et al.* 2009) compounds during fermentation. Sourdough fermentation has been demonstrated also to enhance the sensory quality of rye breads (Rizzello *et al.* 2014b).

Besides the technological and nutritional aspects, sourdough fermentation may also provide several health benefits, such as a decrease of the glycemic response of baked goods, enhancement of the content of bioactive compounds and increase of mineral uptake (Gobbetti *et al.* 2013). Compared with sourdough bread, white wheat bread started with baker's yeast alone shows some drawbacks such as low protein digestibility, high carbohydrate content, high glycemic index, low-resistant starch and low level of dietary fibre (Dhinda *et al.* 2011).

Sourdough microorganisms

The proportions of LAB and yeasts in sourdough are not random, but rather respect a defined relationship. The ratio between LAB and yeasts is generally reported to be 100:1 (Gobbetti *et al.* 1994; Ottogalli *et al.* 1996). This could be the result of direct interaction between these microbial groups.

The study of the social activities exhibited by microorganisms is defined as 'sociomicrobiology' (Parsek and Greenberg 2005). The microorganisms were for a long time believed to exist as single cells in a given environment searching for nutrients to multiply. Bacteria were the first microbes to be studied in order to decipher their code of communication. It has been discovered that they are active in performing a census of their population, as well as in investigating the environment for development and in feeling the presence of competitors (Fuqua *et al.* 1996). Such actions are the result of efficient intercellular communication that is based on the production, release and detection of and reply to small signal molecules, which accumulate and trigger cascade events when a 'quorum' concentration is reached. Hence, the term 'quorum sensing' is used to describe cell–cell communication. Based on this system the bacteria can 'count' one another (Fuqua *et al.* 1994), acting as a group. Recently, it became apparent that fungi, like bacteria, also use quorum regulation to affect population-level behaviours. Furthermore, considering the extent to which quorum-sensing regulation controls important processes in many distantly related bacterial genera, it is not surprising that cell density–dependent regulation also appears to be prevalent in diverse fungal species (Hogan 2006).

The studies regarding cell–cell communication in sourdough ecosystems are so far limited to the group of LAB (Di Cagno *et al.* 2007, 2010), but this topic is under study to elucidate the mechanisms of interdomain communication between LAB and yeasts.

Yeasts

Several sourdoughs are reported to host *S. cerevisiae*. This finding is generally due to introduction through the addition of baker's yeast (Corsetti *et al.* 2001) or to cross-contamination in bakeries where both conventional and sourdough breads are produced (Valmorri *et al.* 2010). Yeasts found in sourdoughs belong to different genera (Rossi 1996; Stolz 1999; Gullo *et al.* 2002). The typical sourdough yeasts are *Saccharomyces exiguus*, *Candida humilis* (formerly described as *Candida milleri*) and *Issatchenkia orientalis* (*Candida krusei*; Garofalo *et al.* 2008; Iacumin *et al.* 2009). Other yeast species detected in sourdough ecosystems are *Pichia anomala* (as *Hansenula anomala*), *Saturnispora saitoi* (as *Pichia saitoi*), *Torulaspora delbrueckii*, *Debaryomyces hansenii*, *Pichia membranifaciens* (Corsetti and Settanni 2007) and *Candida famata* (Mohamed *et al.* 2007). The extensive variability in the number and type of species found depends on several factors, dough yield (weight of dough/weight of flour × 100), type of cereal used, temperature of fermentation and temperature for sourdough maintenance (Gobbetti *et al.* 1994). Yeasts via alcoholic fermentation are primarily responsible for the leavening of dough (Corsetti and Settanni 2007).

Lactic acid bacteria

LAB are involved in the process of acidification of sourdough, but heterofermentative species partly contribute to the mass blowing (Gobbetti *et al.* 1995).

Unlike other fermented foods, where LAB responsible for the transformation of raw materials into final products belong to obligate homofermentative and/or facultative heterofermentative species, obligate heterofermentative species play a major role in sourdough (Salovaara 1998), especially when sourdoughs are prepared in a traditional manner (Corsetti *et al.* 2001, 2003). Typical sourdough LAB mainly belong to the genus *Lactobacillus* and include all three metabolic groups discussed (Hammes and Vogel 1995). However, other LAB belonging to *Leuconostoc*, *Weissella*, *Pediococcus*, *Lactococcus*, *Enterococcus* and *Streptococcus* genera have been isolated from sourdough (Corsetti and Settanni 2007).

Most of the LAB species commonly detected in sourdough (Figure 11.2) have sourdough as their primary and sole source of isolation, probably because no other ecosystem can support their growth. For instance, except sourdough, no other habitat is known for *Lactobacillus sanfranciscensis* (Hammes *et al.* 2005).

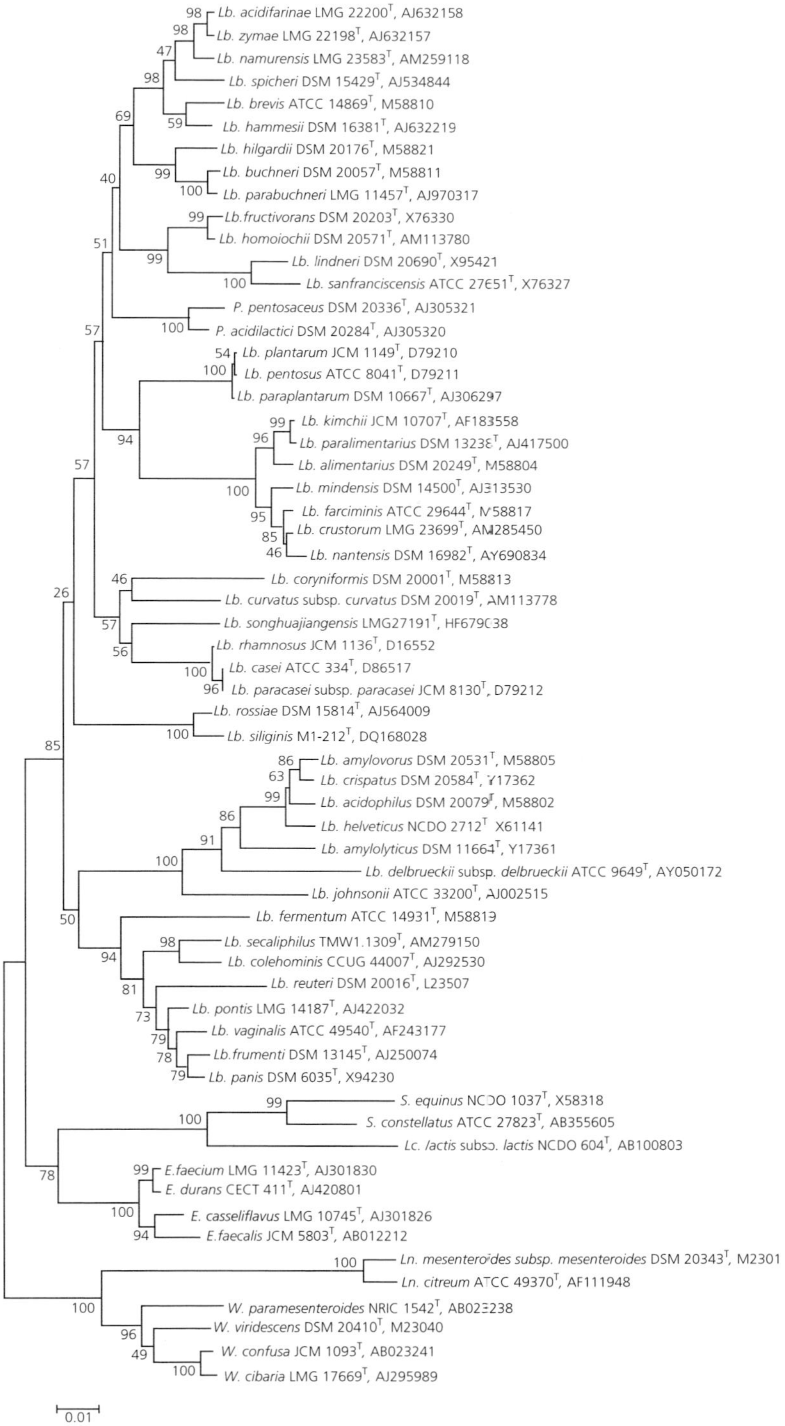

Figure 11.2 Phylogenetic tree of LAB commonly associated with or found in sourdough products based on 16S rRNA gene sequences. Sequence alignment was performed with CLUSTALX (Thompson *et al.* 1997). Sequence and alignment manipulations and calculation of similarity values and nucleotide compositions of sequences were performed with the GeneDoc program version 2.5.000 (K.B. Nicholas and H.B. Nicholas, unpublished data). Positions available for analysis were circa 1150 bp. Phylogenetic and molecular evolutionary analysis was conducted using MEGA version 3.1 (Kumar *et al.* 2004). Bar 0.01 nucleotide substitution per site.

Some of the species associated with this environment have been misidentified because of the lack of application of molecular methods, which were unavailable in the past, or have been found in sourdough due to cross-contamination.

Some non-*Lactobacillus* species isolated from sourdough have only been detected at subdominant levels (Corsetti *et al.* 2007b). As an example, in mature Italian type I sourdough, *E. faecium* and *P. pentosaceus* have been found in the range 10^4–10^6 cfu/g, while lactobacilli were about 2–3 orders of magnitude higher (Valmorri *et al.* 2006). Corsetti *et al.* (2007b) investigated the role of *Ent. faecium* and *Pe. pentosaceus* during sourdough preparation. Strains of both species were followed in dual combination with *Lb. sanfranciscensis*. During the first steps of sourdough preparation, single inocula of *Ent. faecium* and *Pe. pentosaceus* determined a stronger and more rapid acidification of dough than the *Lb. sanfrancisensis* strain used. Subsequently, the behaviour monitored during the co-fermentation of *Ent. faecium*/*Lb. sanfranciscensis* and *Pe. pentosaceus*/*Lb. sanfranciscensis* showed that *Ent. faecium* and *Pe. pentosaceus* prepare the environment for the establishment of the typical species of mature sourdough, including *Lb. sanfranciscensis*, by lowering the pH.

Minervini *et al.* (2014) described the typical LAB population dynamic in sourdough, referred to as 'three-phase evolution', regardless of the type of flour. The three phases indicated are the dominance of LAB species belonging to the genera *Enterococcus*, *Lactococcus* and *Leuconostoc*; the increasingly important presence of sourdough-specific LAB, such as species belonging to the genera *Lactobacillus*, *Pediococcus* and *Weissella*; and the dominance of well-adapted sourdough strains, belonging to obligate heterofermentative species such as *Lb. sanfranciscensis*, *Lb. fermentum* and *Lactobacillus pontis* and to the facultative heterofermentative *Lb. plantarum* (Gänzle *et al.* 2007), although the presence of some *Leuconostoc* spp. is sometimes revealed. This succession of LAB is mainly driven by different tolerance to acidic conditions and to different adaptation mechanisms related to carbohydrate and nitrogen metabolism (Gänzle *et al.* 2007). The establishment of obligate heterofermentative lactobacilli is essential for the optimal fermentation of traditional sourdough (Salovaara 1998).

Regarding cell–cell communication among lactobacilli in sourdough, Di Cagno *et al.* (2007) followed, by a proteomic approach, the growth of *Lb. sanfranciscensis* in mono-culture and co-culture with *Lb. plantarum*, *Lb. brevis* or *Lactobacillus rossiae*. When co-cultured, the *Lb. sanfranciscensis* strain, depending on the combination, overexpressed several proteins during the late stationary phase. The induced polypeptides, only in part common to all co-cultures, were identified as stress proteins, energy metabolism–related enzymes and proline dehydrogenase, GTP-binding protein, *S*-adenosyl-methyltransferase and Hpr phosphocarrier protein. Furthermore, two quorum-sensing genes involved, *luxS* and *metF*, were shown to be expressed in the *Lb. sanfranciscensis* strain studied. Later, the same research group studied the effect of pheromone plantaricin A produced by a *Lb. plantarum* strain (DC400) towards other sourdough LAB, and it was found that this pheromone influenced the growth and survival of the other strains co-cultivated with *Lb. plantarum* DC400 differently (Di Cagno *et al.* 2010).

Starter selection for sourdough production

In the last few years, several research groups have focused their attention on the selection of strains to be used as starter cultures for controlled sourdough fermentation. Analysis of microbial biodiversity and monitoring the dominant microorganisms during fermentation is a key step for the selection of strains able to drive the transformation processes. The starter cultures for fermentation are selected based on their specific technological traits in order to obtain final products with a set of desired characteristics (Figure 11.3).

Generally, LAB are primarily tested for their rapid acidification, whereas yeasts are first tested for their alcoholic fermentation rate. However, depending on the type of bread to be produced, LAB are commonly selected also for other technological performance, such as contribution to the development of the flavour and structure of the dough, reduction of antinutritional factors and, in order to determine microbial stability during fermentation and elongate the

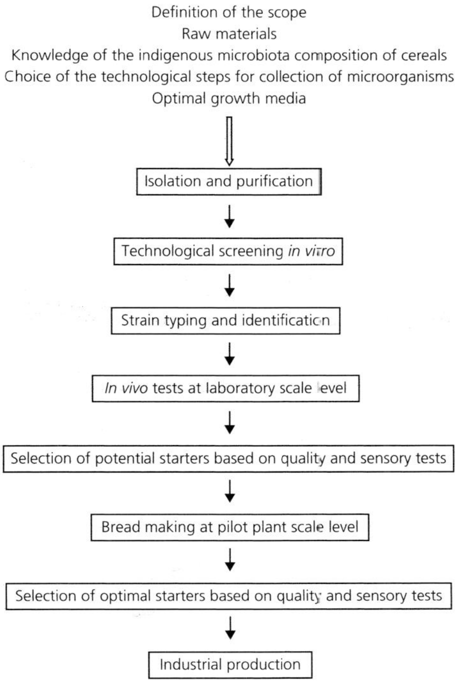

Figure 11.3 Schematic representation of the plan for selection of starter culture(s) to be used for sourdough fermentation.

shelf life of the final products, for their competitiveness with undesired microorganisms based on the production of secondary metabolites, mainly bacteriocins and antifungal compounds (Corsetti and Settanni 2007).

The selection of LAB is performed during fermentation; this allows the collection of dominant strains that have competitive advantages over the native microbiota. The dominant strains will ensure, to a certain extent, the successful transformation of flour into sourdough, thanks to their adaptation to the specific fermentation technology, the environmental conditions and the availability of substrates. Once the dominant isolates are collected, their performance is evaluated *in vitro*, first using the optimal commercial synthetic media and then using sterilized media prepared from raw material extracts that mimic real conditions, without interaction with indigenous microorganisms (Settanni *et al.* 2014). Subsequently, the isolates showing interesting properties are identified at strain and species levels. The different strains are tested *in vivo* using untreated flours, so that their performance is evaluated in the presence of the autochthonous microbiota, and their capacity to dominate the microbial community is monitored (Alfonzo *et al.* 2013). The final products are generally subjected to sensory and quality evaluations to select the best strains, in individual and/or multiple combinations, able to recreate the traditional aromatic profile, in order to obtain breads with the desired characteristics constant over time (Settanni *et al.* 2014).

Fortified fermented cereal-based products

Food fortification refers to the addition of micronutrients to processed foods. It represents a valid technology for reducing micronutrient malnutrition when and where existing food supplies and limited access fail to provide adequate levels of the respective nutrients in the diet. In industrialized countries, food fortification has long been used for the successful control of deficiencies of vitamins A and D, several B vitamins (thiamine, riboflavin and niacin), iodine and iron. From the early 1940s onwards, the fortification of cereal products with thiamine, riboflavin and niacin became common practice (Allen *et al.* 2006). Fortified breads are obtained through the enrichment of flour with nutrients. In fact, the nutritional features of white wheat are quite limited. This is due to the low levels of essential amino acids, such as lysine, and dietary fibre in white flour (Dhinda *et al.* 2011). For years, the most common constituents added to these kind of breads have been folic acid and iron. However, in the last few years the addition of alternative plant-based protein sources has become usual. It is becoming common practice in white bread production to use dietary fibre and ingredients or by-products rich in fibre (De Angelis *et al.* 2007, 2009; Rizzello *et al.* 2012), like a mixture of soy proteins, oat bran and legume flours (Sadowska *et al.* 2003; Kamaljit *et al.* 2010; Dhinda *et al.* 2011; Mohammed *et al.* 2012; Rizzello *et al.* 2014a) to enhance its nutritional value.

These products are gaining importance not only because of the increasing number of people on a vegetarian diet, but also due to the high energy requirements for animal protein production, particularly relevant for people who have no access to an animal protein–rich diet. For example, besides bread, several products are produced in different areas, especially Africa and Asia, with cereals in combination with legumes (Blandino *et al.* 2003), thus improving the overall protein quality of the fermented products. As reported earlier, cereals are deficient in lysine, but are rich in cysteine and methionine. Legumes, on the other hand, are rich in lysine but deficient in sulfur-containing amino acids. Thus, by combining cereals with legumes, the overall protein quality is improved (Campbell-Platt 1994). The main non-bread products obtained from the fermentation of mixtures of cereals and other vegetables are listed in Table 11.2.

Several attempts are being made to fortify the common ingredients used in bread production. Since bread is mainly obtained via the fermentation of wheat flour sugars derived from starch involving chemical interactions of the various food components, these interactions can be adjusted to create desirable products only if the chemical and physical processes are well understood (Sivam *et al.* 2010).

The incorporation of legumes in novel, convenient and healthy food products (Schneider 2002; Gòmez *et al.* 2008) represents a valuable strategy for increasing the global consumption of legumes, which is declining (Kohajdová *et al.* 2013) and is below the recommended amount (McCrory *et al.* 2010). For

Table 11.2 Traditional foods and beverages obtained through fermentation of mixtures of cereals and other vegetable sources.

Products	Cereals and legumes	Countries
Adai	Cereals and legumes	India
Banku	Maize and cassava	Ghana
Chee-fan	Wheat and soybean	China
Dhokla	Rice, wheat and Bengal gram	India
Dosa	Rice and Bengal gram	India
Hamanatto	Wheat and soybean	Japan
Idli	Rice and black gram	India, Sri Lanka
Kanji	Rice and carrots	India
Kecap	Wheat and soybean	Indonesia
Miso	Rice and soybean	China, Japan
Munkoyo	Maize and roots of munkoyo	Africa
Shoyu	Wheat and soybean	China, Japan, Taiwan
Tao-si	Wheat and soybean	Philippines
Taotjo	Wheat, rice and soybean	India
Tarhana	Wheat and vegetables	Turkey
Tauco	Cereals and soybean	Indonesia
Vada	Cereals and legumes	India

this reason, some reports have proposed the addition of legumes to different products such as bread (Dhinda *et al.* 2011; Mohammed *et al.* 2012), biscuits (Eissa *et al.* 2007; Tiwari *et al.* 2011), cakes (Gòmez *et al.* 2008), chapatti (Kadam *et al.* 2012) and crackers (Kohajdová *et al.* 2011).

When legume flours are used for breadmaking, the adjustment of several process parameters is needed to get a high sensory quality that is acceptable to the majority of consumers (Maninder *et al.* 2007; Kohajdová *et al.* 2013). One option to improve the sensory and functional quality of breads containing legume flours is represented by the use of sourdough fermentation. With this in mind, Rizzello *et al.* (2014a) developed a biotechnological protocol for manufacturing a white bread enriched with chickpea, lentil and bean flours through sourdough fermentation. For this purpose, the authors prepared type I sourdough containing legume flours according to traditional protocols routinely used for making typical Italian breads. LAB populations in wheat-legume sourdough included *Lb. plantarum, Lb. sanfranciscensis, Ln. mesenteroides, Lb. fermentum, We. cibaria, Lactobacillus pentosus, Lb. coryniformis, Lb. rossiae, Lb. brevis, Lactobacillus parabuchneri* and *Lactobacillus paraplantarum*, most of which are typically associated with mature sourdough. Compared with wheat-legume breads leavened with commercial yeasts, wheat-legume sourdough breads were characterized by higher quality parameters and acceptability by consumers, thus proving the defining role of sourdough technology in improving the characteristics of bakery products obtained with novel formulas.

Gluten-free cereals

Gluten is toxic for people affected by CS. Gluten causes self-perpetuating mucosal inflammation and subsequent loss of absorptive villi and hyperplasia of the crypts. Proteolytic enzymes of the endoluminal tract acting on prolamins of wheat (a-, b-, c-, and x-gliadin), rye (secalin) and barley (hordein) produce proline- and glycine-rich polypeptides that are responsible for the disease (Silano and De Vincenzi 1999). Triticale also contains gluten (Wolter *et al.* 2014a). The list of proteins that liberate toxic peptides includes the high molecular weight glutenins (Dewar *et al.* 2006). Both gliadins and glutenins are rich in two amino acids, proline (very resistant to the hydrolysis process) and glutamine. The gluten, when ingested, goes through a process of digestion and the protein is not completely degraded by enzymes with prolyl endopeptidase activities (Catassi and Francavilla 2010). The gluten is degraded into small peptides and their further proteolysis is made difficult by the position and abundance of proline residues (Hausch *et al.* 2003). Some of these peptides have an immuno-genetic activity for subjects predisposed to develop CS (Catassi and Francavilla 2010). For these reasons, those affected by CS cannot ingest gluten-containing products.

Most of the food products present on the market that characterize the daily diet in many countries are made with gluten-containing cereals. The number of people affected by CS is high, thus the production of baked goods that can be consumed by these subjects is of paramount importance. The main raw materials used for the production of gluten-free leavened baked are rice (*Oryza sativa* and *Oryza glaberrima*), maize (*Zea mays*), sorghum (*Sorghum bicolor* L. Moench) and millet (*Panicum miliacum*) and pseudo-cereals such as amaranth (species of genus *Amaranthus*), quinoa (species of genus *Chenopodium*) and buckwheat (species of genus *Fagopyrum*; De Angelis and Di Cagno 2010). Rice is one of the most important cereals used for this purpose because the carbohydrate fraction of the flour consists predominantly of starch present in small granules. Amylopectin, easily digested, is present in higher amounts than amylose. When rice flour is used, it provides softness and stability. Maize finds several uses (e.g. polenta, cornflakes, tortillas, snacks etc.) in the diet of coeliac subjects, but it is scarcely used for the production of gluten-free bread. The use of sorghum flour is recommended in the diet of CS patients, as it is the species phylogenetically most distant from the other cereals. Millet flour does not show a good aptitude for baking and for leavened gluten-free production, therefore it is often used in combination with other flours (Schober *et al.* 2003). Quinoa and amaranth have been utilized for the manufacture of different bakery products (Lorenz and Coulter 1991; Nsimba *et al.* 2008) and quinoa flour is also added for the manufacture of enriched gluten-free bakery products (Taylor and Parker 2002).

Most of the gluten-free products present on the market are characterized by a lower quality than conventional products made from gluten-containing flours. Hence, the combination of gluten-free flours and the choice of their ratios is essential for the acceptability of the final products (De Angelis and Di Cagno 2010). Currently, the most utilized gluten-free flours include maize, potato and rice flour and starches, used as base flours due to their bland flavour and neutral effects on baked products. These flours and starches usually tend to be low in nutrition and have very minimal structure-building potential. Chestnut flour presents high potential in the development of gluten-free products (O'Shea *et al.* 2014).

Gluten-free fermented products

The use of gluten-free flours represents a valid alternative to the complete absence of cereals in the diet of CS patients. However, the poor baking performance of these flours (due to the lack of gluten), the low nutritional quality and the poor sensory characteristics of the resulting products determine that there is an important technological challenge to be faced (Wolter *et al.* 2014a). Furthermore, gluten-free products have a limited microbial shelf life (Gallagher 2009; Hager *et al.* 2011).

In the case of bread, the use of sourdough as a bioprocessing ingredient in gluten-free formulations may provide several positive characteristics in the final product. O'Shea *et al.* (2014) summarized the beneficial role of sourdough in gluten-free bread production as follows:

- Production of peptidase able to detoxify the peptides responsible for CS when using wheat and rye flours.
- Activity against mycotoxins produced from fungi found on maize, sorghum and millet.
- Decrease of pH for degradation of phytic acid.
- Extraction of bioactive compounds from the flour and release of biomolecules that are part of the LAB/yeast metabolism.
- Elongation of shelf life.
- Enhancement of the flavour profile.
- Production of exopolysaccarides (EPS), proven to be prebiotic, that are useful to produce breads with a softer texture without the addition of hydrocolloid.

All the positive aspects of gluten-free breads observed with the use of sourdough are due to LAB communities, but their action is strain dependent (Arendt *et al.* 2011).

Among LAB, *Lb. plantarum* is reported to be dominant in several gluten-free sourdoughs produced from amaranth, buckwheat, quinoa, rice and teff flour (Vogelmann *et al.* 2009; Moroni *et al.* 2011a). Strains of this species have been proven to improve the staling rate and crumb hardness of brown rice, buckwheat-based, gluten-free formulations with the addition of sourdough and an inoculum size of 10^8 cfu/g (Moore *et al.* 2007). However, breads made from different gluten-free flours may exhibit an undesirable aroma (Hager *et al.* 2012), especially when compared to wheat sourdough breads, due to the absence or low intensity of a wheat bread–like note. In general, the characteristic odour attributes of gluten-free breads are pea-like with buckwheat, quinoa and teff flours, cooked potato–like with quinoa and teff flours, vomit-like with sorghum and teff flours and mouldy with buckwheat and quinoa flours (Wolter *et al.* 2014a). The odorants and the resulting undesirable notes cause a negative impact on the aroma quality of gluten-free breads.

Selected LAB strains are able to generate very specific volatile organic compounds in wheat sourdough (Settanni *et al.* 2013). Hence, the use of individual metabolic properties of LAB seems to be a promising approach to increase also the aroma quality of gluten-free breads, although the type of flour influences sourdough fermentation, affecting the availability of carbohydrates as primary fermentation substrates, nitrogen sources and growth factors such as vitamins, minerals and the buffering capacity (Hammes *et al.* 2005).

In order to include starter cultures in gluten-free sourdough, it is important to select the correct strain(s) for a given flour type (Rühmkorf *et al.* 2012a). To study the role of LAB, the effect of *Lb. plantarum* sourdough on gluten-free bread quality was evaluated using a composite recipe (Moore *et al.* 2008; Coda *et al.*

2010). This species has been used by Wolter *et al.* (2014a) to ferment single gluten-free (buckwheat, oat, quinoa, sorghum and teff) flours, in order to investigate the influence of the corresponding sourdoughs added for fermentation in the resulting breads. All sourdoughs decreased dough strength, resulting in softer doughs, reduced the staling rate for buckwheat and teff breads and increased the cell volume in sorghum and teff breads, but they did not prolong the shelf life and did not improve the aroma of breads.

Formation of EPS is a positive characteristic of sourdough LAB, since this feature influences the viscosity of sourdough (Vogel *et al.* 2002). Homopolysaccharides (HoPS) are generally applied to improve the structural characteristics of baked goods (Corsetti and Settanni 2007). Sourdough lactobacilli have not been found to produce heteropolysaccharides (HePS; Tieking and Gänzle 2005), which are mainly applied in fermented milk products (Laws and Marshall 2001). However, some authors (Galle *et al.* 2011) have assessed that the utilization of LAB strains producing HePS expands the variety of cultures and the diversity of polysaccharides for applications in gluten-free baking. For this purpose, two LAB, one *Lb. casei* and one *Lactobacillus buchneri,* were tested in sorghum sourdough and the resistance to deformation of the sorghum sourdough started with *Lb. buchneri* was registered at lower levels, due to the presence of HePS. Rühmkorf *et al.* (2012b) evaluated four EPS-producing LAB strains. One *Lb. curvatus* provided the best results, in terms of reduced bake loss, higher crumb moisture content and slower rate of staling, when used in a rice/buckwheat bread formulation (Rühmkorf *et al.* 2012b). Different strains of *Lb. sanfranciscensis, Lb. curvatus, Lactobacillus reuteri, Lactobacillus animalis* and *We. cibaria* did not increase the loaf volume of buckwheat, rice and sorghum sourdough breads (Galle *et al.* 2012; Rühmkorf *et al.* 2012b). Moroni *et al.* (2011b) reported that a multiple LAB strain starter culture comprising species *Lb. brevis, Lactobacillus paralimentarius, Lb. plantarum* and *We. cibaria* determined a small loaf volume of a buckwheat formulation as a result of a decrease in CO_2 production due to the sourdough inclusion.

From the works cited it emerged that, although the application of sourdough might not necessarily lead to improved bread quality, the effect of its inclusion in gluten-free breadmaking is strictly dependent on the flour matrix used (Wolter *et al.* 2014b).

Microbial strategies to reduce gluten content

Gluten-free breads are undoubtedly characterized by lower-quality properties than wheat breads. For this reason, several research groups have studied different strategies to reduce the toxic effects of wheat breads for CS patients. Shan *et al.* (2002) proposed oral therapy with a prolyl-endopeptidase produced by *Flavobacterium meningosepticum* that hydrolyses the 33-mer peptide, reported as one of the most potent peptides involved in triggering CS. This enzyme has also been purified from *Myxococcus xanthus* (Gass *et al.* 2005), *Sphingomonas capsulata* (Shan *et al.* 2004) and *Lactobacillus helveticus* (Chen *et al.* 2003). Stepniak *et al.* (2006) proposed oral supplementation with the prolyl-endopeptidase from *Aspergillus niger* that is stable under gastric conditions. However, some authors found that different LAB can decrease the CS-inducing effects of gluten.

The most significant developments in this field date to the beginning of the 2000s. Di Cagno *et al.* (2002) demonstrated active hydrolysis of various proline-rich peptides by lactobacilli. Following this finding, sourdoughs were prepared from a mixture of wheat (30%) and gluten-free (oat, buckwheat and millet; 70%) flours inoculated with strains of *Lb. alimentarius, Lb. brevis, Lb. sanfranciscensis* and *Lactobacillus hilgardii*. After 24 hours of fermentation, the gliadin fractions of the resulting bread were almost completely hydrolysed and the product was tolerated by CS patients (Di Cagno *et al.* 2004). The probiotic preparation VSL#3 (VSL Pharmaceuticals, Gaithesburg, MD, USA) containing *Streptococcus thermophilus, Lb. plantarum, Lactobacillus acidophilus, Lb. casei, Lactobacillus delbrueckii* spp. *bulgaricus, Bifidobacterium breve, Bifidobacterium longum* and *Bifidobacterium infantis* was also found to hydrolyse gliadin polipeptides (De Angelis *et al.* 2006). The same behaviour was registered for *Ent. faecalis* isolated from fermented wheat doughs (M'hir *et al.* 2008) and Wieser *et al.* (2008) confirmed the degradation of gluten proteins during sourdough fermentation in the presence of lactobacilli and enterococci during the selection of gluten-degrading LAB. Gerez *et al.* (2008) reported the functionality of LAB peptidase activities in the hydrolysis of gliadin-like fragments. In that study, none of the LAB strains alone could hydrolyse 57–89 α-gliadin peptide, while the combination of *Lb. plantarum* and *Pe. pentosaceus* strains led to the hydrolysis of 57% of the peptide in 8 hours.

Rizzello *et al.* (2007) showed that fermentation by selected sourdough lactobacilli and addition of fungal proteases decreased the residual concentration of gluten of wheat flour below the threshold level indicated by the Codex Alimentarius Commissions of the World Health Organization and the Food and Agricultural Organization for gluten-free foods. Thus, the application of this combined approach based on the activities of sourdough lactobacilli and fungal proteases allowed the production of baked goods made from wheat flour that were not toxic to patients with CS (Greco *et al.* 2011). M'hir *et al.* (2009) used this approach, including a pool of selected enterococci and fungal proteases. However, some limitations to the use of sourdough to reduce the intolerance of CD patients derive from the long fermentation time required for complete hydrolysis of the toxic peptides. Under such conditions, stability and dough resistance are decreased as a result of the disruption of the gluten network (Cabrera-Chávez and Calderón de la Barca 2010). Rizzello *et al.* (2014b) developed a protocol for the manufacture of a traditional wheat flour bread with an intermediate content of gluten, enhanced digestibility, high bioavailability of free essential amino acids and high protein nutritional quality, but other studies are needed to obtain breads with complete degradation of gluten during flour fermentation and good structural and sensory features.

Thanks to the properties reported here, sourdough LAB cultures are also useful during processing of gluten-free flours because they can eliminate the risks of cross-contamination by gluten (Di Cagno *et al.* 2008).

Conclusion

Cereal-based foods are key components of the diet of several populations. This chapter has analysed the process of fermentation of different cereal flours carried out by different microorganisms, basically LAB and yeasts, whose activities are responsible for the desirable and typical characteristics of the final products. Sourdough technology is applied worldwide to produce breads and provides a useful strategy to solve the main problems related to special bread production. Fortified and/or gluten-free breads are made with mixtures of ingredients and flours that do not generally result in high-quality products. The use of sourdough might improve several quality characteristics, but the effect depends on the active strains. From this standpoint, the selection of an ad hoc starter culture is defining to improve the sensory notes of a given fermented cereal-based product. Several LAB showed the potential to hydrolyse the toxic peptides responsible for CS during long fermentation, but the complete suitability of sourdough for the production of gluten-free breads from wheat or other flours containing gluten is still under study, in order to guarantee the safety of consumers as well as to decrease the time necessary for production.

References

Alfonzo, A., Ventimiglia, G., Corona, O. *et al.* (2013) Diversity and technological potential of lactic acid bacteria of wheat flours. *Food Microbiology*, **36**, 343–354.

Allen, L.H., de Benoist, B., Dary, O. and Hurrell, R. (2006) Guidelines on Food Fortification with Micronutrients, Geneva: World Health Organization/Food and Agricultural Organization of the United Nations.

Arendt, E.K., Moroni, A. and Zannini, E. (2011) Medical nutrition therapy: Use of sourdough lactic acid bacteria as a cell factory for delivering functional biomolecules and food ingredients in gluten free bread. *Microbial Cell Factories*, **10**, S15.

Berghofer, L.K., Hocking, A.D., Miskelly, D. and Jansson, E. (2003) Microbiology of wheat and flour milling in Australia. *International Journal of Food Microbiology*, **85**, 137–149.

Billings, T. (1998) On fermented foods, *Living and Fermented Foods*, http://www.living-foods.com/articles/fermented.html (accessed 3 September 2014).

Blandino, A., Al-Aseeri, M.E., Pandiella, S.S., Canterob, D. and Webb, C. (2003) Cereal based fermented foods and beverages. *Food Research International*, **36**, 527–543.

Böcker, G., Stolz, P. and Hammes, W.P. (1995) Neue Erkenntnisse zum Ökosystem Sauerteig und zur Physiologie des Sauerteig-Typischen Stämme *Lactobacillus sanfrancisco* und *Lactobacillus pontis*. *Getreide Mehl und Brot*, **49**, 370–374.

Bottega, G., Pagani, M.A. and Mariotti, E. (2010) Tecnologia dei prodotto lievitati da forno. In M. Gobbetti and A. Corsetti (eds), *Biotecnologia dei prodotti lievitati da forno* (pp. 255–287), Milan: Casa Editrice Ambrosiana.

Bushuk, W. (2001) Rye production and uses worldwide. *Cereal Foods World*, **46**, 70–73.

Cabrera-Chávez, F. and Calderón de la Barca, A.M. (2010) Trends in wheat technology and modification of gluten proteins for dietary treatment of coeliac disease patients. *Journal of Cereal Science*, **52**, 337–341.

Campbell-Platt, G. (1994) Fermented foods: A world perspective. *Food Research International*, **27**, 253–257.

Cannella, C., Galterio, G. and Gobbetti, M. (2010) Aspetti nutrizionali dei prodotti lievitati da forno. In M. Gobbetti and A. Corsetti (eds), *Biotecnologia dei prodotti lievitati da forno* (pp. 285–295), Milan: Casa Editrice Ambrosiana.

Cannella, C. and Piredda, M. (2006) I cereali nella dieta mediterranea. *Informatore Agrario*, **12**, 7–8.

Caputo, I., Lepretti, M., Martucciello, S. and Esposito, C. (2010) Enzymatic strategies to detoxify gluten: Implications for celiac disease. *Enzyme Research*, **2010**, 174354.

Carraro, L., Maifreni, M., Bartolomeoli, I. *et al.* (2011) Comparison of culture-dependent and -independent methods for bacterial community monitoring during Montasio cheese manufacturing. *Research in Microbiology*, **162**, 231–239.

Catassi, C. and Francavilla, R. (2010) Intolleranza al glutine. In M. Gobbetti and A. Corsetti (eds), *Biotecnologia dei prodotti lievitati da forno* (pp. 297–316), Milan: Casa Editirice Ambrosiana.

Chavan, J.K. and Kadam, S.S. (1989) Critical reviews in food science and nutrition. *Food Science*, **28**, 348–400.

Chen, Y.S., Christensen, J.E., Broadbent, J.R. and Steele, J.L. (2003) Identification and characterization of *Lactobacillus helveticus* PepO2, an endopeptidase with post-proline specificity. *Applied and Environmental Microbiology*, **69**, 1276–1282.

Coda, R., Rizzello, C.G. and Gobbetti, M. (2010) Use of sourdough fermentation and pseudo-cereals and leguminous flours for the making of a functional bread enriched of γ-aminobutyric acid (GABA). *International Journal of Food Microbiology*, **137**, 236–245.

Corsetti, A., De Angelis, M., Dellaglio, F. *et al.* (2003) Characterization of sourdough lactic acid bacteria based on genotypic and cell-wall protein analyses. *Journal of Applied Microbiology*, **94**, 641–654.

Corsetti, A., Gobbetti, M. and Smacchi, E. (1996) Antibacterial activity of sourdough lactic acid bacteria: Isolation of a bacteriocin-like inhibitory substance from *Lactobacillus sanfranciscensis* C57. *Food Microbiology*, **13**, 447–456.

Corsetti, A., Lavermicocca, P., Morea, M., Baruzzi, F., Tosti, N. and Gobbetti, M. (2001) Phenotypic and molecular identification and clustering of lactic acid bacteria and yeasts from wheat (species *Triticum durum* and *Triticum aestivum*) sourdoughs of Southern Italy. *International Journal of Food Microbiology*, **64**, 95–104.

Corsetti, A. and Settanni, L. (2007) Lactobacilli in sourdough fermentation: A review. *Food Research International*, **40**, 539–558.

Corsetti, A., Settanni, L., Chaves-López, C., Felis, G.E., Mastrangelo, M. and Suzzi, G. (2007a) A taxonomic survey of lactic acid bacteria isolated from wheat (*Triticum durum*) and non-conventional flours. *Systematic and Applied Microbiology*, **30**, 561–571.

Corsetti, A., Settanni, L., Valmorri, S., Mastrangelo, M. and Suzzi, G. (2007b) Identification of subdominant sourdough lactic acid bacteria and their evolution during laboratory-scale fermentations. *Food Microbiology*, **24**, 592–600.

Czerny, M. and Schieberle, P. (2002) Important aroma compounds in freshly ground wholemeal and white wheat flour: Identification and quantitative changes during sourdough fermentation. *Journal of Agricultural and Food Chemistry*, **50**, 6835–6840.

Dahlberg, J. (2007). Grains in human evolution. In K.D. Gifford and S. Baer-Sinnot (eds), *The Oldways Table*, Berkeley, CA: Ten Speed Press.

De Angelis, M., Damiano, N., Rizzello, C.G., Cassone, A., Di Cagno, R. and Gobbetti, M. (2009) Sourdough fermentation as a tool for the manufacture of low-glycemic index white wheat bread enriched in dietary fibre. *European Food Research and Technology*, **229**, 593–601.

De Angelis, M.D. and Di Cagno, R. (2010) I prodotti lievitati da forno senza glutine. In M. Gobbetti and A. Corsetti (eds), *Biotecnologia dei prodotti lievitati da forno* (pp. 317–333), Milan: Casa Editrice Ambrosiana.

De Angelis, M., Gallo, G., Corbo, M.R. *et al.* (2003) Phytase activity in sourdough lactic acid bacteria: Purification and characterization of a phytase from *Lactobacillus sanfranciscensis* CB1. *International Journal of Food Microbiology*, **87**, 259–270.

De Angelis, M., Rizzello, C.G., Alfonsi, G. *et al.* (2007) Use of sourdough lactobacilli and oat fibre to decrease the glycaemic index of white wheat bread. *British Journal of Nutrition*, **98**, 1196–1205.

De Angelis, M., Rizzello, C.G., Fasano, A. *et al.* (2006) VSL#3 probiotic preparation has the capacity to hydrolize gliadin polypeptides responsible for celiac sprue. *Biochimica et Biophysica Acta*, **1762**, 80–93.

De Vuyst, L. and Neysens, P. (2005) The sourdough microflora: Biodiversity and metabolic interactions. *Trends in Food Science and Technology*, **16**, 43–56.

De Vuyst, L., Vrancken, G., Ravyts, F., Rimaux, T. and Weckx, S. (2009) Biodiversity, ecological determinants, and metabolic exploitation of sourdough microbiota. *Food Microbiology*, **26**, 666–675.

Dewar, D.H., Amato, M., Ellis, H.J., Pollock, E.L. and Gonzalez-Cinca, N. (2006) The toxicity of high molecular weight glutenin subunits of wheat to patients with coeliac disease. *European Journal of Gastroenterology and Hepatology*, **18**, 483–491.

Dhinda, F., Lakshmi, J.A., Prakash, J. and Dasappa, I. (2011) Effect of ingredients on rheological, nutritional and quality characteristics of high protein, high fibre and low carbohydrate bread. *Food Bioprocess and Technology*, **5**, 2998–3006.

Di Cagno, R., De Angelis, M., Auricchio, S. *et al.* (2004) Sourdough bread made from wheat and nontoxic flours and started with selected lactobacilli is tolerated in celiac sprue patients. *Applied and Environmental Microbiology*, **70**, 1088–1096.

Di Cagno, R., De Angelis, M., Calasso, M. *et al.* (2010) Quorum sensing in sourdough *Lactobacillus plantarum* DC400: Induction of plantaricin A (PlnA) under co-cultivation with other lactic acid bacteria and effect of PlnA on bacterial and Caco-2 cells. *Proteomics*, **10**, 2175–2190.

Di Cagno, R., De Angelis, M., Lavermicocca, P. *et al.* (2002) Proteolysis by sourdough lactic acid bacteria: Effects on wheat flour protein fractions and gliadin peptides involved in human cereal intolerance. *Applied and Environmental Microbiology*, **68**, 623–633.

Di Cagno, R., De Angelis, M., Limitone, A. *et al.* (2007) Cell-cell communication in sourdough lactic acid bacteria: A proteomic study in *Lactobacillus sanfranciscensis* CB1. *Proteomics*, **7**, 2430–2446.

Di Cagno, R., Rizzello, C.G., De Angelis, M. *et al.* (2008) Use of selected sourdough strains of Lactobacillus for removing gluten and enhancing the nutritional properties of gluten-free bread. *Journal of Food Protection*, **71**, 1491–1495.

Di Giandomenico, M. (2010) Storia e sociologia dei prodotti lievitati da forno. In M. Gobbetti and A. Corsetti (eds), *Biotecnologia dei prodotti lievitati da forno* (pp. 1–15), Milan: Casa Editrice Ambrosiana.

Eissa, A., Hussein, A.S. and Mostafa, B.E. (2007) Rheological properties and quality evaluation of Egyptian balady bread and biscuits supplemented with flours of ungerminated and germinated legume seeds or mushroom. *Polish Journal of Food and Nutrition Sciences*, **57**, 487–496.

Flander, L., Salmenkallio-Marttila, M., Suortti, T. and Autio, K. (2007) Optimization of ingredients and baking process for improved wholemeal oat bread quality. *LWT – Food Science Technology*, **40**, 860–870.

Francis, F.J. (2000) *The Wiley Encyclopedia of Food Science and Technology*, Hoboken, NJ: John Wiley & Sons.

Fuqua, W.C., Winans, S.C. and Greenberg, E.P. (1994) Quorum sensing in bacteria: The LuxR-LuxI family of cell density-responsive transcriptional regulators. *Journal of Bacteriology*, **176**, 269–275.

Fuqua, W.C., Winans, S.C. and Greenberg, E.P. (1996) Census and consensus in bacterial ecosystems: The *LuxR–LuxI* family of quorum-sensing transcriptional regulators. *Annual Review of Microbiology*, **50**, 727–751.

Galati, A., Oguntoyinbo, F.A., Moschetti, G., Crescimannoa, M. and Settannia, L. (2014) The cereal market and the role of fermentation in cereal-based food production in Africa. *Food Reviews International*, **30**, 317–337.

Gallagher, E. (2009) *Gluten-free Food Science and Technology*, Dublin: Ashtown Food Research Centre/John Wiley & Sons.

Galle, S., Schwab, C., Arendt, E.K. and Gänzle, M.G. (2011) Structural and rheological characterisation of heteropolysaccharides produced by lactic acid bacteria in wheat and sorghum sourdough. *Food Microbiology*, **28**, 547–553.

Galle, S., Schwab, C., Dal Bello, F., Coffey, A., Gänzle, M.G. and Arendt, E.K. (2012) Influence of *in-situ* synthesized exopolysaccharides on the quality of gluten-free sorghum sourdough bread. *International Journal of Food Microbiology*, **155**, 105–112.

Galli, A. and Franzetti, L. (1987) Ricerche sulla composizione microbiologica della farina di grano tenero. *Annals of Microbiology*, **37**, 73–80.

Gänzle, M.G., Vermeulen, N. and Vogel, R.F. (2007) Carbohydrate, peptide and lipid metabolism of lactic acid bacteria in sourdough. *Food Microbiology*, **24**, 128–138.

Garofalo, C., Silvestri, G., Aquilanti, L. and Clementi, F. (2008) PCR-DGGE analysis of lactic acid bacteria and yeast dynamics during the production processes of three varieties of Panettone. *Journal of Applied Microbiology*, **105**, 243–254.

Gass, J., Ehren, J., Strohmeier, G., Isaacs, I. and Khosla, C. (2005) Fermentation, purification, formulation, and pharmacological evaluation of a prolyl endopeptidase from *Myxococcus xanthus*: Implications for celiac sprue therapy. *Biotechnology and Bioengineering*, **92**, 674–684.

Gerez, C.L., De Valdez, G.F. and Rollán, G.C. (2008) Functionality of lactic acid bacteria peptidase activities in the hydrolysis of gliadin-like fragments. *Letters in Applied Microbiology*, **47**, 427–432.

Giardini, A. and Vecchiettini, M. (2000). Mais o Granturco (*Zea mays* L.). In R. Baldoni and L. Giardini (eds), *Coltivazioni erbacee* (pp. 155–199), Bologna: Pàtron Editore.

Gifford, K.D. and Baer-Sinnot, S. (eds) (2007) *The Oldways Table*, Berkeley, CA: Ten Speed Press.

Gobbetti, M., Corsetti, A., Rossi, J., La Rosa, F. and De Vincenzi, S. (1994) Identification and clustering of lactic acid bacteria and yeasts from wheat sourdoughs of central Italy. *Italian Journal of Food Science*, **6**, 85–94.

Gobbetti, M., Rizzello, C.G., Di Cagno, R. and De Angelis, M. (2013) How the sourdough may affect the functional features of leavened baked goods. *Food Microbiology*, **37**, 30–40.

Gobbetti, M., Simonetti, M.S., Corsetti, A., Santinelli, F., Rossi, J. and Damiani, P. (1995) Volatile compound and organic acid production by mixed wheat sour dough starters: Influence of fermentation parameters and dynamics during baking. *Food Microbiology*, **12**, 497–507.

Gòmez, M., Oliete, B., Rosell, C.M., Pando, V. and Fernández, E. (2008) Studies on cake quality made of wheat–chickpea flour blends. *LWT – Food Science and Technology*, **41**, 1701–1709.

Greco, L., Gobbetti, M., Auricchio, R. *et al.* (2011) Safety for patients with celiac disease of baked goods made of wheat flour hydrolyzed during food processing. *Clinical Gastroenterology and Hepatology*, **9**, 24–29.

Gullo, M., Romano, A. D., Pulvirenti, A. and Giudici, P. (2002) *Candida humilis*-dominant species in sourdoughs for the production of durum wheat bran flour bread. *International Journal of Food Microbiology*, **80**, 55–59.

Haard, N.F., Odunfa, S.A., Lee, C.H., Quintero-Ramírez, R., Lorence-Quiñones, A. and Wacher-Radarte, C. (1999) Fermented cereals: A global perspective. *FAO Agricultural Services Bulletin*, 138.

Hager, A.S., Axel, C. and Arendt, E.K. (2011) Status of carbohydrates and dietary fiber in gluten free diets. *Cereal Food World*, **56**, 109–114.

Hager, A.S., Wolter, A., Czerny, M. *et al.* (2012) Investigation of product quality, sensory profile and ultrastructure of breads made from a range of commercial gluten-free flours compared to their wheat counterparts. *European Food Research and Technology*, **235**, 333–344.

Hamad, S.H., Dieng, M.C., Ehrmann, M.A. and Vogel, R.F. (1997) Characterization of the bacterial flora of Sudanese sorghum flour and sorghum sourdough. *Journal of Applied Microbiology*, **83**, 764–770.

Hammes, W.P., Brandt, M.J., Francis, K.L., Rosenheim, J., Seitter, M.F.H. and Vogelmann, S.A. (2005) Microbial ecology of cereal fermentations. *Trends in Food Science and Technology*, **16**, 4–11.

Hammes, W.P. and Gänzle, M.G. (1998) Sourdough breads and related products. In B.J.B. Wood (ed.), *Microbiology of Fermented Foods* (pp. 199–216). London: Blackie Academic and Professional.

Hammes, W.P. and Vogel, R.F. (1995) The genus Lactobacillus. In B.J.B. Wood and W.H. Holzapfel (eds), Genera *of* Lactic Acid Bacteria (pp. 19–54), London: Blackie Academic and Professional.

Hansen, A. and Schieberle, P. (2005) Generation of aroma compounds during sourdough fermentation: Applied and fundamental aspects. *Trends in Food Science and Technology*, **16**, 85–94.

Harlander, S. (1992) Food biotechnology. In J. Lederberg (ed.), *Encyclopedia of Microbiology* (pp. 191–207), New York: Academic Press.

Hausch, F., Shan, L., Santiago, N.A., Gray, G.M. and Khosla, C. (2003) Intestinal digestive resistance of immunodominant gliadin peptides. *American Journal of Physiology*, **283**, 996–1003.

He, H. and Hoseney, R.C. (1991) Gas retention on different cereal flours. *Cereal Chemistry*, **68**, 334–336.

Hogan, D.A. (2006) Talking to themselves: Autoregulation and quorum sensing in fungi. *Eukaryotic Cell*, **5**, 613–619.

Iacumin, L., Cecchini, F., Manzano, M. *et al.* (2009) Description of the microflora of sourdoughs by culture-dependent and culture-independent methods. *Food Microbiology*, **26**, 128–135.

INSOR (Istituto Italiano di Sociologia Rurale) (2012) *Atlante dei prodotti tipici: Il pane*, Rome: AGRA-RAI-ERI Editrice.

Jenson, I. (1998) Bread and baker's yeast. In B.J.B. Wood (ed.), *Microbiology of Fermented Foods* (pp. 172–198), London: Blackie Academic and Professional.

Kadam, M.L., Salve, R.V., Mehrajfatema, Z.M. and More, S.G. (2012) Development and evaluation of composite flour for Missi roti/chapatti. *Food Processing and Technology*, **3**, 1000134.

Kamaljit, K., Baljeet, S. and Amarjeet, K. (2010) Preparation of bakery products by incorporating pea flour as a functional ingredient. *American Journal of Food Technology*, **5**, 130–135.

Kohajdová, Z., Karovičová, J. and Magala, M. (2011) Utilisation of chickpea flour for cracker production. *Acta Chimica Slovaca*, **4**, 98–107.

Kohajdová, Z., Karovičová, J. and Magala, M. (2013) Effect of lentil and bean flours on rheological and baking properties of wheat dough. *Chemical Papers*, **67**, 398–407.

Kumar, S., Tamura, K. and Nei, M. (2004) MEGA3: Integrated software for molecular evolutionary genetics analysis and sequence alignment. *Briefings in Bioinformatic*, **5**, 150–163.

Laws, A. and Marshall, V.M. (2001) The relevance of exopolysaccharides to the rheological properties in milk fermented with ropy strains of lactic acid bacteria. *International Dairy Journal*, **11**, 709–721.

Lorenz, K. and Coulter, L. (1991) Quinoa flour in baked products. *Plant Foods for Human Nutrition*, **41**, 213–223.

M'hir, S., Aldric, J.M., El-Mejdoub, T. *et al.* (2008) Proteolytic breakdown of gliadin by *Enterococcus faecalis* isolated from Tunisian fermented dough. *World Journal of Microbiology and Biotechnology*, **24**, 2775–2781.

M'hir, S., Rizzello, C.G., Di Cagno, R., Cassone, A. and Hamdi, M. (2009) Use of selected enterococci and *Rhizopus oryzae* proteases to hydrolyse wheat proteins responsible for celiac disease. *Journal of Applied Microbiology*, **106**, 421–431.

Macrae, R., Robinson, R.H. and Sadler, M.J. (eds) (1993) *Encyclopedia of Food Science, Food Tecnology and Nutrition*, London: Academic Press.

MacRitchie, F. (1992) Physiochemical properties of wheat proteins in relation to functionality. *Advances in Food Nutritional Research*, **36**, 71–76.

Maninder, K., Sandhu, K.S. and Singh, N. (2007) Comparative study of the functional, thermal and pasting properties of flours from different field pea (*Pisumsativum* L.) and pigeon pea (*Cajanus cajan* L.) cultivars. *Food Chemistry*, **104**, 259–267.

Mariotti, M., Lucisano, M. and Pagani, M.A. (2006) Development of a baking procedure for the production of out supplemented wheat bread. *International Journal of Food Science Technology*, **41**, 151–157.

Mariotti, M., Lucisano, M., Pagani, M.A. and Iametti, S. (2008) Macromolecular interactions and rheological properties of buckwheat dough obtained from differently processed grain. *Journal Agricultural and Food Chemistry*, **56**, 4258–4267.

McCrory, M.A., Hamaker, B.R., Lovejoy, J.C. and Eichelsdorfer, P.E. (2010) Pulse consumption, satiety, and weight management. *Advances in Nutrition*, **1**, 17–30.

Mensah, P. (1997) Fermentation: The key to food safety assurance in Africa. *Food Control*, **8**, 271–278.

Meroth, C.B., Hammes, W.P. and Hertel, C. (2004) Characterisation of the microbiota of rice sourdoughs and description of *Lactobacillus spicheri* sp. nov. *Systematic and Applied Microbiology*, **27**, 151–159.

Minervini, F., De Angelis, M., Di Cagno, R. and Gobbetti, M. (2014) Ecological parameters influencing microbial diversity and stability of traditional sourdough. *International Journal of Food Microbiology*, **171**, 136–146.

Mohamed, L., Zakaria, M., Ali, A. *et al.* (2007) Optimization of growth and extracellular glucoamylase production by *Candida famata* isolate. *African Journal of Biotechnology*, **22**, 2590–2595.

Mohammed, I., Ahmed, A.R. and Senge, B. (2012) Dough rheology and bread quality of wheat–chickpea flour blends. *Industrial Crops and Products*, **36**, 196–202.

Moore, M.M., Bello, F.D. and Arendt, E.K. (2008) Sourdough fermented by *Lactobacillus plantarum* FST 1.7 improves the quality and shelf life of gluten-free bread. *European Food Research and Technology*, **226**, 1309–1316.

Moore, M.M., Juga, B., Schober, T.J. and Arendt, E.K. (2007) Effect of lactic acid bacteria on properties of gluten-free sourdoughs, batters, and quality and ultrastructure of gluten-free bread. *Cereal Chemistry*, **84**, 357–364.

Moroni, A.V., Arendt, E.K. and Bello, F.D. (2011a) Biodiversity of lactic acid bacteria and yeasts in spontaneously-fermented buckwheat and teff sourdoughs. *Food Microbiology*, **28**, 497–502.

Moroni, A.V., Bello, F.D., Zannini, E. and Arendt, E.K. (2011b) Impact of sourdough on buckwheat flour, batter and bread: Biochemical, rheological and textural insights. *Journal of Cereal Science*, **54**, 195–202.

Mosca, G. and Toniolo, L. (2000) Orzo (*Hordeum* sp. pl.). In R. Baldoni and L. Giardini (eds), *Coltivazioni erbacee*, Bologna: Pàtron Editore.

Neumann, H. and Bruemmer, J.M. (1997) Investigations with the production of gluten free bread and roll specialities. *Getreide Mehl und Brot*, **51**, 50–55.

Nsimba, R.Y., Kikuzaki, H. and Konishi, Y. (2008) Antioxidant activity of various extracts fractions of *Chenopodium quinoa* and *Amaranthus* spp. seeds. *Food Chemistry*, **106**, 760–766.

O'Shea, N., Arendt, E. and Gallagher, E. (2014). State of the art in gluten-free research. *Journal of Food Science*, **79**, R1067–R1076.

Ottogalli, G., Galli, A. and Foschino, R. (1996) Italian bakery products obtained with sour dough: Characterization of the typical microflora. *Advances in Food Science*, **18**, 131–144.

Owczarek, L., Jasińska, U., Osińka, M. and Skapska, S. (2004) Juices and beverages with a controlled phenol content and antioxidant capacity. *Polish Food Nutritional Sciences*, **5**, 261–268.

Pagani, M.A., Lucisano, M. and Mariotti, M. (2006) Italian bakery products. In Y.H. Huy, H. Corke, I. De Leyn, W.K. Nip and N. Cross (eds), *Baking Science and Technology* (pp. 527–560), Ames, IA: Blackwell.

Parsek, M.R. and Greenberg, E.P. (2005) Sociomicrobiology: The connections between quorum sensing and biofilms. *Trends in Microbiology*, **13**, 27–33.

Perricone, M., Bevilacqua, A., Corbo, M.R. and Sinigaglia, M. (2014) Technological characterization and probiotic traits of yeasts isolated from Altamura sourdough to select promising microorganisms as functional starter cultures for cereal-based products. *Food Microbiology*, **38**, 26–35.

Pogna, N.E., Tusa, P. and Boggini, G. (1996) Genetic and biochemical aspects of dough quality in wheat. *Advances in Food Science*, **18**, 145–151.

Poutanen, K., Flander, L. and Katina, K. (2009) Sourdough and cereal fermentation in a nutritional perspective. *Food Microbiology*, **26**, 693–699.

Prescott, L.M., Harley, J.P. and Klein, D.A. (eds) (2005) *Microbiology*, New York: McGraw-Hill.

Raffo, A., Pasqualone, A., Sinesio, F., Paoletti, F., Quaglia, G. and Simeone, R. (2003) Influence of durum wheat cultivar on the sensory profile and staling rate of Altamura bread. *European Food Research and Technology*, **218**, 49–55.

Rizzello, C.G., Calasso, M., Campanella, D., De Angelis, M. and Gobbetti, M. (2014a) Use of sourdough fermentation and mixture of wheat, chickpea, lentil and bean flours for enhancing the nutritional, texture and sensory characteristics of white bread. *International Journal of Food Microbiology*, **180**, 78–87.

Rizzello, C.G., Coda, R., Mazzacane, F., Minervini, D. and Gobbetti, M. (2012). Micronized by-products from debranned durum wheat and sourdough fermentation enhanced the nutritional, textural and sensory features of bread. *Food Research International*, **46**, 304–313.

Rizzello, C.G., Curiel, J.A., Nionelli, L. *et al.* (2014b) Use of fungal proteases and selected sourdough lactic acid bacteria for making wheat bread with an intermediate content of gluten. *Food Microbiology*, **37**, 59–68.

Rizzello, C.G., De Angelis, M., Di Cagno, R. *et al.* (2007) Highly efficient gluten degradation by lactobacilli and fungal proteases during food processing: New perspectives for celiac disease. *Applied and Environmental Microbiology*, **73**, 4499–4507.

Rocha, J.M. and Malcata, F.X. (1999) On the microbiological profile of traditional Portuguese sourdough. *Journal of Food Protection*, **62**, 1416–1429.

Rossi, J. (1996) The yeasts in sourdough. *Advances in Food Science*, **18**, 201–211.

Rühmkorf, C., Jungkunz, S., Wagner, M. and Vogel, R.F. (2012a) Optimization of homoexopolysaccharide formation by lactobacilli in gluten-free sourdoughs. *Food Microbiology*, **32**, 286–294.

Rühmkorf, C., Rübsam, H., Becker, T. *et al.* (2012b) Effect of structurally different microbial homoexopolysaccharides on the quality of gluten-free bread. *European Food Research and Technology*, **235**, 139–146.

Ryan, L.A.M, Dal Bello, F. and Arendt, E.K. (2008) The use of sourdough fermented by antifungal LAB to reduce the amount of calcium propionate in bread. *International Journal of Food Microbiology*, **125**, 274–278.

Sadowska, J., Blaszczak, W. and Fornnal, J. (2003) Changes of wheat dough and bread quality and structure as a result of germinated pea flour addition. *European Food Research and Technology*, **216**, 46–50.

Salovaara, H. (1998) Lactic acid bacteria in cereal-based products. In S. Salminen and A. von Wright (eds), *Lactic Acid Bacteria Microbiology and Functional Aspects* (pp. 115–138), New York: Marcel Dekker.

Schneider, A.V.C. (2002) Overview of the market and consumption of pulses in Europe. *British Journal of Nutrition*, **88**, S243–S250.

Schober, T.J. and Bean, S.R. (2008) Sorghum and maize. In E. Arendt (ed.), *Gluten-Free Cereal Products and Beverages* (pp. 101–118), Food Science and Technology International Series, Amsterdam: Elsevier.

Schober, T.J., O'Brien, C.M., McCarthy, D., Darnedde, A. and Arendt, E.K. (2003) Influence of gluten-free flour mixes and fat powders on the quality of gluten-free biscuits. *European Food Research and Technology*, **216**, 369–376.

Schoenlechner, R., Scelsi, C., Mariotti, M., Lucisano, M. and Berghofer, E. (2006) Production and optimiation of bread from wheat and amaranth flour blends, World Grains Summit: Foods and Beverages, San Francisco, CA, 17–20 September.

Settanni, L., Conterno, L. and Cavazza, A. (2008) Il microbiota lattico degli impasti acidi e delle materie prime usate in panificazione. *Tecnica Molitoria*, **59**, 631–638.

Settanni, L., Guarcello, R., Gaglio, R. *et al.* (2014) Production, stability, gene sequencing and in situ anti-*Listeria* activity of mundticin KS expressed by three *Enterococcus mundtii* strains. *Food Control*, **35**, 311–322.

Settanni, L., Massitti, O., Van Sinderen, D. and Corsetti, A. (2005) *In situ* activity of a bacteriocin-producing *Lactococcus lactis* strain. Influence on the interactions between lactic acid bacteria during sourdough fermentation. *Journal of Applied Microbiology*, **99**, 670–681.

Settanni, L. and Moschetti, G. (2010) Biodiversity of sourdough lactic acid bacteria. In V. Rescigno and S. Maletta (eds), *Biodiversity Hotspots* (pp. 37–79), New York: Nova Science.

Settanni, L. and Moschetti, G. (2014) New trends in technology and identity of traditional dairy and fermented meat production processes. *Trends in Food Science and Technology*, **37**, 51–58.

Settanni, L., Ventimiglia, G., Alfonzo, A., Corona, O., Miceli, A. and Moschetti, G. (2013) An integrated technological approach to the selection of lactic acid bacteria of flour origin for sourdough production. *Food Research International*, **54**, 1569–1578.

Shan, L., Marti, T., Sollid, L.M., Gray, G.M. and Khosla, C. (2004) Comparative biochemical analysis of three bacterial prolyl endopeptidases: Implications for celiac sprue. *Biochemistry Journal*, **383**, 311–318.

Shan, L., Molberg, O., Parrot, I. *et al.* (2002) Structural basis for gluten intolerance in celiac sprue. *Science*, **297**, 2275–2279.

Shewry, P.R. (2003) Improving wheat quality: The role of biotechnology. In S.P. Cauvain (ed.), *Bread Making: Improving Quality* (pp. 168–186), Cambridge: Woodhead.

Shinohara, M., Aoyama, C., Fujiwara, K. *et al.* (2011) Microbial mineralization of organic nitrogen into nitrate to allow the use of organic fertilizer in hydroponics. *Soil Science and Plant Nutrition*, **57**, 190–203.

Silano, M. and De Vincenzi, M. (1999) Bioactive antinutritional peptides derived from cereal prolamins: A review. *Nahrung*, **43**, 175–184.

Simango, C. (1997) Potential use of traditional fermented foods for weaning in Zimbabwe. *Journal of Social Science and Medicine*, **44**, 1065–1068.

Sivam, A.S., Sun-Waterhouse, D., Quek, S. and Perera, C.O. (2010) Properties of bread dough with added fiber polysaccharides and phenolic antioxidants: A review. *Journal of Food Science*, **75**, R163–R174.

Soni, S.K. and Sandhu, D.K. (1990) Indian fermented foods: Microbiological and biochemical aspects. *Indian Journal of Microbiology*, **30**, 135–157.

Spicher, G. (1999) Zur Geschichte des Sauerteiges. In G. Spicher and H. Stephan (eds), *Handbuch Sauerteig: Biologie, Biochemie, Technologie* (pp. 3–7), Hamburg: Behr's.

Spiertz, J.H.J. and Ewert, F. (2009) Crop production and resource use to meet the growing demand for food, feed and fuel: Opportunities and constraints. *NJAS Wageningen Journal of Life Science*, **56**, 281–300.

Stepniak, D., Spaenij-Dekking, L., Mitea, C. *et al.* (2006) Highly efficient gluten degradation with a newly identified prolyl endoprotease: Implications for celiac disease. *American Journal of Physiology – Gastrointestinal and Liver Physiology*, **291**, G621–G629.

Stolz, P. (1999) Mikrobiologie des Sauerteiges. In G. Spicher and H. Stephan (eds), *Handbuch Sauerteig: Biologie, Biochemie, Technologie* (pp. 35–60), Hamburg: Behr's.

Tack, G.J., Verbeek, W.H., Schreurs, M.W. and Mulder, C.J. (2010) The spectrum of celiac disease: Epidemiology, clinical aspects and treatment. *Nature Reviews Gastroenterology and Hepatology*, **7**, 204–213.

Taylor, J.R.N. and Emmambux, N. (2008) Gluten-free foods and beverages from millets. In E. Arendt and F. Dal Bello (eds), *Gluten-Free Cereal Products and Beverages* (pp. 119–148), Food Science and Technology International Series, Amsterdam: Elsevier.

Taylor, J.R.N. and Parker, M.L. (2002) Quinoa. In P.S. Belton and J.R.N. Taylor (eds), *Pseudocereals and Less Common Cereals: Grain Properties and Utilization*, Berlin: Springer.

Thompson, J.D., Gibson, T.J., Plewniak, F., Jeanmougin, F. and Higgins, D.G. (1997) The CLUSTAL-X windows interface: Flexible strategies for multiple sequence alignment aided by quality analysis tools. *Nucleic Acids Research*, **25**, 4876–4882.

Tieking, M. and Gänzle, M.G. (2005) Exopolysaccharides from cerealassociated lactobacilli. *Trends in Food Science and Technology*, **16**, 79–84.

Tiwari, B.K., Brennan, C.S., Jaganmohan, R., Surabi, A. and Alagusundaram, K. (2011) Utilisation of pigeon pea (*Cajanus cajan* L) by-products in biscuit manufacture. *LWT – Food Science and Technology*, **44**, 1533–1537.

Toderi, G. (1989). Frumento (*Triticum* spp.). In R. Baldoni and L. Giardini (eds), *Coltivazioni erbacee* (pp. 15–89), Bologna: Pàtron Editore.

Toderi, G. (1993) Aspetti generali sulle colture cerealicole. In R. Baldoni and L. Giardini (eds), *Coltivazioni erbacee* (pp. 26–42), Bologna: Pàtron Editore.

Topping, D. (2007) Cereal complex carbohydrates and their contribution to human health. *Journal of Cereal Science*, **46**, 220–229.

Valmorri, S., Settanni, L., Suzzi, G., Gardini, F., Vernocchi, P. and Corsetti, A. (2006) Application of a novel polyphasic approach to study the lactobacilli composition of sourdoughs from the Abruzzo region (central Italy). *Letters in Applied Microbiology*, **43**, 343–349.

Valmorri, S., Tofalo, R., Settanni, L., Corsetti, A. and Suzzi, G. (2010) Yeast microbiota associated with spontaneous sourdough fermentations in the production of traditional wheat sourdough breads of the Abruzzo region (Italy). *Antonie Van Leeuwenhoek*, **97**, 119–129.

Vogel, R.F., Ehrmann, M.A. and Gänzle, M. G. (2002) Development and potential of starter lactobacilli resulting from exploration of the sourdough ecosystem. *Antonie van Leeuwenhoek*, **81**, 631–638.

Vogelmann, S.A., Seitter, M., Singer, U., Brandt, M.J. and Hertel, C. (2009) Adaptability of lactic acid bacteria and yeasts to sourdoughs prepared from cereals, pseudocereals and cassava and use of competitive strains as starters. *International Journal of Food Microbiology*, **130**, 205–212.

Wieser, H., Vermeulen, N., Gaertner, F. and Vogel, R.F. (2008) Effects of different *Lactobacillus* and *Enterococcus* strains and chemical acidification regarding degradation of gluten proteins during sourdough fermentation. *European Food Research Technology*, **226**, 1495–1502.

Wolter, A., Hager, A.S., Zannini, E., Czerny, M. and Arendt, E.K. (2014a) Impact of sourdough fermented with *Lactobacillus plantarum* FST 1.7 on baking and sensory properties of gluten-free breads. *European Food Research and Technology*, **239**, 1–12.

Wolter, A., Hager, A.S., Zannini, E., Czerny, M. and Arendt, E.K. (2014b) Influence of dextran-producing *Weissella cibaria* on baking properties and sensory profile of gluten-free and wheat breads. *International Journal of Food Microbiology*, **172**, 83–91.

CHAPTER 12

The role of starter cultures and spontaneous fermentation in traditional and innovative beer production

Antonietta Baiano and Leonardo Petruzzi
Department of the Science of Agriculture, Food and Environment, University of Foggia, Italy

The knowledge of ingredients, operations and flavour-formation reactions in beer processing is of fundamental importance to get a product with the desired olfactory characteristics and without off-flavours.

Brewing technologies are generally based on a recipe including water, malted barley, hops, and yeasts. The Bavarian Beer Purity Law of 1516, known as *Reinheitsgebot*, listed malt, hops and water as the allowable materials for brewing. At that time, yeasts were unknown, and only between 1855 and 1876 Louis Pasteur published his fermentation theory and distinguished between aerobic and anaerobic utilization of sugars by yeast. Today and worldwide, brewers have more flexibility and can choose, for example, other source of starch than barley or wheat malt, and further flavouring agents than hops.

Water is quantitatively the main ingredient, since it comprises more than 90% of beer (Baiano *et al.* 2012). The mineral content and composition of the water used in brewing affect starch conversion into sugars and thus, indirectly, the beer flavour profile. Nevertheless, once the sugars have been produced, their effect on beer flavour is negligible.

Barley malt is the most widely used raw material, followed by wheat malt. Barley is suitable for brewing because its fibrous hull remains attached to the grain during threshing, thus making easier the separation of the solid parts from the wort immediately after mashing. Other sugar sources such as cereals different from wheat and malt (corn, rice, sorghum, rye), pseudo-cereals (quinoa, buckwheat, amaranth), legumes and starchy fruits (chestnuts) can be used because of their lower costs or the peculiar flavour that they grant. The addition of small amounts of unmalted cereals supplies starch that is then hydrolysed by enzymes of malted cereals, while roasted cereals grant a particular flavour to

Starter Cultures in Food Production, First Edition. Edited by Barbara Speranza, Antonio Bevilacqua, Maria Rosaria Corbo and Milena Sinigaglia.

the beer. Malted and unmalted grains supply sugars as the main source of carbon compounds and energy and nitrogen compounds for yeasts. The sugars mainly present in wort and readily utilized by yeast in fermentation are maltose (50%), maltotriose (18%), glucose (10%), sucrose (8%) and fructose (2%). Other complex carbohydrates include dextrins (ca. 12%). The sequence of use is always the same: first glucose, fructose and sucrose, then maltose, and finally maltotriose. Sucrose is hydrolysed into glucose and fructose by invertase in the yeast cell wall. The permeases responsible for the transport of maltose and maltotriose through the yeast cell membrane are inhibited by monosaccharides, thus their uptake occurs when the concentrations of glucose and fructose decrease. Once within the yeast cell, maltose and maltotriose are converted to glucose by maltase. Only glucose and fructose may be assimilated into the glycolytic (or Embden–Myerhof–Parnas) pathway.

Hops are used as a flavouring and preservative agent. The lupulin glands located on the bracteoles of the hop umbel are of concern for brewers due to their content of bitter resins and aroma compounds (Krottenthaler 2009). The contribution of hops to beer flavour is due to more than 300 volatile substances (Kammhuber and Hagl 2001). The monoterpene myrcene (17–37% of oils) is responsible for the typical taste. Linalool and geraniol are oxidized terpenes. Linalool has a floral-like aroma, while geraniol has a rose-like aroma. Esters can contribute to a fruity flavour. Fatty acids can be responsible for a cheese-like aroma. Hops also contain epoxides formed by autoxidation of sesquiterpene hydrocarbons and sulfur compounds such as thioester, sulfides and sulfur heterocycles (Lermusieau *et al.* 2001).

Today, fermentation is generally carried out by the addition of standard yeast cultures, while in the past, before the role of yeast was understood, fermentation involved microorganisms naturally present in the air (Baiano *et al.* 2012). In fact, in the Middle Ages it was already known that the best beers were produced next to bakeries.

Traditionally, brewers have distinguished two types of brewer's yeasts, ale (top-fermenting) yeasts and lager (bottom-fermenting) yeasts (Hansen and Piškur 2004). Initially, lager brewing yeasts were classified as *Saccharomyces carlsbergensis*, while ale brewing yeasts were categorized as *Saccharomyces cerevisiae*. According to more recent and sophisticated taxonomic techniques, the species *S. carlsbergensis* was included in the *Saccharomyces pastorianus* taxon. On the basis of genetic studies, *S. pastorianus* strains are now described as allopolyploid interspecies hybrids of *S. cerevisiae* and *Saccharomyces bayanus* (Berlowska *et al.* 2014). Apart from the technological traits of lager yeasts such as the production of sulfur-like flavours, their ability to ferment at lower temperatures (10–15 °C) than ale yeasts (20–28 °C), and their property to sediment at the bottom instead of rising to the surface of the fermentation broth, the major taxonomic distinction between the two groups of yeasts is the inability of ale yeasts to ferment the disaccharide melibiose (Verbelen and Delvaux 2009).

Despite the clear advantages of *Saccharomyces* spp., their dominant position as starters in industrial fermentation limits the spectrum of fermentation characteristics, and more generally bioconversion processes, available to producers. Many non-*Saccharomyces* yeasts are able to produce large ranges of aroma compounds that contribute to the general aroma profile of the products obtained (Steensels and Verstrepen 2014). Conversely, great improvements in brewing yeast performance have been achieved by using classic genetic techniques such as hybridization and mutagenesis followed by selection. Another possible approach is genetic engineering. Progress has been made in improving technologically important properties of brewing yeasts, such as carbohydrate utilization, fermentation of dextrins, flocculation and filtration, reduction of hydrogen sulfide (H_2S) and diacetyl production, and osmotolerance (Deák 2008). Recently, the concept of the creation of novel strains to produce 'healthy' beers, such as those with less alcohol and/or sugar, has gained increased attention (Stewart *et al.* 2013).

This chapter provides a summary of the field of fermentation systems and processes, with an emphasis on parameters affecting the aroma of beer, but it is also intended as a research update on technological yeast applications in traditional and innovative beer brewing.

Batch brewing process

Batch fermentation

The conventional brewing process consists of a series of batch processes and operations including malting, grinding, mashing, separation of non-soluble components, boiling with hops or hop extracts, fermentation (primary or main fermentation, secondary fermentation, and eventually refermentation in the bottle), filtration, stabilization, bottling, and eventually pasteurization (Baiano *et al.* 2012).

The malting process allows the grain to partially germinate, making starch and proteins available for the successive brewing operations thanks to the synthesis of new hydrolytic enzymes (α–amylase, β–amylase, cellulose etc.). These enzymes partially break down the endosperm matrix composed of large and small starch granules and also the cell walls within the matrix holding the starch granules, which are mainly made of β-glucans, pentosans, and proteins.

Grinding is an operation necessary to extract malt. On the other hand, the husks should be kept intact to obtain a permeable filter cake. In fact, with a severe milling, the pores of the filter cake clog up very rapidly. Currently, hammer mills and roller mills are mostly employed.

Mashing is the step in which the solubilization of the components of malt and other cereals occurs by enzymatic, physical and chemical reactions. The main processes are starch and protein conversion into fermentable sugarsa and peptides, respectively. The first aim of mashing is the conversion of starch (non-fermentable) into simple sugars. Starch is a polymer composed of amylose (17–24%) and amylopectin (76–83%). The first is a single chain of glucose molecules linked together with α-1,4 bonds, while amylopectin has a branched structure with the branches formed by bonds between the two molecules of 2 glucose (α-1,6 links). In order to convert starch into water-soluble sugars (fermentable and non-fermentable), first starch is gelatinized to become water soluble (starch of barley and malt gelatinizes at 60 °C while other starches gelatinize only above 90 °C) and then amylases break down the long-chained starch molecules into shorter chains. β-amylase is able to completely convert amylose into maltose, the main wort sugar, by splitting two glucose molecules from the non-reducing end of a glucose chain. β-amylase is not able to break down the branch links, thus amylopectin cannot completely be converted to maltose. The optimal temperature range for β-amylase is 60–65 °C, while at 70 °C it is inactivated. α-amylase is able to split the α-1,4 bonds, thus exposing additional non-reducing ends for the β-amylase and allowing the further conversion of amylopectin to maltose. The optimal temperature range is between 72 and 75 °C, while the enzyme is rapidly denatured at 80 °C (Narziss 2005). A high heating rate can be required during mashing if certain enzymatic reactions are to be inhibited, such as maltose production for the production of low-alcohol beers. Another goal of mashing is the partial conversion of malt protein into a balanced mix of short (amino acids) and medium-chained proteins. Amino acids are necessary yeast nutrients, while medium-chained proteins are important for the body of a beer as well as for foam stabilization. The proteolytic activity shows a maximum between 50 °C and 55 °C, but even at higher temperatures protein breakdown is significant (Narziss 2005). Rest temperatures closer to 50 °C determine a higher production of amino acids, while those closer to 55 °C allow the formation of medium-chained proteins.

After mashing, the spent grains are separated from the liquid part through a lauter tun or a mash filter, and the wort is boiled with hops or hop extracts. The wort boiling is divided into two steps, hot holding and evaporation. During the first, different physical and chemical reactions take place: hop isomerization, development of aroma substances and colour (Maillard reactions occur at 80 °C and generate new aroma substances, for example the Strecker aldehydes developed from amino acids that influence the taste stability of the beer), dissolution processes, inactivation of enzymes, formation and precipitation of protein-tannin complexes (break), and sterilization. Evaporation removes undesired aroma substances such as myrcene from hops, different carbonyl as well as sulfur substances, especially dimethylsulfide, and aroma substances from lipid metabolism. In addition, it adjusts the original extract, in order to ensure the constancy and legal requirements/marketability of the beer (Yamashita *et al.* 2006).

After cooling, the wort is submitted to fermentation. Fermentation is divided into primary or main fermentation, secondary fermentation, and refermentation in the bottle. The primary fermentation starts with yeast inoculation and ends when most of the sugar has been fermented. At this time, wort beer is transferred to another tank, thus the sediment is filtered out. The secondary fermentation allows for the complete transformation of sugars and for the refinement of taste and flavour.

After fermentation, maturation occurs. During maturation, diacetyl is reduced to acetoin.

Filtration consists in removing suspended materials from the green beer while stabilization consists in avoiding potential turbidity formers. Filtration can be classified as surface and depth filtration. In the surface filtration, the particles are retained on the surface of the filter material. In the depth filtration, the separation takes place inside the filter material. A variant is represented by the so-called 'cake filtration', where a filter cake is built up by the separated solid materials during surface filtration and the outer layer of the filter cake takes over the separation. Stabilization can be carried out in different ways (adsorptive, sedimentative and enzymatic). Two key compounds leading to turbidity are proteins and tannins. Proteins are removed by adsorption using colloidal silica, while tannins are mainly removed by adsorption using polyvinylpyrrolidone (Lindemann 2009).

Concerning beer bottling, four main categories of packaging are used worldwide: glass bottles, aluminium or tinplate cans, bottles made of polyethylene terephthalate (PET) or polyethylene naphthalate (PEN) and kegs.

In order to establish the proper conditions of pasteurization, it is important to have a focus on the beer's characteristics. Beer has a low pH (around 4.5) and contains carbon dioxide (CO_2; anaerobic atmosphere), alcohol (an average of 5 vol%), bittering substances deriving from hops, and low concentrations of readily utilizable sugars and amino acids. Due to these characteristics, pathogens and heat-resistant microorganisms cannot grow in beer. Therefore, temperatures of 62–72 °C can be used during the pasteurization of bottled beer to achieve microbiological safety (Back 2009). Flash pasteurization is an alternative approach to the traditional pasteurization. Flash pasteurization is a method of microbial stabilization of beer and other beverages prior to filling into containers. In this process the product is heat treated at 71.5–74 °C for 15–30 seconds.

Flavour formation

The conversion of sugar to ethanol is the primary function of yeast during beer brewing and is directly involved in a number of biological processes, such as the transport of sugar across the plasma membrane, sugar catabolism, reserve carbon storage, energy generation and ethanol formation (Yu *et al.* 2012). However, in beverages such as beer and wine, flavour is a predominant quality aspect for producers and consumers (Daenen *et al.* 2008a). Flavour has been defined as 'the sum of perceptions resulting from stimulation of the sense ends that are grouped together at the entrance of the alimentary and respiratory tracts'. In practice, 'flavour' can be considered to comprise four different components: odour, aroma, taste and mouthfeel (Baert *et al.* 2012).

The main flavour-active compounds produced by yeast during fermentation can be classified into six groups, including esters, fusel alcohols, ketones, various phenolics and fatty acids on metabolic sideways. Beer flavour is largely affected by the concentrations of some compounds, such as vicinal diketones (VDKs), higher aliphatic and aromatic alcohols, multivalent alcohols, esters, organic acids and carbonyl compounds. VDKs include two compounds, 2,3-butanedione (diacetyl, the more flavour-active one, with a disagreeable butterscotch aroma) and 2,3-pentanedione. Diacetyl is the product of the chemical oxidative decarboxylation of excess α-acetolactate leaked from the valine biosynthetic pathway to the extracellular environment, while 2,3-pentanedione comes from α-acetohydroxybutyrate. Diacetyl is reassimilated and reduced by yeast to acetoin and 2,3-butanediol. The great contribution to an alcoholic or solvent-like aroma and to a warm mouthfeel is given by higher alcohols such as n-propanol, iso-butanol and 2-methyl and 3-methyl butanol (isoamyl alcohols; Brányik *et al.* 2008). They can be produced through anabolic and catabolic routes. In the first, the 2-oxo acids deriving from carbohydrate metabolism are decarboxylated to aldehydes, which are reduced to the corresponding alcohols. In the catabolic (Ehrlich) route, the 2-oxo acids come from amino acid utilization (Chen 1978). The final concentration of higher alcohols depends on the rate of these reactions.

The esters of beer are responsible for fruity flowery flavours and can be divided into two groups: acetate esters and ethyl or medium-chain fatty acid esters. Esters are synthesized by yeasts thanks to the acyltransferase activities that catalyse the condensation reaction between acetyl/acyl-CoA and alcohols (Brányik *et al.* 2008). Concerning organic acids, around 50% derives from wort while the other part is a product of yeast metabolism. They can be divided into two groups: organic acids with a short carbon skeleton (pyruvate, acetate, lactate, citrate, succinate, malate, oxo-acids), which derive from the incomplete turnover of the Krebs cycle occurring during anaerobic growth and also from the catabolism of amino acids; and medium-chain fatty acids (C6–C12) deriving from long-chain fatty acid anabolism under anaerobic conditions. The short-chain acids confer a ‘sour’ taste on beer, while medium-chain fatty acids are toxic to yeast cells since they cause cell membrane disruption (Brányik *et al.* 2008). Concerning carbonyl compounds, molecules such as 3-methyl butanal, 2-methyl butanal, hexanal and heptanal are present in wort and contribute to a worty off-flavour mainly detected in low-alcohol beer produced by limited fermentation. Beer aldehydes are produced mainly during mashing and boiling and are also synthesized from yeast oxo-acid pools via both anabolic and catabolic processes, from a carbon source and exogenous amino acids, respectively (Peppart and Halsey 1981). Yeast enzymatic systems such as alcohol dehydrogenase isoenzymes, branched-chain alcohol dehydrogenase, aldehyde dehydrogenase and aldo-keto reductases, both NADH and NADPH dependent, are involved in aldehyde transformation during fermentation (Brányik *et al.* 2008).

Continuous brewing process

Continuous fermentation

The first fully continuous process for beer fermentation was patented in 1906. In the late 1950s, Morton Coutts of Dominion Breweries (now DB Breweries), a factory located in New Zealand, introduced the concept of continuous fermentation, whose main attractive aspect is the accelerated transformation of wort into beer.

Continuous fermentation involves the recycling of part of the fermented beer back to the wort at the start of fermentation. It requires a continuous supply of wort into the system and the result is a continuous flow of beer out of the process (Campbell 2014). Many different systems for continuous beer fermentation have been designed, especially during the 1950s and 1960s. These systems can be classified as stirred versus unstirred tanks; single-vessel systems versus a number of in-series vessels; and vessels that allow yeasts to overflow freely with the beer (open system) versus closed or semi-closed system vessels that have abnormally high yeast concentrations (Willaert 2012). Continuous main fermentation can be performed with free or immobilized yeast cells. In the latter case it occurs in bioreactors, where yeast cells are entrapped or adsorbed on the surface of a carrier material.

The main benefits of using immobilized cells are enhanced fermentation productivity due to higher biomass densities, improved cell stability, easier implementation of continuous operation, improved operational control and flexibility, cell recovery and reuse and simplified downstream processing. Key parameters of this technology are selection of the carrier material and the method of immobilization, together with the bioreactor design. The determination of these parameters is directed by operational conditions such as temperature, pH, substrate composition and fluid dynamics, where special attention should be paid to mass transfer properties since a limited nutrient supply can result in changes in yeast metabolism, leading to inadequate flavour in the final product (Nedović *et al.* 2005).

Immobilized cell technology allows the production of lager beer (main and secondary fermentation) in less than 2 days, while modern batch fermentation technology cannot reduce the production time below 10–12 days (Leskosek-Cukalovic and Nedovic 2005). Continuous fermentation performed with immobilized yeast cells ensures high volumetric cell densities of yeasts and thus higher volumetric productivities and shorter residence times, 2–3 days according to Tata *et al.* (1999). Thus bioreactors used in continuous fermentation with immobilized yeasts are smaller and require lower capital costs (Willaert and Nedovic 2006). However, while immobilized cell technology has been successfully designed for different stages in beer production (it greatly decreases the maturation period from 3–4 weeks to 2 hours and reduces the problem of the undesirable wort aroma deriving from wort aldehydes), it shows some limitations. In particular, a challenge is the design of a successful approach to combine main and secondary fermentation. In fact, the main fermentation is complex and has various side reactions that are important for beer quality. For example, insufficient free amino nitrogen consumption by immobilized yeast cells, coupled with reduced cell growth, causes an unpleasant flavour profile (Leskosek-Cukalovic and Nedovic 2005). In fact, the low amino nitrogen consumption leads to lower amounts of higher alcohols and esters and to a higher pH in the final beer. The first carrier material used in beer fermentation was wood: yeast cells enter the walls of the wooden vessel and remain there until the fermentation starts (Virkajärvi and Pohjala 2000). Matrices most commonly used for yeast immobilization include gel-type materials such as alginate, calcium pectate and carrageenan; porous structures such as glass beads, silicon carbide gluten pellets and diatomaceous earth; and cellulose-based materials (Brányik *et al.* 2002).

Continuous fermentation allows non-negligible financial savings and reduction of the environmental impact. Capital costs are lower than those invested in batch fermentation, since less equipment is required and consequently costs of energy, labour and testing are reduced. In continuous brewing, CO_2 can be immediately collected, purified and continuously supplied for reuse in other parts of the process or sold for a profit to gas suppliers, while in batch plants the amount of CO_2 produced during the first days of fermentation is relatively low (Campbell 2014).

The differences between batch and continuous fermentation concern not only the way they occur, but mainly yeast metabolic changes. During batch fermentation, yeasts adapt their metabolism and their growth curve to changes in the external environment. The metabolic changes associated with the individual growth phases are exerted at the level of gene expression (induction or repression of specific genes) and enzyme activity (the modulation of metabolic pathways is mediated through the stimulatory or inhibitory effects of intracellular metabolites). Thus beer flavour results from a mixture of products from both aerobic and anaerobic conditions during the various yeast growth phases. In a continuous culture, which remains in steady-state conditions, yeast cells are not exposed to significant environmental changes and thus they do not experience the different growth phases typical of a batch culture (Brányik *et al.* 2008). The number of reactors in multistage systems is a compromise between flavour requirements (better flavour quality with a higher number of stages), investment and operational costs (lower for a lower number of stages; Van De Winkel *et al.* 1993; Yamauchi *et al.* 1994; Virkajärvi and Kronlöf 1998). One of the first immobilized cell systems was a multistage process for fermentation and maturation of lager beer within 3–5 days designed by the Japanese Kirin Brewery Company (Leskosek-Cukalovic and Nedovic 2005). The system consisted of three bioreactors: the first an aerated, continuously stirred tank for yeast growth; the second represented by two packed bed reactors, in series, for the main fermentation and heat treatment for the conversion of α-acetolactate into diacetyl; and the last a packed-bed reactor with immobilized yeast for maturation.

Differences in the characteristics of beer can be attributed to the type of carrier used to immobilize yeasts. The carrier materials must meet the requirements of high cell load, stability, food grade, suitability for regeneration and sterilization, and direct contact of the yeast biomass surrounding the carrier surface with the bulk liquid, thus reducing the mass transfer problems associated with other immobilized systems (Brányik *et al.* 2008). Age, surface charge, composition of cell wall and hydrophobicity affect the adsorption, flocculation and immobilization of yeast (Virkajärvi and Pohjala 2000). The first carrier was constituted by alginate beads, but they were rapidly replaced because of some drawbacks like decreased fermenting capacity, insufficient mechanical strength and swelling. Ethanol productivity is higher when yeasts are immobilized onto a cellulose carrier with a small positive surface charge with respect to a negatively charged cellulose carrier. The lower bitterness of the beer from immobilized fermentation was related to the adsorption of isohumulones onto alginate beads (Hsu and Bernstein 1985). The nature of the carrier can affect the concentrations of flavour compounds in combination with yeast. Virkajärvi and Pohjala (2000) tested four yeast strains (very flocculent, flocculent, weakly flocculent and non-flocculent) immobilized on three different carriers (porous glass, DEAE-cellulose based – a non-porous composite carrier of irregular shape – and a kieselguhr-based cylindrical carrier). They found that higher amounts of propanol and 2-methyl butanol were produced by strongly flocculent yeasts on porous glass and DEAE-cellulose-based carriers, respectively. The beers produced using the porous glass carrier contained higher concentrations of 3-methyl butyl acetate. The ethyl acetate concentration was higher in the beer produced with the kieselguhr-based cylindrical carrier. With the porous glass carrier both total diacetyl and total pentanedione concentrations were lower than with the DEAE-cellulose-based carrier. Although these carriers share the same immobilization technique (adsorption), they differ by nature of the phenomena. In fact, the DEAE-cellulose-based carrier contains anion exchange groups and is a non-porous system, while the kieselguhr-based cylindrical and porous glass carriers base their adsorption on van der Waals forces and hydrophobicity and are porous materials.

Polyvinyl alcohol (PVA) is a promising support for cell entrapment due to the absence of toxicity and mechanical stability. PVA particles can be produced by multiple freezing and thawing of PVA solutions, the jet-cutting technique, cross-linking by UV radiation or boric acid, and a technology that, including controlled partial drying, gives them desirable properties such as good diffusion (due to the small lens thickness), good mechanical properties and simple separation (typical for large carriers). Cells immobilized on polyvinyl alcohol particles obtained through this technology had high fermentation activity, and in three successive gas-lift reactor fermentations, the apparent attenuation of 80% was reached after only 2 days. Polyvinyl alcohol particles obtained through this technology showed a shelf life of 6 months without significant change in cell activity

(Durieux *et al.* 2000; Bezbradica *et al.* 2007). These findings suggest that the choice of carrier should be made taking into account the nature of the carrier, the kind of yeast and the desired beer characteristics. In addition, the combination of carrier material and bioreactor design influences the quality of the final product. Compared to packed-bed reactors, an airlift reactor with pneumatically forced circulation offers the advantages of higher CO_2 removal, absence of channelling and clogging, and better mass and heat transfer (Linko *et al.* 1998). K-carrageenan beads can be used as a carrier material since their density is close to that of water, thus minimizing the energy required for fluidization (Pilkington *et al.* 1999). Some researchers developed a tubular matrix of sintered silicon carbide installed into a loop bioreactor to be used for alcohol-free beer production main fermentation and maturation (Van De Winkel 1995; Andries *et al.* 2000).

However, as the cost of the immobilization carrier is the major economic limitation for implementation of this system on an industrial scale, the use of low-cost materials for cell immobilization is required (Dragone *et al.* 2008), as well as the use of a support that is cheap, hygienic and abundant in nature (Bekatorou *et al.* 2002). Brewer's spent grain, the main brewery by-product, is a low-cost material and possesses a high capacity for yeast cell retention (Dragone *et al.* 2008). Similarly, yeast cells immobilized on food-grade supports such as gluten (Bardi *et al.* 1997; Bekatorou *et al.* 2001) and delignified cellulosic material (Bekatorou *et al.* 2002), corn cobs (Brányik *et al.* 2006) or dried figs (*Ficus carica*; Bekatorou *et al.* 2002) can serve as a potential substrate for the production of value-added products. In particular, beers with excellent taste and aroma could be produced using low-temperature brewing, psychrotolerant and ethanol-resistant yeasts immobilized on food-grade supports (Kopsahelis *et al.* 2007). Bardi *et al.* (1996) reported that in beers produced by the fermentation of wort by cells immobilized on gluten pellets, higher alcohols were reduced as the temperature was decreased and ethyl acetate was higher for immobilized cells at low temperatures as compared to free cells.

Wort acidification using immobilized lactic acid bacteria (LAB) has also been proposed. In general, the controlled use of thermophilic lactobacilli to perform biological acidification has technological advantages, since it improves the sensory quality of beer with regard to flavour, colour and foam stability (Hammes *et al.* 2005). Pittner *et al.* (1993) applied this approach to produce acidified wort using a *Lactobacillus amylovorus* strain immobilized on a DEAE-cellulose carrier.

Flavour formation

In fermentation carried out with immobilized yeast cells, chemical changes in cellular composition were associated with an increased resistance to stress. It has been shown that free and immobilized yeast cells differ in chemical composition and ploidy. Immobilized cells have a higher content of glycogen, trehalose, structural polysaccharides (glucan and mannan), fatty acids and DNA. Immobilization also causes changes in the proteome of a cell, in the level of gene expression, and has a significant impact on the quantitative composition and organization of the cytoplasmic membrane and cell wall structures. Many studies have also reported an increase in metabolic activity (increased rate of sugar uptake and productivity of selected metabolites; Berlowska *et al.* 2013). The alterations of plasma membrane composition have significant effects on several enzymes, transporters and membrane fluidity, thus increasing ethanol tolerance and altered sugar and amino acid uptake (Brányik *et al.* 2008). These changes, together with ageing and mutation of yeasts in continuous cultures, are responsible for the changes of flavour from batch to continuous brewing. The amount of total diacetyl formed during continuous main fermentation is higher than that produced during traditional batch fermentation. Immobilization could accelerate the expression of the acetohydroxy acid synthase responsible for the formation of α-acetolactate from pyruvic acid (Shindo *et al.* 1994). Nevertheless, an alternate amino acid metabolism could also be the leading factor for a lower amino acid uptake by immobilized cells (Brányik *et al.* 2008). Another reason could be an enhanced anabolic formation of amino acid precursors due to rapid yeast growth induced by overaeration (Brányik *et al.* 2004). In order to repress excessive VDKs formation during continuous brewing, it is possible to add bacterial α-acetolactate decarboxylase, thus converting α-acetolactate to acetoin (Godtfredsen *et al.* 1984); use genetically modified yeast encoding an α-acetolactate decarboxylase (Yamano *et al.* 1995); maintain an optimum wort-free amino nitrogen; increase the concentration of immobilized cells and prolong the residence time (Brányik *et al.* 2006); or accelerate the conversion of all α-acetolactate to diacetyl between primary and secondary fermentation by heating (10 min at 90 °C; Yamauchi *et al.* 1995).

The behaviour of immobilized yeasts depends on the kind of immobilization. When cells are immobilized by entrapment (e.g. in carriers such as alginate, carrageenan and calcium pectate), the decrease of higher alcohol is limited to the decrease of free amino nitrogen utilization (Dömény *et al.* 1998). In the case of cells immobilized by attachment (e.g. in DEAE-cellulose and stainless-steel cloth), there is a slight increase of higher alcohols (Shen *et al.* 2003). Conditions that promote yeast cell growth (high levels of nutrients such as amino acids, oxygen, lipids, zinc, increased temperature or agitation) enhance the production of higher alcohols, while the conditions for yeast growth (lower temperature and higher CO_2 pressure) reduce higher alcohol production (Renger *et al.* 1992; Landaud *et al.* 2001). Generally, esters are less in a continuous process than in a batch system. This behaviour can be due to insufficient aeration of the bioreactor. In fact, although ethanol production is an anaerobic process, oxygen is necessary for yeast growth and for unsaturated fatty acid and sterol synthesis (Masschelein 1997). Different rates of ester formation can be observed when entrapped or adsorbed cells are used. Entrapment reduces the diffusion of nutrients, while adsorbed cells show behaviours analogous to those of the free ones (Šmogrovičová and Dömény 1999). For example, *Saccharomyces uvarum* cells entrapped in calcium pectate beads show a suitable flavour, with a low level of diacetyl, an optimum ratio of higher alcohols to ester content and maximum specific rate of saccharide utilization (Šmogrovičová *et al.* 1998). The most common strategy to control organic acid biosynthesis in continuous fermentation includes the regulation of cell growth and nutrient consumption rate (Yamauchi *et al.* 1995).

Beers produced in a continuous fermentation system shows higher amounts of acetaldehyde; acetaldehyde biosynthesis can be controlled by proper oxygen supply and can be reduced by prolonging the maturation time. Yeast immobilization on carbonyl does not affect or improve its capacity to reduce carbonyl compounds. The increased alcohol dehydrogenase activity in immobilized yeast is correlated with an immobilization-induced (DEAE-cellulose) higher glucose flux in cells and with efficient NADH/NADPH regeneration during faster glycolysis and the pentose phosphate pathway (Van Iersel *et al.* 2000).

Temperature, wort gravity and alcohol content, feed volume and aeration play a major role in controlling the production of flavour-active by-products (Willaert and Nedovic 2006). The influence of the temperature between 5 and 20 °C has been investigated for bottom-fermented yeast entrapped in calcium pectate or κ-carrageenan and adsorbed on DEAE-cellulose (Šmogrovičová and Dömény 1999).

High and very high gravity fermentation

Although continuous beer fermentation is considered a promising technology, the number of industrial applications is still limited and only a low number of industrial producers successfully produce full-strength beers by using this approach. In fact, beer contains hundreds of different components and even little changes in technology can result in undesired changes in flavour. The major issue hindering the industrial application of this technology is the difficulty in achieving the development of the desired flavour compounds and the reduction of the undesired ones during the short time typical for continuous brewing (Brányik *et al.* 2008). Other issues are the higher complexity of operation, lower yeast viability, the low flexibility in the ability to change beer type, the carrier price and inconvenience of immobilization, and the higher hygiene requirements of a continuous brewing system (Ault 1965).

A promising and interesting way is high gravity (HG) technology (Yu *et al.* 2012). Traditionally, normal gravity (NG, 12 °P, i.e. 12 g extract per 100 g liquid) fermentation of wort yields beer with 5% (v/v) ethanol concentration. Nowadays, HG wort fermentation has been adopted as standard practice in many modern breweries throughout the world. The use of such a technology can improve yeast fermentation performance with faster and more complete conversion of sugars to alcohol, as well as lower production time (Ekberg *et al.* 2013).

In recent years, the concept of HG fermentation was further extended to very high gravity (VHG; ≥18 °P) fermentation to maximize the benefits of HG fermentation (Puligundla *et al.* 2011). However, intensification of the brewing process poses further issues for brewing yeast, the evolutionary history of which has not been towards the extreme conditions associated with higher gravity brewing (Gibson 2011). Osmotic stress, ethanol toxicity, viscosity, carbon dioxide concentration and nutrient limitations can result in a slow or stuck fermentation, a decrease in both cell growth and the production of ethanol (Hou 2010; Yu *et al.* 2012). Stress-responsive element sequences have recently been identified in promoter sections of many stress-induced genes, including heat shock protein genes (e.g. *HSP12* and *HSP104*; Lei *et al.* 2012).

The disadvantages of VHG brewing could be overcome by using more robust yeast strains that can resist the environmental stresses under VHG conditions (Blieck *et al.* 2007; Huuskonen *et al.* 2010). For example, Sanchez *et al.* (2012) generated new lager yeast and *S. cerevisiae* hybrids by classic genetics, which were improved regarding resistance to high osmolarity conditions.

Similarly, the brewing industry also showed interest in genetically engineered yeasts with improved maltose utilization, in order to accelerate the rate of fermentation. For example, overexpression of the gene *MTY1* conferred the ability to ferment maltose and maltotriose on an *S. cerevisiae* Mal^{-} strain (Duong 2009), whereas constitutive expression of the maltose transporter gene (*MALT*) with high-copy-number plasmids in a lager yeast strain has been found to accelerate the fermentation of maltose during high gravity (24 °P) brewing (Willaert 2012).

However, the problem of unbalanced flavour profiles due to the relative overproduction of acetate esters has been associated with HG fermentation. This overproduction results in overfruity and solvent-like beers (Verstrepen *et al.* 2003a). Verstrepen *et al.* (2003b) have demonstrated that overexpression of the gene *ATF1* in a commercial brewer's strain leads to significantly increased concentrations of isoamyl acetate and ethyl acetate in the beers produced; overexpression of *ATF2* leads to smaller increases in isoamyl acetate formation and no significant changes in ethyl acetate levels.

Different supplements have been used or proposed to maintain yeast performance under stressful conditions, like metal ions, lipids and lipid components such as fatty acids and sterols and free amino nitrogen, usually supplied in the form of a complex yeast food (Gibson 2011). Piddocke *et al.* (2011) showed that nitrogen supplementation generated by addition of the multicomponent enzyme (Flavourzyme; Novozymes, Bagsværd, Denmark) with both endo- and exopeptidase activities resulted in the best fermentation performance in terms of higher ethanol yield, specific growth rate and specific ethanol productivity in addition to high free amino nitrogen (FAN) utilization.

Finally, a possibility of reducing detrimental factors acting in HG and VHG wort could be yeast immobilization. Dragone *et al.* (2008) evaluated flavour compound formation and fermentative parameters of continuous HG brewing with yeasts immobilized on spent grains at different temperatures. According to Tran *et al.* (2010), sugar assimilation and ethanol production rates of immobilized yeasts in calcium alginate beads in HG wort were significantly higher than those of free yeast. Similarly, Pátková *et al.* (2000) found that by using calcium alginate-entrapped yeast, 24% (w/w) wort was successfully fermented within 8 days. Virkajärvi *et al.* (2002) used porous glass beads as the carrier for wort fermentation, and found that the optimal wort gravity in regard to ethanol productivity was between 18 and 21 °P.

Brewer's yeast in action

Specialty and traditional beers

Saccharomyces is the unique starter culture for at last 99% of worldwide beers. However, other beers, which are gaining increased popularity worldwide, incorporate secondary, non-*Saccharomyces* starter cultures, uncharacterized 'natural' starter cultures or autochthonous, non-starter microbiota during fermentation or maturation, leading to distinctive, unusual products. Belgian *lambic* beers, produced in a single area close to Brussels, are a well-known example of mixed-inoculum beer fermentations (Bokulich and Bamforth 2013). The unique aspect of these beers is that fermentation is spontaneous. The wort is allowed to cool in large, shallow open vessels, which are located in rooms where good ventilation provides maximum opportunity for contamination with the microbial flora of the room. After being held overnight in the cooling vessel, the wort is transferred to wooden casks, where it receives a further natural inoculum. The fermentation takes place over a period of several months in the casks, thus different microorganisms play an important role (Boulton and Quain 2001). Namely, there is a succession of enterobacteria, *Saccharomyces*, LAB and *Brettanomyces* (Vanderhaegen *et al.* 2003). Since these beers are fermented and matured in the same vessel *sur lies*, the unique flavor profile is likely influenced by microbial autolysis (Bokulich and Bamforth 2013).

In Africa several 'traditional' beers, made from cereals and especially from sorghum, involve contributions from a range of yeasts and bacteria (Hansen and Piškur 2004). These beers are known as *kaffir* in South Africa, *otika* or *burukutu* in Nigeria, *mtama* in Tanzania, *doro* or *chibuku* in Zimbabwe, *bili bili* in Tchad, *dolo* in Burkina Faso, *ikigage* in Rwanda and *tchoukoutou* in Togo and Benin (Djegui *et al.* 2014). These products are fermented by backslopping flocculent yeast slurry from a previous batch. Thus, *S. cerevisiae* dominates the fermentation of these beers, similar to other spontaneous beer fermentations. LAB are the second most prominent category of microorganisms in most of these beers, and they carry out mash acidification, which is an important processing step. The most commonly observed LAB are *Lactobacillus fermentum*, *Lactobacillus buchneri*, *Lactobacillus delbrueckii, Pediococcus acidilacti, Leuconostoc lactis* and *Lactococcus lactis* (Bokulich and Bamforth 2013). Interestingly, African beers are consumed in an active state of fermentation; hence they are effervescent in appearance (Atter *et al.* 2014).

Many mixed-inoculum beer fermentations have traditionally been brewed in Belgium, with the most renowned group being the acid beers of Flanders. These beers are inoculated with a mixture of *S. cerevisiae*. *Lactobacillus* spp. and *Pediococcus* spp. and are fermented in steel tanks for 7–8 weeks to create a fruity, refreshingly tart beer (Bokulich and Bamforth 2013). Similar to acid beers, American coolship ale (ACA) is a sour ale produced in the United States using production practices adopted from the *lambic* brewers of Belgium. ACA spontaneous fermentations were shown to follow a consistent fermentation progression, initially dominated by *Enterobacteriaceae* and a range of oxidative yeasts in the first month, then by *Saccharomyces* spp. and *Lactobacillales* for the following year. After one year of fermentation, *Brettanomyces bruxellensis* was the dominant yeast (occasionally accompanied by *Candida* spp., *Pichia* spp. and other yeasts) and *Lactobacillales* remained dominant, although various aerobic bacteria became more prevalent (Bokulich *et al.* 2012).

Bottle refermentation

Bottle refermentation, also known as bottle conditioning or bottle krausening, involves an extra fermentation process by adding fermentable carbohydrates and yeast in the bottle. The resulting beers are appreciated for their organoleptic traits and the visual aspect of the yeast sediment in the bottle. Additionally, some authors suggest that yeast offers a natural protection against oxygen, as it can act as an oxygen scavenger, making beer less sensitive to oxidation (Saison *et al.* 2010).

In Belgium, many specialty beers, including *Trappist* and many abbey and strong blond beers, are produced in this way. These beers are economically important since their consumption and export are still increasing (Van Landschoot *et al.* 2005). Following the active refermentation phase, a long storage period in the bottle of refermented beer in contact with yeast may result in yeast autolysis, with the excretion in the beer of intracellular compounds as amino acids, peptides, nucleotides, fatty acids and enzymes, which may affect the flavour profile (Vanderhaegen *et al.* 2003).

For some beers, such as *Berliner Weisse*, a mixed-starter culture of yeasts and LAB is used for refermentation. Other beers, such as *gueuze* beers, are the refermented products of mixtures of spontaneously fermented *lambic* beers. For the production of *gueuze* beer, a young (typically one-year-old) *lambic* beer with residual dextrin carbohydrates is mixed with old (typically three-year-old) *lambic* beer, which contains the microbiota that can convert the dextrin carbohydrates to more simple fermentable compounds. Once mixed, the beer referments spontaneously, without the addition of energy sources, yeast or bacterial cells. *Dekkera/Brettanomyces* spp. and LAB species are the dominant microorganisms in the refermenting beer, although after 14 months of refermentation only LAB are isolated (Spitaels *et al.* 2014).

The flavor of *gueze* beers is somewhat different from that of most beers, because of the high concentrations of organic acids (mainly lactic acid; Spitaels *et al.* 2014). On the other hand, the impact of *Brettanomyces* on beer flavour is complex. The presence of a cell-bound esterase changes the ester profile. Additionally, *Brettanomyces* affects beer flavour by releasing isovaleric acid and volatile phenols derived from hydroxycinammic acids (Vanderhaegen *et al.* 2003).

Bioflavouring

An improvement in the organoleptic quality of beverages and the design of new beverages can be attained through bioflavouring. One method relies on the possibility of enhancing or modifying the flavour profile through an enzymatic hydrolysis of flavour precursors such as glycosidically bound flavour compounds. Glycosides extracted from hops can play an important role. In *lambic* and *gueuze* beers, glycosides from added sour cherries or raspberries are extracted during secondary fermentation (Daenen *et al.* 2008a). The addition of whole fruit to beer is traditionally practised in Belgium for the production of *cherry lambic* ('Kriek') or *raspberry lambic* ('Framboise') by adding, respectively, sour cherries (*Prunus cerasus* L.) or raspberries (*Rubus idaeus* L.) to fermenting *lambic* in casks (Daenen *et al.* 2008b).

Interestingly, yeasts are not only responsible for the direct production of aroma compounds, but they can also mediate in the bioconversion of covalently bound, non-volatile and odourless flavour precursors into flavour-active compounds. Some microbes produce glycosidases that catalyse the liberation of the volatile aroma-active aglycons. Although some industrial *S. cerevisiae* strains show glycosidase activity, the incidence is very low and the activity relatively weak. However, several non-conventional yeasts, such as *Brettanomyces* spp., *Debaryomyces* spp. and *Issatchenkia terricola*, can produce high levels of β-glycosidase (Steensels and Verstrepen 2014). Indeed, the exploration of non-*Saccharomyces* yeasts represents an attractive alternative to the development of more complex beer aromas (Yeo and Liu 2014).

Fermentation conditions and nutritional supplements are important in beer brewing due to their influence on fermentation performance and on the characteristics of beer. However, yeasts respond differently to various nutritional and fermentation conditions (Hiralal *et al.* 2014). Indeed, the genome associated with each strain is unique and will ultimately define the final aroma profile of the product (Pires *et al.* 2014). Saerens *et al.* (2008) found that there is a good correlation between flavour production and the expression level of specific genes involved in the biosynthesis of aroma compounds.

The development of gene technology has opened up the possibility of metabolism engineering and thus flavour production (Vanderhaegen *et al.* 2003).

The inactivation of the gene encoding sulfite reductase (*MET10*) in a brewer's yeast resulted in increased sulfite accumulation during beer production and increased flavour stability, and no sign of H_2S production (Hansen and Kielland-Brandt 1996). H_2S is a well-known, volatile sulfur compound that strongly masks desirable characters in beer (Oka *et al.* 2008). For example, overexpression of the *CYS4* gene, encoding cystathionine β-synthetase, was shown to reduce H_2S production (Tezuka *et al.* 1992). To increase isoamyl acetate levels, Hirata *et al.* (1992) introduced extra copies of the *LEU4* gene into the *S. cerevisiae* genome. This approach results in increased production of isoamyl alcohol and its corresponding acetate ester, isoamyl acetate. Vanderhaegen *et al.* (2003) reported that overexpression of the alcohol acetyltransferase (*ATF1*) gene results in a strong increase in the levels of ethyl acetate, isoamyl acetate and 2-phenylethyl acetate. On the other hand, the expression level of the *ATF1* gene was significantly raised in the immobilized cells, resulting in a twofold increase in isoamyl acetate (Willaert and Nedovic 2006).

Acceleration of maturation

The basic principle of acceleration of maturation is to force yeast to remove diacetyl (Hansen and Piškur 2004). This step in the lager beer production process is time consuming and energy demanding; it is of great concern as it can decrease maturation time without affecting quality (Krogerus and Gibson 2013).

There have been many attempts, including improvement in the craft of fermenting and genetic modification (GM), to decrease the diacetyl content during the production of beer; for example, the α-acetolactate decarboxylase (*ALDC*) gene, the expression of a heterologous gene in brewer's yeast, could reduce the diacetyl content in beer (Zhang *et al.* 2008). Genes encoding *ALDC* from different bacteria, for instance *Enterobacter aerogenes, Klebsiella terrigena, Lc. lactis* and *Acetobacter aceti*, were also expressed in yeast using either episomal plasmids or genomic integrations (Duong *et al.* 2011). The *ALDC* gene derived from *Ac. aceti* has been brought under the transcriptional control of the constitutive yeast *PGK* promoter and introduced into the *S. cerevisiae* genome. The mutant strain has been used in pilot-scale brewing trials and produced beers of good quality, with low-level diacetyl (Vanderhaegen *et al.* 2003). These new types of brewer's yeast can be used in conventional batch fermentation or in a continuous maturation process. For example, continuous fermentation with immobilized genetically modified yeast without any subsequent maturation is a realistic option (Virkajärvi 2006). Nowadays, the combination of low-temperature fermentation and cell immobilization has been found to lead to a reduction in fermentation time and elimination of the maturation stage due to faster reduction in diacetyl (Kanellaki *et al.* 2014).

Finally, a promising approach to decreasing diacetyl production during fermentation, without using GM strains, is control of the valine content of the wort. By supplementing valine to brewer's wort, it was possible to decrease both the maximum diacetyl concentration observed during fermentation and the diacetyl concentration at the end of fermentation (Krogerus and Gibson 2013).

Low-alcohol and alcohol-free beers

There is an increasing trend towards low-alcohol beer (<2.5% alcohol content) and alcohol-free beer (<0.5% alcohol content) to meet worldwide demand or the needs of Muslims (Sohrabvandi *et al.* 2010).

The classic technology to produce alcohol-free or low-alcohol beer is based on the suppression of alcohol formation by arrested batch fermentation. However, the resulting beers are characterized by an undesirable wort aroma, since the wort aldehydes have only been reduced to a limited degree (Willaert 2012). In this regard, a careful selection of appropriate yeast strains can provide a useful alternative. *Saccharomycodes ludwigii*, a species unable to ferment maltose and maltotriose, is sometimes used to produce low-alcohol beer (Steensels and Verstrepen 2014). In general, non-*Saccharomyces* yeasts tend to have low fermentative power and produce low levels of ethanol, but the resultant beer would contain higher levels of residual sugars than conventional beer (Yeo and Liu 2014).

A reduction of ethanol production can be achieved by metabolic engineering of carbon flux in yeast, resulting in increased formation of other fermentation product (Willaert 2012). For example, Nevoigt *et al.* (2002) decreased the ethanol content in beer by overexpressing the *GPD1* gene, encoding the key enzyme in glycerol formation, glycerol-3-phosphate dehydrogenase, in an industrial lager brewing yeast. However, besides glycerol, other undesirable fermentation by-products, such as acetoin, diacetyl and acetaldehyde, were also produced in elevated concentrations. Similarly, mutants of *S. cerevisiae* strains deficient in tricarboxylic acid cycle genes or strains deficient in fumarase and α-ketoglutarate dehydrogenase have been proposed as suitable strains for the production of alcohol-free beer (Willaert 2012).

Finally, a possible strategy relies on the use of LAB in beers produced by the method of stopped fermentation to provide a fresh character and improved drinkability compared with yeast-driven fermentation (Vriesekoop *et al.* 2012).

Low-carb, light or diet beers

Nowadays, there is a rising trend towards low-calorie beverages due to the increased awareness of health in alcoholic beverages (Liu *et al.* 2004). Beer is of great commercial interest here, but *S. cerevisiae* lacks the ability to produce extracellular depolymerizing enzymes that can efficiently liberate fermentable sugar from abundant, polysaccharide-rich substrates (Zhang *et al.* 2008). Non-conventional strains with amylase activity can be used to lower the amount of residual sugars in beer. In particular, species such as *Br. bruxellensis* could be used to obtain highly attenuated beverages (Steensels and Verstrepen 2014).

Instead of adding an exogenous enzyme during beer fermentation, starch hydrolysis and oligosaccharide reduction can be achieved by introducing an amylolytic enzyme gene into brewing yeast (Liu *et al.* 2004). The α-amylase gene (*ALP1*) from *Saccharomycopsis fibuligera*, the dextranase gene (*LSD1*) from *Lipomyces starkeyi* and the glucoamylase genes (*STA*, *SGA1* and *GLA*) from *Saccharomyces diastaticus*, *S. cerevisiae* and *Sac. fibuligera* have been used to develop brewer's yeasts able to reduce the calorie content of beer. However, these recombinant yeasts cannot completely degrade wort dextrins to fermentable sugars (Park *et al.* 2014).

To avoid any problems caused by the transfer of bacterial antibiotic-resistance markers from GM yeasts, integrating cassettes, carrying the amylolytic enzyme gene and lacking bacterial DNA sequences, have been constructed with the aureobasidin A resistance gene (*AUR*1-C) as a selective marker (Liu *et al.* 2004). This enzyme from *Aspergillus niger*, however, is heat stable. The presence of active glucoamylase produced a sweet taste during storage. To overcome this problem, the *GAM1* gene from *Schwanniomyces occidentalis* was introduced into brewer's yeast. The resulting glucoamylase is heat sensitive and possesses debranching activity, and the mutant can ferment dextrin efficiently (Duong 2009).

References

Andries, M., Van Beveren, P.C., Goffin, O., Rajotte, P. and Masschelein, C. A. (2000) Results on semi-industrial continuous top fermentation with the Meura-Delta immobilized yeast fermenter. *Brauwelt International*, **2**, 134–136.

Atter, A., Obiri-Danso, K. and Amoa-Awua, W.K. (2014) Microbiological and chemical processes associated with the production of *burukutu* a traditional beer in Ghana. *International Food Research Journal*, **21**, 1769–1776.

Ault, R.F. (1965) Spoilage bacteria in brewing. *Journal of the Institute of Brewing*, **71**, 376–391.

Back, W. (2009) Microbiology. In H.M. Eßlinger (ed.), *Handbook of Brewing* (2nd edn, pp. 477–490), Weinheim: Wiley-VCH.

Baert, J.J., De Clippeleer, J., Hughes, P.S., De Cooman, L. and Aerts, G. (2012) On the origin of free and bound staling aldehydes in beer. *Journal of Agricultural and Food Chemistry*, **60**, 11449–11472.

Baiano, A., Conte, A. and Del Nobile, M.A. (2012) Beer: Advances in processing and preservation. *Recent Patents on Engineering*, **6**, 83–95.

Bardi, E., Koutinas, A.A. and Kanellaki, M. (1997) Room and low temperature brewing with yeast immobilized on gluten pellets. *Process Biochemistry*, **32**, 691–696.

Bardi, E., Soupioni, M., Koutinas, A.A. and Kannelaki, M. (1996) Effect of temperature on the formation of volatile byproducts in brewing by immobilized cells. *Food Biotechnology*, **10**, 203–217.

Bekatorou, A., Koutinas, A.A., Psarianos, K. and Kannelaki, M. (2001) Low-temperature brewing by freeze-dried immobilized cells on gluten pellets. *Journal of Agricultural and Food Chemistry*, **49**, 373–377.

Bekatorou, A., Sarellas, A., Ternan, N.G. *et al.* (2002) Low-temperature brewing using yeast immobilized on dried figs. *Journal of Agricultural and Food Chemistry*, **50**, 7249–7257.

Berlowska, J., Kregiel, D. and Ambroziak, W. (2013) Physiological tests for yeast brewery cells immobilized on modified chamotte carrier. *Antonie van Leeuwenhoek*, **104**, 703–714.

Berlowska, J., Kregiel, D. and Rajkowska, K. (2014) Biodiversity of brewery yeast strains and their fermentative activities. *Yeast*, **32**, 289–300.

Bezbradica, D., Obradovic, B., Leskosek-Cukalovic, I., Bugarski, B. and Nedovic, V. (2007) Immobilization of yeast cells in PVA particles for beer fermentation. *Process Biochemistry*, **42**, 1348–1351.

Blieck, L., Toye, G., Dumortier, F. *et al.* (2007) Isolation and characterization of brewer's yeast variants with improved fermentation performance under high-gravity conditions. *Applied and Environmental Microbiology*, **73**, 815–824.

Bokulich, N.A. and Bamforth, C.W. (2013) The microbiology of malting and brewing. *Microbiology and Molecular Biology Reviews*, **77**, 157–172.

Bokulich, N.A., Bamforth, C.W. and Mills, D.A. (2012) Brewhouse-resident microbiota are responsible for multi-stage fermentation of American coolship ale. *PLoS ONE*, **7**(4), e3550.

Boulton, C. and Quain, B. (2001) *Brewing Yeast and Fermentation*. Oxford: Blackwell.

Brányik, T., Silva, D.P., Vicente, A.A. *et al.* (2006) Continuous immobilized yeast reactor system for complete beer fermentation using spent grains and corn cobs as carrier materials. *Journal of Industrial Microbiology and Biotechnology*, **33**, 1010–1018.

Brányik, T., Vicente, A.A., Dostálek, P. and Teixeira, J.A. (2008) A review of flavour formation in continuous beer fermentations. *Journal of the Institute of Brewing*, **114**, 3–13.

Brányik, T., Vicente, A., Machado Cruz, J. and Teixeira, J. (2002) Continuous primary beer fermentation with brewing yeast immobilized on spent grains. *Journal of the Institute of Brewing*, **108**, 410–415.

Brányik, T., Vicente, A.A., Machado Cruz, J.M. and Teixeira, J.A. (2004) Continuous primary fermentation of beer with yeast immobilized on spent grains: The effect of operational conditions. *Journal of the American Society of Brewing Chemists*, **62**, 29–34.

Campbell, S. L. (2014) *The Continuous Brewing of Beer*. http://nzic.org.nz/ChemProcesses/food/6A.pdf (accessed 17 June 2014).

Chen, E.C.H. (1978) The relative contribution of Ehrlich and biosynthetic pathways to the formation of fusel alcohols. *Journal of the American Society of Brewing Chemists*, **36**, 39–43.

Daenen, L., Saison, D., Sterckx, F., Delvaux, F.R., Verachtert, H. and Derdelinckx, G. (2008a) Screening and evaluation of the glucoside hydrolase activity in *Saccharomyces* and *Brettanomyces* brewing yeasts. *Journal of Applied Microbiology*, **104**, 478–488.

Daenen, L., Sterckx, F., Delvaux, F.R., Verachtert, H. and Derdelinckx, G. (2008b) Evaluation of the glycoside hydrolase activity of a *Brettanomyces* strain on glycosides from sour cherry (*Prunus cerasus* L.) used in the production of special fruit beers. *FEMS Yeast Research*, **8**, 1103–1114.

Deák, T. (2008) Yeasts in specific types of foods. In T. Deák (ed.), *Handbook of Food Spoilage Yeasts* (2nd edn, pp. 117–201), Boca Raton, FL: CRC Press.

Djegui, K.Y., Gachomo, E.W., Hounhouigan, D.J., Kayodé, A.P. and Kotchoni, S.O. (2014) Biochemical characterization and growth patterns of new yeast isolates. *Molecular Biology Reports*, **41**, 5199–5206.

Dömény, Z., Šmogrovičová, D., Gemeiner, P., Sturdík, E., Pátková, J. and Malovíková, A. (1998) Continuous secondary fermentation using immobilised yeast. *Biotechnology Letters*, **20**, 1041–1045.

Dragone, G., Mussatto, S.I. and Almeida e Silva, J.B. (2008) Influence of temperature on continuous high gravity brewing with yeasts immobilized on spent grains. *European Food Research and Technology*, **228**, 257–264.

Duong, C.T. (2009) An integrative approach to identify novel target genes for reduction of diacetyl production in lager yeast. PhD thesis, Berlin University of Technology.

Duong, C.T., Strack, L., Futschik, M. *et al.* (2011) Identification of Sc-type ILV6 as a target to reduce diacetyl formation in lager brewers' yeast. *Metabolic Engineering*, **13**, 638–647.

Durieux, A., Nicolay, X. and Simon, J.P. (2000) Continuous malolactic fermentation by *Oenococcus oeni* entrapped in Lentikats. *Biotechnology Letters*, **22**, 1679–1684.

Ekberg, J., Rautio, J., Mattinen, L., Vidgren, V., Londesborough, J. and Gibson, B.R. (2013) Adaptive evolution of the lager brewing yeast *Saccharomyces pastorianus* for improved growth under hyperosmotic conditions and its influence on fermentation performance. *FEMS Yeast Research*, **13**, 335–349.

Gibson, B.R. (2011) 125th Anniversary Review: Improvement of higher gravity brewery fermentation via wort enrichment and supplementation. *Journal of the Institute of Brewing*, **117**, 268–284.

Godtfredsen, S.E., Rasmussen, A.M., Ottesen, M., Rafn, P. and Peitersen, N. (1984) Occurrence of α-acetolactate decarboxylases among lactic-acid bacteria and their utilization for maturation of beer. *Applied Microbiology and Biotechnology*, **20**, 23–28.

Hammes, W.P., Brandtet, M.J., Francis, K.L., Rosenheim, J., Seitter, M.F.H. and Vogelmann, S.A. (2005) Microbial ecology of cereal fermentations. *Trends in Food Science and Technology*, **16**, 4–11.

Hansen, J. and Kielland-Brandt, M.C. (1996) Inactivation of MET10 in brewer's yeast specifically increases SO formation during beer production. *Nature Biotechnology*, **14**, 1587–1591.

Hansen, J. and Piškur, J. (2004) Fungi in brewing. In D.K. Arora (ed.), *Biodiversity and Biotechnology Perspectives* (2nd edn), New York: Marcel Dekker.

Hiralal, L., Olaniran, A.O. and Pillay, B. (2014) Aroma-active ester profile of ale beer produced under different fermentation and nutritional conditions. *Journal of Bioscience and Bioengineering*, **117**, 57–64.

Hirata, D., Aoki, S., Watanabe, K., Tsukiokaa, M. and Suzukia, T. (1992) Stable overproduction of isoamyl alcohol by *S. cerevisiae* with chromosome integrated copies of the LEU4 genes. *Bioscience, Biotechnology and Biochemistry*, **56**, 1682–1683.

Hou, L. (2010) Improved production of ethanol by novel genome shuffling in *Saccharomyces cerevisiae*. *Applied Biochemical Biotechnology*, **160**, 1084–1093.

Hsu, W.P. and Bernstein, L. (1985) Fermentation Symposium: Part II A new type of bioreactor employing immobilized yeast. *Technical Quarterly of the Master Brewers' Association of the Americas*, **22**,159–161.

Huuskonen, A., Markkula, T., Vidgren, V. *et al.* (2010) Selection from industrial lager yeast strains of variants with improved fermentation performance in very-high-gravity worts. *Applied Environmental Microbiology*, **76**, 1563–1573.

Kammhuber, K. and Hagl, S. (2001) Statistische Untersuchungen zur Korrelation von Hopfenölkomponente. *Monatsschrift für Brauwissenschaft*, **54**, 100–103.

Kanellaki, M., Bekatorou, A. and Koutinas, A.A. (2014). Low-temperature production of wine, beer, and distillates using cold-adapted yeasts. In P. Buzzini and R. Margesin (eds), *Cold-Adapted Yeasts* (pp. 3–22), Berlin: Springer,.

Kopsahelis, N., Kanellaki, M. and Bekatorou, A. (2007) Low temperature brewing using cells immobilized on brewer's spent grains. *Food Chemistry* **104**, 480–488.

Krogerus, K. and Gibson, B.R. (2013) Influence of valine and other amino acids on total diacetyl and 2,3-pentanedione levels during fermentation of brewer's wort. *Applied Microbioliology and Biotechnology*, **97**, 6919–6930.

Krottenthaler, M. (2009) Hops. In H.M. Eßlinger (ed.), *Handbook of Brewing* (2nd edn, pp. 85–101), Weinheim: Wiley-VCH.

Landaud, S., Latrille, E. and Corrieu, G. (2001) Top pressure and temperature control the fusel alcohol/ester ratio through yeast growth in beer fermentation. *Journal of the Institute of Brewing*, **107**, 107–117.

Lei, H., Zhao, H., Yu, Z. and Zhao, M. (2012) Effects of wort gravity and nitrogen level on fermentation performance of brewer's yeast and the formation of flavor volatiles. *Applied Biochemistry and Biotechnology*, **166**, 1562–1574.

Lermusieau, G., Bulens, M. and Collin, S. (2001) Use of GC-olfactometry to identify the hop aromatic compounds in beer. *Journal of Agricultural and Food Chemistry*, **49**, 3867–3874.

Leskosek-Cukalovic, I.J. and Nedovic, V.A. (2005) Immobilized cell technology in beer brewing: Current experience and results. *Proceedings for Natural Sciences*, **109**, 129–141.

Lindemann, B. (2009) Filtration and stabilization. In H.M. Eßlinger (ed.), *Handbook of Brewing* (2nd edn, pp. 225–234), Weinheim: Wiley-VCH.

Linko, M., Haikara, A., Ritala, A. and Penttilä, M. (1998) Recent advances in the malting and brewing industry. *Journal of Biotechnology*, **65**, 85–98.

Liu, Z., Zhang, G. and Liu, S. (2004) Constructing an amylolytic brewing yeast *Saccharomyces pastorianus* suitable for accelerated brewing. *Journal of Bioscience and Bioengineering*, **98**, 414–419.

Masschelein, C.A. (1997) Realistic view on the role of research in the brewing industry today. *Journal of the Institute of Brewing*, **103**, 103–113.

Narziss, L. (2005) *Abriss der Bierbrauerei*, Weinheim: Wiley-VCH.

Nedović, V., Willaert, R., Leskošek-Čukalović, I., Obradović, B. and Bugarski, B. (2005) Beer production using immobilised cells. In V. Nedovi and R. Willaert (eds), *Applications of Cell Immobilisation Biotechnology* (pp. 259–273), Berlin: Springer.

Nevoigt, E., Pilger, R., Mast-Gerlash, E. *et al.* (2002) Genetic engineering of brewing yeast to reduce the content of ethanol in beer. *FEMS Yeast Research*, **2**, 225–232.

Oka, K., Hayashi, T., Matsumoto, N. and Yanase, H. (2008) Decrease in hydrogen sulfide content during the final stage of beer fermentation due to involvement of yeast and not carbon dioxide gas purging. *Journal of Bioscience and Bioengineering*, **106**, 253–257.

Park, J.Y., Leeet, J.Y., Choi, S.H. *et al.* (2014) Construction of dextrin and isomaltose-assimilating brewer's yeasts for production of low-carbohydrate beer. *Biotechnology Letters*, **36**, 1693–1699.

Pátková, J., Šmogrovičová, D., Dömenyet, Z. and Bafrncová, P. (2000) Very high gravity wort fermentation by immobilized yeast. *Biotechnology Letters*, **22**, 1173–1177.

Peppart, T.L. and Halsey, S.A. (1981) Malt flavour: Transformation of carbonyl compounds by yeast during fermentation. *Journal of the Institute of Brewing*, **87**, 386–390.

Piddocke, M.P., Fazio, A., Vongsangnak, W. *et al.* (2011) Revealing the beneficial effect of protease supplementation to high gravity beer fermentations using '-omics' techniques, *Microbial Cell Factories*, **10**, 27.

Pilkington, H., Maragaritis, A., Mensour, N., Sobczak, J., Hancock, I. and Russell, I. (1999) Kappa-carrageenan gel immobilization of lager brewing yeast. *Journal of the Institute of Brewing*, **105**, 398–404.

Pires, E.J., Teixeira, J.A., Brányik, T. and Vicente, A.A. (2014) Yeast: The soul of beer's aroma – a review of flavour-active esters and higher alcohols produced by the brewing yeast. *Applied Microbiology and Biotechnology*, **98**, 1937–1949.

Pittner, H., Back, W., Swinkels, W., Meersman, E., van Dieren, B. and Lommi, H. (1993) Continuous production of acidified wort for alcohol-free beer using immobilised lactic acid bacteria. *Proceedings of the European Brewery Convention Congress, Oslo* (pp. 323–329), Oxford: IRL Press.

Puligundla, P., Šmogrovičová, D., Obulam V.S.R. and Ko, S. (2011) Very high gravity (VHG) ethanolic brewing and fermentation: A research update. *Journal of Industrial Microbiology and Biotechnology*, **38**, 1133–1144.

Renger, R.S., Vanhateren, S.H. and Luyben, K. (1992) The formation of esters and higher alcohols during brewery fermentation: The effect of carbon dioxide pressure. *Journal of the Institute of Brewing*, **98**, 509–513.

Saerens, S.M.G., Verbelen, P.J., Vanbeneden, N., Thevelein, J.M. and Delvaux, F.R. (2008) Monitoring the influence of high-gravity brewing and fermentation temperature on flavour formation by analysis of gene expression levels in brewing yeast. *Applied Microbiology and Biotechnology*, **80**, 1039–1051.

Saison, D., De Schutter, D.P., Vanbeneden, N, Daenen, L., Delvaux, F. and Delvaux, F.R. (2010) Decrease of aged beer aroma by the reducing activity of brewing yeast. *Journal of Agricultural and Food Chemistry*, **58**, 3107–3115.

Sanchez, R.G., Solodovnikova, N. and Wendland, J. (2012) Breeding of lager yeast with *Saccharomyces cerevisiae* improves stress resistance and fermentation performance. *Yeast*, **29**, 343–355.

Shen, H.Y., Moonjai, N., Verstrepen, K.J. and Delvaux, F.R. (2003) Impact of attachment immobilization on yeast physiology and fermentation performance. *Journal of the American Society of Brewing Chemists*, **61**, 79–87.

Shindo, S., Sahara, H. and Koshino, S. (1994) Suppression of α-acetolactate formation in brewing with immobilized yeast. *Journal of the Institute of Brewing*, **100**, 69–72.

Šmogrovičová, D. and Döméný, Z. (1999) Beer volatile by-product formation at different fermentation temperature using immobilized yeasts. *Process Biochemistry*, **34**, 785–794.

Šmogrovičová, D., Domeny, Z. and Svitel, J. (1998) Effect of immobilized cell concentration on beer quality in continuous fermentation. *Food Biotechnology*, **12**, 123–137.

Sohrabvandi, S., Razavi, S.H., Mousavi, S.M. and Mortzavian, A.M. (2010) Characteristics of different brewer's yeast strains used for non-alcoholic beverage fermentation in media containing different fermentable sugars. *Iranian Journal of Biotechnology*, **8**, 178–185.

Spitaels, F., Van Kerrebroeck, S., Wieme, A.D. *et al.* (2014) Microbiota and metabolites of aged bottled gueuze beers converge to the same composition. *Food Microbiology*, **47**, 1–11.

Steensels, J. and Verstrepen, K.J. (2014) Taming wild yeast: Potential of conventional and non-conventional yeasts in industrial fermentations. *Annual Review of Microbiology*, **68**, 61–80.

Stewart, G.G., Hill, A.E. and Russel, I. (2013) 125th Anniversary Review: Developments in brewing and distilling yeast strains. *Journal of the Institute of Brewing*, **119**, 202–220.

Tata, M., Bower, P., Bromberg, S. *et al.* (1999) Immobilized yeast bioreactor systems for continuous beer fermentation. *Biotechnology Progress*, **15**, 105–113.

Tezuka, H., Mori, T., Okumura, Y., Kitabatake, K. and Tsumura, Y. (1992) Cloning of a gene suppressing hydrogen sulfide production by *Saccharomyces cerevisiae* and its expression in a brewing yeast. *Journal of the American Society of Brewing Chemists*, **50**, 130–133.

Tran, Q.H., Nguyen, T.T., Le, V.V.M. and Hoang, K.A. (2010) Effect of Tween 80 and ergosterol supplementation on fermentation performance of the immobilized yeast in high gravity brewing. *International Food Research Journal*, **17**, 309–318.

Vanderhaegen, B., Neven, H., Coghe, S., Verstrepen, K.J., Derdelinckx, G. and Verachtert, H. (2003) Bioflavoring and beer refermentation. *Applied Microbiology and Biotechnology*, **62**, 140–150.

Van De Winkel, L. (1995) Design and optimization of a multipurpose immobilized yeast bioreactor system for brewery fermentations. *Cerevisia*, **20**, 77–80.

Van De Winkel, L., Van Beveren, P.C., Borremans, E., Goossens, E. and Masschelein, C.A. (1993) High performance immobilized yeast reactor design for continuous beer fermentation. *Proceedings of the European Brewing Convention Congress, Oslo* (pp. 307–314), Oxford: IRL Press.

Van Iersel, M.F.M., Brouwer-Post, E., Rombouts, F.M. and Abee, T. (2000) Influence of yeast immobilization on fermentation and aldehyde reduction during the production of alcohol-free beer. *Enzyme and Microbial Technology*, **26**, 602–607.

Van Landschoot, A., Vanbeneden, N. and Machtelinckx, M. (2005) Peculiarities of seven refermented Belgian strong ales and their corresponding industrial yeasts. *Cerevisia*, **30**, 181–188.

Verbelen, P.J. and Delvaux, F.R. (2009) Brewing yeast in action: Beer fermentation. In M. Rai and P.D. Bridge (eds), *Applied Mycology* (pp. 110–135), Wallingford: CAB International.

Verstrepen, K.J., Derdelinckx, G., Dufour, J.P. *et al.* (2003a) Flavor-active esters: Adding fruitiness to beer. *Journal of Bioscience and Bioengineering*, **96**, 110–118.

Verstrepen, K.J., Van Laere, S.D.M., Vanderhaegen, B.M.P. *et al.* (2003b) Expression levels of the yeast alcohol acetyltransferase genes ATF1, Lg-ATF1, and ATF2 control the formation of a broad range of volatile esters. *Applied and Environmental Microbiology*, **69**, 5228–5237.

Virkajärvi, I. (2006) Accelerated processing of beer. In C.W. Bamforth (ed.), *Brewing New Technologies* (pp. 254–274), Boca Raton, FL: CRC Press.

Virkajärvi, I. and Kronlöf, J. (1998) Long-term stability of immobilized yeast columns in primary fermentation. *Journal of the American Society of Brewing Chemists*, **56**, 70–75.

Virkajärvi, I. and Pohjala, N. (2000) Primary fermentation with immobilized yeast: Some effects of carrier materials on the flavour of the beer. *Journal of the Institute of Brewing*, **106**, 311–318.

Virkajärvi, I., Vainikka, M., Virtanen, H. and Home, S. (2002) Productivity of immobilized yeast reactors with very-high-gravity worts. *Journal of the American Society of Brewing Chemists*, **60**, 188–197.

Vriesekoop, F., Krahl, M., Hucker, B. and Menz, G. (2012) 125th Anniversary Review: Bacteria in brewing: The good, the bad and the ugly. *Journal of the Institute of Brewing*, **118**, 335–345.

Willaert, R. (2012) Biochemistry of beer fermentation. In B.K. Simpson (ed.), *Food Biochemistry and Food Processing* (2nd edn, pp. 627–653), Ames, IA: John Wiley & Sons.

Willaert, R. and Nedovic, V.A. (2006) Primary beer fermentation by immobilised yeast: A review on flavour formation and control strategies. *Journal of Chemical Technology and Biotechnology*, **81**, 1353–1367.

Yamano, S., Tomizuka, K., Sone, H. *et al.* (1995) Brewing performance of a brewer's yeast having α-acetolactate decarboxylase from *Acetobacter aceti* subsp. *xylinum*. *Journal of Biotechnology*, **39**, 21–26.

Yamashita, H., Kühbeck, F., Hohrein, A., Herrmann, M., Back, W. and Krottenthaler, M. (2006) Fractionated boiling technology: Wort boiling of different lauter fractions. *Monatsschrift für Brauwissenschaft*, **59**, 130–147.

Yamauchi, Y., Okamoto, T., Murayama, H. *et al.* (1995) Rapid maturation of beer using an immobilized yeast bioreactor. 1. Heat conversion of α-acetolactate. *Journal of Biotechnology*, **38**, 101–108.

Yamauchi, Y., Okamoto, T., Murayama, H. *et al.* (1994) Beer brewing using an immobilized yeast bioreactor design of an immobilized yeast bioreactor for rapid beer brewing system. *Journal of Fermentation and Bioengineering*, **78**, 443–449.

Yeo, H.Q. and Liu, S.Q. (2014) An overview of selected specialty beers: Developments, challenges and prospects. *International Journal of Food Science and Technology*, **49**, 1607–1618.

Yu, Z., Zhao, M., Li, H. *et al.* (2012) A comparative study on physiological activities of lager and ale brewing yeasts under different gravity conditions. *Biotechnology and Bioprocess Engineering*, **17**, 818–826.

Zhang, Y., Wang, Z.Y., He, X.P., Liu, N. and Zhang, B.R. (2008) New industrial brewing yeast strains with ILV2 disruption and LSD1 expression. *International Journal of Food Microbiology*, **123**, 18–24.

CHAPTER 13

Wine microbiology

Patrizia Romano and Angela Capece
Scuola di Scienze Agrarie, Forestali, Alimentari ed Ambientali, Università degli Studi della Basilicata, Italy

Humans have been producing and consuming wine for more than 7000 years, making it one of the first processed agricultural products. At one level wine is simple to make: crushing ripened grapes and letting the indigenous microflora do the rest is how wine has been made for millennia.

Wine production involves a succession of biological processes, including alcoholic fermentation by yeasts and malolactic fermentation by lactic acid bacteria.

Alcoholic fermentation

The conversion of sugar to ethanol is the foundation of the transformation of grapes into wine. Even if grape must is a substrate that is rich enough in nutrients, certain factors, such as low pH and high sugar content, exert a strong selective action on microorganisms, and therefore only some yeast species and a few bacterial species survive and are able to grow. In particular, alcoholic fermentation is conducted by yeasts that, by using a range of enzymes, convert the glucose, fructose and sucrose found in grape must and juice into ethanol and carbon dioxide (CO_2) via the process of fermentation. In addition to CO_2 and alcohol, many other secondary compounds are produced by yeasts, and it must be underlined that the efficiency of yeast and fermentation conditions alter the proportions of various by-products and consequently wine quality. The transformation of grape must in wine can occur by spontaneous fermentation, which is carried out by the action of different yeast genera and species present in grapes and must, or by inoculated fermentation, which is performed at an industrial level by the use of starter cultures (Table 13.1).

Starter Cultures in Food Production, First Edition. Edited by Barbara Speranza, Antonio Bevilacqua, Maria Rosaria Corbo and Milena Sinigaglia.

Table 13.1 Characteristics of spontaneous and inoculated fermentation.

Fermentation	Advantages	Disadvantages
Spontaneous	High microbial complexity High aromatic complexity	Low reproducibility Production of off-flavours
Inoculated	Reproducible process Control of spoilage microorganisms	Complexity reduction Reduction of varietal characteristics

Spontaneous fermentation

It has been known for a long time that freshly crushed grape juice is a non-sterile substrate that contains several types of microorganisms, in particular various yeasts. The main diffused yeasts belong to the genera *Hanseniaspora* (anamorph *Kloeckera*), *Pichia, Candida, Metschnikowia, Kluyveromyces* and *Saccharomyces*. Occasionally, species in other genera such as *Zygosaccharomyces, Saccharomycodes, Torulaspora, Dekkera* and *Schizosaccharomyces* may be present (reviewed in Fleet 2003). These yeasts originate from the microbial communities of the grape berry and winery environment. Many of these non-*Saccharomyces* yeasts (especially species of genera *Hanseniaspora, Candida, Pichia* and *Metschnikowia*) initiate spontaneous alcoholic fermentation of the juice, but they are very soon overtaken by the growth of *Saccharomyces cerevisiae* as a consequence of their weak ethanol tolerance. After 3–4 days, *S. cerevisiae* becomes the dominant yeast and most often it is the only species found in the fermenting juice from the middle to final stages of the process. It is generally considered that the successional evolution of strains and species throughout fermentation is largely determined by their different susceptibilities to the increasing concentration of ethanol, the non-*Saccharomyces* species dying off earlier in the process because they are more sensitive to ethanol than *S. cerevisiae*. In addition to ethanol, other phenomena such as temperature of fermentation, dissolved oxygen content, killer factors, quorum-sensing molecules and spatial density are known to affect the competitive interaction between yeast species and strains in wine fermentation (Perez-Nevado *et al.* 2006).

The non-*Saccharomyces* species usually achieved maximum populations of 10^7 colony forming units (cfu) per mL or more in the early stages of fermentation before dying. The amount of biomass resulting is sufficient to have an impact on the chemical composition of the wine and the contribution of these yeasts to the overall wine character has been recognized as much more significant than previously thought. Nowadays, the important contribution of non-*Saccharomyces* species has been admitted in both spontaneous and inoculated fermentations.

Inoculated fermentation

Spontaneous fermentations give inconsistent results from vintage to vintage and there is a risk of spoilage by undesirable yeasts and/or bacteria. The practice of spontaneous fermentations remained prevalent in 'Old World' wine-producing areas until the 1980s, because of the popular belief that the superior yeast strains associated with specific vineyards gave a distinctive style and quality to the wine. Even today, winemakers at many 'boutique' wineries accept the potentially staggering risks involved in spontaneous fermentations to achieve stylistic distinction and vintage variability. In large-scale wine production, however, where rapid and reliable fermentations are essential for consistent wine flavour and predictable quality, the use of selected pure yeast inocula is preferred. These large wineries will be the main beneficiaries of programmes aimed at selecting new yeast strains with even more reliable performance, reducing processing inputs and facilitating the production of affordable high-quality wines.

Hundreds of strains of *S. cerevisiae* have been isolated from successful wine fermentations and have been used over decades as starter cultures in winemaking, thus taking some of the risk out of the process. With the commercial availability of active dry cultures of *S. cerevisiae,* the inoculation of grape must has become attractive and convenient. At present, the use of selected yeast cultures is widespread in both the newer wine-producing countries, such as the United States, South Africa and Australia, and the more traditional wine-producing countries, such as Italy, Germany and France. In this context, extensive use of starter cultures in all winemaking areas around the world represents an important advance in wine biotechnology.

The traditional role of wine yeasts, which is the transformation of grape sugars into ethanol, has been significantly widened by modern oenological microbiology. Nowadays, it is widely recognized that the metabolic peculiarities and physiological properties of a particular yeast may lead to the formation of metabolites and the transformation of grape molecules that may sensorially enrich a wine (Figure 13.1). As a consequence, yeast selection is addressed in the search for strains that might improve wines in terms of their colour, aroma, structure and other technological properties (Pretorius 2000).

The selection of yeasts for winemaking consists of identifying those strains that can ferment grape juice efficiently and produce good-quality wines. This process is generally carried out within the genus *Saccharomyces*. Generally, the sources for the isolation of wine yeast candidates as starter cultures are two ecological habitats: the vineyard (primarily the grapes) and spontaneous fermentations that have given wines of an acceptable or unique quality. Although the grapes are always considered a potential source of new wine yeasts, *Saccharomyces* yeasts are present in very low numbers on grapes. As a consequence, it is difficult to isolate *Saccharomyces* species from mature, undamaged grapes by direct culture on agar media, but they are frequently found by enrichment culture methods. This method consists in the isolation of *Saccharomyces* yeasts from spontaneously fermented grapes, aseptically harvested from vines and crushed in the laboratory. The high concentration of ethanol that accumulates in grape juice during fermentation is the main factor favouring the growth of *Saccharomyces* strains.

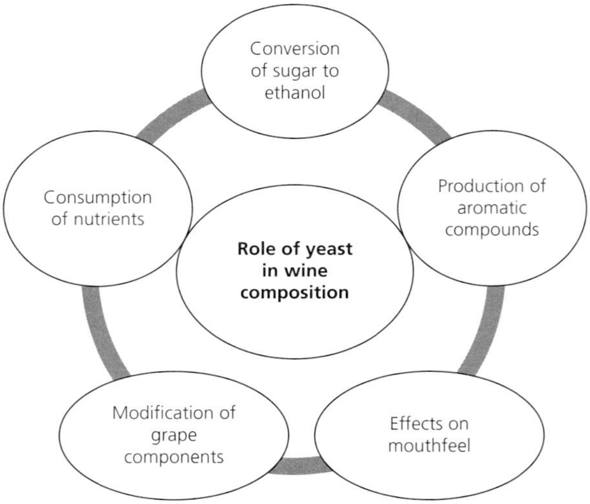

Figure 13.1 Impact of yeast on wine characteristics.

Criteria for the selection of yeast starter

In essence, the criteria for the selection and design of yeast starter for wine fermentation can be considered under three categories (Figure 13.2):

- Properties that affect the performance of the fermentation process ('technological or conventional traits').
- Properties that determine wine quality and character (metabolic or non-'conventional' traits).
- Properties associated with the commercial production of wine yeasts.

Within each category, there are properties of varying degrees of significance and importance, some being essential and some being desirable (Fleet 2008).

Commercial strains have been selected on the basis of specific attributes: ethanol tolerance, sulfur dioxide (SO_2) resistance, predictability of fermentation behaviour, fermentation to dryness (reduction of sugar below 0.2%), absence of production of spoilage characteristics, such as hydrogen sulfide (H_2S) formation, production of specific desirable esters, ability to dominate at diverse fermentation conditions, tolerance of other microorganisms and neutrality (minimal impact on grape varietal character).

Technological or conventional traits

Fermentation performance

Fermentation predictability and wine quality are directly dependent on wine yeast attributes that guarantee the rapid establishment of numerical and metabolic dominance in the early phase of wine fermentation, and determine the ability to conduct efficient fermentation with a desirable residual sugar level.

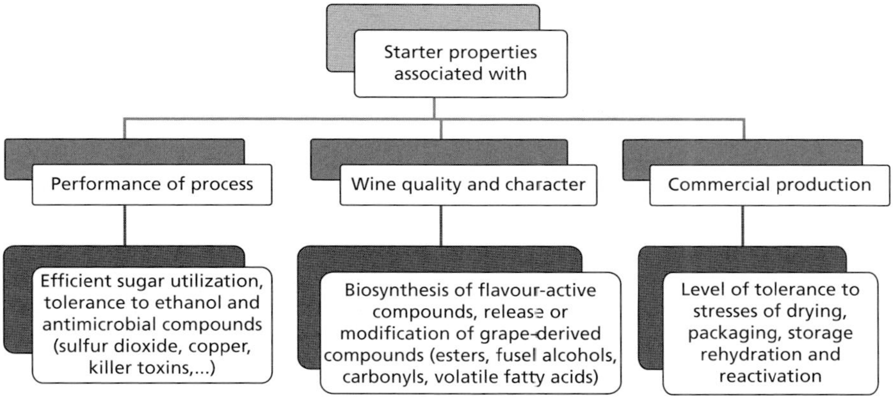

Figure 13.2 Selection criteria for yeast starters.

A wide range of factors affect the fermentation performance of wine yeasts. Apart from successful inoculation with the appropriate starter culture strain, the physiological condition of active dried wine yeast culture and its ability to adapt to and cope with nutritional deficiency, and the presence of inhibitory substances are vital to fermentation performance. Yeasts are subjected to numerous stress factors during the fermentation of grape juice, such as the initial high osmolarity (22–26% sugar or more), the generation of high ethanol concentrations (11–17%), the low pH (3.0–3.9), competition from other microorganisms, the lack of nutrients, extremes of temperature and the presence of inhibitory compounds, such as acetate. If the strain is not able to adapt to these conditions, fermentation will slow and can arrest. These arrested or 'stuck fermentations' are notoriously difficult to restart and lead to losses in production.

Efficient sugar utilization

S. cerevisiae metabolizes glucose and fructose, the main sugars present in grape must, in ethanol. The primary criteria applied during the selection programme are to find strains that are able to complete the fermentation, with the conversion of grape sugar to alcohol and CO_2, at a controlled rate and without the development of off-flavours. Fast, vigorous and complete fermentation of grape juice sugars to high ethanol concentrations (>8% v/v) are essential requirements of wine yeasts. The rate of fermentation and the amount of alcohol produced per unit of sugar during the transformation of grape must into wine is of considerable commercial importance. During winemaking and yeast glycolysis, one molecule of glucose or fructose yields two molecules of ethanol and CO_2. However, the theoretical conversion of 180 g sugar into 92 g ethanol (51.1%) and 88 g CO_2 (48.9%) could only be expected in the absence of any yeast growth, production of other metabolites and loss of ethanol as vapour.

Ethanol tolerance

The ethanol production and fermentation rates are closely linked to ethanol tolerance. In fact, although ethanol is the major desired metabolic product of grape juice fermentation, it is also a potent chemical stress factor that is often the cause of sluggish or stuck fermentations. The production of excessive amounts of ethanol, coming from the harvest of overripe grapes, is known to inhibit the uptake of solutes (e.g. sugars and amino acids) and to inhibit yeast growth rate, viability and fermentation capacity.

Wine strains usually possess survival factors, such as unsaturated long-chain fatty acids and sterols, which confer ethanol tolerance during vinification. These factors are present in higher levels in wine yeasts than in non-wine *Saccharomyces* strains. Furthermore, wine yeasts possess defensive adaptations, conferring enhanced ethanol tolerance, which range from alterations in membrane fluidity to the synthesis of detoxification enzymes. The physiological response to ethanol challenge is also greater in wine yeasts than in non-wine strains.

Tolerance to antimicrobial compounds

During wine fermentation, other antimicrobial compounds produced by yeasts can be present, such as acetic acid, ethanol, medium-chain fatty acids (e.g. decanoic acid) and killer toxins, chemical preservatives (in particular sulfite) and agrochemicals containing heavy metals (e.g. copper). *S. cerevisiae* strains show wide variability in their ability to resist or tolerate these compounds. SO_2 is widely used in wineries to inhibit the growth of unwanted microorganisms, and the tolerance to sulfite assures the implantation of yeast starter cultures into grape must. Although *S. cerevisiae* tolerates higher levels of sulfite than most unwanted yeasts and bacteria, strain variability was found in the maximum levels tolerated and excessive SO_2 dosages may cause sluggish or stuck fermentations.

Improper application of copper formulates to control fungal infections in the vineyard could lead to high levels of copper residues in musts that may cause lagging fermentation. In fact, it is well known that low amounts of copper play a key role in microbial activities, whereas elevated concentrations can be toxic to yeasts, affecting cell growth or causing the acquisition of tolerance to the metal. *S. cerevisiae* species exhibit high variability in the expression of 'copper resistance' character and the acquisition of this trait has been suggested as a consequence of environmental adaptation.

There are several compounds produced by yeasts during alcoholic fermentations that may become inhibitory to other yeast species or strains. One of these inhibitory mechanisms is the so-called yeast killer activity. This phenomenon consists in the production of specific extracellular glycoproteins, the killer toxins, by certain yeast strains (killer yeasts) that are able to kill other yeast strains (sensitive yeasts). There is an extensive literature on the isolation of killer toxin-producing strains, killer-sensitive strains and killer-neutral strains of *S. cerevisiae* from fermenting grape juice. Although many winemaking variables affect the expression of killer and killer-sensitive phenotypes, there is good evidence that killer interactions may determine species and strain evolution during fermentation. Killer strains of *S. cerevisiae* sometimes predominate at the completion of fermentation, suggesting that they have asserted their killer property and taken over the fermentation. Different ecological studies report that the killer phenomenon confers a competitive edge on the producer strain by excluding other yeasts from its habitat (Yap *et al.* 2000), although some studies (Capece *et al.* 2013) indicated that the competitive advantage of one strain over other strains is not always related to killer activity.

Metabolic or non-'conventional' traits

The organoleptic quality of the final product is a very important factor in winemaking. The wine bouquet is determined by the presence of desirable flavour compounds and metabolites in a well-balanced ratio, and the absence of undesirable ones. While some volatile aroma compounds arise directly from the chemical components of the grapes, many grape-derived compounds are released and/or modified by the action of flavour-active yeasts. The importance of yeast in the development of wine flavour is central. During alcoholic fermentation, yeasts do not only convert sugars to ethanol and CO_2, they also produce a range of minor, but sensorially important, volatile metabolites that give wine its character. Today, wine yeast can be selected to optimally biosynthesize flavour-active compounds, and to release grape-derived flavour compounds and/or modify grape-derived flavour compounds, without affecting the general fermentation performance. One of the numerous tools that can assist winemakers in producing wines with specific flavour profiles and market specifications is the choice of microbial starter culture. With respect to wine quality and character, it is essential that any yeast produces a balanced array of flavour metabolites, without excesses of undesirable volatiles.

The capacity to form aroma depends not only on yeast species but also on the particular strain of the individual species. Different strains of *S. cerevisiae* can produce significantly different flavour profiles when fermenting the same must. This is a consequence of both the differential ability of wine yeast strains to release varietal volatile compounds from grape precursors and the differential ability to synthesize *de novo* yeast-derived volatile compounds. Therefore, selecting the proper yeast strain can be critical for the development of the desired wine style (Mauriello *et al.* 2009). Ethyl esters, acetate esters, fusel alcohols, carbonyls and volatile fatty acids are secondary metabolites synthesized by yeasts, which contribute to the so-called yeast bouquet (Cordente *et al.* 2012). The use of different yeast strains during fermentation contributes considerably to variations in profiles and concentrations of higher alcohol in wine.

Another character thought to have beneficial sensory effects is glycerol, which is believed to have a positive impact on mouthfeel and to add sweetness to the wine. Wine yeast strains producing a high level of glycerol would therefore be of considerable value in improving the organoleptic quality of wine, in particular for starter addressed to the production of red wine, in which usually glycerol concentrations are higher than in white wines.

S. cerevisiae wine strains produce variable amounts of esters, such as isoamyl acetate, hexyl acetate, ethyl hexanoate and ethyl octanoate, which have a potential impact on the aroma profile. For the production of young wines, the wineries select yeast strains that produce both the high levels of esters and acetates, needed for the desirable fruity taste, and the low levels of higher alcohols, that in high amounts contribute negatively to the aroma.

Some compounds such as sulfur-containing compounds are undesirable from a sensory perspective in the finished wine. The formation of sulfite and sulfide by wine yeasts greatly affects the quality of wine. Owing to their high volatility, reactivity and potency at very low threshold levels, sulfur-containing compounds have a great effect on the flavour of wine. Health concerns and an unfavourable public perception of sulfite have led to demands for restriction of its use and reassessment of all aspects of sulfite accumulation in wine. The formation of sulfite and sulfide is affected by many factors, including the composition of the fermentation medium and the strain's ability to produce sulfur-containing compounds. Yeast strains differ widely in their ability to produce sulfite and sulfide. Selecting a yeast with a low propensity to produce sulfur derivatives can minimize the production of off-flavours.

Properties associated with the commercial production of wine yeasts

During the industrial production of yeast starter for the wine industry, some basic criteria have been considered. The cost of production needs to be contained so that the final product is affordable to the wine industry. Consequently, the yeast must be amenable to large-scale cultivation on relatively inexpensive substrates such as molasses. The manufacturers of active dried wine yeast starter cultures can positively influence the degree of viability and vitality, as well as the subsequent fermentation performance of their cultures, by the way in which they cultivate their yeasts. Industrial cultivation of wine yeasts can have a profound effect on microbiological quality and fermentation rate, as well as on tolerance to drying and rehydration. Therefore, the cultivated yeast needs to be tolerant to the stresses of drying, packaging, storage and, finally, rehydration and reactivation by the winemaker (Soubeyrand *et al.* 2006). These requirements need to be achieved without loss of the essential and desirable winemaking properties. Some authors report that the starter formulation significantly affects the content of some volatile compounds influencing wine aroma (Romano *et al.* 2015).

Novel traits

Although the selection procedure and the production of wine yeasts with basic oenological criteria are well defined, in fact there are increasing requests for a wine industry that is more efficient and more sustainable, and there are growing demands from consumers for wines with more distinctive and specific styles. For example, in the last few years increasing attention has been addressed to the improvement of healthy attributes in wine, such as the reduction of ethanol content or the increase in antioxidant compounds. For this purpose, the selection protocol will involve the definition of specific attributes required in the finished wine and the individuation of yeast properties related to these specific attributes, without compromising the essential oenological criteria.

Colour improvement

One of the most important sensorial variables of red wines is the colour, which is related to the extraction of polyphenols from grapes during maceration. Wine phenolic contents depend mainly on grape variety, vintage and winemaking conditions. The choice of a yeast strain is also an important factor that affects wine colour, since these microorganisms have different capacities to retain or adsorb phenolic compounds. Different studies have shown that anthocyanin adsorption is affected by numerous factors, but also by yeast activity. Yeast strains expressing β-glucosidase activities promote anthocyanin degradation. On the other hand, yeast may contribute to stabilizing wine colour, as a result of participating in the formation of vitisins (formed by the reaction between pyruvic acid, acetaldehyde and anthocyanins) during fermentation. A strong correlation has been reported between the quantity of metabolites formed by different yeast strains and the quantity of vitisins produced in the wine. Yeasts may affect the wine's chromatic profile and phenolic content, during both fermentation and wine ageing, and may be used as a tool for obtaining stable and highly coloured wines.

Wine wholesomeness

Nowadays, two controversial tendencies are related to wine consumption. Although in moderate amounts, wine has been shown to protect from heart disease and type 2 diabetes; however, consumers are paying more attention to the negative effects of alcohol on health, leading the World Health Organization to introduce a global strategy to reduce the use of alcohol. In this context, it is therefore of the utmost importance to focus on health aspects and to select yeasts that may reduce the risks and enhance the benefits, for instance yeasts that give less ethanol, decreased formation of ethyl carbamate and biogenic amines, and increased production of resveratrol and antioxidants.

Currently more studies are focused on improving technologies that remove ethanol from wines, but these approaches increase production costs and can have a negative impact on wine quality. Yeast selection, however, can provide glycolytically inefficient strains that consume more sugar per degree of alcohol produced. The use of such yeasts would provide a natural method of reducing the ethanol content of wines, potentially increasing at the same time the formation of derived metabolites with a positive impact on quality. However, it is very difficult to select low-ethanol-producing yeast strains; until now, the simplest and most economical biological approaches to produce wine with low ethanol concentrations has been the use of genetically modified yeasts, in which the carbon metabolism was partially diverted away from ethanol production during fermentation.

Ethyl carbamate (also known as urethane) is a suspected carcinogen that occurs in most fermented foods and beverages and, given the potential health hazard, there is a growing demand to reduce the allowable limits of ethyl carbamate in wines. This compound forms by reaction between urea and ethanol. Numerous practices in the vineyard and in the cellar can lead to high urea levels in wine, but *S. cerevisiae* strains also vary widely with regard to their urea-forming ability. In *S. cerevisiae* urea is formed during the breakdown of arginine, one of the main amino acids in grape juice, by the *CAR1*-encoded arginase. Certain yeast strains secrete urea into wine and, depending on fermentation conditions, may be unable to metabolize the external urea further. Although all *S. cerevisiae* strains secrete urea, the extent to which they reabsorb it differs. Strain selection is the only way of reducing the accumulation of urea in wine.

Other toxic compounds present in the wine are biogenic amines, such as histamine, tyramine and tryptamine, implicated as causative agents of 'wine headache'. Some *Saccharomyces* strains have been associated with a total biogenic amine production of 12.14 mg/L, but 0.01 mg/L is the average (Caruso *et al.* 2002). However, the production of biogenic amines is a problem mainly caused by the metabolism of lactic acid bacteria (LAB) during malolactic fermentation (MLF) and wine maturation.

Ochratoxin A (OTA) is a toxic and carcinogenic compound produced by fungal growth on the grapes; this compound has several toxic effects and the European Commission has proposed a limit of 2 μg/L. Some *S. cerevisiae* strains are able to remove OTA from wines by adsorption mechanisms of OTA on yeast cell walls during fermentation. Therefore, yeast selection offers an interesting option for the reduction of OTA content in wine.

In addition to the influence on wine colour, the polyphenols present in the wine possess pharmacological characteristics; in fact, red wine phenolics, such as resveratrol, have been shown to reduce the risk of coronary heart disease. By acting as an antioxidant and an antimutagen, resveratrol shows cancer chemopreventive activity. Since yeasts can influence the wine's phenolic content, starter cultures performing the fermentation process can affect the total antioxidant capacity of the wine (Brandolini *et al.* 2011).

As regards the characterization of *Saccharomyces* yeasts, the International Organization of Vine and Wine (OIV) has adopted the resolution 'Guidelines for the characterization of wine yeasts of the genus *Saccharomyces* isolated from vitivinicultural environments' (OIV-OENO 370-2012), which contains a collection of criteria useful in the process of isolation and characterization of *Saccharomyces* wine yeasts suitable for the production of high quality wine.

New wine yeasts

The search for new yeast strains, in order to improve the characteristics of wine or to facilitate specific stages of the production process, relies on the isolation and screening of new yeast strains from grape and wine samples, and this is indeed the origin of the vast majority of wine yeast strains currently on the market. However, finding wine yeast strains possessing an ideal combination of oenological characteristics is highly improbable, and therefore strain selection is extended to the improvement of *S. cerevisiae* strains or to the use of mixed-starter cultures.

This unstated conclusion has led wine microbiologists to look for alternative ways to exploit yeast's natural genetic diversity, or even to genetically manipulate yeast strains in order to improve specific properties.

Improvement of *S. cerevisiae* wine yeast strains

Due to the demanding nature of modern winemaking practices, there is a continuously growing quest for specialized *S. cerevisiae* strains possessing a wide range of optimized or novel oenological properties. The need for new wine yeast strains derives from both producer- and consumer-oriented requirements. Winemakers require cost-competitive production of wine with minimum resource inputs. These include improved fermentation performance and processing efficiency, such as the increase of ethanol and stress tolerance, efficient nitrogen assimilation, resistance to antimicrobial compounds, and improvement of the biological control of spoilage microorganisms through the expression of antimicrobial enzymes and peptides or through the metabolic production of SO_2. Consumer-oriented tendencies include the requirement for new healthful wines, that is wines with higher antioxidant compounds (resveratrol); wines with lower concentrations of toxic substances, such as ethyl carbamate and biogenic amines; and also wines with a lower ethanol content. Furthermore, consumers request an improvement of the organoleptic quality of wines, such as enhanced production of desirable volatile esters, optimization of phenolic content and reduced production of higher alcohols, sulfites and sulfides.

The starting point for the genetic improvement of wine yeasts is always the isolation from grape musts and wines of a high number of yeast strains, which are then submitted to the analysis of their oenological properties, a process named 'clonal selection' (Giudici *et al.* 2005). Although this process rarely produces pure strain clones possessing all the desired traits for winemaking, it allows the constitution of a biodiversity background, which is very useful for successive genetic improvement programmes. The genetic improvement of industrial strains has traditionally relied on classic genetic techniques, such as variant selection, mutagenesis, hybridization (mating, spore-cell mating), rare mating, cytoduction and spheroplast fusion, followed by selection for broad traits, such as fermentation capacity, ethanol tolerance and absence of off-flavours (e.g. H_2S). Genetic improvement of wine yeasts is rarely based on the sexual cycle, either because the yeasts lack it, it is difficult to manipulate or there are faster or cheaper alternatives. An alternative to the sexual cycle in industrially important microorganisms is parasexual hybridization, in the form of protoplast (or spheroplast) fusion. Protoplast fusion can be intraspecific (two strains from the same species) or interspecific (fusing cells from two more or less distant species). There is increasing interest on this technology for the genetic improvement of wine yeast, especially since several authors (De Barros Lopes *et al.* 2002; Gonzalez *et al.* 2006) showed that a number of industrially important strains are the result of natural interspecific hybridization events. Random mutagenesis with chemical or physical agents is perhaps the simplest way to genetically improve industrial microorganisms. One of the main limitations to the usefulness of this technique in wine yeast strains is related to their genomic structure: since most genes will be present in two or more copies, selecting recessive mutations is difficult. As a consequence, only a few examples of random mutagenesis have been applied to wine yeast improvement. The classic genetic methods are especially advantageous to improve and combine traits under polygenic control, but the major limitation of these techniques is the difficulty of adding or removing features from a strain without altering its performance.

Genetic improvement of industrial strains by classic genetics was followed in the last 20 years by the use of recombinant DNA technologies. Publication of the complete *S. cerevisiae* genome (Goffeau *et al.* 1996), together with the growing availability of recombinant DNA technologies, made the construction of specialized commercial strains possible, mainly by heterologous gene expression or by altered gene dosage (overexpression or deletion). In the last few decades, gene technology has broadened the possibilities of adding new traits and improved characteristics into the target cell, and currently numerous research laboratories worldwide have obtained genetically modified yeast (GMY) strains. These have been reviewed by different authors (Pizzarro *et al.* 2007). Some of these advances include glycerol overproduction, lower ethanol yields, increased release of phenolic compounds, maloethanolic and malolactic yeast strains, decreased ethyl carbamate production, production of the grape antioxidant resveratrol and increased ester production.

Strains improved with classic genetic techniques do not give rise to products that are included in the definition of genetically modified organisms (GMOs), are not treated with the same level of public suspicion and are not subject to the same strict regulations that pertain to GMOs. In general, all genetic materials applied for the construction of microorganisms used for food fermentation should be derived from the host species (self-cloning) or GRAS (generally regarded as safe) organisms with a history of safe food use, while the use of DNA sequences from species taxonomically closely related to pathogenic species should be avoided.

The shortest path to commercial implementation of GMY will probably lie in strains developed through self-cloning techniques that are based on the use of host-derived material (Coulon *et al.* 2006). Recently, a saké yeast strain improved through self-cloning was approved by the Japanese government and does not need to be treated as a GMO (Akada 2002; Schuller and Casal 2005). This technology has been used to develop commercial wine yeast strains that produce significantly lower amounts of the carcinogen ethyl carbamate.

Mixed-starter cultures

It is now widely accepted that many wines, whether produced with or without inoculated yeasts, are the outcome of mixed fermentation that involves contributions from many species and strains. With the current understanding of the yeast ecology of wine fermentation, winemakers are seeking to enhance the flavour diversity and appeal of wines and predictability by controlled fermentation with multiple yeast species or strains (Figure 13.3). The use of mixed and multistarter fermentations has been proposed as a tool to take advantage of spontaneous fermentation, while avoiding the risks of stuck fermentation.

S. cerevisiae multistarters

Different studies have investigated the effect of simultaneously inoculating multiple *Saccharomyces* strains to conduct fermentation (co-inoculation). This technique has been used to investigate the volatile profiles of wines produced

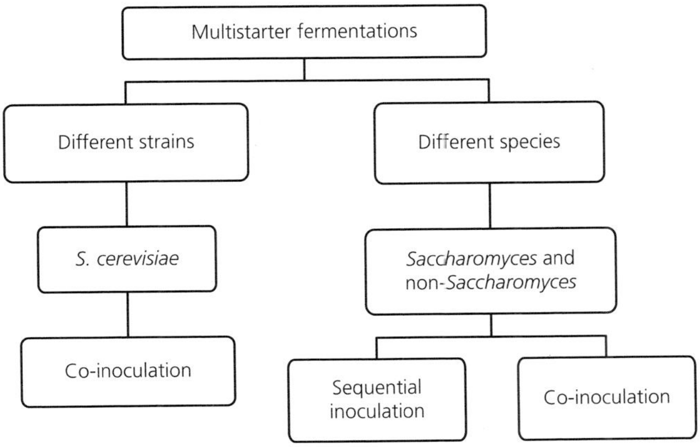

Figure 13.3 Typologies of inoculated mixed fermentations.

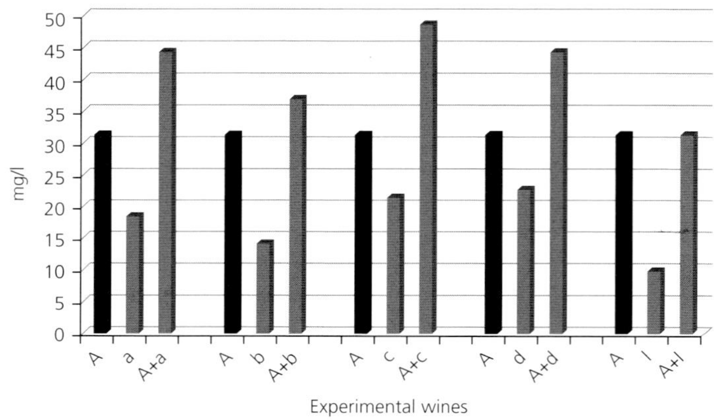

Figure 13.4 (a) Content of terpenic compounds in experimental wines produced by single (A, a, b, c, d, I) and mixed cultures (A+a, A+b, A+c, A+d, A+I).

using multiple *Saccharomyces* strains in an attempt to control the production of desirable metabolites and potentially enhance aroma complexity in wines. In some cases one yeast strain dominated the yeast population towards the end of fermentation, whereas other research (Howell *et al.* 2006; Grossmann *et al.* 1996) showed that each of the strains in the co-inoculations had an effect on the volatile composition of the co-inoculated wines. In Figure 13.4a, the content of terpenic compounds in experimental wines produced by single cultures is compared with the levels detected in wines from mixed cultures. Different authors (Howell *et al.* 2006; Capece *et al.* 2013) showed that the volatile composition of the co-inoculated fermentations could not be replicated by blending wines from the single-strain fermentations. In fact, wines made with mixed cultures of the yeasts gave a combination of volatile aroma metabolites different from that obtained by blending together monoculture wines, made with the same yeast strains composing mixed starter (Figure 13.4b). These results might confirm the hypothesis that yeasts can modify the products of fermentation when grown in mixed cultures and that different yeasts growing together in wine fermentations interact to change the volatile outcome. The modification of wine flavour observed in mixed-culture wines arises from complex, largely unknown interactions between wine yeasts. It has been demonstrated that the co-inoculation of different *S. cerevisiae* commercial strains also affects the sensory profile of wines (King *et al.* 2008) when complementary yeast strains are combined. Due to the influence on the volatile profile of wines, co-inoculated fermentation is a promising tool for the wine industry, assisting winemakers to tailor wine to market specifications for increased competitiveness, and providing novel products to increase the diversity of wine styles.

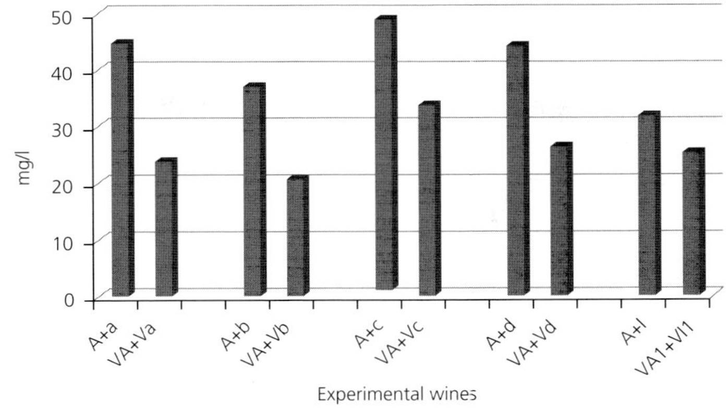

Figure 13.4 (b) Content of terpenic compounds in experimental wines produced by mixed cultures (A+a, A+b, A+c, A+d, A+I) and blended wines (VA+Va, VA+Vb, VA+Vc, VA+Vd, VA+VI) obtained by mixing the corresponding single-culture wines; i.e. the VA+Va blended wine, corresponding to the A+a mixed culture wine, was obtained by mixing wine A and wine a.

Non-*Saccharomyces* multistarters

Non-*Saccharomyces* yeasts are usually considered as spoilage organisms and have been recommended only with the use of a *S. cerevisiae* starter. Furthermore, several studies carried out over the last 25 years have shown the quantitatively significant presence of non-*Saccharomyces* yeasts even during the various stages of fermentation inoculated with *S. cerevisiae* strains. Although the potential of using stand-alone non-*Saccharomyces* yeasts in the development of new styles of wine has been explored, their use as single-starter cultures is not feasible. These species are generally limited in their ability to fully ferment the grape juice sugars and in their ability to produce sufficient concentrations of ethanol. Some may grow too slowly in comparison with other indigenous yeasts and not establish themselves. By manipulating some conditions, such as temperature, sulfur dioxide addition, inoculum levels and time of inoculation, it is possible to enhance the extent of their survival and contribution to the overall fermentation. Inoculating ethanol-sensitive or slower-growing non-*Saccharomyces* yeasts into the grape juice several days before inoculating *S. cerevisiae* (sequential inoculation) is one strategy for enhancing their contribution to fermentation. Various authors have reported results on the deliberate inoculation of selected non-*Saccharomyces* yeasts for wine production, due to their specific metabolic characteristics. An example of the influence of non-*Saccharomyces* yeasts on wine aroma is reported in Figure 13.5, where wines produced inoculating the same *S. cerevisiae* strain (Sc1) together different non-*Saccharomyces* yeasts are significantly different from wine obtained by inoculating only Sc1 starter. Until now, the yeast species tested belonged to the genera *Torulaspora*, *Candida*, *Hanseniaspora*, *Zygosaccharomyces*, *Schizosaccharomyces*, *Lachancea* (formerly *Kluyveromyces*; Lachance and Kurtzman 2011) and *Pichia*.

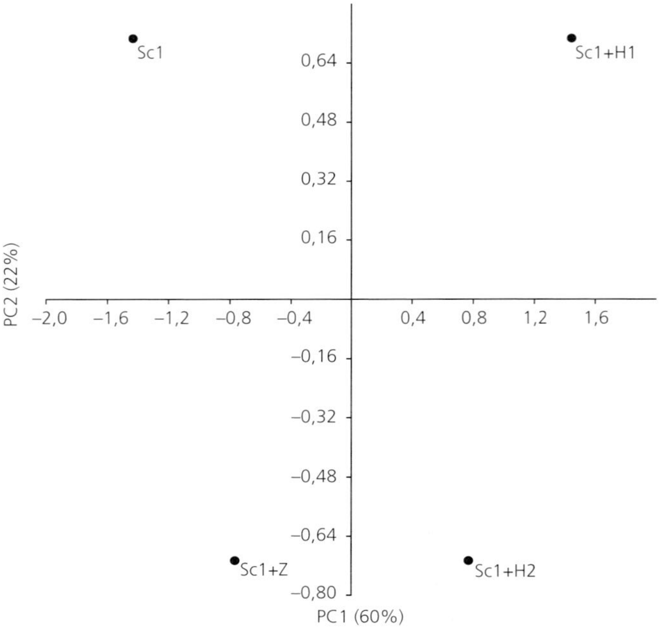

Figure 13.5 Principal component analysis (PCA) biplot of secondary compounds determined in wines produced by inoculating *Saccharomyces cerevisiae* pure culture (Sc1), mixed cultures of *S. cerevisiae* and two *Hanseniaspora uvarum* strains (Sc1+H1, Sc1+H2) and mixed cultures of *S. cerevisiae* and *Zygosaccharomyces bailii* (Sc1+Z).

All these yeasts are poor fermenters, therefore *S. cerevisiae* is always needed to complete wine fermentation. Typically, non-*Saccharomyces* yeasts are used in sequential fermentation where these yeasts are allowed to grow or ferment for between one hour and several days before inoculation with *S. cerevisiae* (Ciani and Ferraro 1998; Jolly *et al.* 2006; Ciani *et al.* 2010).

Torulaspora delbrueckii

Torulaspora delbrueckii (anamorph *Candida colliculosa*) was one of the first commercial non-*Saccharomyces* yeasts to be released. *T. delbrueckii* has previously been suggested for the vinification of musts with low content of sugar and acid, and was used for the commercial production of red and rosé wines in Italy (Castelli 1955). Recently, pure cultures of *T. delbrueckii* have been shown to produce lower levels of volatile acidity than *S. cerevisiae* strains. Thus, *T. delbrueckii* has been useful in the production of wines from high-sugar musts derived from botrytized grapes (Bely *et al.* 2008). Other metabolites produced by *T. delbrueckii* include succinic acid and, for particular strains, linalool, which adds the varietal aroma of Muscat-type wines. As *T. delbrueckii* affects wine composition, it also modulates wine flavour and aroma. Following a co-inoculated strategy with T. *delbrueckii* and *S. cerevisiae*, Sauvignon Blanc and Chenin Blanc wines were both judged to be better than their respective *S. cerevisiae* reference wines 5 and 18 months after production (Jolly *et al.* 2006).

Similarly, Amarone wines produced by sequential inoculation with *T. delbrueckii* and *S. cerevisiae* were judged to have increased aroma intensity, including 'ripe red fruit' aroma, increased sweetness and astringency and decreased intensity for vegetal attributes (Azzolini *et al.* 2012). In 2003, the first commercial release of *T. delbrueckii* was as a component of a yeast blend (Vinoflora® Melody.nsac and Vinoflora® Harmony.nsac) with *S. cerevisiae* and *Klyveromyces thermotolerans* (now classified as *Lachancea thermotolerans*). Subsequently, the *T. delbrueckii* component was released on its own. A further two *T. delbrueckii* strains are also available, indicating that some winemakers are eager to experiment with carefully selected and tested non-*Saccharomyces* yeasts (Jolly *et al.* 2014).

Candida pulcherrima

Metschnikowia pulcherrima (anamorph *C. pulcherrima*) is another yeast that is commercially available. This commercial strain produces an extracellular α-arabinofuranosidase that has an impact on the concentration of varietal aromas, such as terpenes and volatile thiols. Wines of different grape varieties (Sauvignon Blanc, Chenin Blanc, Muscat d'Alexandrie, Debina) obtained by sequential fermentation with *C. pulcherrima* and *S. cerevisiae* showed higher quality scores than control wines obtained by fermentation with *S. cerevisiae* (Jolly *et al.* 2014). However, other studies reported that sequential inoculation with *C. pulcherrima* and *S. cerevisiae* was judged to be of inferior quality than the control wine.

Candida zemplinina/Candida stellata

In 2011, specific strains of *C. stellata* were reclassified to *C. zemplinina* (Kurtzman *et al.* 2011). *C. stellata* is known as a high glycerol producer. Unlike *S. cerevisiae*, which favours glucose utilization (glucophilic yeast), *C. stellata* consumes fructose preferentially to glucose (fructophilic yeast). Wines obtained by mixed cultures or following a sequential inoculation strategy with *C. stellata* and *S. cerevisiae* showed no residual sugar, due to the complementary utilization of fructose and glucose by both species. Furthermore, the resulting wines showed increased concentrations of glycerol and succinic acid and reduced concentrations of acetic acid and higher alcohols. The use of sequential inoculation for the production of Sauvignon Blanc or co-inoculation of Macabeo grape juice (Andorra *et al.* 2010) produced wines with very different volatile profiles than wines fermented with *S. cerevisiae* monocultures. Different influences on flavour profile and the sensory attributes of wines produced by both co-inoculation and sequential inoculation were found, therefore it seems that the use of *C. zemplinina* for wine production might involve a role in increasing wine complexity rather than increasing the perception of particular 'desirable' sensory attributes.

Hanseniaspora species

Hanseniaspora spp. generally show low fermentative power, but they are important for the production of wine volatile compounds, and the chemical composition of wines made with *Hanseniaspora* spp./*S. cerevisiae* combinations differs from that of wines produced with *S. cerevisiae* monoculture. *Hanseniaspora vineae* and *Hanseniaspora guilliermondii* have been reported to produce increased amounts of 2-phenyl-ethyl acetate (associated with 'rose', 'honey', 'fruity' and 'flowery' aroma descriptors) and higher 'fruity' sensory scores than wines produced with *S. cerevisiae* monoculture (Viana *et al.* 2009). Furthermore, wines produced with *H. guilliermondii* and *S. cerevisiae* have shown higher concentrations of hexyl acetate, ethyl acetate and isoamyl acetate than wines produced with *S. cerevisiae* (Moreira *et al.* 2008). In these wines, the production of heavy sulfur compounds was also affected by *H. guilliermondii*. Although some of these compounds are associated with unpleasant sensory descriptors, they might have a role in increasing wine complexity. Apiculate yeasts are also known as high producers of acetic acid (0.75–2.25 g/L) and ethyl acetate, making them less attractive for wine production, although high-strain variability for this character was found. Some reports have observed that the initial growth of *Hanseniaspora* had a retarding effect on the subsequent growth of *S. cerevisiae*. This phenomenon could have further implications as a cause for lagging or stuck fermentations. Therefore, a cautionary approach would have to be taken when considering using *Hanseniaspora* spp. in wine production.

Zygosaccharomyces species

Zygosaccharomyces spp. are considered to be winery contaminants as a consequence of high production levels of acetic acid and are especially a problem in wineries producing sweet and sparkling wines (Loureiro and Malfeito-Ferreira 2003). However, wines produced by mixed fermentation with combinations of *Zygosaccharomyces bailii*/*S. cerevisiae* and *Zygosaccharomyces florentinus*/*S. cerevisiae* have shown increased production of polysaccharides, which can have a positive influence on wine taste (Domizio *et al.* 2011). A commercial *Zygosaccharomyces* yeast was released specifically for restarting stuck fermentations due to its fructophilic nature. This may also be beneficial in fermentations of grape musts from riper grapes, where the fructose concentration can exceed that of glucose at the start of fermentation and this can affect *S. cerevisiae* growth.

Schizosaccharomyces species

Schizosaccharomyces spp. can degrade organic acids such as malic acid and gluconic acid. *Schiz. pombe* was used in mixed fermentations with *S. cerevisiae* to remove malic acid and total acidity in Arien grape juice (Benito *et al.* 2013). Although wines obtained by mixed fermentation showed increased concentration of acetaldehyde, propanol and 2,3-butanediol and slightly decreased concentration of esters, they received a more favourable sensory score from the judging panel than wine produced by a *S. cerevisiae* monoculture (Benito *et al.* 2013). Dry immobilized cells of *Schiz. pombe* for malic acid consumption in winemaking have been proposed, and a commercial yeast strain of *Schiz. pombe* is now available in an immobilized form (ProMalic; Proenol, Canelas, Portugal). In addition to *Schiz. pombe*, a strain of *Issatchenkia orientalis* can degrade malic acid rapidly, and has been proposed to reduce the malic acid content in wine as a mixed culture with *S. cerevisiae* (Kim *et al.* 2008).

Lachancea thermotolerans (Kluyveromyces thermotolerans)

Lachancea thermotolerans (formerly *K. thermotolerans*; Kurtzman *et al.* 2011) has been described as a yeast species able to produce wines with increased concentrations of lactic acid, glycerol and 2-phenylethanol during mixed fermentations of grape musts (Comitini *et al.* 2011; Gobbi *et al.* 2013). In addition, commercial-scale fermentations (10,000 L) of Sangiovese grape must with *Lac. thermotolerans*/*S. cerevisiae* produced wines that were scored higher in 'spicy' and 'acidity' attributes than *S. cerevisiae* monoculture wines (Gobbi *et al.* 2013). However, the effect of *Lac. thermotolerans* on wine chemical composition and therefore on wine flavour depends on the time of inoculation with *S. cerevisiae*, and the choice of *S. cerevisiae* strain can also influence the wine aroma profile in mixed fermentations. A commercial active dried yeast blend of *Lac. thermotolerans* (marketed as *K. thermotolerans*) and *S. cerevisiae* (Viniflora® Symphony.nsac) was previously commercially available. This combination was developed for enhancement of aroma and flavour in white and red grape varieties. According to the product information sheet, the use of this yeast in simultaneous inoculation could lead to enhancement of floral and tropical fruit aromas and more complex and rounded flavours in white and red wine, respectively. In 2012, the *Lac. thermotolerans* component of Viniflora® Symphony.nsac was released on its own as a single-active dried yeast.

Other non-*Saccharomyces* species

Other non-*Saccharomyces* yeasts have also been investigated for their potential contribution to wine.

Co-fermentation with *Pichia kluyveri* has been reported to lead to higher levels of varietal thiols and in fact it is available as a commercial yeast product that is reported to extract flavour precursors from grape juice at higher levels than other yeasts tested. Mixed fermentations with *Pichia fermentans* and *S. cerevisiae* produced wines with increased concentrations of some volatile compounds, such as acetaldehyde, ethyl acetate, 1-propanol, n-butanol, 1-hexanol, ethyl octanoate, 2,3-butanediol and glycerol, and showed an increased concentration of polysaccharides, which can improve wine taste and body (Domizio *et al.* 2011). The use of *Hansenula anomala* in mixed fermentation with *S. cerevisiae* increased the formation of some higher alcohols as well as acetate and ethyl esters.

Other species include *Williopsis saturnus*, *Candida cantarellii*, *Issatchenkia orientalis* and *Saccharomycodes ludwigii*.

Malolactic fermentation

Spontaneous malolactic fermentation

Malolactic fermentation (MLF) is a biological process that usually follows the completion of alcoholic fermentation by yeasts. Although MLF is regarded as a secondary fermentation process, it plays an integral role in the production of the majority of red wines, as well as some white wines (Lerm *et al.* 2010). The MLF reaction is defined as the conversion by lactic acid bacteria (LAB) strains of L-malic acid, a dicarboxylic acid, to L-lactic acid, a monocarboxylic acid, with the production of CO_2.

The desirable effects of MLF in wine are:

- Deacidification of the wine with a concomitant increase in pH.
- Contribution to microbial stability by the removal of malic acid as a possible carbon substrate.
- Modification of the wine aroma profile, linked to the enzymatic activity of LAB, such as a richer bouquet, which would include buttery and nutty attributes and traits of honey, vanilla, leather, spices and smoother tannins. Furthermore, MLF generally reduced vegetative character, modified fruitiness and improved mouthfeel and flavour persistence.

In cooler climates the deacidification process is regarded as the most important modification associated with MLF, while the change in the sensory profile of the wine is a more important consideration in countries where deacidification is less important, such as warmer regions where lower concentrations of malic acid are present in the grapes.

The species of LAB associated with spontaneous MLF belong to the genera *Oenococcus*, *Leuconostoc*, *Lactobacillus* and *Pediococcus*, but *Oenococcus oeni* is the most widespread and dominant species thanks to its tolerance of the harsh wine environment. The stress factors present in the wine include conditions of high ethanol content, low pH, presence of SO_2 and other inhibiting compounds, such as the fatty acids released by yeasts, tannins and various chemical residues (Table 13.2).

A successional growth of several species of LAB from the vineyard to the final vinification stages has been observed, although considerable variability was found due to region, cultivar and vinification procedures. *O. oeni* is the main LAB

Table 13.2 Factors affecting malolactic fermentation.

Promotion	Inhibition
Low sulfur dioxide content	Use of sulfur dioxide
Warm temperatures	Low temperatures
pH adjustment	pH reduction
Late racking	Early racking
Low ethanol content	Filtration/fining
Inoculum	Use of antimicrobials
Nutrient addition	(lysozyme, fumaric acid)

species associated with wine; *Pediococcus damnosus*, *Pe. parvulus* and *Pe. pentosaceus* mostly occur after MLF and in higher pH wines, and several *Lactobacillus* species also occur after MLF. At the start of alcoholic fermentation, the major species of LAB present include *Lactobacillus plantarum*, *Lb. casei*, *Leuconostoc mesenteroides* and *Pe. damnosus*, as well as *O. oeni* to a lesser extent. After the completion of alcoholic fermentation, the surviving bacterial cells, most commonly *O. oeni*, start to multiply. This phase is characterized by vigorous bacterial growth. The pH of the wine is imperative in determining which species of LAB are present, with values above pH 3.5 favouring the growth of *Lactobacillus* and *Pediococcus* species, whereas the *O. oeni* population prevails at lower pH values. When MLF is complete, the remaining LAB are still able to metabolize the residual sugar, which could result in wine spoilage, in particular in high pH wines, where *Lactobacillus* and *Pediococcus* may occur and contribute to wine spoilage.

Inoculated malolactic fermentation

The use of selected cultures of malolactic bacteria (MB) for MLF instead of native bacteria is becoming a common oenological practice. Winemakers are starting to recognize the benefits of inoculating grape must or wine with commercial starter cultures of LAB to ensure the successful completion of MLF (Krieger-Weber 2009) and to reduce the risks associated with spontaneous MLF. Potential risks include the presence of unidentified/spoilage bacteria that can produce undesirable compounds or off-flavours, or toxic metabolites, such as ethyl carbamate and biogenic amines, a delay in the onset or completion of MLF and the development of bacteriophages, all of which contribute to a decrease in the quality of the wine. Due to the risks associated with spontaneous or uncontrolled MLF, it is important for the winemaker to realize the benefits associated with inoculating MLF with a starter culture, as well as inoculating according to the directions of the manufacturer.

Most of the malolactic starter cultures available on the market contain strains of *O. oeni*. Recently, *Lb. plantarum* has also been considered for application in a commercial starter culture.

The selection and characterization of strains for possible use in a commercial culture are crucial, due to the fact that LAB strains differ in their fermentation capabilities and growth characteristics. Strict criteria are used for the selection of bacteria to be used as starter cultures (Krieger-Weber 2009). Apart from its ability to convert malic acid, *O. oeni* is considered an ideal starter culture for MLF thanks to its many oenological properties, such as low production of acetic acid, presence of enzymatic activities that increase the aroma and reduced risk of wine spoilage. However, many ecological and technological investigations suggest that these positive features are not common to all strains of the species. Moreover, the same characteristics can also be found in strains of other species. Great care is therefore essential when selecting strains and exploiting them in the winemaking process, in order to drive the MLF in the desired direction.

Selection criteria for malolactic bacteria

Numerous criteria should be taken into account for the selection of wine bacterial starter culture and these factors are specific for different types of wine.

These criteria can be subdivided as follows (Torriani *et al.* 2011):

- **Stress resistance**: resistance to ethanol, low pH, high SO_2 concentrations and low temperatures.
- **Technological performance**: high malolactic activity, ability to perform MLF in different types of wine, satisfactory growth in a synthetic medium, production of desirable flavours.
- **Impact on wine wholesomeness**: no or very low production of biogenic amines and ethyl carbamate.

In addition, not only the individual effects of the different factors have to be taken into account, but the interactive and synergistic effects. In fact, these influencing factors do not only affect the growth and malolactic activity of LAB, but also the effect that the LAB will have on wine aroma.

Stress resistance

LAB have to tolerate the stress factors emerging during winemaking, such as resistance to high levels of ethanol, tolerance to a pH of 3.0, resistance to high SO_2 concentrations and low temperatures and bacteriophage resistance (not lysogenic).

Ethanol is the main yeast metabolite formed during alcoholic fermentation and, due to its adverse effect on LAB growth and metabolic activity, plays an important role in the ability of LAB to survive in the wine environment and accomplish MLF. As with most LAB inhibitory factors, ethanol also demonstrates synergistically inhibiting effects with temperature. The optimal growth temperature of LAB decreases at high ethanol concentrations, and elevated temperatures lower the ability of LAB to withstand increased ethanol concentrations. It is generally acknowledged that *O. oeni* strains are able to survive and proliferate in 10% (v/v) ethanol at pH 4.7. However, the degree to which LAB are able to tolerate ethanol concentrations is strain dependent, as well as being contingent on the activation steps before inoculation in the wine.

LAB species also differ in their ability to tolerate SO_2. Some authors (Larsen *et al.* 2003) found that *O. oeni* strains were less tolerant to high total SO_2 concentrations than strains of *Pediococcus*.

Besides the addition of SO_2 during the vinification process, it must be underlined that yeasts are also able to produce significant amounts of SO_2. Therefore, when MLF is required, it is essential for the winemaker to consider not only the SO_2 added at different stages of the winemaking process, but also the possible levels of SO_2 produced by the yeast. The combined SO_2 concentration from these two sources will influence bacterial survival and proliferation as well as MLF initiation. It is important to choose a yeast strain that does not produce significant amounts of SO_2, and if sulfur is required to make only small additions at crushing. If larger amounts (>30 mg/L) of sulfur are required (e.g. damaged grapes), MLF inoculation should take place after alcoholic fermentation has been completed. Due to the large influence of wine pH and individual strain tolerance to SO_2, the effects of different SO_2 concentrations are variable. The type of SO_2 present (free or bound) also influences the effect on LAB, with both a reduction in malolactic activity and a reduction in LAB growth. Henick-Kling (1993) reported a 13% reduction in malolactic activity with 20 mg/L of bound SO_2, a 50% reduction at 50 mg/L and no malolactic activity at 100 mg/L of bound SO_2, while a concentration of 30 mg/L of bound SO_2 delayed LAB growth.

The pH of the wine plays a crucial role in determining the success of MLF and it also has a direct effect on the growth rate of bacteria. Although the optimum pH for the growth of *O. oeni* is pH 4.3–4.8, G-Alegría et al. (2004) found that *O. oeni* and *Lb. plantarum* are able to grow at pH 3.2. A further effect of pH is the influence on malolactic activity (Henick-Kling 1993), with the highest malolactic activity exhibited between pH 3.5 and pH 4. The pH is also critical to the commencement of MLF as well as the time taken to complete MLF.

Temperature has a severe effect on bacterial growth and it affects the growth rate, length of the lag phase and population of LAB. The optimum growth temperature for *O. oeni* is reported at 27–30 °C, but due to the presence of alcohol the optimum growth temperature in wine decreases to 20–23 °C. The optimum temperature for both *O. oeni* growth as well as malic acid metabolism in wine is 20 °C. To ensure the rapid initiation and completion of MLF, it is essential to control the fermentation temperature, which should be kept at 18–22 °C. Other selection parameters are the resistance to stress emerging during production steps and the ability to survive and retain viability after the production process, such as freezing and freeze drying, and stress related to the inoculation procedure, such as hydration and inoculation into wine. In order to guarantee complete and successful MLF, it is imperative that winemakers follow the directions for the reactivation of freeze-dried starter cultures as recommended by the manufacturer, as this minimizes some of the potential loss in viability due to direct inoculation in the wine.

Technological performance

The success of the inoculated bacterial culture in initiating and completing MLF is also influenced by the timing of inoculation and the concomitant interaction between the yeast and bacterial cultures. As regards the timing of inoculation, there are three possible inoculation scenarios for MLF: simultaneous inoculation (co-inoculation), inoculation during alcoholic fermentation and inoculation after the completion of alcoholic fermentation (sequential inoculation). The possible risks of simultaneous inoculation are the development of undesirable/antagonistic interactions between yeast and/or bacteria, stuck alcoholic fermentation and the production of possible off-odours. Other authors propose simultaneous inoculation as a tool to induce MLF in high-alcohol wines. Co-inoculation of the bacterial cells results in complete MLF in a shorter time period compared to that of sequential inoculation. Co-inoculation also has the advantage of reducing overall fermentation duration. Other advantages include more efficient MLF in 'difficult' wines (e.g. low pH) due to low levels of ethanol and higher nutrient concentrations. Wines are also immediately available for racking, fining and SO_2 addition. More recent results on co-inoculation highlight this practice as a viable option if care is taken regarding the strain selection of both the bacteria and the yeast. Inoculation during alcoholic fermentation is not a common practice as a consequence of the strong antagonism between yeasts and bacteria. Bacterial populations show drastic decreases with this type of inoculation and this could be attributed to various factors, including the removal of nutrients by the yeast, accumulation of SO_2, ethanol production, toxic metabolite and acid production by the yeast that decreases the pH. Sequential inoculation has several advantages, such as the lack of adverse interactions between yeasts and bacteria as well as a reduced risk of acetic acid production due to smaller residual sugar concentrations. In spite of these advantages, the risks related to sequential inoculation are linked mainly to loss in viability, attributed to the presence of high ethanol concentration, low pH, SO_2, other antimicrobial compounds produced by the yeast as well as nutrient depletion.

The interaction between bacteria and yeasts during alcoholic fermentation and/or MLF will have a direct effect on LAB growth and malolactic activity. Alexandre *et al.* (2004) proposed that the degree and complexity of these interactions are due to three factors. The first factor affecting the extent to which inhibition between these microorganisms occurs is largely dependent on the selected strains of yeast and bacteria. The second factor is the uptake and release of nutrients by the yeast, which will in turn affect the nutrients available for the LAB. Yeast autolysis plays a vital role in the release of essential nutrients for LAB proliferation and survival. The third factor to consider is the ability of the yeast to produce metabolites that can have either a stimulatory or inhibitory/toxic effect on LAB. There are a number of yeast-derived inhibitory compounds, including ethanol, SO_2, medium-chain fatty acids and proteins. Furthermore, the composition of the must and the vinification practices influence the interaction.

Another factor to be taken into account when developing bacterial cultures is the production of aroma compounds that could potentially contribute to the wine aroma profile. Wine-associated LAB have been shown to induce a range of enzymatic activity that has the potential to affect or produce a range of volatile compounds. The use of different starter cultures affects wine aroma and flavour by the modification or production of flavour-active compounds. Bartowsky and Henschke (1995) proposed three mechanisms by which LAB are able to modify wine aroma and flavour: the bacteria production of volatile compounds by metabolizing grape constituents, such as sugars and nitrogen-containing compounds like amino acids; the modification of grape or yeast-derived secondary metabolites by the bacteria; and adsorption to the cell wall or metabolism of flavour compounds. The LAB can produce some aromatic compounds, such as diacetyl, acetoin, butanediol and acetate through the metabolism of citric acid. Diacetyl is responsible for one of the most evident flavour changes that occur during MLF and confers a 'buttery' trait on wine. In addition to the buttery aroma, it has been reported that MLF has enhanced the nutty, vanilla, fruity, vegetative and toasty aromas and reduced the vegetative, green and grassy aromas, possibly due to the catabolism of aldehydes (Liu 2002).

The presence of 'ropy phenotypes' is a negative traits in LAB, negatively affecting the quality of wine, and it is linked to the production of wine-spoiling exopolysaccharides.

Impact on wine wholesomeness

The strains selected as malolactic starters must be safe. In this respect, possible hazards to human health are linked to the release of toxic compounds, namely biogenic amines and ethyl carbamate.

Biogenic amines are a group of organic nitrogen-containing compounds that are formed by certain LAB via the substrate-specific enzymatic decarboxylation of naturally occurring amino acids. The main biogenic amines associated with wine are putrescine, histamine, tyramine and cadaverine, followed by phenylethylamine, spermidine, spermine, agmatine and tryptamine (Lonvaud-Funel 2001). The importance of these compounds in wine is due to their potential toxicological effects in sensitive humans. The ingestion of biogenic amines, histamine in particular, can lead to various health reactions in sensitive humans. These include symptoms like headaches, hypo- or hypertension, cardiac palpitations and in extreme cases even anaphylactic shock. Putrescine and cadaverine, besides being able to enhance the toxicity of histamine, tyramine and phenylethylamine, can also have a detrimental effect on wine quality by imparting flavours of putrefaction and rotten meat, respectively. The presence of alcohol, SO_2 and other amines could potentially amplify the toxic effect of certain biogenic amines. Various factors influence the biogenic amine content, such as the amino acid composition, the microflora present in the wine and the ability of the microflora to decarboxylate amino acids. All parameters that favour bacterial growth will favour biogenic amine formation (Volschenk *et al.* 2006) and it is imperative to be able to identify strains with the potential to produce biogenic amines. It is generally accepted that spoilage LAB are responsible for the formation of biogenic amines, specifically species of *Pediococcus* and *Lactobacillus*. However, *O. oeni* was also identified as a possible biogenic amine producer. In an investigation of the biogenic amine–producing capability of several strains of *O. oeni*, more than 60% were able to produce histamine in concentrations ranging from 1.0 to 33 mg/L. An additional 16% had the added capability of producing putrescine and cadaverine (Guerrini *et al.* 2002). The ability to produce biogenic amines is used as a screening criterion in the selection of LAB starter cultures.

Inoculation for MLF with a starter culture that does not have the ability to produce biogenic amines will eliminate the risk of biogenic amine formation associated with spontaneous MLF.

Ethyl carbamate is subjected to international regulation due to its potential health implications, as it is a suspected carcinogen. Urea, which is yeast derived, and citrulline, derived from LAB metabolism, are the main precursors for ethyl carbamate formation. This molecule is usually synthesized as a side product of arginine catabolism, which contributes to LAB growth due to the generation of adenosine triphosphate, but two of the intermediates formed, citrulline and carbamyl phosphate, are able to react with ethanol to form ethyl carbamate. Strains of *O. oeni* and *Lactobacillus hilgardii* showed the ability to contribute to the ethyl carbamate concentration.

References

Akada, R. (2002) Genetically modified industrial yeast ready for application. *Journal of Bioscience and Bioengineering*, **94**, 536–544.

Alexandre, H., Costello, P.J., Remize, F., Guzzo, J. and Guilloux-Benatier, M. (2004) *Saccharomyces cerevisiae-Oenococcus oeni* interactions in wine: Current knowledge and perspectives. *International Journal of Food Microbiology*, **93**, 141–154.

Andorra, I., Berradre, M., Rozes, N., Mas, A., Guillamon, J.M. and Esteve-Zarzoso, B. (2010) Effect of pure and mixed cultures of the main wine yeast species on grape must fermentations. *European Food Research and Technology*, **231**, 215–224.

Azzolini, M., Fedrizzi, B., Tosi, E. *et al.* (2012) Effects of *Torulaspora delbrueckii* and *Saccharomyces cerevisiae* mixed cultures on fermentation and aroma of Amarone wine. *European Food Research and Technology*, **235**, 303–313.

Bartowsky, E.J. and Henschke, P.A. (1995) Malolactic fermentation and wine flavour. *Australian Grapegrower Winemaker Journal*, **378**, 83–94.

Bely, M., Stoeckle, P., Masnuef-Pomarède, I. and Dubourdieu, D. (2008) Impact of mixed *Torulaspora delbrueckii–Saccharomyces cerevisiae* culture on high-sugar fermentation. *International Journal of Food Microbiology*, **122**, 312–320.

Benito, S., Palomero, F., Morata, A., Calderón, F., Palmero, D. and Suárez-Lepe, J.A. (2013) Physiological features of *Schizosaccharomyces pombe* of interest in making of white wines. *European Food Research and Technology*, **236**, 29–36.

Brandolini, V., Maietti, A., Tedeschi, P., Capece, A. and Romano, P. (2011) Influence of *Saccharomyces cerevisiae* wine strains on total antioxidant capacity. *Annals of Microbiology*, **61**, 125–130.

Capece, A., Siesto, G., Romaniello, R. and Romano, P. (2013) Assessment of competition in wine fermentation among wild *Saccharomyces cerevisiae* strains isolated from Sangiovese grapes in Tuscany region. *LWT – Food Science and Technology*, **54**, 485–492.

Caruso, M., Fiore, C., Contursi, M. and Romano, P. (2002) Formation of biogenic amines as criteria for the selection of wine yeasts. *World Journal of Microbiology and Biotechnology*, **18**, 159–163.

Castelli, T. (1955) Yeasts of wine fermentations from various regions of Italy. *American Journal of Enology and Viticulture*, **6**, 18–20.

Ciani, M. and Ferraro, L. (1998) Combined use of immobilized *Candida stellata* cells and *Saccharomyces cerevisiae* to improve the quality of wines. *Journal of Applied Microbiology*, **85**, 247–254.

Ciani, M., Comitini, F., Mannazzu, I. and Domizio P. (2010) Controlled mixed culture fermentation: A new perspective on the use of non-*Saccharomyces* yeasts in winemaking. *FEMS Yeast Research*, **10**, 123–133.

Comitini, F., Gobbi, M., Domizio, P. *et al.* (2011) Selected non-*Saccharomyces* wine yeasts in controlled multistarter fermentations with *Saccharomyces cerevisiae*. *Food Microbiology*, **28**, 873–882.

Cordente, A.G., Curtin, C.D., Varela, C. and Pretorius, I.S. (2012) Flavour-active wine yeasts. *Applied Microbiology and Biotechnology*, **96**, 601–618.

Coulon, J., Husnik, J.I., Inglis, D.L. *et al.* (2006) Metabolic engineering of *Saccharomyces cerevisiae* to minimize the production of ethyl carbamate in wine. *American Journal of Enology and Viticulture*, **57**, 113–124.

De Barros Lopes, M., Bellon, J.R., Shirley, N.J. and Ganter, P.F. (2002) Evidence for multiple interspecific hybridization in *Saccharomyces* sensu stricto species. *FEMS Yeast Research*, **1**, 323–331.

Domizio, P., Romani, C., Lencioni, L. *et al.* (2011) Outlining a future for non-*Saccharomyces* yeasts: Selection of putative spoilage wine strains to be used in association with *Saccharomyces cerevisiae* for grape juice fermentation. *International Journal of Food Microbiology*, **147**, 170–180.

Fleet, G.H. (2003) Yeast interactions and wine flavor. *International Journal of Food Microbiology*, **86**, 11–22.

Fleet, G.H. (2008) Wine yeasts for the future. *FEMS Yeast Research*, **8**, 979–995.

G-Alegría, E., López, I., Ruiz, J.I. *et al.* (2004) High tolerance of wild *Lactobacillus plantarum* and *Oenococcus oeni* strains to lyophilisation and stress environmental conditions of acid pH and ethanol. *FEMS Microbiology Letters*, **230**, 53–61.

Giudici, P., Solieri, L., Pulvirenti, A.M. and Cassanelli, S. (2005) Strategies and perspectives for genetic improvement of wine yeasts. *Applied Microbiology and Biotechnology*, **66**, 622–628.

Gobbi, M., Comitini, F., Domizio, P. *et al.* (2013) *Lachancea thermotolerans* and *Saccharomyces cerevisiae* in simultaneous and sequential co-fermentation: A strategy to enhance acidity and improve the overall quality of wine. *Food Microbiology*, **33**, 271–281.

Goffeau, A., Barrell, B.G., Bussey, H. *et al.* (1996) Life with 6000 genes. *Science*, **274**, 546–567.

Gonzalez, S.S., Barrio, E., Gafner, J. and Querol A. (2006) Natural hybrids from *Saccharomyces cerevisiae, Saccharomyces bayanus* and *Saccharomyces kudriavzevii* in wine fermentations. *FEMS Yeast Research*, **6**(8), 1221–1234.

Grossmann, M., Linsenmeyer, H., Muno, H. and Rapp, A. (1996) Use of olig-strain yeast cultures to increase complexity of wine aroma. *Viticulture and Enology Science*, **51**, 175–179.

Guerrini, S., Mangani, S., Granchi, L. and Vincenzini, M. (2002) Biogenic amine production by *Oenococcus oeni*. *Current Microbiology*, **44**, 374–378.

Henick-Kling, T. (1993) Malolactic fermentation. In G.H. Fleet (ed.), *Wine Microbiology and Biotechnology* (pp. 289–326), Chur: Harwood Academic.

Howell, K.S., Cozzolino, D., Bartowsky, E., Fleet, G.H. and Henschke, P.A. (2006) Metabolic profiling as a tool for revealing *Saccharomyces* interactions during wine fermentation. *FEMS Yeast Research*, **6**, 91–101.

Jolly, N.P., Augustyn, O.P.H. and Pretorius, I.S. (2006) The role and use of non-*Saccharomyces* yeasts in wine production. *South African Journal for Enology and Viticulture*, **27**, 15–38.

Jolly, N.P., Varela, C. and Pretorius, I.S. (2014) Not your ordinary yeast: Non-*Saccharomyces* yeasts in wine production uncovered. *FEMS Yeast Research*, **14**, 215–237.

Kim, D.H., Hong, Y.A. and Park, H.D. (2008) Co-fermentation of grape must by *Issatchenkia orientalis* and *Saccharomyces cerevisiae* reduces the malic acid content in wine. *Biotechnology Letters*, **30**, 1633–1638.

King, E.S., Swiegers, J.H., Travis, B., Francis, I.L., Bastian, S.E. and Pretorius, I.S. (2008) Coinoculated fermentations using *Saccharomyces* yeasts affect the volatile composition and sensory properties of *Vitis vinifera* L. cv. Sauvignon Blanc wines. *Journal of Agricultural and Food Chemistry*, **56**, 10829–10837.

Krieger-Weber, S. (2009) Application of yeast and bacteria as starter cultures. In H. König, G. Unden and J. Fröhlich (eds), *Biology of Microorganisms on Grapes, in Must and in Wine* (pp. 498–511), Berlin: Springer.

Kurtzman, C.P., Fell, W.F. and Boekhout, T. (2011) *The Yeasts: A Taxonomic Study* (5th edn), Amsterdam: Elsevier Science.

Lachance, M.A. and Kurtzman, C.P. (2011) Lachancea Kurtzman (2003). In C.P. Kurtzman, J.W. Fell and T. Boekhout, (eds), *The Yeasts: A Taxonomic Study* (5th edn), pp. 511–519, Amsterdam: Elsevier Science.

Larsen, J.T., Nielsen, J.C., Kramp, B. *et al.* (2003) Impact of different strains of *Saccharomyces cerevisiae* on malolactic fermentation by *Oenococcus oeni*. *American Journal of Enology and Viticulture*, **54**, 246–251.

Lerm, E., Engelbrecht, L. and du Toit, M. (2010) Malolactic fermentation: The ABC's of MLF. *South African Journal for Enology and Viticulture*, **31**(2), 1836–1212.

Liu, S.Q. (2002) A review: Malolactic fermentation in wine – beyond deacidification. *Journal of Applied Microbiology*, **92**, 589–601.

Lonvaud-Funel, A. (2001) Biogenic amines in wine: Role of lactic acid bacteria. *FEMS Microbiology Letters*, **199**, 9–13.

Loureiro, V. and Malfeito-Ferreira, M. (2003) Spoilage yeasts in the wine industry (review). *International Journal of Food Microbiology*, **86**, 23–50.

Mauriello, G., Capece, A., D'Auria, M., Garde-Cerdán, T. and Romano, P. (2009) SPME-GC method as a tool to differentiate VOC profiles in *Saccharomyces cerevisiae* wine yeasts. *Food Microbiology*, **26**, 246–252.

Moreira, N., Mendes, F., de Pinho, R.G., Hogg, T. and Vasconcelos, I. (2008) Heavy sulphur compounds, higher alcohols and esters production profile of *Hanseniaspora uvarum* and *Hanseniaspora guilliermondii* grown as pure and mixed cultures in grape must. *International Journal of Food Microbiology*, **124**, 231–238.

Perez-Nevado, F., Albergaria, H., Hogg, T. and Girio, F. (2006) Cellular death of two non-*Saccharomyces* wine-related yeasts during mixed fermentation with *Saccharomyces cerevisiae*. *International Journal of Food Microbiology*, **108**, 336–345.

Pizarro, F., Vargas, F.A. and Agosin, E. (2007) A systems biology perspective of wine fermentations. *Yeast*, **24**, 977–991.

Pretorius, I.S. (2000) Tailoring wine yeast for the new millennium: Novel approaches to the ancient art of winemaking. *Yeast*, **16**, 675–729.

Romano, P., Pietrafesa, R., Romaniello, R., Zambuto, M., Calabretti, A. and Capece, A. (2015) Impact of yeast starter formulations on the production of volatile compounds during wine fermentation. *Yeast*, **32**, 245–256.

Schuller, D. and Casal, M. (2005) The use of genetically modified *Saccharomyces cerevisiae* strains in the wine industry. *Applied Microbiology and Biotechnology*, **68**, 292–304.

Soubeyrand, V., Julien, A. and Sablayrolles, J.M. (2006) Rehydration protocols for active dry wine yeasts and search for early indicators of yeast activity. *American Journal of Enology and Viticulture*, **57**, 474–480.

Torriani, S., Felis, G.E. and Fracchetti, F. (2011) Selection criteria and tools for malolactic starters development: An update. *Annals of Microbiology*, **61**(1), 33–39.

Viana, F., Gil, J.V., Valles, S. and Manzanares, P. (2009) Increasing the levels of 2-phenylethyl acetate in wine through the use of a mixed culture of *Hanseniaspora osmophila* and *Saccharomyces cerevisiae*. *International Journal of Food Microbiology*, **135**, 68–74.

Volschenk, H., van Vuuren, H.J.J. and Viljoen-Bloom, M. (2006) Malic acid in wine: Origin, function and metabolism during vinification. *South African Journal for Enology and Viticulture*, **27**, 123–136.

Yap, N.A., de Barros Lopes, M., Langridge, P. and Henscke, P.A. (2000) The incidence of killer activity of non-*Saccharomyces* yeasts towards indigenous yeast species of grape must: Potential application in wine fermentation. *Journal of Applied Microbiology*, **89**, 381–389.

CHAPTER 14

Starter cultures in vegetables with special emphasis on table olives

Francisco Noé Arroyo-López, Antonio Garrido-Fernández and Rufino Jiménez-Díaz

Food Biotechnology Department, Instituto de la Grasa (CSIC), Spain

Fermented vegetables

Vegetables are fundamental in the human diet because they are sources of water-soluble vitamins, phytosterols, dietary fibre, phytochemicals and minerals (Gebbers 2007). Various works have encouraged the consumption of vegetables to prevent chronic pathologies such as hypertension, coronary heart diseases and the risk of stroke (Dauchet *et al.* 2007; He *et al.* 2007). Many vegetables are consumed as fresh or minimally processed foods (dried, canned, salads etc.). They tend to have short shelf lives and are prone to rapid microbial spoilage, and in some cases to contamination by pathogens. Pasteurization, cooking or the addition of preservatives are the main technological options that guarantee safe and high-quality vegetables, but these practices could induce certain undesirable changes in their physico-chemical properties (Zia-ur-Rehman *et al.* 2003; Zhang and Hamauzu 2004). In contrast, among the diverse technologies currently available, fermentation is still one of the oldest and cheapest methods to preserve vegetables. The production of fermented vegetables is lost in the mists of history and the processes have been adapted over generations to favour the most robust procedures. Therefore, fermentation is responsible for many favourable effects on manufactured vegetables, improving the flavour, shelf life, texture and safety of the final product, and also removing antinutritional components.

The processing of vegetables on a large scale has allowed for the establishment of important industrial activities around the world and also the development of starter cultures to produce improved products. The cucumber is one of the most popular fermented vegetables. Its worldwide production reached around 65,134,000 tons in the 2012 season (Helgi Library 2014), with China, the European Union (EU), Turkey and Iran as the main producers. It would be difficult to distinguish the proportion of production devoted to fermentation, but it should still be high (Helgi Library 2014). Germany, China, Russia and Ukraine

---

Starter Cultures in Food Production, First Edition. Edited by Barbara Speranza, Antonio Bevilacqua, Maria Rosaria Corbo and Milena Sinigaglia.

are in the top ten in the list of white cabbage–producing countries. A substantial proportion of white cabbage production is fermented as sauerkraut. A recent overview of the production and marketing of sauerkraut has estimated the annual production to be almost 58,000,000 tons, mainly in non-industrial settings. However, pasteurized sauerkraut is made industrially in Germany, Austria and France (Freitag 2012). Capers and caper berries are typical in the Mediterranean basin and their production goes from the Canary Islands and Morocco to the Caspian Sea and Iran. Their worldwide production is estimated at 10,000 tons/year (mainly in Turkey, Morocco and Spain). Another well-known fermented product is kimchi, which is one of the most popular ethnic fermented vegetables in Korea and other eastern countries (China, Japan etc.). The term is used to denote a group of fermented cabbage, radish and garlic foods. Due to its nutritional properties, kimchi was recently included in the list of the top five 'World's Healthiest Foods'. Its preparation is also progressively done in processing factories. In Korea alone, the production of kimchi may reach around 500,000 tons/year (Tamang 2012).

Therefore, many types of vegetable products, among them table olives, are produced by fermentation throughout the world, all of them sharing a general process that requires salting and acidification by lactic acid production. Salt, mainly sodium chloride (NaCl), has four primary roles in the preservation of vegetables: it determines the flavour of the final product; it has low water activity and consequently influences the type and extent of microbial metabolism; it indirectly helps to prevent the softening of vegetable tissue; and it assists with the breakdown of fruit membranes allowing the diffusion of components into the brines (sugars, vitamins, aminoacids etc.). The concentration of salt used varies widely for different vegetables, depending on the softening tendency of the vegetable during brine storage (Perez-Díaz *et al.* 2013).

The manufacturing of fermented vegetables has traditionally taken advantage of the beneficial microbiota that is spontaneously established in the raw material, mainly composed of lactic acid bacteria (LAB) (Fleming and McFeeters 1981; Chiu *et al.* 2008; Di Cagno *et al.* 2013) and yeasts (Arroyo-López *et al.* 2012c). If the fermentation proceeds to completion and good manufacturing practices are applied, sugars are completely converted into lactic acid, which drops the pH below 4.6, and prevents the growth of or even destroys spoilage and pathogens of public health significance (*Enterobacteriaceae*, *Clostridium* etc.). Microbial growth during the natural fermentation of brined vegetables may follow a sequential process characterized by four stages: initiation, primary fermentation, secondary fermentation and postfermentation. During initiation, the various Gram-positive and Gram-negative bacteria that colonize fresh vegetables or are present in the processing water and the fermentation environment compete for predominance. *Enterobacteriaceae*, aerobic spore-formers, LAB and other groups of bacteria and yeasts may be active for several days or even weeks. Then, the LAB population gains predominance, lowering the pH by the production of lactic acid through sugar consumption, and primary lactic acid fermentation occurs (Garrido-Fernández *et al.* 1997; Perez-Díaz *et al.* 2013). Yeasts are also present during fermentation and contribute to the sensory attributes of the final product, aiding in sugar consumption as well (Arroyo-López *et al.* 2012c). The secondary fermentation and postfermentation phases are not considered desirable. These phases begin at the end of lactic acid fermentation and can take place inside the fermentation or packaging containers. Spoilage microorganisms (oxidative yeasts, *Propionibacterium* etc.) can consume the lactic acid produced by LAB, which originates a pH rise with the consequent health or spoilage risk (Garrido-Fernández *et al.* 1997; Perez-Díaz *et al.* 2013).

In many cases, the fermentation of vegetables remains in its artisanal form and is not adequate for the utilization of starter cultures due to the high salt and acid levels used in preservation, the presence of inhibitory compounds or the low scale of processing (Ruiz-Barba *et al.* 1993; Pérez Pulido *et al.* 2012; Lanza 2013). In this chapter, only those vegetables for which the use of starter cultures has potential application will be mentioned, with a special emphasis on table olives.

Table olives

The olive fruit is a drupe. It has a bitter component (the glucoside oleuropein), a high oil content (10–30%) and moderate sugar concentration (2.6–6.0%), depending on variety and ripeness of fruits. These natural characteristics prevent olives from being consumed directly from the tree and promote diverse processes to make them eatable, which can slightly differ among regions (Garrido-Fernández *et al.* 1997). The Trade Standard Applying to Table Olives (IOC 2004) defines this food as 'the product obtained from suitable olive cultivars (*Olea europaea* var. *sativa*, L.), processed to remove their original bitterness, and preserved (by natural fermentation, heat treatment or preservatives) with or without brine until consumption'.

Table olives are a traditional fermented vegetable with many centuries of history, particularly in the Mediterranean basin, where this food has had a great influence on the culture and diet of many countries. The oldest written reference on the preparation of table olives is by Lucius Columela in his treatise *De Re Rustica*, in 54 BCE. The production of table olives originally started at a small domestic scale millennia ago and it was a craft process until the middle of the 20th century, when mass fermentation and storage systems were introduced together with pitting and stuffing automation. Progressive mechanization in the industry has led today to a rapid expansion of the sector and the search for starter cultures.

At present, the table olive is one of the major fermented vegetables, with an overall production above 2,400,000 tons/year. Most table olives are produced in the EU, with Spain, Greece and Italy as the main contributors. There are also important production quantities in Egypt, Turkey, Syria, Algeria, Morocco, Peru, the United States and Argentina. Thus, table olive processing occurs worldwide and represents an important economic source for olive-growing countries.

According to the International Olive Council (IOC), table olives can be classified as a function of the type of olive fruit used (green, turning colour and natural black olives) and the method of processing used. Processing alludes to the system used for fruit debittering, the way in which they are kept and the preservation procedure. The Trade Standard includes treated olives (fruits are debittered using lye treatments, followed by fermentation in brine); natural olives (fruits are brined directly and the debittering process occurs only by dilution or natural hydrolysis); dehydrated or shriveled olives (fruits, regardless of lye treatment or not, are subjected to dehydration by salt, heat or any other technological process); and olives darkened by oxidation (fruits are darkened by oxidation and preserved by sterilization). Therefore, the most important table olive industrial processing methods are alkali-treated green olives (the so-called Spanish style), which represents about 50–60% of production; ripe olives by alkaline oxidation (the so-called Californian style); and untreated or directly brined olives (green, turning colour or naturally black; Garrido-Fernández *et al.* 1997; Arroyo-López *et al.* 2012b). However, there are also many other traditional/industrial ways of processing table olives distributed around the world (Garrido-Fernández *et al.* 1997).

Among all table olive industrial preparations, the one most prone to the use of LAB starter cultures is the so-called green Spanish style, which is the only product that follows typical lactic acid fermentation. Ripe olives are only able to support microbial activity during prior storage in brine or any other solution, but their inoculation is of little interest in this case because of the later successive lye treatments, washing, conditioning and, mainly, sterilization, which destroy any microbial load on the fruits or brine. Consequently, the final product can hardly carry any favourable characteristic provided by any possible starter culture. In the case of directly brined olives (green or naturally black), inoculation with LAB also has limited interest because the environmental conditions (high salt and acid levels, presence of inhibitory compounds) are not adequate for the proper support of any LAB starter culture.

Application of starter cultures to table olive fermentation

The process of transforming the olive drupe into an edible food is the result of complex biochemical reactions that are influenced by interactions among the indigenous microbiota of the olives together with a variety of contaminating microorganisms from different industrial sources (pumps, fermenters, pipes, water etc.); the compositional characteristics of the fruits (mainly the presence of inhibitory compounds and availability of nutrients; and environmental factors (such as temperature, pH, the addition of organic acids or NaCl etc.). Traditionally, olive fermentation has occurred spontaneously, but the process is not fully predictable and can lead to spoilage in products with low quality or health risks (Lanza 2013). If good hygiene and technological practices are performed, table olives can be considered a safe product. However, suspected cases of intoxication caused by the growth of *Clostridium botulinum* type B in both green and black olives have been reported (Fenicia *et al.* 1992; Cawthorne *et al.* 2005). Also, the presence of *Listeria monocytogenes* was reported in one commercial (thermally treated) sample of green table olives (Caggia *et al.* 2004).

To prevent these problems, the fermentation processes can be controlled through physico-chemical (the addition of acids, salt, temperature management etc.) or microbiological approaches. To improve fermentation and consistently produce high-quality, safe final products, many authors have recommended strict process control of the parameters discussed in addition to the use of starter cultures (see Corsetti *et al.* 2012 for a complete review). LAB are classified as GRAS (generally recognized as safe) microorganisms and considered as the main microorganisms responsible for olive fermentation due to the production of lactic acid and bacteriocins, which originate the rapid and safe acidification of brines (Jiménez-Díaz *et al.* 1993; De Castro *et al.* 2002; Hurtado *et al.* 2012; Ruiz-Barba and Jiménez-Díaz 2012). The main genus of LAB isolated from table olive fermentations is *Lactobacillus*, but *Enterococcus*, *Pediococcus*, *Leuconostoc* and *Lactococcus* have also been isolated to a lesser extent. *Lactobacillus plantarum* and *Lactobacillus pentosus* are the predominant species in most fermentation processes but, depending on the olive cultivar, the processing method and the geographical origin, other lactobacilli or genera can play an essential role or even be the main species (Hurtado *et al.* 2012; Heperkan 2013).

For many years, the search for starters with application in olive fermentation and vegetables in general has in practice been strictly focused on the activity of LAB and their technological applications. However, several recent publications have emphasized the importance of the role that selected yeasts can play when used as starter cultures during table olive processing (Arroyo-López *et al.* 2012c; Bevilacqua *et al.* 2012). Although yeasts can sometimes cause spoilage of the product due to the production of carbon dioxide (CO_2), bad odours and flavours, the clouding of brines or softening of fruits, which is especially harmful in olive packaging or storage, they also shown some desirable features such as lipase, esterase, β-glucosidase or catalase activities, and have been used as bio-control agents or to improve LAB growth and the organoleptic profile of fruits (Arroyo-López *et al.* 2012c). These microorganisms could be especially effective in diverse olive preparations such as directly brined olives, where LAB growth is partially inhibited due to the presence of high concentrations of polyphenol compounds (Ruiz-Barba *et al.* 1993).

In the past, the selection of LAB starters in olive fermentation and vegetables in general has been exclusively based on diverse technological criteria, including homo-fermentative metabolism, high acidification rate and fast consumption of fermentable substrates, organic acids, polyphenols, high pH and salt tolerance, flavour development, wide temperature range for growth, oleuropein-splitting capability, minimum nutritional requirements and bacteriocin production (Duran-Quintana *et al.* 1999; Sánchez *et al.* 2001; Delgado *et al.* 2005; Corsetti *et al.* 2012; Hurtado *et al.* 2012; Di Cagno *et al.* 2013; Heperkan 2013). Another important characteristic of a starter culture must be its ability to dominate the indigenous microbiota as well as to be resistant to bacteriophages (Zago *et al.* 2013). Dominance of the starter culture would be exerted by its fast and predominant growth under fermentation conditions and/or its ability to produce antagonistic substances (in the case of LAB bacteriocins; Ruiz-Barba and Jiménez-Díaz 2012). In addition, for commercial purposes, it is necessary for starter cultures to resist freezing or freeze-drying processes.

Table 14.1 shows the main LAB and yeast starters selected from diverse table olive processing methods according to their technological traits.

New challenges for the development of starter cultures in table olives and vegetables in general

A great opportunity for the development of non-dairy probiotic products has arisen from the demands of people who are lactose intolerant, and also because of the high cholesterol levels of dairy foods (Granato *et al.* 2010). The increasing demand for vegetarian products poses an extra challenge to the food industry in its effort to manufacture high-quality and functional foods (Heenan *et al.* 2004). This opportunity for diversification of the sources of beneficial microorganisms includes such traditional non-dairy raw materials as cereals, legumes, fruits and vegetables (Luckow and Delahunty 2004); following appropriate technological handling, they may also serve as probiotic carriers (Mattila-Sandholm *et al.* 2002). Unfortunately, scarce knowledge is available of the behaviour of vegetables, specifically in terms of the delivery of probiotic strains (Lavermicocca *et al.* 2005; Peres *et al.* 2012).

Recently, researchers have focused their attention on the study of LAB species with probiotic potential isolated from fermented vegetables (for complete reviews of this topic see Peres *et al.* 2012 and Di Cagno *et al.* 2013). In this regard, authors have found putative probiotic LAB strains in sauerkraut, kimchi, york cabbage, carrot, cauliflower and fresh beans (Chang *et al.* 2010; Beganović *et al.* 2011; Lee *et al.* 2011; Jaiswal *et al.* 2012; Vitali *et al.* 2012; Xiong *et al.* 2012). These findings suggest that specific LAB strains collected from spontaneous plant fermentations and raw material hold great promise as probiotics (Di Cagno *et al.* 2013). Moreover, it has been suggested that the survival rates of LAB probiotic strains in

Table 14.1 Studies carried out to obtain or validate starter cultures in table olives based exclusively on technological features.

Microbial group	Microorganisms	Type of elaboration	Reference
Yeast	*S. cerevisiae*	Directly brined olives	Papoff *et al.* (1996)
	D. hansenii	Greek olive juice	Tsapatsaris and Kotzekidou (2004)
	W. anomalus, K. marxianus, S. cerevisiae, C. maris	Portuguese seasoned olives	Hernández *et al.* (2007)
	C. guilliermondi, C. famata	Italian NaOH-treated olives	Bevilacqua *et al.* (2009)
	W. anomalus	Spanish directly brined and NaOH-treated olives	Bautista-Gallego *et al.* (2011)
	W. anomalus, C. boidinii, C. diddensiae	Spanish directly brined and NaOH-treated olives	Rodriguez-Gómez *et al.* (2012)
	S. cerevisiae	Italian natural black olives	Pistarino *et al.* (2012)
	W. anomalus, K. lactis, C. norvegica	Italian NaoH-treated olives	Bevilacqua *et al.* (2013)
	P. galeiformis, S. cerevisiae, W. anomalus	Italian directly brined olives	Tofalo *et al.* (2013)
LAB	*Lactobacilli sp.*	Spanish-style olives	De la Borbolla y Alcalá *et al.* (1964)
	Strep. thermophilus, Lb. bulgaricus	Spanish-style olives	Balatsouras *et al.* (1971)
	Lb. plantarum	Spanish-style olives	Pelagatti and Brighigna (1981)
	Lactobacilli sp.	Spanish-style olives	Roig and Hernández (1991)
	Lb. plantarum	Spanish-style olives	Ruiz Barba *et al.* (1994)
	Lb. plantarum, Strep. faecium	Directly brined olives	Papoff *et al.* (1996)
	Lb. plantarum	Spanish-style olives	Duran-Quintana *et al.* (1999)
	Lb. pentosus	Spanish-style olives	Sánchez *et al.* (2001)
	Lb. pentosus, Ent. casseliflavus	Spanish-style olives	De Castro *et al.* (2002)
	Lb. plantarum	Spanish-style olives	Leal-Sánchez *et al.* (2003)
	Lb. plantarum	Greek olive juice	Tsapatsaris and Kotzekidou (2004)
	Lb. plantarum, Lb. paracasei	Italian directly brined olives	Randazzo *et al.* (2004)
	Lb. plantarum	Spanish-style olives	Mokhbi *et al.* (2009)
	Lb. plantarum	Spanish-style olives	Perricone *et al.* (2010)
	Lb. pentosus	Spanish-style olives	Bevilacqua *et al.* (2010)
	Lb. pentosus	Spanish directly brined olives	Hurtado *et al.* (2010)
	Lb. plantarum	Italian natural black olives	Pistarino *et al.* (2012)
	Lb. pentosus	Spanish-style olives	Aponte *et al.* (2012)

Notes: C. = *Candida; D.* = *Debaryomyces; Ent.* = *Enterococcus; K.* = *Kluyveromyces; Lb.* = *Lactobacillus;* NaOH = sodium hydroxide; *P.* = *Pichia; S.* = *Saccharomyces; Strep.* = *Streptococcus; W.* = *Wickerhamomyces.*

fermented plant-based materials (e.g. artichokes and table olives) are comparable to, or even higher than, those of milk-originated probiotics (Lavermicocca *et al.* 2005; Ranadheera *et al.* 2010; Arroyo-López *et al.* 2014). This can be attributed to the composition of the microorganism, since its cell wall is forced to become more solid and thicker in order to allow adaptation to the harsh environmental conditions prevailing in raw materials (the presence of antibacterial compounds and high osmotic pressure, poor nutrient profile etc.; Masuda *et al.* 2010). Additionally, the microarchitecture of the vegetable surface (e.g. its roughness) and the presence of natural prebiotic compounds (e.g. oligosaccharides) in plant materials are likely to contribute to cell protection, and thus to improve the survival of bacteria (Ranadheera *et al.* 2010). Bear in mind that plant tissues are multiphase systems, with an intricate internal microstructure formed by cells, intercellular spaces, capillaries and pores. The edible portions of most vegetables are in fact composed of flesh parenchyma cells that can store nutrients and metabolites (Alzamora *et al.* 2005). Therefore, all the probiotic strains isolated from plants and vegetables, their intrinsic properties and chemical composition and the physical structure of the plant matrices themselves contribute to the efficacy of those matrices as probiotic carriers (Peres *et al.* 2012).

In recent years, several researchers have tried to turn olives into a delivery vehicle of probiotic microorganisms to the human body. As occurs in other vegetables, two main options may be pursued: the use of allochthonous/exogenous or autochthonous/native microorganisms (Di Cagno *et al.* 2013). In table olives, the first study in this direction was carried out by inoculating olives using exogenous lactobacilli strains with probiotic features as starters, as was the case of the human-origin strain *Lactobacillus paracasei* IMPC2.1 (Lavermicocca *et al.* 2005; De Bellis *et al.* 2010). This microorganism showed high adaptability to the fermentation process, adhering to the surface of the olives and being recovered from the faeces of people who were fed with olives fermented in the presence of this bacterium for ten days. Therefore, the Italian health ministry allowed its application as a starter culture in table olives. Various strains of *Lb. paracasei* have also been used for the fermentation of green olives, but always with bacteria exogenous to the olive fermentation process (Saravanos *et al.* 2008).

Great developments in this area have been obtained recently thanks to the European project ProBiolives (FP7-SME-243471). In the framework of this research project, diverse native LAB strains were isolated from the fermentation of different types of table olive elaborations and regions, and molecularly identified and characterized according to their potential probiotic characteristics such as resistance to gastric and pancreatic digestion, auto-aggregation ability, hydrophobicity, inhibition of pathogens and production of exopolysaccharides, among others (Argyri *et al.* 2013; Bautista-Gallego *et al.* 2013; Botta *et al.* 2014; Peres *et al.* 2014). Although the project was originally aimed at obtaining LAB strains from brine fermentations with probiotic characteristics that could potentially be used as starter cultures during fermentation, the results obtained have opened

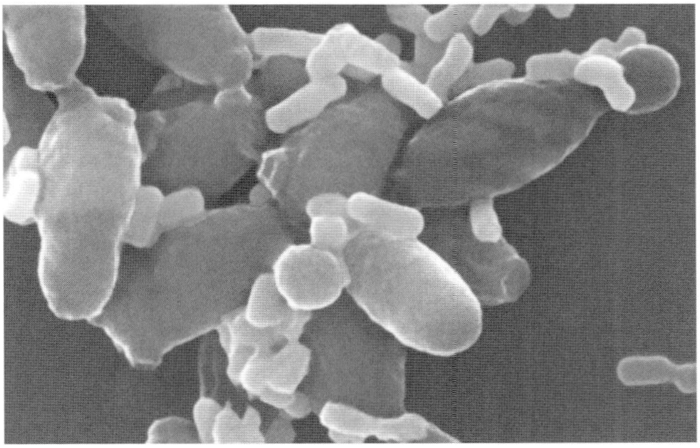

Figure 14.1 Mixed biofilms formed by yeasts and LAB obtained from fermented olives.

new and interesting research lines. In this regard, recent studies using scanning electron microscopy techniques have proven that the ability of fruits to act as a carrier was due to the formation of biofilm by microorganisms on the olive surface (Arroyo-López *et al.* 2012a; Domínguez-Manzano *et al.* 2012). This biofilm was a poly-microbial ecosystem, consisting essentially of lactobacilli and yeasts, which were attached to the epidermis of the fruits, forming a community characterized by the excretion of a protective adhesive extracellular matrix mainly composed of exopolysaccharide (Figure 14.1). These exopolysaccharides produced by native microorganisms and isolated from olives have also shown the ability to inhibit the adhesion of the enterohaemorrhagic strain *Escherichia coli* K88 to the intestinal mucosa of pigs (González-Ortiz *et al.* 2013). Therefore, it could have application in both animals and humans for the treatment of diarrhoea.

The study of mixed biofilms between yeast and LAB on the olive surface is a recent issue, with a direct application in starter development because it could make this fermented vegetable an excellent vehicle of beneficial microorganisms to the final consumers. Recent studies conducted with table olives found that the major species found in these mixed biofilms during fermentation were *Lb. pentosus* among LAB, and *Pichia galeiformis, Candida sorbosa, Geotrichum candidum, Saccharomyces cerevisiae* and *Wickerhamomyces anomalus* among yeasts, reaching population levels on the olives of up to 8 $\log_{10}$ colony forming units (cfu) per gram (Arroyo-López *et al.* 2012a; Domínguez-Manzano *et al.* 2012). Moreover, these microorganisms are able to survive for long periods of time in olive packing without a cold chain (Rodriguez-Gómez *et al.* 2014b). However, current knowledge of the genes regulating the different steps during biofilm formation, the interaction among microorganisms, population dynamics and the composition of the matrix surrounding the microorganisms is still scarce, despite the important role that these yeast and LAB species play during table olive processing.

Table 14.2 Studies carried out to obtain or validate starter cultures in table olives based on their multifunctional features (both technological and probiotic characteristics).

Microbial group	Microorganisms	Native to olives	Type of elaboration	Reference
LAB	*Lb. paracasei*, *Lb. rhamnosus*, *Bifidobacterium*	No	Italian black and green pitted olives	Lavermicocca *et al.* (2005)
	Lb. paracasei	No	Italian NaOH-treated green olives	De Bellis *et al.* (2010)
	Lb. pentosus	Yes	Spanish NaOH-treated olives	Bevilacqua *et al.* (2010)
	Lb. pentosus	Yes	Spanish directly brined olives	Abriouel *et al.* (2012)
	Lb. pentosus	Yes	Spanish directly brined and NaOH-treated olives	Bautista-Gallego *et al.* (2013)
	Lb. pentosus, *Lb. plantarum*, *Lb. paracasei*	Yes	Greek natural table olives	Argyri *et al.* (2013)
	Lb. pentosus, *Lb. plantarum*	Yes	Spanish NaOH-treated olives	Argyri *et al.* (2014)
	Lb. plantarum	Yes	Italian NaOH-treated olives	Botta *et al.* (2014)
	Lb. pentosus	Yes	Spanish NaOH-treated olives	Rodriguez-Gómez *et al.* (2014a)
	Lb. rhamnosus	No	Italian directly brined olives	Randazzo *et al.* (2014)
	Lb. plantarum, *Lb. paraplantarum*	Yes	Portuguese directly brined olives	Peres *et al.* (2014)
	Lb. pentosus, *Lb. plantarum*	Yes	Greek NaoH-treated olives	Blana *et al.* (2014)
Yeast	*T. delbrueckii*, *D. hansenii*	Yes	Greek-style black olives	Psani and Kotzekidou (2006)
	P. fermentans, *C. oleophila*	Yes	Portuguese directly brined olives	Silva *et al.* (2011)
	P. guilliermondii	Yes	Greek natural black olives	Bonatsou *et al.* (2015)

Notes: *C.* = *Candida*; *D.* = Debaryomyces; *Lb.* = *Lactobacillus*; NaOH = sodium hydroxide; *P.* = *Pichia*; *T.* = *Torulaspora*.

In addition to technological characteristics, further studies on the development of starter cultures for table olives must be directed towards the study of the probiotic potential of native microorganisms from table olives, with special attention paid to determining the following characteristics: degradation of cholesterol; phytase activity; inhibition and exclusion of pathogens; resistance to gastric and pancreatic digestion; antioxidant activity; immunostimulatory activity; auto-aggregation and hydrophobicity; biofortification with folates and vitamins; biodegradation/bioabsorption of toxic compounds (biogenic amines, mycotoxins etc.); ability to form biofilms; lack of antibiotic resistance; and production of exopolysaccharides. These studies must be focused on both LAB and yeasts for the development of a mixed multifunctional starter in order to improve and expand the mode of action of the starter by the use of two complementary microorganisms with different properties.

Table 14.2 shows, as a summary, the most promising microorganisms with probiotic potential currently isolated in table olive processing.

Acknowledgements

The authors wish to thank the Junta de Andalucía Regional Government, Spanish Government and the EU's Seventh Framework for funding their research through projects AGR7755 (PrediAlo: www.predialo.science.com.es), AGL2013-48300-R (OliFilm: www.olifilm.science.com.es) and FP7/2007-2013 grant agreement n°243471 (PROBIOLIVES: www.probiolives.eu), respectively. FNAL wishes also to thank the Spanish Government and CSIC for his Ramón y Cajal postdoctoral research contract.

References

Abriouel, H., Benomar, N., Cobo, A. *et al.* (2012) Characterization of lactic acid bacteria from naturally-fermented Manzanilla Aloreña green table olives. *Food Microbiology*, **32**, 308–316.

Alzamora, S.M., Salvatori, D., Tapia, M.S., López-Malo, A., Welti-Chanes, J. and Fito, P. (2005) Novel functional foods from vegetable matrices impregnated with biologically active compounds. *Journal of Food Engineering*, **67**, 205–214.

Aponte, M., Blaiotta, G., La Croce, F. *et al.* (2012) Use of selected autochthonous lactic acid bacteria for Spanish-style table olive fermentation. *Food Microbiology*, **30**, 8–16.

Argyri, A.A., Nisiotou, A.A., Malauchos, A. and Tassou, C.C. (2014) Performance of two potential probiotic *Lactobacillus* strains from the olive microbiota as starters in the fermentation of heat shocked green olives. *International Journal of Food Microbiology*, **171**, 68–76.

Argyri, A.A., Zoumpopoulou, G., Karatzas, K.A. *et al.* (2013) Selection of potential probiotic lactic acid bacteria from fermented olives by in vitro tests. *Food Microbiology*, **33**, 282–291.

Arroyo-López, F.N., Bautista-Gallego, J., Domínguez-Manzano, J. *et al.* (2012a) Formation of lactic acid bacteria-yeasts communities on the olive surface during Spanish-style Manzanilla fermentations. *Food Microbiology*, **32**, 295–301.

Arroyo-López, F.N., Bautista-Gallego, J., Romero-Gil, V. *et al.* (2012b) Fermentation of olive fruit. In Y.H. Hui (ed.), *Handbook of Plant-Based Fermented Food and Beverage Technology* (pp. 307–326), Boca Raton, FL: CRC Press.

Arroyo-López, F.N., Romero-Gil, V., Bautista-Gallego, J. and Garrido-Fernández, A. (2012c) Yeasts in table olive processing: Desirable or spoilage microorganisms. *International Journal of Food Microbiology*, **160**, 42–49.

Arroyo-López, F.N., Blanquet-Diot, S., Denis, S. *et al.* (2014) Survival of pathogenic and lactobacillus species of fermented olives during simulated human digestions. *Frontiers Microbiology*, **5**, Art 540.

Balatsouras, G., Papamichael, V. and El Din, N. (1971) Fermentation of green olives by thermophilic lactic acid bacteria at a temperature of 37–40 °C. *Grasas y Aceites*, **22**, 357–366.

Bautista-Gallego, J., Rodríguez-Gómez, F., Barrio, E. and Arroyo-López, F.N. (2011) Exploring the yeast biodiversity of green table olive industrial fermentations for technological applications. *International Journal of Food Microbiology*, **147**, 89–96.

Bautista-Gallego, J., Arroyo-López, F.N., Rantsiou, K. and Cocolin, L. (2013) Screening of lactic acid bacteria isolated from fermented table olives with probiotic potential. *Food Research International*, **50**, 135–142.

Beganović, J., Leboš Pavunc, A., Gjuračić, K., Spoljarec, M., Šušković, J. and Kos, B. (2011) Improved sauerkraut production with probiotic strain *Lactobacillus plantarum* L4 and *Leuconosctoc mesenteroides* LMG 7954. *Journal of Food Science*, **76**, M124–M129.

Bevilacqua, A., Perricone, M., Cannarsi, M., Corbo, M.R. and Sinigaglia, M. (2009) Technological and spoiling characteristics of the yeast microflora isolated from Bella di Cerignola table olives. *International Journal of Food Science and Technology*, **44**, 2198–2207.

Bevilacqua, A., Altieri, C., Corbo., M.R., Sinigaglia, M. and Ouoba, L.I. (2010) Characterization of lactic acid bacteria isolated from Italian Bella di Cerignola table olives: Selection of potential multifunctional starter cultures. *Journal of Food Science*, **75**, M536–M544.

Bevilacqua, A., Corbo, M.R. and Sinigaglia, M. (2012) Selection of yeasts as starter cultures for table olives: A step-by-step procedure. *Frontiers Microbiology*, **3**, Art. 194.

Bevilacqua, A., Beneduce, L., Sinigaglia, M. and Corbo, M.R. (2013) Selection of yeasts as starter cultures for table olives. *Journal of Food Science*, **78**, M742–M751.

Blana, V., Grounta, A., Tassou, C.C., Nychas, G.J. and Panagou, E.Z. (2014) Inoculated fermentation of green olives with potential probiotic *Lactobacillus pentosus* and *Lactobacillus plantarum* starter cultures isolated from industrially fermented olives. *Food Microbiology*, **38**, 208–218.

Bonatsou, S., Benítez, A., Rodríguez-Gómez, F., Panagou, E.Z. and Arroyo-López, F.N. (2015) Selection of yeasts with multifunctional features for application as starter in natural black table olive processing. *Food Microbiology*, **46**, 66–73.

Botta, C., Langerholc, T., Cencič, A. and Cocolin, L. (2014) In vitro selection and characterization of new probiotic candidates from table olive microbiota. *PLoS ONE*, **9**(4), e94457.

Caggia, C., Randazzo, C.L., Di Salvo, M., Romeo, F. and Giudici, P. (2004) Occurrence of *Listeria monocytogenes* in green table olives. *Journal of Food Protection*, **67**, 2189–2194.

Cawthorne, A., Celentano, L.P., D'Anconce, F. *et al.* (2005) Botulism and preserved green olives. *Emerging Infectious Diseases*, **11**, 781–782.

Chang, J.H., Shim, Y.Y., Cha, S.K. and Chee, K.M. (2010) Probiotic characteristics of lactic acid bacteria isolated from kimchi. *Journal of Applied Microbiology*, **106**, 220–230.

Chiu, H.H., Tsai, C.C., Hsih, H.Y. and Tsen, H.Y. (2008) Screening from pickled vegetables the potential probiotic strains of lactic acid bacteria able to inhibit the *Salmonella* invasion in mice. *Journal of Applied Microbiology*, **104**, 605–612.

Columela, L.J.M. (54) *De Re Rustica*, vol **II**. Santander: Nestlé.

Corsetti, A., Perpetuini, G., Schirone, M., Tofalo, R. and Suzzi, G. (2012) Application of starter cultures to table olive fermentation: An overview on the experimental studies. *Frontiers Microbiology*, **3**, Art. 248.

Dauchet, L., Kesse-Guyot, E., Czernichow, S. *et al.* (2007) Dietary patterns and blood pressure change over 5-y follow-up in the SU.VI.MAX cohort. *American Journal Clinical and Nutrition*, **85**, 1650–1656.

De Bellis, P., Valerio, F., Sisto, A., Lonigro, S.L. and Lavermicocca, P. (2010) Probiotic table olives: Microbial populations adhering on olive surface in fermentation sets inoculated with the probiotic strain *Lactobacillus paracasei* IMPC2.1 in an industrial plant. *International Journal of Food Microbiology*, **140**, 6–13.

De Castro, A., Montaño, A., Casado, F.J. and Rejano, L. (2002) Utilization of *Enterococcus casseliflavus* and *Lactobacillus pentosus* as starter cultures for Spanish-style green olive fermentation. *Food Microbiology*, **19**, 637–644.

De la Borbolla y Alcalá, J.R., Fernández Díez, M.J. and González Cancho, F. (1964) Emploi de cultures pures de lacto-bacilles pour la préparation des olives vertes. *Informations Oleicoles International*, **28**, 119–129.

Delgado, A., Brito, D., Peres, C. and Garrido-Fernández, A. (2005) Bacteriocin production by *Lactobacillus pentosus* B96 can be expressed as a function of temperature and NaCl concentration. *Food Microbiology*, **22**, 521–528.

Di Cagno, R., Coda, R., De Angelis, M. and Gobbetti, M. (2013) Exploitation of vegetables and fruits through lactic acid fermentation. *Food Microbiology*, **33**, 1–10.

Domínguez-Manzano, J., Olmo-Ruiz, C., Bautista-Gallego, J. *et al.* (2012) Biofilm formation on abiotic and biotic surfaces during Spanish style green table olive fermentation. *International Journal of Food Microbiology*, **157**, 230–238.

Duran-Quintana, M.C., García-García, P. and Garrido-Fernández, A. (1999) Establishment of conditions for green table olive fermentation at low temperature. *International Journal of Food Microbiology*, **51**, 133–143.

Fenicia, L., Ferrini, A.M. and Aureli, P. (1992) Epidemia di botulismo da olive nere. *Industrie Alimentari*, **31**, 55–58.

Fleming, H.P. and McFeeters, R.F. (1981) Use of microbial cultures: Vegetable products. *Food Technology*, **35**, 84–88.

Freitag, C. (2012) Overview about production and market of sauerkraut. Presentation for the 4th International Kimchi Conference, Washington, DC.

Garrido-Fernández, A., Fernández-Díez, M.J. and Adams, R.M. (1997) *Table Olives Production and Processing*, London: Chapman and Hall.

Gebbers, J.O. (2007) Atherosclerosis, cholesterol, nutrition, and statins: A critical review. *German Medical Science*, **5**, 1–11.

González-Ortiz, G., Pérez, J.F., Hermes, R.G., Molist, F., Jiménez-Díaz, R. and Martín-Orúe, S.M. (2013) Screening the ability of natural feed ingredients to interfere with the adherence of enterotoxigenic *Escherichia coli* (ETEC) K88 to the porcine intestinal mucus. *British Journal of Nutrition*, **11**, 633–642.

Granato, D., Branco, G.F., Nazzaro, F., Cruz, A.G. and Faria, J.A.F. (2010) Functional foods and nondairy probiotic food development: Trends, concepts, and products. *Comprehensive Reviews in Food Science and Food Safety*, **9**, 292–302.

He, F.J., Nowson, C.A., Lucas, M. and MacGregor, G.A. (2007) Increased consumption of fruit and vegetables is related to a reduced risk of coronary heart disease: Metaanalysis of cohort studies. *Journal Human Hypertension*, **21**, 717–728.

Heenan, C.N., Adams, M.C., Hosken, R.W. and Fleet, G. (2004) Survival and sensory acceptability of probiotic microorganisms in a non-fermented frozen vegetarian dessert. *LWT – Food Science and Technology*, **37**, 461–466.

Helgi Library (2014) FAOSTAT. *Cucumber and gherkin production.* http://www.helgilibrary.com/indicators/index/cucumber-and-gherkin-production (accessed October 2014).

Heperkan, D. (2013) Microbiota of table olive fermentation and criteria of selection for their use as starters. *Frontiers Microbiology*, **4**, Art. 143.

Hernández, A., Martin, A., Aranda, E., Pérez-Nevado, F. and Córdoba, M.G. (2007) Identification and characterization of yeast isolated from the elaboration of seasoned green table olives. *Food Microbiology*, **24**, 346–351.

Hurtado, A., Reguant, C., Bordons, A. and Rozès, N. (2010) Evaluation of a single and combined inoculation of *Lactobacillus pentosus* starter for processing cv. Arbequina natural green olives. *Food Microbiology*, **27**, 731–740.

Hurtado, A., Reguant, C., Bordons, A. and Rozès, N. (2012) Lactic acid bacteria from fermented olives. *Food Microbiology*, **31**, 1–8.

IOC (International Olive Council) (2004) Trade standard applying to table olives. COI/NC n° 1, December 2004, Madrid, Spain.

Jaiswal, A., Gupta, S. and Abu-Ghannam, N. (2012) Optimization of lactic acid fermentation of York cabbage for the development of potential probiotic products. *International Journal of Food Science and Technology*, **47**, 1605–1612.

Jiménez-Díaz, R., Ríos-Sánchez, R.M., Desmazeaud, M., Ruiz-Barba, J.L. and Piard, J.C. (1993) Plantaricins S and T, two new bacteriocins produced by *Lactobacillus plantarum* LPCO10 isolated from a green olive fermentation. *Applied and Environmental Microbiology*, **59**, 1416–1424.

Lanza, B. (2013) Abnormal fermentation in table-olive processing: Microbial origin and sensory evaluation. *Frontiers Microbiology*, **4**, Art. 91.

Lavermicocca, P., Valerio, F., Lonigro, S.L. *et al.* (2005) Study of the adhesion and survival of Lactobacilli and Bifidobacteria on table olives with the aim of formulating a new probiotic food. *Applied and Environmental Microbiology*, **71**, 4233–4240.

Leal-Sánchez, M.V., Ruiz-Barba, J.L., Sánchez, A.H. and Garrido-Fernández, A. (2003) Fermentation profile and optimization of green olive fermentation using *Lactobacillus plantarum* LPCO10 as a starter culture. *Food Microbiology*, **20**, 421–430.

Lee, H., Yoon, H., Ji, Y. *et al.* (2011) Functional properties of *Lactobacillus* strains from kimchi. *International Journal of Food Microbiology*, **145**, 155–161.

Luckow, T. and Delahunty, C. (2004) Which juice is healthier? A consumer study of probiotic non-dairy juice drinks. *Food Quality Preferences*, **15**, 751–759.

Masuda, T., Kimura, M., Okada, S. and Yasui, H. (2010) *Pediococcus pentosaceus* Sn26 inhibits IgE production and the occurrence of ovalbumin-induced allergic diarrhoea in mice. *Bioscience Biotechology Biochemistry*, **74**, 329–335.

Mattila-Sandholm, T., Myllarinen, P., Crittenden, R., Mogensen, G., Fondén, R. and Saarela, M. (2002) Technological challenges for future probiotic foods. *International Dairy Journal*, **12**, 173–182.

Mokhbi, A., Kaid-Harche, M., Lamri, K., Rezki, M. and Kacem, M. (2009) Selection of *Lactobacillus plantarum* strains for their use as starter cultures in Algerian olive fermentations. *Grasas y Aceites*, **60**, 82–88.

Papoff, C.M., Agabbio, M., Vodret, A. and Farris, G.A. (1996) Influenza di combinazioni biotechologiche sulla qualitá sensoriale di olive verdi da mensa. *Industri Alimentari*, **35**, 375–381.

Pelagatti, O. and Brighigna, A. (1981) Rivitalizzazione e impiego di fermenti lattice essicati por la preparazione delle olive da tavola col sistema Sivibliano. *Industri Alimentari*, March, 182–184.

Peres, C.M., Alves, M., Hernandez-Mendoza, A. and Malcata, F.X. (2014) Novel isolates of lactobacilli from fermented Portuguese olive as potential probiotics. *LWT – Food Science and Technology*, **59**, 234–246.

Peres, C.M., Peres, C., Hernández-Mendoza, A. and Malcata, F.X. (2012) Review on fermented plant materials as carriers and sources of potentially probiotic lactic acid bacteria with an emphasis on table olives. *Trends Food Science and Technology*, **26**, 31–42.

Perez-Díaz, I.M., Breidt, F., Buescher, R.W. *et al.* (2013) Fermented and acidified vegetables. In P. Downes and K. Ito (eds), *Compendium of Methods for the Microbiological Examination of Foods* (4th edn, pp. 521–532), Washington, DC: American Public Health Association.

Pérez Pulido, R., Benomar, N. and Martínez Cañamero, M. (2012) Fermentation of caper products. In Y.H. Hui and E. Özgül Evranuz (eds), *Handbook of Plant Based Fermented Food and Beverage Technology* (2nd edn, pp. 201–207), Boca Raton, FL: CRC Press.

Perricone, M., Bevilacqua, A., Corbo, M.R. and Sinigaglia, M. (2010) Use of *Lactobacillus plantarum* and glucose to control the fermentation of 'Bella di Cereignola' table olives, a traditional variety of Apulian region (southern Italy). *Journal of Food Science*, **75**, M430–M436.

Pistarino, E., Aliakbarian, B., Casazza, A.A., Paini, M., Cosulich, M.E. and Perego, P. (2012) Combined effect of starter culture and temperature on phenolic compounds during fermentation of Taggiasca black olives. *Food Chemistry*, **138**, 2043–2049.

Psani, M. and Kotzekidou, P. (2006) Technological characteristics of yeast strains and their potential as starter adjuncts in Greek-style black olive fermentation. *World Journal of Microbiology and Biotechnology*, **22**, 1329–1336.

Ranadheera, R.D.C.S., Baines, S.K. and Adams, M.C. (2010) Importance of food in probiotic efficacy. *Food Research International*, **43**, 1–7.

Randazzo, C.L., Restuccia, C., Romano, A.D. and Caggia, C. (2004) *Lactobacillus casei*, dominant species in naturally fermented Sicilian green olives. *International Journal of Food Microbiology*, **90**, 9–14.

Randazzo, C.L., Todaro, A., Pinoa, A. *et al.* (2014) Giarraffa and Grossa di Spagna naturally fermented table olives: Effect of starter and probiotic cultures on chemical, microbiological and sensory traits. *Food Research International*, **62**, 1154–1164.

Rodríguez-Gómez, F., Romero-Gil, V., Bautista-Gallego, J., Garrido-Fernández, A. and Arroyo-López, F.N. (2012) Multivariate analysis to discriminate yeast strains with technological applications in table olive processing. *World Journal of Microbiology and Biotechnology*, **28**, 1761–1770.

Rodriguez-Gómez, F., Romero-Gil, V., Bautista-Gallego, J., García-García, P., Garrido-Fernández, A. and Arroyo-López, F.N. (2014a) Production of potential probiotic Spanish-style Green table olives at pilot scale using multifunctional starters. *Food Microbiology*, **44**, 278-287.

Rodriguez-Gómez, F., Romero-Gil, V., Garcia-Garcia, P., Garrido-Fernández, A. and Arroyo-López, F.N. (2014b). Fortification of table olive packing with the potential probiotic bacteria *Lactobacillus pentosus* TOMC-LAB2. *Frontiers Microbiology*, **5**, Art. 467.

Roig, J.M. and Hernández, J.M. (1991) El uso de microorganismos iniciadores ('Starters') en la fermentación de aceitunas de mesa. *Olivae*, **37**, 20–28.

Ruiz-Barba, J.L., Brenes, M., Jiménez Díaz, R. and Garrido-Fernández, A. (1993) Inhibition of *Lactobacillus plantarum* by polyphenols extracted from two different kinds of olive brines. *Journal of Applied Bacteriology*, **74**, 15–19.

Ruiz-Barba, J.L., Cathcart, D.P., Warner, P.J. and Jiménez-Díaz, R. (1994) Use of *Lactobacillus plantarum* LPCO10, a bacteriocin producer, as a starter culture in Spanish-style green olive fermentations. *Applied and Environmental Microbiology*, **60**, 2059–2064.

Ruiz-Barba, J.L. and Jiménez-Díaz, R. (2012) A novel *Lactobacillus pentosus*-paired starter culture for Spanish-style green olive fermentation. *Food Microbiology*, **30**, 253–259.

Sánchez, A.H., Rejano, L., Montaño, A. and de Castro, A. (2001) Utilization at high pH of starter cultures of lactobacilli for Spahish-style green olive fermentation. *International Journal of Food Microbiology*, **67**, 115–122.

Saravanos, E., Kagli, D., Zoumpopoulou, G., Panagou, E. and Tassou, C. (2008) Use of probiotic lactic acid bacteria as starter cultures in Spanish-style green olive fermentation and determination of their survival using PFGE. *Food Microbiology Congress*, Aberdeen, 1–4 September.

Silva, T., Reto, M., Sol, M. *et al.* (2011) Characterization of yeasts from Portuguese brined olives, with a focus on their potentially probiotic behaviour. *LWT – Food Science and Technology*, **44**, 1349–1354.

Tamang, J.P. (2012) Plant-based fermented foods and beverages in Asia (kimchi). In Y.H. Hui and E. Özgül Evranuz (eds), *Handbook of Plant Based Fermented Food and Beverage Technology* (2nd edn, pp. 51–80), Boca Raton, FL: CRC Press.

Tofalo, R., Perpetuini, G., Schirone, M., Suzzi, G. and Corsetti, A. (2013) Yeast biota associated to naturally fermented table olives from different Italian cultivars. *International Journal of Food Microbiology*, **161**, 203–208.

Tsapatsaris, S. and Kotzekidou, P. (2004) Application of central composite design and response surface methodology to the fermentation of olive juice by *Lactobacillus plantarum* and *Debaryomyces hansenii*. *International Journal of Food Microbiology*, **95**, 157–168.

Vitali, B., Minervini, G., Rizzello, C.G. *et al.* (2012) Novel probiotic candidates for humans isolated from raw fruits and vegetables. *Food Microbiology*, **31**, 116–125.

Xiong, T., Song, S., Huang, X. *et al.* (2012). Screening and identification of functional *Lactobacillus* specific for vegetable fermentation. *Journal of Food Science*, **78**, M84–M89.

Zago, M., Lanza, B., Rossetti, L., Muzzalupo, I., Carminati, D. and Giraffa, G. (2013) Selection of *Lactobacillus plantarum* strains to use as starters in fermented table olives: Oleuropeinase activity and phage sensitivity. *Food Microbiology*, **34**, 81–87.

Zhang, D.L. and Hamauzu, Y. (2004) Phenolics, ascorbic acid, carotenoids and antioxidant activity of broccoli and their changes during conventional and microwave cooking. *Food Chemistry*, **88**, 503–509.

Zia-ur-Rehman, Z., Islam, M. and Shah, W.H. (2003) Effect of microwave and conventional cooking on insoluble dietary fibre components of vegetables. *Food Chemistry*, **80**, 237–240.

CHAPTER 15

New trends in dairy microbiology: Towards safe and healthy products

Ana Rodríguez, Beatriz Martínez, Pilar García, Patricia Ruas-Madiedo and Borja Sánchez

Instituto de Productos Lácteos de Asturias–Consejo Superior de Investigaciones Científicas (IPLA-CSIC), Spain

Biopreservation: Bacteriocins and bacteriophages

Biopreservation is based on the rational exploitation of antimicrobials produced by microorganisms with a long history of safe use in foods to extend shelf life and enhance safety. Among antimicrobial weapons, bacteriocins produced by lactic acid bacteria (LAB) have long been studied as natural food preservatives and nisin is used worldwide as a food additive. More recently, bacteriophages have been focused on as biocontrol agents in foods (García *et al.* 2008).

Bacteriocins produced by lactic acid bacteria

LAB have historically been involved in the fermentation of different food matrices (milk, meat, vegetables etc.), where they inhibit pathogenic and spoilage microorganisms, due to the production of organic acids (mainly lactic acid) and the concomitant pH reduction. Their inhibitory activity is enhanced by the production of bacteriocins, ribosomally synthesized antimicrobial peptides. As metabolites produced by LAB, bacteriocins also have GRAS (generally recognized as safe) status and become more attractive food preservatives than chemical ones for consumers who demand high-quality natural foods (Balciunas *et al.* 2013). Quite a few studies on bacteriocins provide information on their chemical, structural and genetic characteristics, mode of action and biotechnological applications. Nowadays, over 100 bacteriocins are known, most of them thermo-stable cationic peptides with a length of 20–100 amino acid residues and a simple structure (http://bactibase.pfba-lab-tun.org/main.php).

Bacteriocin-producing bacteria export these peptides across the cell membrane by a dedicated membrane-associated ATP-binding cassette (ABC) transporter (Havärstain *et al.* 1995), but some of them show a typical N-terminal sequence of a sec-dependent type and consequently are secreted through the

Starter Cultures in Food Production, First Edition. Edited by Barbara Speranza, Antonio Bevilacqua, Maria Rosaria Corbo and Milena Sinigaglia.

general secretory pathway (Worobo *et al.* 1995). To protect themselves against bacteriocin action, bacteriocinogenic strains produce dedicated immunity proteins. Bacteriocin biosynthesis is usually organized in operon clusters, encoding both production and immunity genes and located on plasmids, chromosomes and transposons (Sahl and Bierbaum 1998).

Different classifications of bacteriocins have been proposed by several authors over time based on biochemical, genetic and activity properties (Table 15.1). Bacteriocins were divided into four classes by Klaenhammer (1993): Class I (lantibiotics) (<5 kDa) that are modified following translation and show high thermostability. They harbour thioether-based intramolecular rings of lanthionine and ß-methyl-lanthionine. Among them, nisin (E-234), used as a food biopreservative in over 50 countries, lacticin 3147, lacticin 481 and plantaricin C are outstanding representatives. Class II (<10 kDa) is composed of non-modified cationic peptides with high isoelectric points. This class is also divided into IIa (pediocin-like), particularly relevant in food preservation because of its antilisteria activity; IIb (two-peptides), comprising two-component bacteriocins showing separate peptide for full activity (lactococcin G, lactacin F); and IIc (non-pediocin-like single peptide), including thiol-containing bacteriocins secreted through a

Table 15.1 Examples of bacteriocins produced by lactic acid bacteria.

Classification	Bacteriocin	Producer
Class I Lantibiotics	Nisin A and natural variants (Z, F, Q)[1]	*Lc. lactis*
	Lacticin 481	*Lc. lactis*
	Lacticin 3147 (two component)	*Lc. lactis*
	Plantaricin C	*Lb. plantarum*
	Lactocin S	*Lb. sakei*
Class II		
Pediocin-like (IIa)	Pediocin PA1/AcH	*Pe. acidilatici*
	Enterocin A	*Ent. faecium*
	Sakacin P	*Lb. sakei*
	Leucocin A	*Leuconostoc spp.*
Two-component (IIb)	Lactococcin G	*Lc. lactis*
	Lacticin F	*Lb. johnsonii*
	Plantaricin S	*Lb. plantarum*
Circular (IIc)	Enterocin AS-48	*Ent. faecalis/faecium*
	Gassericin A	*Lb. gasseri*
	Garvicin ML	*Lc. garviae*
Others (IId)	Lactococcin 972 (non-pore forming)	*Lc. lactis*
	Enterocins L50 (leaderless)	*Ent. faecium*
	Divergicin A (sec-dependent)	*Ca. divergens*
Class III (large)	Helveticin J	*Lb. helveticus*

Notes: [1] Nisin variants U and U2 are produced by *Streptococcus uberis* and *Streptococcus agalactiae*, respectively.
Ca.= *Carnobacterium*; *Ent.*= *Enterococcus*; *Lc.*= *Lactococcus*; *Lb.*= *Lactobacillus*; *Pe.* = *Pediococcus*.

general sec pathway that lack a leader sequence (lactocin B). Class III (>30 kDa) comprises heat-labile high molecular weight peptides, with very limited food preservation ability (helveticin J). Class IV includes complex bacteriocins carrying lipid or carbohydrate moieties. More recently, bacteriocin classification has been simplified and two major classes proposed: Class I (lantibiotics) and Class II (non-lantibiotics), in which four subclasses are included that match the previous IIa and IIb, while subclass IIc includes cyclic bacteriocins, characterized by the head-to-tail peptide bond (enterocin AS-48) and with potential application as food preservatives, and subclass IId is composed of miscellaneous bacteriocins (Cotter *et al.* 2005; Heng and Tagg 2006).

Regarding mode of action, most bacteriocins act as cell membrane permeabilizers due to the electrostatic interactions that occur between the antimicrobial peptides and the cell membrane (Figure 15.1). Formation of pores disrupts the proton motive force and causes the leakage of ions, resulting in cell death. It has also been reported that a single bacteriocin possesses more than one mode of action. This is the case for nisin and other lantibiotics (i.e. lacticin 3147) that target the membrane-bound cell wall precursor lipid II as a docking molecule to form further pores on bacterial membranes (Breukink *et al.* 1999). Thus, the ability to inhibit cell wall biosynthesis and pore formation is involved in their dual mechanism of action and allows inhibitory activity at nM concentrations (Wiedemann *et al.* 2001). Nisin also interacts with the cell wall teichoic acid (WTA) precursors (lipid III and lipid IV) to block WTA biosynthesis, as reported by Müller *et al.* (2012). Mannose-phosphotransferase system (man-PTS) components (i.e. EII permease) act as receptors for several Class II bacteriocins in sensitive cells to further provoke the dissipation of proton motive force by formation of pores in the cell membrane (Diep *et al.* 2007). The accumulation of hydroxyl radicals has been identified as the selective antimicrobial activity of lacticin Q (Li *et al.* 2013). Besides pore formation, other modes of action have been described. The non-pore-forming bacteriocin lactococcin 972 inhibits cell division by specific binding to lipid II at the division septum in lactococci (Martínez *et al.* 2008) and

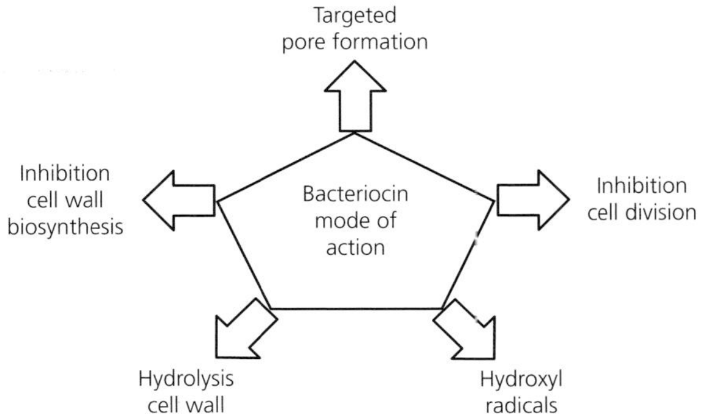

Figure 15.1 Mode of action of bacteriocins produced by lactic acid bacteria.

garvicin A may have a similar mode of action (Maldonado-Barragán *et al.* 2013). Moreover, some Class III bacteriocins (bacteriolysins) are endowed with peptidoglycan-degrading activities (Roces *et al.* 2012).

Preservation of dairy products with bacteriocins

Bacteriocins produced by LAB have several attributes that make them suitable as food biopreservatives: ability to inhibit Gram-positive pathogenic and spoilage bacteria (*Listeria monocytogenes, Staphylococcus aureus, Bacillus cereus, Clostridium botulinum, Clostridium tyrobutyricum* etc.); sensitivity to (human) digestive proteases; stability in a wide range of temperature and pH; not affecting the organoleptic properties of food as they have no taste, odour or colour; non-toxic to eukaryotic cells; and easy to produce at a large scale (Chen and Hoover 2003). Of note is that the spectrum of inhibition, stability, interaction with the food matrix or the specific activity of bacteriocins can be modified by genetic engineering. This is the case, for instance, for a nisin A derivative (glycine at position 29 was replaced by serine) with an extended inhibitory spectrum towards Gram-negative foodborne pathogens such as *Enterobacter sakazakii, Escherichia coli* and *Salmonella enterica* serovar. Tiphymurium (Field *et al.* 2012).

Bacteriocins can be used in combination with different methods of preservation (heat, chelating agents, carbon dioxide, biocides etc.) as part of hurdle technology to enhance their effectiveness. The combination of barriers improves the control of microbial proliferation. Among a number of examples compiled in the literature, some are focused on broadening the inhibition spectrum, on reducing the concentration of bacteriocins or on controlling the proliferation of sublethally injured cells (Gálvez *et al.* 2007). A synergistic effect against foodborne pathogens was shown when pediocin PA-1/AcH was combined with a carbon dioxide (CO_2) atmosphere (Nilsson *et al.* 2000), while nitrites enhance nisin activity against *Clostridium sporogenes* (Rayman *et al.* 1981). Inactivation of this microorganism in milk was enhanced by a combination of nisin and high pressure (Black *et al.* 2005; Arqués *et al.* 2005). Nisin was also combined with pulsed electrical fields to inactivate *Ls. monocytogenes* in skimmed milk (Calderón-Miranda *et al.* 1999), *Listeria innocua* in whey (Gallo *et al.* 2007) and *Staph. aureus* in milk (Sobrino-López and Martín-Belloso 2006). The cyclic bacteriocin AS-48 showed enhanced bactericidal activity against *Ls. monocytogenes* in combination with chemical preservatives, essential oils and natural bioactive compounds (Cobo-Molinos *et al.* 2009).

The incorporation of bacteriocins in packaging systems allows the slow release of these antimicrobials in food products, thereby extending their shelf life. The antimicrobial efficacy can be even higher than that resulting from the addition of bacteriocins to the food matrix. In this regard, the immobilization of nisin on polyamide and cellulose material was able to reduce *Staph. aureus* in cheese slices under cold conditions for three months (Scannell *et al.* 2000a) and in Cheddar cheese (Scannell *et al.* 2000b). Low-density polyethylene (LDPE) films coated with enterocin 416K1 showed antilisterial activity on cheeses during cold storage (Iseppi *et al.* 2008), or with nisin-inhibited *Micrococcus luteus* in raw milk (Mauriello *et al.* 2005), while soy-based films impregnated with nisin and lauric acid reduced up to 6 log units of *Ls. monocytogenes* counts on turkey bologna after 21 days at 4 °C (Dawson *et al.* 2002).

Bacteriocin producers may be used for *ex situ* production of bacteriocins, but culture ferments are also authorized as food ingredients in some countries. At present, several commercial fermentation products are available containing bacteriocins/small antimicrobial peptides along with organic acids, such as ALTA™ 2431 (pediocin PA-1/AcH, manufactured by Quest International, IL, USA), MICROGARD™ (manufactured by Dupont Nutrition Biosciences ApS, Denmark) and Bactoferm F-LC (pediocin and sakacin, manufactured by Christian Hansen A/S, Denmark). In addition, semi-purified preparations of Nisin A (Nisaplin® and Chrisin) are manufactured by Dupont Nutrition Biosciences ApS and Christian Hansen A/S, respectively, as food additives that show high efficacy in processed cheeses by controlling the outgrowth of spores from *Cl. botulinum* and *Cl. tyrobutyricum* (Nisaplin dosage 200–600 mg/kg=nisin 5–15 mg/kg), the development of *Ls. monocytogenes* in Ricotta cheese (Nisaplin 100–200 mg/kg=nisin 2.5–5 mg/kg) and in stirred yoghurt (Nisaplin 20–50 mg/kg=nisin 0.5–1.25 mg/kg) to avoid overacidification throughout the shelf life (Delves-Broughton 2007). Bacteriocins can also be applied in food by *in situ* production. This requires bacteriocinogenic strains (protective cultures) well adapted to the particular food matrix (i.e. milk) in such a way that they are able to grow and produce bacteriocin under processing, ripening and storage conditions (Gálvez *et al.* 2007). This system is cost effective and not affected by legal regulations.

Wild-type bacteriocin producers have been combined with starters to improve the safety of fermented dairy products. Nisin-producing strains of *Lactococcus lactis* inhibited *Ls. monocytogenes* in Camembert cheese (Maisnier-Patin *et al.* 1992), *Staph. aureus* in Afuega'l Pitu cheese (Rilla *et al.* 2004), and *Ls. innocua* in semi-hard cheese made with raw milk (Rodríguez *et al.* 2000). The strain *Lc. lactis* ssp. *lactis* IPLA 729 (Nisin Z producer) was able to inhibit *Cl. tyrobutyricum*, a late-blowing agent, in semi-hard Vidiago cheese (Rilla *et al.* 2003). Regarding lacticin 3147, a two-component lantibiotic, several commercial strains with improved technological properties have been produced by transferring the lacticin 3147-encoding conjugative plasmid pMRC01 to Cheddar cheese starter strains (Coakley *et al.* 1997). Lacticin 3147 producers inhibited *Ls. monocytogenes* in smear-surface cheese by spraying the bacteriocin producer on the cheese surface (O'Sullivan *et al.* 2006) and during cold storage of cottage cheese (McAuliffe *et al.* 1999). These bacteriocinogenic starters are also able to control cheese quality by inhibiting the growth of non-starter lactic acid bacteria (NSLAB) populations during ripening. Likewise, bacteriocin-producing cultures accelerate cheese ripening by increasing the lysis of the starter strains, with the consequent release of intracellular enzymes to the cheese matrix (Morgan *et al.* 2002). On the other hand, bacteriocinogenic enterococci have been used as adjunct cultures in both cheeses and dairy fermented beverages. *Enterococcus faecalis* AS 48-32, an AS-48 producer, clearly inhibited *Bc. cereus* growth in non-fat hard cheese throughout ripening (Muñoz *et al.* 2004). The inhibitory efficacy of *Ent. faecalis* AS 48-32 and *Enterococcus faecium* UJA32-81 (electro-transformant harbouring the plasmid encoding for AS 48), however, was weaker against *Staph. aureus* in unripened cheese (Muñoz *et al.* 2007). The use of *Ent. faecium* MMRA, an enterocin A producer, as a protective adjunct culture halted the growth of *Ls. monocytogenes*, but viability was hardly affected during cold storage of Rayeb (Rehaiem *et al.* 2012).

Bacteriophages

Bacteriophages (phages) are the natural enemies of bacteria, being harmless for humans, animals and plants. They are ubiquitous in nature and can be isolated in all the habitats colonized by bacteria, including foods. Indeed, they have been isolated from dairy products, meat products and fresh produce. They have a great advantage in being used as biocontrol agents due to their specificity. This means that is it possible to use preparations of phages not harmful for beneficial microbiota, such as starter cultures or commensal microbiota of the human gut. Phages are classified according their morphology, nucleic acid type and the presence/absence of envelope. So far, 17 different families have been described, most of them belonging to the order *Caudovirales* (tailed phages, double-strain DNA), divided into three families: *Siphoviridae* (long and flexible tails), *Myoviridae* (long, straight and contractile tails) and *Podoviridae* (very short tails). Other phages, either filamentous or pleomorphic with double-strand or single-strand DNA, and double-strand or single-strand RNA, have also been isolated (Ackermann 2007).

To proliferate, phages depend on the genome replication machinery of the host. In tailed phages, the first step of the life-cycle is adsorption mediated by tail proteins binding specific bacterial receptors (Figure 15.2). Phage genome is

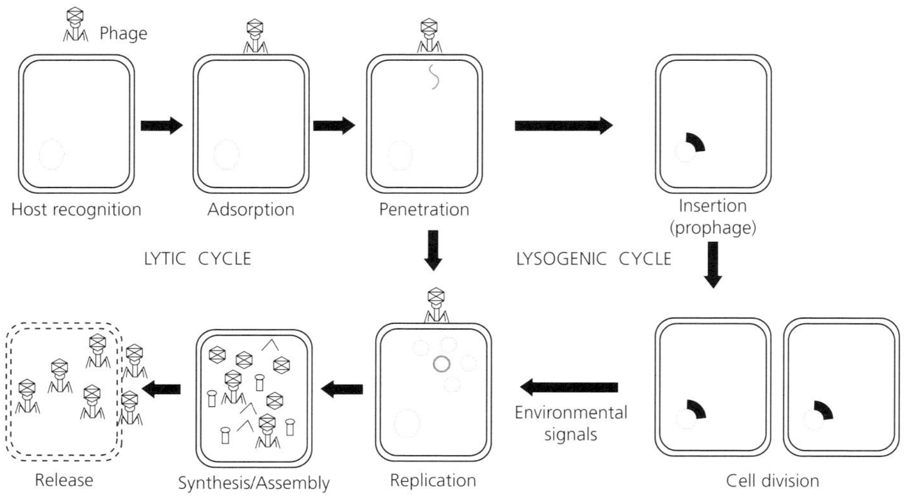

Figure 15.2 Life-cycles of bacteriophage (lytic and lysogenic).

thereafter injected into the cell. Genome replication occurs followed by the synthesis of viral structural proteins. After the assembling process, the new viral progeny is released through the action of the cell wall hydrolases (i.e. endolysins) and the consequent lysis of the cell host. For application purposes, the virulent phages, showing exclusively a lytic cycle, must be chosen. However, the phenomenon of lysogeny is observed in temperate phages in which the phage genome is integrated into the chromosome of the host cell (prophage), replicates along with the cell chromosome and is passed to the bacterial progeny. The lysogenic cells are resistant to infection by the same phage (superinfection immunity). Under certain environmental conditions, the prophage can be activated and starts the lytic cycle.

Application of bacteriophages as biocontrol agents and biopreservatives

Immediately after their discovery by Twort (1915) and D'Herelle (1917), phages were used in therapy to treat bacterial infections in humans, reaching a great degree of development in the former Soviet Union, although they declined in the West with the advent of antibiotics (Sulakvelidze *et al.* 2001). The increasing resistance to antibiotics of pathogenic microorganisms has renewed interest in bacteriophages as antimicrobials. Apart from their use in phage therapy, they can be exploited as food preservatives. A number of studies support the efficacy of bacteriophages in inhibiting the growth of pathogens in dairy foods, but also in meat products and ready-to-eat (RTE) foods. A mixture of the virulent phages phiIPLA35 and phiIPLA88, derived from their lysogenic counterparts isolated from mastitic milk, completely inhibited or reduced *Staph. aureus* in acid and enzymatic curds (García *et al.* 2007) and in cheeses (Bueno *et al.* 2012). These phages showed a synergistic effect in combination with high hydrostatic pressure in pasteurized raw milk subjected to a cold chain break (Tabla *et al.* 2012). However, the proliferation of *Staph. aureus*–infecting phage K in raw milk is significantly reduced in comparison with pasteurized milk, hence limiting its application in dairy products (O'Flaherty *et al.* 2005). Thus, it is necessary specifically to optimize the phage concentration for each food system to ensure passively diffusing phage virions with their host cells in order to get the complete elimination of pathogens in foods (Hagens and Offerhaus 2008). In addition, phage efficacy should be evaluated in each food system to guarantee proper performance in the food environment and under the particular processing conditions (García *et al.* 2008). Recently, the virulent *Myoviridae* phage SPW specific to *Staph. aureus* has been shown to lyse efficiently several strains of this pathogen, revealing its potential as a tool for prophylaxis and/or therapy in cattle suffering mastitis (Li and Zhang 2014). *En. sakazakii* (formerly *Cronobacter sakazakii*) was removed by phage in reconstituted infant formula milk (Kim *et al.* 2007). In addition, *Sal. enterica* serovar. Enteritidis was killed by phage SJ2 in both raw and pasteurized milk Cheddar cheese (Modi *et al.* 2001). The virulent phage A511 was used for controlling the surface contamination of Camembert and Limburger-type cheeses by *Ls. monocytogenes*. It was observed that the higher the phage dose, the higher the efficacy shown by the phage in different types of cheeses (Guenther and Loessner 2011). This supports the need to identify the concentration of phages that should be added to the food matrix to ensure the killing of a particular pathogen (Obeso *et al.* 2010).

In order to have a look at the possibility of avoiding the presence of pathogens in foods, the use of phages as therapeutic and prophylactic tools in livestock has been evaluated (Table 15.2). In these cases the data do not refer to dairy products, but there are a significant number of trials focused on zoonotic pathogens (*E. coli*, *Salmonella* ssp., *Campylobacter* ssp. and *Listeria* ssp.,) colonizing food-producing animals, which have been performed to reduce the microbial load of animals before the derived meat products were consumed (Atterbury 2009). At present, several bacteriophage-based products for use by the food industry have been commercialized. For instance, OmniLytics, Inc. (Sandy, UT, USA) has developed products suitable for external application to livestock (poultry, pigs, sheep and cattle), both prior to and after slaughtering, to reduce or eliminate contamination by *E. coli*, *Salmonella* and *Listeria* (http://www.omnilytics.com). IntraLytix, Inc. (Baltimore, MD, USA) has commercialized three phage-based products. ListShield™ focuses on reducing or eliminating the risk of listeriosis. In 2006, EcoShield™ was approved by the US Food and Drug Administration (FDA) as a food additive for RTE foods that is effective against *E. coli* O157:H7, reducing the contamination of different types of foods (fruits, vegetables and ground beef) by this pathogen. SalmoFresh™ is effective in reducing or eliminating the risk of foodborne salmonellosis provoked by the most common *Salmonella* serotypes; it received GRAS status from the FDA in 2013 for direct applications on fresh and processed fruits, vegetables, poultry, fish and shellfish. It has also been approved as a processing aid in the production of poultry products (http://www.intralytix.com/). Micreos Food Safety (Wageningen, The Netherlands) has commercialized a broad-spectrum product against *Listeria* named Listex P100, recognized as GRAS by the FDA and used as a processing aid in cheese, meat, fish and vegetable production. As it is quickly inactivated, it is not active in the final product. This company (http://www.micreosfoodsafety.com/) has also developed a new product against *Salmonella* named Salmonelex™.

In the infection cycle of phages, bacterial lysis occurs by the action of phage-encoded endolysins, proteins with hydrolytic activity on the peptidoglycan (PG). These proteins are accumulated in the bacterial cytoplasm until the assembly of the viral particles is complete. Endolysins need the action of holins that form pores in the cytoplasmic membrane to be allowed to get the PG. Despite the fact that the natural mode of action of endolysins is from within, they are able to degrade the bacterial PG when applied exogenously, supporting their use as food preservatives (Rodríguez-Rubio *et al.* 2016). The effectiveness of several endolysins has been tested in food but, as far as we know, no endolysin has yet

Table 15.2 Application of bacteriophages and endolysins as biocontrol agents and biopreservatives.

	Target	Bacteria	Phage/protein	Mode of application	Effectiveness	Reference
Phage therapy/ prophylaxis	Chicken	*E. coli* O157:H7	Phage mixture	Oral	Total removal Diarrhoea reduction	Li *et al.* (2012)
	Cattle	*Staph. aureus*	Phage mixture	Topic	Reduction	O'Flaherty *et al.* (2005)
	Fish	*Ps. plecoglossicida*	Phage PPp-W4	Oral	Contamination prevention	Park and Nakai (2003)
	Chicken	*Salmonella*	P22 Tailspike protein	Oral	Intestinal colonization reduction	Waseh *et al.* (2010)
	Broiler chicken	*Camp. jejuni*	Phage mixture	Oral	Dose and time effect reduction	Loc Carrillo *et al.* (2005)
Food decontamination	Pork, chicken, eggs, lettuce	*Sal. enterica*	Phage mixture	Spray Immersion	Bacteria reduction No efficacy on eggs	Spricigo *et al.* (2013)
	Lettuce Spinach	*E. coli* 0157:H7	Phage mixture	Aerosol	Total removal	Viazis *et al.* (2011a)
	Raw/ cooked meat	*Camp. jejuni*	Phage Cj6	Surface application	Reduction after treatment (8 days)	Bigwood *et al.* (2008)
Surface disinfection	Inox	*E. coli* 0157:H7	Phage mixture	Immersion	Biofilm removal	Viazis *et al.* (2011b)
	Inox	*Ls. monocytogenes*	Phage mixture	Immersion	Biofilm reduction	Ganegama-Arachchi *et al.* (2013)
	Inox/glass	*Y. pestis*	Phage mixture	Immersion	Biofilm removal	Rashid *et al.* (2012)
Food preservation	Cheese	*Salmonella*	PhageSJ2	Addition	Total removal	Modi *et al.* (2001)
	Cheese	*Ls. monocytogenes*	PhageA511	Spray on surface	Reduction	Guenther and Loessner (2011)
	Beer	*Lb. brevis*	Phage SA-C12	Addition	*Lactobacillus* growth control	Deasy *et al.* (2011)
	Raw/pasteurized milk	*Staph. aureus*	Hydrolase HydH5	Addition	Partial/total reduction	Rodriguez-Rubio *et al.* (2013)
	Pasteurized milk	*Staph. aureus*	Endolysin LysH5	Addition	Total removal	Obeso *et al.* (2008)
	Soya milk	*Ls. monocytogenes*	Endolysin LysZ5	Addition	Reduction	Zhang *et al.* (2012)

Notes: Camp. = Campylobacter; E. = Escherichia; Lb.= Lactobacillus; Ls.= Listeria; Ps. = Pseudomonas; Staph.= Staphylococcus; Y. = Yersinia.

been approved for use in foods for human consumption. For instance, LysH5 killed *Staph. aureus* in pasteurized milk (Obeso *et al.* 2008) and its use in combination with nisin enhanced its antistaphylococcal activity (García *et al.* 2010). Control of staphylococcal strains causing mastitis by phage endolysin phi 11 has been obtained in milk (Donovan *et al.* 2006). Contamination of soya milk with *Ls. monocytogenes* has been reduced by addition of the endolysin LysZ5 at refrigeration temperature (Zhang *et al.* 2012). Chimeric endolysins (endopeptidase domain of streptococcal phage SA2 fused to cell wall–binding domains of lysostaphin and LysK endolysin) combined with lysostaphin reduced *Staph. aureus* in a mouse model of mastitis (Schmelcher *et al.* 2012). Likewise, chimeric proteins composed of the catalytic domains of virion-associated PG hydrolase HydH5 and the cell-binding protein domain of lysostaphin also reduced the *Staph. aureus* load in raw and pasteurized milk (Rodríguez-Rubio *et al.* 2013). Partially purified endolysin ctp1 from the *Cl. tyrobutyricum* phage phiCTP1 cloned and expressed in *E. coli* induced bacterial cells in milk, supporting its potential as an anti–late blowing agent in cheese (Mayer *et al.* 2010). Endolysins also have great potential as disinfectants by removing biofilms developed on the surfaces of food-processing equipment. In particular, the lytic activity of the endolysin LysH5 against *Staph. aureus* and *Staphylococcus epidermidis* biofilms was observed in *in vitro* assays (Gutiérrez *et al.* 2014). Other examples of the application of phages and endolysins through the food chain are provided in Table 15.2. The use of phage endolysins as food preservatives offers some advantages over phages: a wider host spectrum, generally including all species from the genus; no transmission of virulence genes; and no reports of resistant bacteria that have been published so far.

Healthy products: Probiotics and prebiotics

The interest of consumers in functional foods containing probiotics and prebiotics is increasing due to both the positive perception and the growing scientific evidence associating their intake to benefits for human health. Among the different food delivery vehicles, probiotics have traditionally been consumed in fermented dairy products (Prasanna *et al.* 2014). In addition, milk is a source for obtaining some specific prebiotics; indeed, human milk is rich in HMO (human milk oligosaccharides), which can be considered as natural prebiotic substrates (Al-Sheraji *et al.* 2013). Dietary interventions with products containing probiotics and/or prebiotics are currently being explored for their potential to modulate intestinal microbiota, notably in the framework of inflammatory and autoimmune diseases (Blottiere *et al.* 2013), as well as to restore microbial dysbiosis after infection and antibiotic treatment (Reid *et al.* 2011). Additionally, probiotics can be regarded as prophylactic and/or therapeutic agents for non-intestinal-associated diseases (Tojo *et al.* 2014).

Probiotics

One well-accepted definition of probiotics is 'live microorganisms, which when administered in adequate amounts confer a health benefit on the host' (Hill *et al.* 2014). Most probiotic strains used for human consumption belong to the genera *Bifidobacterium* and *Lactobacillus*, the latter being a member of the LAB group. Some *Bacillus* species and the yeast *Saccharomyces cerevisiae* are also being used as probiotics. Indeed, specific species of these genera are included in the QPS (qualified presumption of safety) list of the European Food Safety Authority (EFSA), which means that their use in foods can be regarded as safe (EFSA 2007). Recent advances in the study of the human gut microbiota suggested that commensal bacteria inhabiting our intestine could be a source of new probiotic strains (Qin *et al.* 2010). However, without considering food security issues, commensal bacteria are often difficult to grow at an industrial scale due to their sensitivity to manufacturing processes, thus limiting their inclusion in food products (Leroy and De Vuyst 2004).

According to the roadmap proposed by a group of experts brought together by the Food and Agriculture Organization and the World Health Organization (FAO-WHO 2006), there are guidelines that should be followed in order to obtain a probiotic food with claimed health benefits (Figure 15.3). In short, the probiotic strain candidate must be well identified (at a species level) and its phenotypical and genetic fingerprint well known. Usually, evidence of the safety and functionality of the candidate can be obtained by means of *in vitro* tests; both characteristics should be further confirmed by means of *in vivo* (animal model)

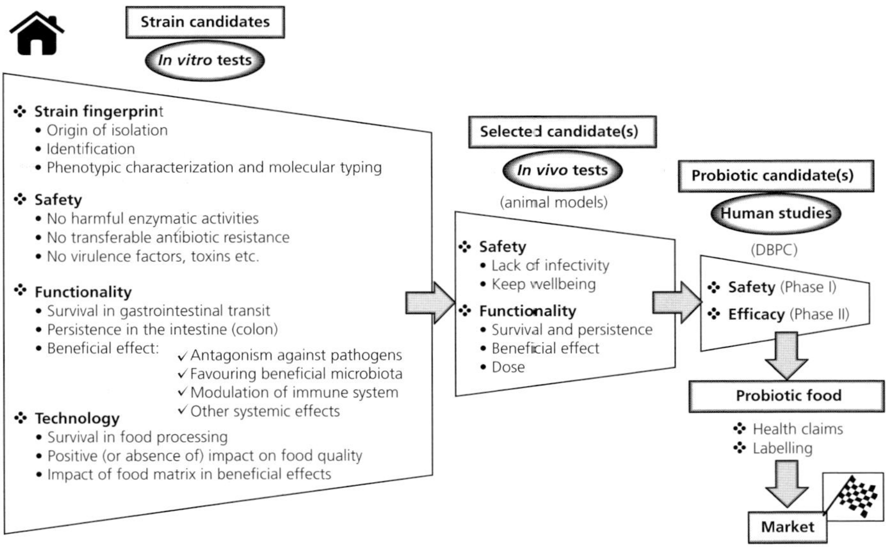

Figure 15.3 Guidelines proposed by experts brought together by FAO-WHO (2006) for selection of probiotic strains intended for food applications.

studies, although, due to ethical and cost concerns, only strains having great probiotic potential should arrive at this point. Finally, well-designed human intervention studies are required to demonstrate the lack of side effects (Phase I studies) and to prove the efficacy of the claimed benefit (Phase II studies). In addition, if the strain(s) is included in the formulation of a food, the 'probiotic food' must accomplish this final step, since the food carrier can modify the performance of the probiotic.

It has been indicated that there are three main mechanisms of action by which probiotics could achieve a beneficial effect (Lebeer *et al.* 2010): antagonism against pathogens, either by direct action (antimicrobials and niche competition) or through modulation of the commensal microbiota; enhancement of the innate epithelial barrier function (inducing production of mucins, defensins, proteins of tight junctions etc.); and modulating the mucosal and systemic immune responses. Bacteriocin production by probiotic strains fits in the first category, and this trait could have a great impact on the ability to modulate the complex intestinal microbial community. The probiotic effect of bacteriocin production *in vivo* is attributed to their action as colonizing peptides that allow probiotics to compete with the indigenous microbiota; the elimination of pathogens by antimicrobial peptides; and their action as signaling peptides that recruit other bacteria and/or the immune system to eliminate the infective microorganism (Dobson *et al.* 2012). Up to 56 species of *Lactobacillus* species have been found in the human intestinal microbiota, most of them producing bacteriocins *in vitro*, and some of them also *in vivo* (Gillor *et al.* 2008), with activity against both Gram-positive and Gram-negative pathogens. An interesting study demonstrated that *Lactobacillus salivarius* UC118 strain produces *in vivo* the potent bacteriocin Abp118, which inhibits *Ls. monocytogenes* in mice (Claesson *et al.* 2006). Similarly, *Lactobacillus casei* L26 LAFTI, a potential bacteriocin producer, was able to inhibit enterohaemorrhagic *E. coli* and *Ls. monocytogenes* strains in mice (Su *et al.* 2007). Inhibition activity against *Helicobacter pylori* bound to intestinal epithelial cells was provided by *Lactobacillus johnsonii* (Gotteland and Cruchet 2003) and *Lactobacillus acidophilus* strains in mice (Coconnier *et al.* 1998). This is particularly interesting, as bacteriocins produced by Gram-positive bacteria have the limitation of usually being inactive against classic enteropathogenic bacteria.

Probiotics included in dairy products

The success of dairy products as probiotic vehicles resides in the high abundance of nutrients in milk, as well as in their pH values close to neutrality (around 6.5) and high water activity (Ritter *et al.* 2009). The mild acidic conditions typical of fermented milks can serve as preadaptation for the lower pH values that probiotics will further face within the stomach; this could facilitate probiotic survival through gastrointestinal transit. In the case of cheeses, those with long ripening times are not suitable for probiotic inclusion, as they show lower water activity values and a very competitive and complex microbiota, which must compromise probiotic survival (Karimi *et al.* 2011). However, fresh short-ripening cheeses, which have physico-chemical characteristics closer to those of fermented milks, could constitute a good matrix for probiotic delivery (Cárdenas *et al.* 2014). In dairy products, probiotics can be partially responsible for milk fermentation or just be added as adjunct cultures. For instance, bifidobacteria produce relatively high amounts of acetic acid, a characteristic that is not desirable in the case of those strains included in fermented milks. In this sense, random mutagenesis, using UV radiation, has been successfully used for isolating a *Bifidobacterium animalis* subps. *lactis* strain (CECT 7953) that decreased the ability to produce acetate. In turn, this strain also produced less ethanol but increased the production of desirable volatile compounds such as diacetyl, thus making it more suitable for dairy fermentation (Margolles and Sánchez 2012).

During the life-cycle of a probiotic included in a fermented milk, it must cope first with the manufacturing conditions and secondly with the stress factors associated with the human gastrointestinal tract (Sánchez *et al.* 2010). Successful survival in these processes and a certain persistence in the human gut will allow probiotics to exert their beneficial effects on human health. During the last few years, the development of new molecular techniques has allowed scientists to define markers of resistance to different stress conditions in probiotic strains. For instance, acid-adapted bifidobacteria, such as those present in dairy products, have induced the membrane-bound F_1F_0-ATPase and thus have persisted in high viable numbers during the shelf life of the product (Sánchez *et al.* 2006). Furthermore, the same enzyme will allow bifidobacteria to counteract the acidic conditions found in the human stomach (Sánchez *et al.* 2007). Together with mild acidic conditions, exposure to oxygen is the other main factor affecting the viability of probiotics included in a fermented dairy product. Aerobic conditions are present during both manufacturing and storage of dairy foods; thus, good oxygen tolerance is a key requisite for a probiotic strain to be included in the formulation of the product. Aerotolerance is a species/strain-dependent trait; for instance, one of the reasons underlying the choice to include *Bif. animalis* subsp. *lactis* in dairy products is its better tolerance to oxygen in comparison to other *Bifidobacterium* species (Andriantsoanirina *et al.* 2013).

Choosing the best food vehicle is paramount from the microbiological point of view, as it may condition later stability and survival. Milk protein and fat are good protectants enhancing probiotic survival during gastrointestinal transit (Ritter *et al.* 2009). In this sense it has been shown that the components of skimmed milk protected two *Bifidobacterium* strains during simulated gastric digestion, mainly by buffering the acidic conditions of the stomach (Sánchez *et al.* 2010). Dairy products traditionally have been a model for the study of the interactions between probiotic bacteria and starters, as well as between probiotic and the food matrix itself; usually most studies target the mechanism of probiotic survival (Ashraf and Shah 2011). Raw milk may contain natural antimicrobial compounds affecting probiotic viability and stability; indeed, if the starter is affected it will end in failed fermentations (Clare *et al.* 2008). Also, the amount of fat can have an impact on probiotic survival (Vinderola *et al.* 2002), whereas including prebiotic carbohydrates may result in enhanced probiotic stability and an increase of health-promoting metabolites through bacteria metabolism (Oliveira *et al.* 2009). Other microorganisms present in dairy products may influence the behaviour and stability of probiotics, either negatively by, for instance, producing hydrogen peroxide (H_2O_2) or bacteriocins (Ng *et al.* 2011), or positively, such as via the co-cultivation of probiotic bifidobacteria with *Lc. lactis* prior to the manufacture of fermented milks (Odamaki *et al.* 2011). On the other hand, care should be taken in choosing a food carrier for probiotics, as the matrix can affect the probiotic pathways involved in host colonization. In this way, adhesion of *Lactobacillus rhamnosus* GG to a colonocyte cell layer was higher when grown in yogurt in comparison to ice cream (Deepika *et al.* 2011), and *Bifidobacterium bifidum* growing in kefir induced the expression of genes involved in host–microbe interaction, among them a pili operon (Serafini *et al.* 2014).

Prebiotics

The definition of a prebiotic was first proposed by Gibson and Roberfroid as a 'non-digestible food ingredient that beneficially affects the host by selectively stimulating the growth and/or activity of one or a limited number of bacteria in the colon, and thus improves host health' (Gibson and Roberfroid 1995). Later, a group of experts brought together by the FAO indicated that a prebiotic is 'a non-viable food component that confers a health benefit on the host associated with modulation of the microbiota' (FAO 2007). Although there is no consensus definition, a recent one states that a prebiotic is a substrate that achieves 'the selective stimulation of growth and/or activity(ies) of one or a limited number of microbial genus(era)/species in the gut microbiota that confer(s) health benefits to the host' (Roberfroid *et al.* 2010). A review on the evolution of the concept of prebiotics, as well as their functions and applications, has been published after a meeting of experts on this topic (Blatchford *et al.* 2013). Thus, in order to consider a substrate as a prebiotic candidate it must resist gastric hydrolysis and gastrointestinal absorption; be fermented by members of the intestinal microbiota; and selectively stimulate the growth and/or activity of a limited number of beneficial bacteria (Roberfroid *et al.* 2010). Combining probiotics and prebiotics in a single functional food leads to the concept of 'synbiotics'. Indeed, it is noteworthy that the dual use of prebiotics together with probiotics could improve the performance of the latter, since prebiotics have the capability of conferring bacterial protection during freeze drying applied for biomass production (Tymczyszyn *et al.* 2011). Apart from this protection within the manufacture of the functional food, combination prebiotic/bacteria could later promote the survival of probiotics through the upper gastrointestinal transit as well as their growth in the colon (Avila-Reyes *et al.* 2014).

Often, prebiotics are carbohydrates that consist of non-starch polysaccharides and oligosaccharides that are not processed by human enzymes. These carbohydrates reach the colon, where they promote the growth of beneficial microorganisms that harbour the necessary enzymatic machinery enabling complete prebiotic digestion (Flint *et al.* 2012). Short-chain fatty acids (SCFA), such as propionate, butyrate and acetate, are the main end products obtained from the microbial fermentation of prebiotics. Moreover, these metabolites or the prebiotics themselves can directly (without microbiota modulation) exert beneficial effects for host health, among others increasing mineral absorption (Rastall 2010); controlling metabolic syndrome and cardiovascular risk by reducing blood pressure, levels of glucose, triglycerides and/or cholesterol (Manning and Gibson 2004); or decreasing levels of blood ammonia in patients with liver cirrhosis (Kajihara *et al.* 2000).

The highest scientific evidence of health-promoting activity was obtained only for a few well-recognized carbohydrates that are currently commercialized as prebiotics: FOS (fructooligosaccharides), inulin-like fructans, GOS (galactooligosaccharides) and lactulose (Gibson *et al.* 2010). There are other substrates currently under study, such as xylooligosaccharides (XOS), isomaltooligosaccharides (IMOS), glucooligosaccharides and soya-oligosaccharides (Al-Sheraji *et al.* 2013). Most of these substrates are obtained from the hydrolysis of natural products (mainly vegetables), but they can also be synthesized by enzymatic or chemical reactions from the simplest sugars, such as those found in milk (Mussatto and Mancilha 2007).

Prebiotics from milk

Some prebiotic substrates can be obtained from milk components, for instance those synthetized from lactose, but they can also be naturally present in milk, such as human milk oligosaccharides (HMO; Table 15.3).

Lactose (β-D-Gal-(1 → 4)-D-Glc) is the main carbohydrate present in milk from different mammals and it can be easily, and relatively costlessly, purified from the whey obtained as a residual product during cheese manufacture. This natural disaccharide has a wide range of industrial applications, one of them being its use as substrate for the production of prebiotics (Villamiel *et al.* 2014). Some of the synthetic oligosaccharides (OS) obtained from lactose are lactulose, lactosucrose and GOS, which account for the majority of production in the sector of non-digestible OS functional food ingredients. With the exception of lactulose, which can also be obtained from lactose isomerization, most of these OS are obtained through enzymatic synthesis carried out by trans-glycosylation reactions. Basically, lactose acts as a galactosyl donor that is transferred to diverse acceptor carbohydrates (having galactosyl, glucosyl or fructosyl moieties) through enzymes, finally rendering OS (Gänzle 2012). The enzymes often used at an industrial level are of microbial origin and have traditionally been named glycosyltransferases, but currently they are included in the family of glycosyl

Table 15.3 Some prebiotics obtained from milk components or naturally present in milk.

Source	Prebiotic	Method(s)
Lactose (and fructose as acceptor)	**Lactulose** (synthetic disaccharide) (β-D-Gal-(1→4)-D-Fru)	Isomerization or enzymatic transglycosylation
Lactose or sucrose	**Lactosucrose** (synthetic trisaccharide) (β-D-Gal-(1→4)-α-D-Glc-(1→2)-β-D-Fru)	Enzymatic transglycosylation
Lactose	**2-α-Glucosyl-lactose** (synthesic trisaccharide)	Enzymatic transglycosylation
Lactose or glucose or galactose or lactulose	**GOS** (synthetic oligosaccharide, DP 2-8) Galactose linked by β-(1→4), β-(1→6) and β-(1→3) bonds and terminal glucose	Enzymatic transglycosylation
Human milk (animal milk)	**HMO** (natural oligosaccharides) Mainly composed of N-acetylglucosamine, fucose, sialic acid, galactose and glucose	Enzymatic transglycosylation

Notes: DP = degree of polymerization; Fru = fructose; Gal = galactose; Glc = glucose; GOS = galactooligosaccharides; HMO = human milk oligosaccharides.

hydrolases (GH) that are able to catalyze the hydrolysis and/or produce rearrangements of the glycosidic bonds (Díez-Municio *et al.* 2014). A catalogue of microbial GH used for OS production as food ingredients has recently been collected by Díez-Municio and co-workers (2014). As an example, β-glactosidases from GH1, GH2 and GH42 families are involved in the hydrolysis of lactose for OS formation (Gänzle 2012). The way of using these microbial enzymes, for example in a soluble or immobilized state, is an issue of special relevance to optimizing the cost of lactose-derived OS at an industrial level (Gosling *et al.* 2010). Furthermore, due to the growing interest in obtaining bioactive food ingredients, the analysis and structural characterization of novel OS derived from lactose, as well as the study of their beneficial effects using a range of biological models, is gaining importance in this research area (Moreno *et al.* 2014).

Currently it is well recognized that breastfeeding is the gold-standard diet for newborns due to the protective effect exerted by human milk. Apart from immunoglobulins, milk glycoconjugates such as HMO may play a relevant role in the prevention of disease (Peterson *et al.* 2013). In addition, these glycans shape the development and composition of the gut microbiota at an earlier stage of life, thus acting as 'natural prebiotic' substrates (Koropatkin *et al.* 2012). Indeed, the *Bifidobacterium* genus, which is a predominant member of breastfed babies' microbiota, is well adapted to using these carbohydrates, which have been considered 'bifidogenic factors' (Sánchez *et al.* 2013). Some of the bifidobacteria species most often found in the infant gut, as well as in human milk, are *Bif. bifidum*, *Bifidobacterium breve* and *Bifidobacterium longum* subsp. *infantis*, which have genomes highly specialized for the use of these glycans (Xiao *et al.* 2010). In fact, there is an HMO cluster of genes encoding proteins involved in regulation, transport and hydrolysis of these molecules (Sela and Mills 2010; Ventura *et al.* 2012). HMO are a mixture of OS that have a skeleton composed of glucose (Glc), galactose (Gal) and N-acetyl-glucosamine (GlcNAc; Table 15.3); so far, more than 200 HMO have been described due to the presence of different linkages and side modifications, such as fucose and sialic acid residues (Wu *et al.* 2010). The core present in type 1 chains, which are predominant especially in colostrum, is LNT or lacto-*N*-tetraose (Gal-β-1,3-GlcNAc-β-1,3Gal-β-1,4Glc); other abundant examples of HMO are lacto-*N*-fucopentaose I (Fuc-α-1,2Gal-β-1,3GlcNAc-β-1,3Gal-β-1,4Glc), lacto-*N*-difucohexaose I (Fuc-α-1,2Gal-β-1,3(Fuc-α-1,4)GlcNAc-β-1,3Gal-β-1,4Glc) and 2′-fucosyllactose (Fuc-α-1,2Gal-β-1,4Glc; Wada *et al.* 2008). With respect to the synthesis of biomimetic, functional, synthetic HMO, which could be regarded as prebiotic additives for infant formula milk, it can also be achieved by transglycosylation reactions. However, due to the number of 'rare enzymes' that are involved in the synthesis of these molecules, this synthetic process still deserves progress in research and development (Zeuner *et al.* 2014). Additionally, although animal milk has a much lower concentration of these OS types that have a less complex structure than human milk, non-human milk could also be regarded as a source for obtaining HMO analogues (Zivkovic and Barile 2011).

Finally, it is of note that dairy products, with the exception of infant formula, do not constitute the main matrix for the delivery of prebiotics in foods. Nevertheless, some dairy-based products have been formulated with different prebiotic substrates (Charalampopoulus and Rastall 2012).

Conclusion

Consumer demand for safe foods with minimal amounts of chemically synthesized additives makes the food industry interested in natural food preservatives. However, despite the overwhelming scientific data on the successful use of several bacteriocins as biocontrol tools, only nisin has been authorized as a food additive along with a few culture fermentates. This is undoubtedly due to the high cost–benefit ratio linked to the use of purified bacteriocins and their approval by the legal authorities. Further research on the production of bacteriocins at lower costs should be undertaken. Bacteriocin resistance is also an issue of concern, as well as the impact in complex ecosystems, for instance on the intestinal microbiota, which has hardly been addressed so far. On the other hand, the use of bacteriophages and endolysins as biocontrol agents and biopreservatives has been approached more recently, and some commercial products are already available. However, despite no adverse effects of oral administration of phages having been described, the most important drawback in the use of phage-derived products might be their acceptance by consumers, which must be accompanied by a suitable regulatory status for these products.

The functional foods market is another growing area of interest for the food industry. However, foods with the claimed beneficial effects must be supported by scientific evidence of their functionality. Dietary intervention with probiotics and/or prebiotics, towards modulating the intestinal microbiota, is a hot research topic. More basic knowledge is needed in order to establish the mechanisms of probiotic and prebiotic action on human health, either through interaction with the intestinal microbiota or directly with the host cells. To date most scientific effort has been devoted to this field, but less attention has been paid to the influence of the food matrix used as delivery for the performance of both players. Therefore, together with the selection of the best strain/substrate candidates for survival/stability in a given food, the efficacy of the probiotic or prebiotic food itself must be proven in the target (human) population and for a specific health claim.

Acknowledgements

The authors acknowledge financial support for their research activities by projects from the Spanish Ministry of Economy and Competiveness (MINECO) and the Regional 'Plan de Ciencia, Tecnología e Innovación' (PCTI, Principado de Asturias), both partially supported by FEDER funds of the European Union.

References

Ackermann, H.W. (2007) 5500 Phages examined in the electron microscope. *Archives in Virology*, **152**, 227–243.

Al-Sherajia, S.H., Ismaila, A., Manap, M.Y., Mustafa, S., Yusof, R.M. and Hassan, F.A. (2013) Prebiotics as functional foods: A review. *Journal of Functional Foods*, **5**, 1542–1553.

Andriantsoanirina, V., Allano, S., Butel, M.J. and Aires, J. (2013) Tolerance of *Bifidobacterium* human isolates to bile, acid and oxygen. *Anaerobe*, **21**, 39–42.

Arqués, J.L., Rodríguez, E., Gaya, P. and Nuñez, M. (2005) Effect of combinations of high-pressure treatment and bacteriocin-producing lactic acid bacteria on the survival of *Listeria monocytogenes* in raw milk cheese. *International Dairy Journal*, **15**, 893–900.

Ashraf, R. and Shah, N.P (2011) Selective and differential enumerations of *Lactobacillus delbrueckii* subsp. *bulgaricus*, *Streptococcus thermophilus*, *Lactobacillus acidophilus*, *Lactobacillus casei* and *Bifidobacterium* spp. in yoghurt: A review. *International Journal of Food Microbiology*, **149**(3), 194–208.

Atterbury, R.J. (2009) Bacteriophage biocontrol in animals and meat products. *Microbial Biotechnology*, **2**, 601–612.

Avila-Reyes, S.V., Garcia-Suarez, F.J., Jiménez, M.T., San Martín-Gonzalez, M.F. and Bello-Perez, L.A. (2014) Protection of *L. rhamnosus* by spray-drying using two prebiotics colloids to enhance the viability. *Carbohydrate Polymers*, **102**, 423–430.

Balciunas, E.M., Castillo Martínez, F.A., Todorov, S.D., Gombossy de Melo Franco, B.D., Converti, A. and Pinheiro de Souza Oliveira, P. (2013) Novel biotechnological applications of bacteriocins: A review. *Food Control*, **32**, 134–142.

Bigwood, T., Hudson, J.A., Billington, C., Carey-Smith, G.V. and Heinemann, J.A. (2008) Phage inactivation of foodborne pathogens on cooked and raw meat. *Food Microbiology*, **25**, 400–406.

Black, E.P., Kelly, A.L. and Fitzgerald, G.F. (2005) The combined effect of high pressure and nisin on inactivation of microorganisms in milk. *Innovation Food Science Emergent Technologies*, **6**, 286–292.

Blatchford, P., Ansell, J., De Godoy, M.R.C. *et al.* (2013) Prebiotic mechanisms, functions and applications: A review. *International Journal of Probiotics and Prebiotics*, **8**, 109–131.

Blottiere, H.M., de Vos, W.M., Ehrlich, S.D. and Doré, J. (2013) Human intestinal metagenomics: State of the art and future. *Current Opinion in Microbiology*, **16**, 234–239.

Breukink, E., Wiedemann, I., van Kraaij, C., Kuipers, O.P., Sahl, H.G. and de Kruijff, B. (1999) Use of the cell wall precursor lipid II bya pore-forming peptide antibiotic. *Science*, **286**, 2361–2364.

Bueno, E., García, P., Martínez, B. and Rodríguez, A. (2012). Phage inactivation of *Staphylococcus aureus* in fresh and hard-type cheeses. *International Journal of Food Microbiology*, **158**, 23–27.

Calderón-Miranda, M.L., Barbosa-Cánovas, G.V. and Swanson, B.G. (1999) Transmission of electron microscopy of *Listeria innocua* treated by electric pulse fields and nisin in skimmed milk. *International Journal of Food Microbiology*, **51**, 31–38.

Cárdenas, N., Calzada, J., Peirotén, Á. *et al.* (2014) Development of a potential probiotic fresh cheese using two lactobacillus salivarius strains isolated from human milk. *BioMed Research International*, **2014**, 801918.

Charalampopoulos, D. and Rastall, R.A. (2012) Prebiotics in foods. *Current Opinion in Biotechnology*, **23**, 187–191.

Chen, H. and Hoover, D. (2003) Bacteriocins and their food applications. *Comprehensive Reviews of Food Science and Food Safety*, **2**, 82–100.

Claesson, M.J., Li, Y., Leahy, S. *et al.* (2006) Multireplicon genome architecture of *Lactobacillus salivarius*. *Proceedings of the Natural Academy of Science USA*, **103**, 6718–6723.

Clare, D.A., Zheng, Z., Hassan, H.M., Swaisgood, H.E. and Catignani, G.L. (2008) Antimicrobial properties of milkfat globule membrane fractions. *Journal of Food Protection*, **71**(1), 126–133.

Coakley, M., Fitzgerald, G. and Ross, R.P. (1997) Application and evaluation of the phage resistance- and bacteriocin-encoding plasmid pMRC01 for the improvement of dairy starter cultures. *Applied and Environmental Microbiology*, **63**, 1434–1440.

Cobo-Molinos, A., Abriouel, H., Lucas López, R., Omar, N.B., Valdivia, E. and Gálvez, A. (2009) Enhanced bactericidal activity of enterocin AS-48 in combination with essential oils, natural bioactive compounds, and chemical preservatives against *Listeria monocytogenes* in ready-to-eat salads. *Food Chemical Toxicology*, **47**, 2216–2223.

Coconnier, M.H., Lievin, V., Hemery, E. and Servin, A.L. (1998) Antagonistic activity against Helicobacter infection *in vitro* and *in vivo* by the human *Lactobacillus acidophilus* strain LB. *Applied and Environmental Microbiology*, **64**, 4573–4580.

Cotter, P.D., Hill, C. and Ross, R.P. (2005) Food microbiology: Bacteriocins: Developing innate immunity for food. *Nature Reviews Microbiology*, **3**, 777–788.

Dawson, P.L., Carl, G.D., Acton, J.C. and Han, I.Y. (2002) Effect of lauric acid and nisin-impregnated soy-based films on the growth of *Listeria monocytogenes* on turkey bologna. *Poultry Science*, **81**, 721–726.

Deasy, T., Mahony, J., Neve, H., Heller, K.J. and van Sinderen, D. (2011) Isolation of a virulent *Lactobacillus brevis* phage and its application in the control of beer spoilage. *Journal of Food Protection*, **74**, 2157–2161.

Deepika, G., Rastall, R.A. and Charalampopoulos, D. (2011) Effect of food models and low-temperature storage on the adhesion of *Lactobacillus rhamnosus* GG to Caco-2 cells. *Journal of Agricultural and Food Chemistry*, **59**, 8661–8666.

Delves-Broughton, J. (2007) Nisin as biopreservative. *Food Australia*, **57**, 525–527.

Diep, D.B., Skaugen, M., Salehian, Z., Holo, H. and Nes, I.F. (2007) Common mechanisms of target cell recognition and immunity for class II bacteriocins. *Proceedings of the Natural Academy of Science USA*, **104**, 2384–2389.

Díez-Municio, M., Herrero, M., Olano, A. and Moreno, F.J. (2014) Synthesis of novel bioactive lactose-derived oligosaccharides by microbial glycoside hydrolases. *Microbial Biotechnology*, **7**, 315–331.

Dobson, A., Cotter, P.D., Ross, R.P. and Hill, C. (2012) Bacteriocin production: A probiotic trait? *Applied and Environmental Microbiology*, **78**, 1–6.

Donovan, D.M., Lardeo, M. and Foster-Frey, J. (2006) Lysis of staphylococcal mastitis pathogens by bacteriophage phi11 endolysin. *FEMS Microbiological Letters*, **265**,133–139.

EFSA (2007) Introduction of a Qualified Presumption of Safety (QPS) approach for assessment of selected microorganisms referred to EFS. *EFSA Journal*, **587**, 1–16.

FAO (2007) FAO Technical Meeting on Prebiotics. Rome: Food and Agriculture Organization of the United Nations.

FAO-WHO (2006) Probiotics in Food: Health and Nutritional Properties and Guidelines for Evaluation. FAO Food and Nutrition paper, No. 85. Rome: World Health Organization.

Field, D., Begley, M., O'Connor, P.M. *et al.* (2012). Bioengineered nisin A derivatives with enhanced activity against Gram positive and Gram negative pathogens. *PLoS ONE*, **7**, e46884.

Flint, H.J., Scott, K.P., Duncan, S.H., Louis, P. and Forano, E. (2012) Microbial degradation of complex carbohydrates in the gut. *Gut Microbes*, **3**, 289–306.

Gallo, L.I., Pilosof, A.M.R. and Jagus, R.J. (2007) Effect of the sequence of nisin and pulsed electric fields treatments and mechanisms involved in the inactivation of *Listeria innocua* in whey. *Journal of Food Engineering*, **79**, 188–193.

Gálvez, A., Abriouel, H., López, R.L. and Ben Omar, N. (2007) Bacteriocin based strategies for food biopreservation. *International Journal of Food Microbiology*, **120**, 51–70.

Ganegama-Arachchi, G.J., Cridge, A.G., Dias-Wanigasekera, B.M. *et al.* (2013) Effectiveness of phages in the decontamination of *Listeria monocytogenes* adhered to clean stainless steel, stainless steel coated with fish protein, and as abiofilm. *Journal of Industrial Microbiology and Biotechnology*, **40**, 1105–1116.

Gänzle, M.G. (2012) Enzymatic synthesis of galacto-oligosaccharides and other lactose derivatives (hetero-oligosaccharides) from lactose. *International Dairy Journal*, **22**, 116–122.

García, P., Madera, C., Beatriz Martínez, B. and Rodríguez, A. (2007) Biocontrol of *Staphylococcus aureus* by lytic-derived phages in curd manufacturing processes. *International Dairy Journal*, **17**, 1232–1239.

García, P., Martínez, B., Obeso, J.M. and Rodríguez A. (2008) Bacteriophages and their application in food safety. *Letters in Applied Microbiology*, **47**, 479–485.

García, P., Martínez, B., Rodríguez, L. and Rodríguez, A. (2010) Synergy between the phage endolysin LysH5 and nisin to kill *Staphylococcus aureus* in pasteurized milk. *International Journal of Food Microbiology*, **141**, 151–155.

Gibson, G.R. and Roberfroid, M.B. (1995) Dietary modulation of the human colonic microbiota: Introducing the concept of prebiotics. *Journal of Nutrition*, **125**, 1401–1412.

Gibson, G.R., Scott, K.P., Rastall, R.A. *et al.* (2010) Dietary prebiotics: Current status and new definition. *Food Science and Technology Bulletin*, **7**, 1–19.

Gillor, P., Etzion, A. and Riley, M.A. (2008) The dual role of bacteriocins as anti and probiotics. *Applied Microbiology and Biotechnology*, **81**, 591–606.

Gosling, A., Stevens, G.W., Barber, A.R., Kentish, S.E. and Gras, S.L. (2010) Recent advances refining galactooligosaccharide production from lactose. *Food Chemistry*, **121**, 307–318.

Gotteland, M. and Cruchet, S. (2003) Suppressive effect of frequent ingestion of *Lactobacillus johnsonii* La1 on *Helicobacter pylori* colonization in asymptomatic volunteers. *Journal of Antimicrobial Chemotherapy*, **51**, 1317–1319.

Guenther, S. and Loessner, M.J. (2011) Bacteriophage biocontrol of *Listeria monocytogenes* on soft ripened white mold and red-smear cheeses. *Bacteriophage*, **1**, 94–100.

Gutiérrez, D., Ruas-Madiedo, P., Martínez, B., Rodríguez, A. and García, P. (2014) Effective removal of staphylococcal biofilms by the endolysin LysH5. *PLoS ONE*, **9**, e107307.

Hagens, S. and Offerhaus M.L. (2008) Bacteriophages: New weapons for food safety. *Food Technology*, **62**, 46–54.

Havärstein, L.S., Diep, D.B. and Nes, I.F. (1995) A family of bacteriocin ABC transporters carry out proteolytic processing of their substrates concomitant with export. *Molecular Microbiology*, **16**, 229–240.

Heng, N.C. and Tagg, J.R. (2006) What's in a name? Class distinction for bacteriocins. *Nature Reviews Microbiology*, **4**. doi:10.1038/nrmicro1273-c1

Hill, C., Guarner, F., Reid, G. *et al.* (2014) The International Scientific Association for Probiotics and Prebiotics consensus statement on the scope and appropriate use of the term probiotic. *Nature Reviews Gastroenterology and Hepatology*, **11**, 506–514.

Issepi, R., Pilati, F., Marini, M. *et al.* (2008) Anti-listerial activity of a polymeric film coated with hybrid coatings doped with Enterocin 416K1 for use as bioactive food packaging. *International Journal of Food Microbiology*, **123**, 281–287.

Kajihara, M., Kato, S., Konishi, M. and Ishii, H. (2000) Xylooligosaccharide decreases blood ammonia levels in patients with liver cirrhosis. *American Journal of Gastroenterology*, **95**, 2514–2514.

Karimi, R., Mortazavian, A.M. and Da Cruz, A.G. (2011) Viability of probiotic microorganisms in cheese during production and storage, a review. *Dairy Science and Technology*, **91**, 283–308.

Kim, K.P., Klumpp, J. and Loessner, M.J. (2007) *Enterobacter sakazakii* bacteriophages can prevent bacterial growth in reconstituted infant formula. *International Journal of Food Microbiology*, **115**, 195–203.

Klaenhammer, T.R. (1993). Genetics of bacteriocins produced by lactic acid bacteria. *FEMS Microbiology Reviews*, **12**, 39–86.

Koropatkin, N.M., Cameron, E.A. and Martens, E.C. (2012) How glycan metabolism shapes the human gut microbiota. *Nature Reviews Microbiology*, **10**, 323–335.

Lebeer, S., Vanderleyden, J. and De Keersmaecker, S.C.J. (2010) Host interactions of probiotic bacterial surface molecules: Comparison with commensals and pathogens. *Nature Review Microbiology*, **8**, 171–184.

Leroy, F. and De Vuyst, L. (2004) Lactic acid bacteria as functional starter cultures for the food fermentation industry. *Trends in Food Science and Technology*, **15**, 67–78.

Li, H., Ma, M.L., Xie, H.J. and Kong, J. (2012) Biosafety evaluation of bacteriophages for treatment of diarrhea due to intestinal pathogen *Escherichia coli* 3-2 infection of chickens. *World Journal of Microbiology and Biotechnology*, **28**, 1–6.

Li, L. and Zhang, Z. (2014) Isolation and characterization of a virulent bacteriophage *SPW* specific for *Staphylococcus aureus* isolated from bovine mastitis in lactating dairy cattle. *Molecular Biology Reports*, **41**, 4829–4838.

Li, M., Yoneyama, F., Toshimitsu, N., Zendo, T., Nakayama, J. and Sonomoto, K. (2013) Lethal hydroxyl radical accumulation by a lactococcal bacteriocin, lacticin Q. *Antimicrobials Agents and Chemotheraphy*, **57**, 3897–3902.

Loc Carrillo, C., Atterbury, R.J., el-Shibiny, A. *et al.* (2005) Bacteriophage therapy to reduce *Campylobacter jejuni* colonization of broiler chickens. *Applied and Environmental Microbiology*, **71**, 6554–6563.

Maisnier-Patin, S., Deschamps, N., Tatini, S.R. and Richard, J. (1992) Inhibition of *Listeria monocytogenes* in Camembert cheese made with a nisin-producing starter. *Lait*, **72**, 249–263.

Maldonado-Barragán, A., Cárdenas, N., Martínez, B. *et al.* (2013) Garvicin A, a novel class IId bacteriocin from *Lactococcus garvieae* that inhibits septum formation in *L. garvieae* strains. *Applied and Environmental Microbiology*, **79**, 4336–4346.

Manning, T.S. and Gibson, G.R. (2004) Prebiotics. *Best Practice Research Clinical Gastroenterology*, **18**, 287–298.

Margolles, A. and Sánchez, B. (2012) Selection of a *Bifidobacterium animalis* subsp. *lactis* strain with a decreased ability to produce acetic acid. *Applied and Environmental Microbiology*, **78**, 3338–3342.

Martínez, B., Böttiger, T., Schneider, T., Rodríguez, A., Sahl, H.-G. and Wiedemann, I. (2008) Specific interaction of the unmodified bacteriocin Lactococcin 972 with the cell wall precursor lipid II. *Applied and Environmental Microbiology*, **74**, 4666–4670.

Mauriello, G., De Luca, E., La Storia, A., Villani, F. and Ercolini, D. (2005) Antimicrobial activity of a nisin-activated plastic film for food packaging. *Letters in Applied Microbiology*, **41**, 464–469.

Mayer, M.J., Payne, J., Gasson, M.J. and Narbad, A. (2010) Genomic sequence and characterization of the virulent bacteriophage phiCTP1 from *Clostridium tyrobutyricum* and heterologous expression of its endolysin. *Applied and Environmental Microbiology*, **76**, 5415–5422.

McAuliffe, O., Hill, C. and Ross, R.P. (1999) Inhibition of *Listeria monocytogenes* in cottage cheese manufactured with lacticin 3147 producing starter culture. *Journal of Applied Microbiology*, **86**, 251–256.

Modi, R., Hirvi, Y., Hill, A. and Griffiths, M.W. (2001) Effect of phage on survival of *Salmonella* Enteritidis during manufacture and storage of Cheddar cheese made from raw and pasteurized milk. *Journal of Food Protection*, **64**, 927–933.

Moreno, F.J., Montilla, A., Villamiel, M., Corzo, N. and Olano, A. (2014) Analysis, structural characterization, and bioactivity of oligosaccharides derived from lactose. *Electrophoresis*, **35**, 1519–1534.

Morgan, S.M., O'Sullivan, L., Ross, R.P. and Hill, C. (2002) The design of a three strain starter system for Cheddar cheese manufacture exploiting bacteriocin-induced starter lysis. *International Dairy Journal*, **17**, 760–769.

Müller, A., Ulm, H., Reder-Christ, K., Sahl, H.-G. and Schneider, T. (2012) Interaction of type A lantibiotics with undecaprenol-bound cell envelope precursors. *Microbial Drug Resistance*, **18**, 261–270.

Muñoz, A., Ananou, S., Gálvez, A. and Valdivia, E. (2007) Inhibition of *Staphylococcus aureus* in dairy products by enterocin AS-48 produced *in situ* and *ex situ*: Bactericidal synergism through heat and AS-48. *International Dairy Journal*, **17**, 760–769.

Muñoz, A., Maqueda, M., Rodríguez, A. and Valdivia, E. (2004) Control of psychrotrophic *Bacillus cereus* in a non-fat hard type cheese by an enterococcal strain producing enterococcal strain producing enterococin AS-48. *Journal of Food Protection*, **67**, 1517–1521.

Mussatto, S.I., Mancilha, I.M. (2007) Non-digestible oligosaccharides: A review. *Carbohydrate Polymers*, **68**, 587–597.

Ng, E.W., Yeung, M. and Tong, P.S. (2011) Effects of yogurt starter cultures on the survival of *Lactobacillus acidophilus*. *International Journal of Food Microbiology*, **145**(1), 169–175.

Nilsson, L., Chen, Y., Chikindas, M.L., Huss, H.H., Gram, L. and Montville, T.J. (2000) Carbon dioxide and nisin act synergistically on *Listeria monocytogenes*. *Applied and Environmental Microbiology*, **66**, 769–774.

Obeso, J.M., García, P., Martínez, B., Arroyo-López, F.N., Garrido-Fernández, A. and Rodriguez, A. (2010) Use of logistic regression for predicting *Staphylococcus aureus* fate in pasteurized milk in the presence of two lytic phages. *Applied and Environmental Microbiology*, **76**, 6038–6046.

Obeso, J.M., Martínez, B., Rodríguez, A. and García P. (2008) Lytic activity of the recombinant staphylococcal bacteriophage ΦH5 endolysin active against *Staphylococcus aureus* in milk. *International Journal of Food Microbiology*, **128**, 212–218.

Odamaki, T., Xiao, J.Z., Yonezawa, S., Yaeshima, T. and Iwatsuki, K. (2011) Improved viability of bifidobacteria in fermented milk by cocultivation with *Lactococcus lactis* subspecies *lactis*. *Journal of Dairy Science*, **94**(3), 1112–1121.

O'Flaherty, S., Coffey, A., Meaney, W.J., Fitzgerald, G.F. and Ross, R.P. (2005) Inhibition of bacteriophage K proliferation on *Staphylococcus aureus* in raw bovine milk. *Letters in Applied Microbiology*, **41**, 274–279.

Oliveira, R.P., Florence, A.C., Silva, R.C. *et al.* (2009) Effect of different prebiotics on the fermentation kinetics, probiotic survival and fatty acids profiles in nonfat symbiotic fermented milks. *International Journal of Food Microbiology*, **128**(3), 467–472.

O'Sullivan, L., O'Connor, E.B., Ross, R.P. and Hill, C. (2006) Evaluation of live-culture-producing lacticin 3147 as a treatment for the control of *Listeria monocytogenes* on the surface of smear-ripened cheese. *Journal of Applied Microbiology*, **100**, 135–143.

Park, S.C. and Nakai, T. (2003) Bacteriophage control of *Pseudomonas plecoglossicida* infection in ayu *Plecoglossus altivelis*. *Disease of Aquatic Organims*, **53**, 33–39.

Peterson, R., Cheah, W.Y., Grinyer, J. and Packer, N. (2013) Glycoconjugates in human milk: Protecting infants from disease. *Glycobiology*, **23**, 1425–1438.

Prasanna, P.H.P., Grandison, A.S. and Charalampopoulos, D. (2014) Bifidobacteria in milk products: An overview of physiological and biochemical properties, exopolysaccharide production, selection criteria of milk products and health benefits. *Food Research International*, **55**, 247–262.

Qin, J., Li, R., Raes, J. *et al.* (2010) A human gut microbial gene catalogue established by metagenomic sequencing. *Nature*, **464**, 59–65.

Rashid, M.H., Revazishvili, T., Dean, T. *et al.* (2012) A *Yersinia pestis*-specific, lytic phage preparation significantly reduces viable *Y. pestis* on various hard surfaces experimentally contaminated with the bacterium. *Bacteriophage*, **2**, 168–177.

Rastall, R.A. (2010) Functional oilgosaccharides: Application and manufacture. *Annual Reviews in Food Sciences and Technology*, **1**, 305–339.

Rayman, M.K., Aris, B. and Hurst, A. (1981) Nisin: A possible alternative or adjunct to nitrite in the preservation of meats. *Applied and Environmental Microbiology*, **41**, 375–380.

Rehaiem, A., Martínez, B., Manai, M. and Rodríguez, A. (2012). Technological performance of the enterocin A producer *Enterococcus faecium* MMRA as a protective adjunct culture to enhance hygienic and sensory attributes of traditional fermented milk 'Rayeb'. *Food and Bioprocess Technology*, **5**, 2140–2150.

Reid, G., Younes, J.A., Van der Mei, H.C., Gloor, G.B., Knight, R. and Busscher, H.J. (2011) Microbiota restoration: Natural and supplemented recovery of human microbial communities. *Nature Reviews Microbiology*, **9**, 27–38.

Rilla, N., Martínez, B., Delgado, T. and Rodríguez, A. (2003) Inhibition of *Clostridium tyrobutyricum* in Vidiago cheese by *Lactococcus lactis* ssp. *lactis* IPLA 729, a nisin Z producer. *International Journal of Food Microbiology*, **85**, 22–33.

Rilla, N., Martínez, B. and Rodríguez, A. (2004) Inhibition of a methicillin-resistant *Staphylococcus aureus* strain in Afuega'l Pitu cheese by the nisin Z producing strain *Lactococcus lactis* subsp. *lactis* IPLA 729. *Journal of Food Protection*, **67**, 928–933.

Ritter, P., Kohler, C. and Von Ah, U. (2009) Evaluation of the passage of *Lactobacillus gasseri* K7 and bifidobacteria from the stomach to intestines using a single reactor model. *BMC Microbiology*, **9**, 1–9.

Roberfroid, M., Gibson, G.R., Hoyles, L. *et al.* (2010) Prebiotic effects: Metabolic and health benefits. *British Journal of Nutrition*, **104**, S1–S63.

Roces, C., Rodríguez, A. and Martínez, B. (2012) Cell wall active bacteriocins and their application beyond antibiotic activity. *Probiotics and Antimicrobial Proteins*, **4**, 259–272.

Rodríguez, E., Arques, J.L., Gaya, P., Tomillo, J., Nuñez, M. and Medina, M. (2000) Behaviour of *Staphylococcus aureus* in semihard cheese made from raw milk with nisin-producing starter cultures. *Milchwissenschaft*, **55**, 633–635.

Rodríguez-Rubio, L., Gutiérrez, D., Donovan, D. and García, P. (2016) Phage lytic proteins: Biotechnological applications beyond clinical antimicrobials. *Critical Reviews in Biotechnology* **36**(3), 542–552.

Rodríguez-Rubio, L., Martínez, B., Donovan, D.M., García, P. and Rodríguez, A. (2013) Potential of the virion-associated peptidoglycan hydrolase HydH5 and its derivative fusion proteins in milk biopreservation. *PLoS ONE*, **8**, e54828.

Sahl, H.G. and Bierbaum, G. (1998) Lantibiotics: Biosynthesis and biological activities of uniquely modified peptides from gram-positive bacteria. *Annual Reviews in Microbiology*, **52**, 41–79.

Sánchez, B., Champomier-Vergès, M.C., Collado, M.C. *et al.* (2007) Low-pH adaptation and the acid tolerance response of *Bifidobacterium longum* biotype *longum*. *Applied and Environmental Microbiology*, **73**, 6450–6459.

Sánchez, B., de los Reyes-Gavilán, C.G. and Margolles, A. (2006) The F1F0-ATPase of *Bifidobacterium animalis* is involved in bile tolerance. *Environmental Microbiology*, **8**, 1825–1833.

Sánchez, B., Fernández-García, M., Margolles, A., González de los Reyes-Gavilán, C. and Ruas-Madiedo, P. (2010) Technological and probiotic selection criteria of a bile-adapted *Bifidobacterium animalis* subsp. *lactis* strain. *International Dairy Journal*, **20**, 800–805.

Sánchez, B., Ruiz, L., Gueimonde, M., Ruas-Madiedo, P. and Margolles, A. (2013) Adaptation of bifidobacteria to the gastrointestinal tract and functional consequences. *Pharmacological Research*, **69**, 127–136.

Scannell, A.G.M., Hill, C., Ross, R.P. *et al.* (2000a) Development of bioactive food packaging materials using immobilised bacteriocins Lacticin 3147 and Nisaplin®. *International Journal of Food Microbiology*, **60**, 241–249.

Scannell, A.G., Hill, C., Ross, R.P., Marx, S., Hartmeier, W. and Arendt, E.K. (2000b) Continuous production of lacticin 3147 and nisin using cells immobilized in calcium alginate. *Journal of Applied Microbiology*, **89**, 573–579.

Schmelcher, M., Powell, A.M., Becker, S.C., Camp, M.J. and Donovan, D.M. (2012) Chimeric phage lysins act synergistically with lysostaphin to kill mastitis-causing *Staphylococcus aureus* in murine mammary glands. *Applied and Environmental Microbiology*, **78**, 2297–2305.

Sela, S. and Mills, D.A. (2010) Nursing our microbiota: Molecular linkages between bifidobacteria and milk oligosaccharides. *Trends in Microbiology*, **18**, 298–307.

Serafini, F., Turroni, F., Ruas-Madiedo, P. *et al.* (2014) Kefir fermented milk and kefiran promote growth of *Bifidobacterium bifidum* PRL2010 and modulate its gene expression. *International Journal of Food Microbiology*, **178**, 50–59.

Sobrino-López, A. and Martín-Belloso, O. (2006) Enhancing inactivation of *Staphylococcus aureus* in skim milk by combining high-intensity pulsed electric fields and nisin. *Journal of Food Protection*, **69**, 345–353.

Spricigo, D.A., Bardina, C., Cortes, P. and Llagostera, M. (2013) Use of a bacteriophage cocktail to control *Salmonella* in food and the food industry. *International Journal of Food Microbiology*, **165**, 169–174.

Su, P., Henriksson, A. and Mitchell, H. (2007) Survival and retention of the probiotic *Lactobacillus casei* LAFTI (R)L26 in the gastrointestinal tract of the mouse. *Letters in Applied Microbiology*, **44**, 120–125.

Sulakvelidze, A., Alavidze, Z. and Morris, J.G. (2001) Bacteriophage therapy. *Antimicrobial Agents Chemotheraphy*, **45**, 649–659.

Tabla, R., Martínez, B., Rebollo, J.E. *et al.* (2012) Bacteriophage performance against *Staphylococcus aureus* in milk is improved by high hydrostatic pressure treatments. *International Journal of Food Microbiology*, **156**, 209–213.

Tojo, R., Suárez, A., Clemente, M.G. *et al.* (2014) Intestinal microbiota in health and disease: Role of bifidobacteria in gut homeostasis. *World Journal of Gastroenterology*, **20**, 15163–15176.

Tymczyszyn, E.E., Gerbino, E., Illanes, A. and Gómez-Zavaglia, A. (2011) Galacto-oligosaccharides as protective molecules in the preservation of *Lactobacillus delbrueckii* subsp. *bulgaricus*. *Cryobiology*, **62**, 123–129.

Ventura, M., Turroni, F., Motherway, M.O., MacSharry, J. and van Sinderen, D. (2012) Host–microbe interactions that facilitate gut colonization by commensal bifidobacteria. *Trends in Microbiology*, **20**, 467–476.

Viazis, S., Akhtar, M., Feirtag, J. and Diez-Gonzalez, F. (2011a) Reduction of *Escherichia coli* O157:H7 viability on leafy green vegetables by treatment with a bacteriophage mixture and trans-cinnamaldehyde. *Food Microbiology*, **28**, 149–157.

Viazis, S., Akhtar, M., Feirtag, J. and Diez-Gonzalez, F. (2011b) Reduction of *Escherichia coli* O157:H7 viability on hard surfaces by treatment with a bacteriophage mixture. *International Journal of Food Microbiology*, **145**, 37–42.

Villamiel, M., Montilla, A., Olano, A. and Corzo, N. (2014) Production and bioactivity of olisaccharides derived from lactose. In F.J. Moreno and M.L. Sanz (eds), *Food Oligosaccharides: Production, Analysis and Bioactivity* (pp. 137–167), Chichester: John Wiley & Sons.

Vinderola, C.G., Costa, G.A., Regenhardt, S. and Reinheimer, J.A. (2002) Influence of compounds associated with fermented dairy products on the growth of lactic acid starter and probiotic bacteria. *International Dairy Journal*, **12**, 579–589.

Wada, J., Ando, T., Kiyohara, M. *et al.* (2008) *Bifidobacterium bifidum* lacto-N-biosidase, a critical enzyme for the degradation of human milk oligosaccharides with a type 1 structure. *Applied and Environmental Microbiology*, **74**, 3996–4004.

Waseh, S., Hanifi-Moghaddam, P., Coleman, R. *et al.* (2010) Orally administered P22 phage tailspike protein reduces *Salmonella* colonization in chickens: Prospects of a novel therapy against bacterial infections. *PLoS ONE*, **5**, e13904.

Weidemann, I., Breukink, E., van Kraaij, C. *et al.* (2001). Specific binding of nisin to the peptidoglycan precursor lipid II combines pore formation and inhibition of cell wall biosynthesis for potent antibiotic activity. *Journal of Biological Chemistry*, **276**, 1772–1779.

Worobo, R.W., Van Belkum, M.J., Saller, M., Roy, K.L., Vederas, J.C. and Stiles, M.E. (1995) A signal-peptide secretion-dependent bacteriocin from *Carnobacterium divergens*. *Journal of Bacteriology*, **177**, 3143–3149.

Wu, S., Tao, N., German, J.B., Grimm, R. and Lebrilla, C.B. (2010) Development of an annotated library of neutral human milk oligosaccharides. *Journal of Proteome Research*, **9**, 4138–4151.

Xiao, J.Z., Takahashi, S., Nishimoteo, M. *et al.* (2010) Distribution of in *vitro* fermentation ability of lacto-N-biose I, a major building block of human milk oligosaccharides, in bifidobacterial strains. *Applied and Environmental Microbiology*, **76**, 54–59.

Zeuner, B., Jers, C., Mikkelsen, J.D. and Meyer, A.S. (2014) Methods for improving enzymatic trans-glycosylation for synthesis of human milk oligosaccharide biomimetics. *Journal of Agricultural and Food Chemistry*, **62**, 9615–9631.

Zhang, H., Bao, H., Billington, C., Hudson, J.A. and Wang, R. (2012) Isolation and lytic activity of the *Listeria* bacteriophage endolysin LysZ5 against *Listeria monocytogenes* in soya milk. *Food Microbiology*, **31**, 133–136.

Zivkovic, A.M. and Barile, D. (2011) Bovine milk as a source of functional oligosaccharides for improving human health. *Advances in Nutrition*, **2**, 248–289.

CHAPTER 16

Sausages and other fermented meat products

Renata E.F. Macedo[1], Fernando B. Luciano[1], Roniele P. Cordeiro[2] and Chibuike C. Udenigwe[3]

[1]*School of Agricultural Sciences and Veterinary Medicine, Pontifícia Universidade Católica do Paraná, Brazil*

[2]*Department of Food Science, University of Manitoba, Canada*

[3]*Department of Environmental Sciences, Dalhousie University, Canada*

Fermented sausages are defined as a mixture of ground lean meat and minced fat, curing salts, sugar and spices, which is embedded in a casing and subjected to fermentation and drying (Leroy *et al.* 2006; Ammor and Mayo 2007).

The quality of fermented sausages is closely related to the ripening process that gives colour, flavour, aroma and firmness to the product. These characteristics are developed by a complex interaction of chemical and physical reactions associated with the fermentative action of the microbiota present in the mass of the sausage. In handmade production processes, fermentation occurs spontaneously by the action of the indigenous bacteria present on meat. In industrial processes the microbiota, responsible for the fermentation process, is known as the starter culture (Leroy *et al.* 2006). Starter cultures are defined as preparations containing live microorganisms capable of developing desirable metabolic activity in meat. They are used to increase microbiological safety, to maintain stability by inhibiting the growth of undesirable microorganisms and to improve the sensory characteristics of fermented sausages (Työppönen *et al.* 2003b). More recently, starter cultures have also been used in fermented meat products to increase their functional value due to the probiotic properties and to the *in situ* production of nutraceuticals and antibacterial compounds.

This chapter presents the role of starter cultures in fermented meat products, focusing on the applications of these cultures to improve the sensory profile, safety and functional value of fermented meat products.

Starter Cultures in Food Production, First Edition. Edited by Barbara Speranza, Antonio Bevilacqua, Maria Rosaria Corbo and Milena Sinigaglia.

Starter cultures for fermented meat products

In many countries, fermented meat products are unique foods, often represented as elements of culinary heritage and identity (Leroy *et al.* 2013). These products are prepared respecting tradition, resulting in a wide variety of products with different flavours, textures and microbiological quality (Bonomo *et al.* 2008).

In European Union countries, fermented meat products account for 20–40% of total processed meat. Among these countries, Italy along with Germany and Spain manufactures a wide variety of fermented meat products. In some conditions, given the characteristics and the specificity of the production process, these products are protected by the registration of Denomination of Controlled Origin (Di Cagno *et al.* 2008). This great diversity of fermented meat products is related to the type of culture and to the variety of ingredients used in their formulation (Coloretti *et al.* 2014).

The characteristics of fermented meat products are also affected by the physical, biochemical, microbiological and sensorial transformations that occur during the ripening of the product. These transformations include dehydration; reduction in pH; reduction of nitrates to nitrites; formation of nitrosomyoglobin; solubilization and gelation of myofibrillar and sarcoplasmic proteins; proteolytic, lipolytic and oxidative phenomena; and a change in the initial microbiota (Casaburi *et al.* 2007). These metabolic transformations are achieved during the fermentation process promoted by bacteria, and improve the flavour, texture and safety of fermented meat (Toldrá *et al.* 2007). Thus, use of microbial starter cultures has become increasingly important to ensure the consistency of these metabolic activities and improve the characteristics of these products.

Compared to fermented dairy food and alcoholic beverages, the use of starter cultures in fermented meat products is a relatively recent practice (Liepe 1983). While the commercial application of starter cultures in the cheese industry occurred in the 1930s, the use of pure cultures in sausages began after several studies in the 1950s.

Some of the requirements for using starter cultures in fermented meat include the production and preservation of starter cultures in a viable and metabolically active form suitable for commercial distribution; the production of adequate quantities of lactic acid; the ability to grow at salt concentrations up to 6% (w/w); and the capacity to enhance the flavour of the finished products without the production of biogenic amines and slimes.

Starter cultures are formed by mixing different types of microorganisms, where each has a specific function. The most common microorganisms used as starter cultures in meat fermentations are lactic acid bacteria (LAB) and coagulase-negative staphylococci (CNS; Bonomo *et al.* 2009).

LAB and CNS are actively involved in the development of texture, colour and flavour (Hammes *et al.* 1990), as well as in the enhancement of the safety of the fermented product due to suppression of the pathogenic microbiota through the acidification or production of antimicrobial compounds. LAB are Gram-positive microorganisms that include *Lactococcus, Enterococcus, Oenococcus, Pediococcus, Streptococcus, Lactococccus* and *Lactobacillus* species (Makarova *et al.* 2006). The main function of LAB in meat fermentation is to promote a rapid decrease in the pH of the batter, which favours safety by inactivating pathogens; stability and shelf life by inhibiting undesirable changes caused by spoilage microorganisms or abiotic reactions; and adequate biochemical conditions to attain the new sensory properties of the ripe products through modification of the raw materials (Lücke 2000). CNS play a significant role in the development of colour and flavour in fermented meat products and comprise bacteria of the genus *Staphylococcus,* which includes species of Gram-positive, catalase-positive cocci microorganisms. The most common CNS species isolated from dry fermented sausages include *Staphylococcus xylosus, Staphylococcus carnosus, Staphylococcus equorum* and *Staphylococcus saprophyticus*. The most important technological characteristic of CNS is the ability to reduce nitrate to nitrite, leading to the formation of nitrosomyoglobin (Rantsiou and Cocolin 2008), which promotes the desired red colour. The organoleptic quality of meat products also depends on the lipolytic, proteolytic (Casaburi *et al.* 2007) and catalase activity of CNS, where the latter prevents lipid oxidation (Barrière *et al.* 2001).

Other bacteria such as *Kocuria,* yeasts such as *Debaryomyces* and moulds such as *Penicillium* can also be used as starter cultures. Yeasts are usually found internally in the product and are used in the development of the characteristic flavour of cured meat. *Debaryomyces hansenii* is one of the yeast species most frequently isolated in dry-cured meat products (Aquilanti *et al.* 2007). Moulds are present on the surface of the product where there is availability of oxygen (Cocolin *et al.* 2011). Commercial moulds are used on the sausage surface primarily to prevent mycotoxin production (Sunesen and Stahnke 2003), but a more consistent flavour, taste, drying rate and uniform appearance of sausages are also achieved by their use. *Penicillia* is the mould species most widely used as protective in the meat industry (Sunesen and Stahnke 2003).

Table 16.1 shows the microorganism species most commonly used as starter cultures to ferment meat products.

These microorganisms can be used either singly or in multiple-species combinations (Ricke *et al.* 2007). Usually, a large quantity (10^7–10^8 colony forming units (cfu) per gram) of bacterial culture is intentionally added to the meat product in order to ensure an appropriate fermentation process (Resch *et al.* 2008).

The selection of starter cultures for fermented meat products must be carried out according to the product formulation and the technological processing employed, since environmental factors can select a limited number of strains with the ability to compete and overcome in the product. Typically, the species used as the starter culture are selected from strains naturally predominant in meat products and hence well adapted to this environment. Therefore, these

Table 16.1 Species of microorganism most commonly used as starter cultures in fermented meat products.

Microorganism	Genus and species
Lactic acid bacteria	*Lb. acidophilus*,[1] *Lb. alimentarius*,[2] *Lb. brevis*, *Lb. casei*,[1] *Lb. curvatus*, *Lb. fermentum*, *Lb. plantarum*, *Lb. pentosus*, *Lb. sakei*
	Lc. lactis
	Pe. acidilactici, *Pe. pentosaceus*
Actinobacteria	*Kc. varians*[3]
	Sm. griséus
	Bifidobacterium sp.[1]
Staphylococcus	*Staph. xylosus*, *Staph. carnosus subsp. carnosus*, *Staph. carnosus subsp. utilis*, *Staph. equorum*[2]
Halomanadaceae	*Hal. elongata*[2] (tested in dry-cured ham)
Enterobacter	*Aeromonas sp.*
Mould	*Pen. nalgiovense*, *Pen. chrysogeum*, *Pen. camemberti*
Yeast	*D. hansenii*, *C. famata*

Source: Hammes and Hertel (1998); Ruiz-Moyano *et al.* (2008); Rivera-Espinoza and Gallardo-Navarro (2010).

[1] Used as probiotic cultures.

[2] Used in commercial tests on an industrial scale (Laboratorium Wiesby, Niebüll and Rudolf Müller and Co.).

[3] Previously called *Micrococcus varians*.

C. = *Candida*; *D.* = *Debaromyces*; *Hal.* = *Halomonas*; *Kc.* = *Kocuria*; *Lb.* = *Lactobacillus*; *Lc.* = *Lactococcus*; *Pe.* = *Pediococcus*; *Pen.* = *Penicillium*; *Sm.* = *Streptomyces*; *Staph.* = *Staphylococcus*.

species present a tendency to have a greater metabolic capacity, which is reflected in the development of the proper sensory and physico-chemical characteristics in the product (Leroy *et al.* 2006).

Role of starter culture in the flavour, texture and colour formation of fermented meat products

Flavour formation

The characteristic flavour of fermented sausages is a result of the breakdown of carbohydrates, lipids and proteins through the action of microbial and endogenous meat enzymes (Casaburi *et al.* 2008; Sidira *et al.* 2015b). In addition, spices and other condiments also contribute to the distinctive flavour of fermented sausages. While proteolysis in meat takes place due to the action of the cathepsins breaking down sarcoplasmic and myofibrillar proteins, the action of microbial enzymes becomes more prominent during the latter stages of ripening (Casaburi *et al.* 2008). In this context, the selection of starter strains with lipolytic and/or proteolytic activity plays an important role in the breakdown products of lipolysis and proteolysis (i.e. peptides, amino acids, carbonyls and volatile flavour compounds), which in turn contribute to the characteristic flavour and aroma of fermented meat. For instance, the formation of volatile compounds in traditional Spanish dry-fermented sausage was influenced by autochthonous starter cultures (Fonseca *et al.* 2013). Concentrations of volatile compounds derived from microbial esterification were greater than compounds derived from lipid oxidation, carbohydrate fermentation, amino acid catabolism or spices. Among the volatile compounds formed, hexanoic acid, methyl ester, was the most abundant before inoculation. In final products, butanoic acid, 3-methyl, methyl ester, had the highest levels, followed by hexanoic acid, methyl ester.

The profile of volatile compounds formed in fermented sausages may also be influenced by product ingredients and processing conditions such as proteins, temperature, drying and ripening (Hughes *et al.* 2002). Sidira *et al.* (2015b), working with dry-fermented sausages inoculated with *Lactobacillus casei* as starter culture immobilized in wheat grains (300 g/kg of stuffing mixture), found that the content of volatile compounds such as esters, organic acids, alcohols and carbonyl was greater at 45 days than at 28 days of ripening. Besides sausage processing conditions, studies have also reported that the interaction between starter cultures with different proteolytic activity capability can significantly affect the flavour of semi dry-fermented sausages. In a sensorial analysis, the highest scores for flavour of semi dry-fermented sausage stored at refrigerated temperature for 120 days was assigned to samples inoculated with *Lactobacillus brevis* and *Lactobacillus plantarum*; on the other hand, sausages inoculated with *Lactococcus lactis* spp. and *Streptomyces griseus* ssp. had the lowest score for flavour at the same temperature and storage period (Ahmad and Amer 2013). Combinations of starter cultures *Lactobacillus sakei* + *Staph. carnosus* or *Pediococcus pentosaceus* + *Staph. xylosus* caused degradation of sarcoplasmic and myofibrillar proteins during the fermentation and drying stages of sausages (Aro *et al.* 2010). However, quantitative analysis showed that *Staph. xylosus* alone produced significantly higher amino acid content than any other starter culture, which contributed to the large differences in free amino acid contents at the end of the process. Fermented sausages inoculated with proteolytic and lipolytic *Staph. xylosus* coupled with *Lactobacillus curvatus* also showed higher proteolysis and lipolysis than the control treatment (non-inoculated; Casaburi *et al.* 2008). However, these activities in sausage containing only *Staph. xylosus* without lactobacilli were identical to those of the non-inoculated control. Thus, the proteolytic activity of different starter cultures needs to be determined prior to fermentation in order to achieve the desired flavour of fermented sausage.

A strategy to enhance proteolytic activity in dry-fermented sausages and to accelerate ripening and flavour development is the addition of exogenous proteases in fermented sausage formulation (Fernández *et al.* 2000). However, the increase in aroma compounds does not seem to be associated with an increase in the total of free amino acids, since the mechanisms of amino acid degradation must also be favoured in order to yield higher amounts of flavour compounds (Diaz *et al.* 1997). Thus, improvement of the sensory characteristics of dry-fermented sausages by the addition of exogenous protease may depend on its association with starter cultures in order to guarantee product safety and homogeneity in metabolic activities. For instance, dry-fermented sausages ripened with protease EPg222 and *Pediococcus acidilactici* MS200 and *Staphylococcus vitulus* RS34 as starter cultures (P200S34) showed higher amounts of volatile compounds, including 1-propanol, 1-hexanol and 2-propanone, when compared to the control treatment (Casquete *et al.* 2011).

Regarding the use of mould in fermented meat products, the growth of mould on the meat surface is often desirable, as it can be responsible for the development of the specific flavours and aromas of dry meats. Moulds produce enzymes for the degradation of lipid and protein matter; however, proteolytic and lipolytic abilities may differ significantly between strains and species, and are highly dependent on media, pH and temperature (Sunesen and Stahnke 2003). *Penicillium* species are mostly able to colonize salami during seasoning, showing the potential to be used as starters for dry-cured meat (Perrone *et al.* 2015). New species of *Penicillium* such as *Penicillium salami* have proved well adapted to meat-curing environments, with a light green sporulation both on agar media and on the sausages, which is a desirable technological property (Perrone *et al.* 2015). Concerning yeasts, the addition of *Debaryomyces* spp. modifies the flavour pattern of dry-cured sausages through the production of volatile aroma compounds. *D. hansenii* has been found to contribute to higher levels of cyclic and aromatic alcohols during controlled ripening of pork loins (Martín *et al.* 2003).

Texture formation

The texture of fermented sausages is influenced by the acidification of the product, which should be below the isoelectric point of muscle protein in order to affect the coagulation of proteins responsible for the sliceability, firmness and cohesiveness of the final product (Essid and Hassouna 2013). In dry-fermented meat products, LAB are responsible for a rapid fermentation of carbohydrates that causes a decrease in pH, which is responsible for the desired acidification. The dominating LAB are known to produce acids, such as lactic, acetic and formic acids. However, the levels of acid production depend on genus, species and growth conditions (Borch *et al.* 1999). As a result, different cultures of LAB may diversely affect the texture of fermented sausages. A significant difference in pH between sausage samples has been observed due to different starter cultures, which led to significant differences in the texture of the final semi dry-fermented sausage during refrigerated storage (Ahmad and Amer 2013). In addition, increased levels of fat have also been shown to influence texture. Sausage samples inoculated with *Lb. brevis* + *Sm. griseus* ssp. received the highest scores for texture if the fat content was between 20% and 25%. Samples inoculated with

Lb. plantarum + *Lc. lactis* ssp. were given the lowest texture scores even with the addition of a similar fat concentration. Thus, the fat content can influence the metabolic activity of starter cultures.

Colour formation

The typical red colour of fermented sausages is developed through nitrate reductase activity that leads to the formation of nitrosomyoglobin, which results from the interaction between muscle-based myoglobin and nitric oxide. Nitrate has traditionally been used in the curing processes, in which it is added to the meat batter to prompt the formation of nitric oxide. Nitrate is then reduced into nitric oxide during the fermentation process to obtain nitrite. This type of curing agent also plays an essential role in attaining the stability and safety of meat products such as fermented sausages, ham and, more recently, emulsion-type sausages. The most efficient nitrate-reducing organisms are *Staphylococci* (Gøtterup *et al.* 2008). Among several species of CNS, *Staph. xylosus* and *Staph. equorum* have shown the highest nitrate-reductase activity along with several other technological and safety properties relevant for their use as a starter culture in fermented sausages. These properties include catalase, proteolitic and lipolitic activities, as well as antibiotic susceptibility (Mauriello *et al.* 2004; Bonomo *et al.* 2009; Landeta *et al.* 2013). However, *Staph. xylosus* seems to be the predominant species isolated from sausages. In southern Italy, the microflora in traditional fermented sausage (37 strains) was found to be dominated by 17 strains of *Staph. xylosus* (45%; Bonomo *et al.* 2009). This species was also predominantly recommended as a starter culture in the production of artisanal sausages produced in the southern region of Brazil (Fiorentini *et al.* 2009). The *Staph. xylosus* strains selected in this region of Brazil showed nitrate-reductase, catalase and lipase activity, as well as satisfactory growth in the presence of sodium chloride (NaCl) and sodium nitrite ($NaNO_2$).

Although nitrite is responsible for the red colour of cured meat, carcinogenic, teratogenic and mutagenic N-nitrosamines may be formed from nitrite; thus, its use as a curing agent has raised health concerns (Li *et al.* 2013). As an alternative to nitrite in meat products, the reddening ability of some strains of bacteria that are suitable for meat fermentation has been evaluated. In a Chinese-style sausage treated with 10^8 cfu/g meat of *Lactobacillus fermentum*, the cured pink colour closely replicated that of nitrite-cured meat (Zhang *et al.* 2007). *Staph. xylosus* also proved to be a potential alternative for nitrite substitution in meat products (Li *et al.* 2013). *Staph. xylosus* isolated from Chinese dried sausage showed a greater ability to convert metmyoglobin into nitrosylmyoglobin in de Man–Rogosa–Sharp broth model systems and raw pork meat batters without the addition of nitrite when compared to the control samples or samples inoculated with *Pe. pentosaceus*. Thus, the microbial conversion of metmyoglobin into nitrosylmyoglobin has the potential to replace nitrite in sausage production while keeping the desired red colour of meat products.

Biopreservation of fermented meat products by starter cultures

Processes within the food chain change constantly, leading to continuous modifications in the food microbiota. There has been an emergence of foodborne diseases in recent decades that were not recognized in the past. Moreover, globalization has transformed the food market worldwide and most food-processing companies have become exporters. Therefore, any sort of food contamination occurring in a small factory may reach a global scale of foodborne disease, causing burdens on the public health system of several countries (Castellano *et al.* 2008).

The concept of biopreservation has recently gained increasing attention, and fermented food products are some of the best candidates for the use of this technique. Biopreservation refers to the use of microorganisms or natural antimicrobials in order to increase the shelf life and safety of food products. Many microorganisms commonly found in meat and meat products were established to reduce or inhibit the growth of spoilage and pathogenic bacteria. Most of these organisms are LAB, which have generally recognized as safe (GRAS) status for human consumption.

Dry-fermented sausages are traditionally produced with a mixture of pork and/or beef trimmings and pork fat. These ingredients are comminuted to a batter and added to salt, nitrite and/or nitrate (curing salt), sugar, ascorbate, spices and the starter culture (Työppönen *et al.* 2003a). Then, the product is allowed to ferment and dry for weeks or even months. The process and ingredients confer a long shelf life on the final product. Nitrite is a well-known antibacterial agent (mainly against *Clostridium* species) and its production is maintained during the fermentation process through the reduction of nitrate by micrococci species present in the starter culture (Työppönen *et al.* 2003b). High levels of salt also inhibit the growth of various saprophytic and pathogenic bacteria. Moreover, many of the spices used contain essential oils with potent antimicrobial activity (e.g. oregano, thyme, sage, pepper, mustard, garlic; Holley and Patel 2005). Lastly, the starter culture produces several hurdles to the growth of other microorganisms. Usually they are highly adapted to the sausage environment and therefore are strong competitors for the utilization of nutrients. In addition, LAB cause a rapid drop of sausage pH through the production of lactic acid (Gálvez *et al.* 2007) and also produce other bactericidal/bacteriostatic compounds such as hydrogen peroxide, carbon dioxide (CO_2), diacetyl and bacteriocins (Ammor *et al.* 2006).

Some LAB used as starter cultures have the ability to produce antibacterial peptides called bacteriocins. Many of these compounds are able to kill competitor bacteria by forming pores in the cytoplasmic membrane and promoting cell wall lysis, causing the release of metabolites and collapse of the proton motive force (Montville *et al.* 1995). These antimicrobials were found to be more effective against Gram-positive bacteria (Työppönen *et al.* 2003a).

Table 16.2 Bacteriocinogenic lactic acid bacteria used as starter/bioprotective cultures in fermented meat products against foodborne pathogens.

Bacteriocin producer	Bacteriocin	Targeted organism	Reference
Ent. faecalis	Enterocin AS-48	*Staph. aureus*	Ananou *et al.* 2005b
Ent. faecium	Enterocin A and Enterocin B	*Cl. botulinum, Cl. perfringens, Ls. monocytogenes* and *Staph. aureus*	Herranz *et al.* 2001
Lb. sakei	Sakacin	*Ls. monocytogenes* and *Staph. aureus*	Castellano *et al.* 2012
Pe. acidilactici	Pediocin PA-1	*Ls. monocytogenes* and *Clostridium perfringens*	Nieto-Lozano *et al.* 2010
Pe. pentosaceus	Pediocin PA-1/AcH	*Ls. monocytogenes*	Kingcha *et al.* 2012

Notes: Cl. = Clostridium; Ent. = Enterococcus; Lb. = Lactobacillus; Ls. = Listeria; Pe. = Pediococcus; Staph. = Staphylococcus.

Nisin, which is synthesized by *Lc. lactis*, is the only bacteriocin approved as a food additive (Al-Holy *et al.* 2012). Numerous other bacteria also produce bacteriocins, as presented in Table 16.2. However, none of these bacteriocins was approved for commercialization as food additives in its purified form.

Dry-fermented sausages have several characteristics that prevent the growth of common food pathogens such as *Staphylococcus aureus, Bacillus cereus* and *Clostridium botulinum* (Työppönen *et al.* 2003a). However, important pathogens such as *Listeria monocytogenes* (Farber and Peterkin 1991) and *Escherichia coli* O157:H7 (Hinkens *et al.* 1996) are found to survive the acidic and highly osmotic environment of fermented sausages. These bacteria present very low infectious doses and the consumption of contaminated sausages may present a serious risk to public health. In 1994, *E. coli* O157:H7 present in dry-fermented sausages sickened 18 people in the United States, leading the US and Canadian health authorities to demand a minimum of 5 log cfu/g reduction of *E. coli* O157:H7 during the dry-fermented sausage manufacturing process (Hinkens *et al.* 1996; Palanichamy *et al.* 2008).

No outbreak has been correlated with the presence of *Ls. monocytogenes* in dry-fermented sausages so far (Leroy and De Vuyst 1999), but the ability of this pathogen to survive in the food matrix has been proven and it represents a potential hazard to humans (Gálvez *et al.* 2007). Other metabolites produced by bacteria also present antimicrobial activity; among these substances, reuterin (β-hydroxypropinaldehyde) – produced by *Lactobacillus reuteri* – is one of the most studied (Arqués *et al.* 2008; El-Ziney and Jakobsen 2009; Shaefer *et al.* 2010). However, there are very few studies in the literature using purified reuterin or *Lb. reuteri* to eliminate foodborne pathogens in fermented sausages. Some authors (Kuelasan and Çakmakçi 2002) showed that reuterin could significantly reduce the population of *Ls. monocytogenes* after 7 days, but it did not eliminate the pathogen on the surface of traditional Turkish sausages. No effect on *Salmonella* spp. growth was observed instead. Muthukumarasamy and Holley (2007) also used *Lb. reuteri* as an adjuvant to a commercial starter culture composed of *Pe. pentosaceus* and *Staph. carnosus* to produce salami artificially contaminated with five strains of *E. coli* O157:H7. The control group – containing only commercial starter culture – was able to reduce the population of *E. coli* O157:H7 by 0.7 log cfu/g after 30 days, whereas the addition of *Lb. reuteri* to the sausage caused an extra 2.7 log cfu/g reduction in the pathogen count. Reuterin seems to be a good antibacterial agent *in vitro* and for application in dairy products (El-Ziney and Jakobsen 2009; Langa *et al.* 2013), but its use to eliminate this foodborne pathogen in sausages is still very limited.

Utilization of bacteriocinogenic LAB as meat starter cultures to control pathogenic bacteria

Although the utilization of purified bacteriocins to eliminate foodborne pathogens from fermented sausage does not seem to be very promising, *in situ* bacteriocin production has been extensively studied with relative success. Ananou *et al.* (2005a) demonstrated that both bacteriocinogenic *Enterococcus faecalis* A-48-32 and *Enterococcus faecium* S-32-81 were able to reduce the population of *Ls. monocytogenes* below detection levels in a sausage model after 9 and 6 days of incubation, respectively. The concentration of bacteriocin ranged between 60 and 80 AU/g, which was well below the level of purified bacteriocin needed to cause the same listerial reduction. In addition, bacteriocin-producing *Lb. curvatus* rapidly (12 hours) caused the reduction of *Ls. monocytogenes* cell counts below detectable levels, while a non-bacteriociogenic commercial starter culture needed 19 days to produce similar results (Benkerroum *et al.* 2005). Other bacteriocin-producing strains of *Pediococcus* and *Lactobacillus* were also reported to inhibit the growth of *Ls. monocytogenes* in dry and semi dry sausages (Hugas *et al.* 1995; Albano *et al.* 2007; Abrams *et al.* 2011).

Nedelcheva *et al.* (2010) demonstrated the ability of the bacteriocinogenic *Lb. plantarum* NBIMCC 2415 to inhibit the growth of undesirable microorganisms, namely *E. coli* ATCC 25922, *E. coli* ATCC 8739, *Proteus vulgaris* G, *Salmonella* sp., *Salmonella abony* NTCC 6017, *Staph. aureus* ATCC 25093, *Staph. aureus* ATCC 6538 P and *Ls. monocytogenes* in meat products ripened at 15–18 °C.

Production of bacteriocins was found to be influenced by a variety of environmental parameters such as temperature, pH, competing bacteria, levels of CO_2, salt, nitrite and others (Gálvez *et al.* 2007). Therefore, the utilization of a certain bacteriocinogenic LAB as a starter culture will not guarantee the production of bacteriocins in the meat matrix. Apparently, conditions offered to LAB have to meet some requirements in order to achieve high synthesis rates of bacteriocin. Optimum growth conditions were suggested to induce a greater production of nisin by *Lc. lactis*, pediocin AcH by *Pe. acidilactici*, leuconocin Lcm1 by *Leuconostoc carnosum* Lm1 and sakacin A by *Lb. sakei* Lb 706 (Yang and Ray 1994). Conversely, stress caused by moderate doses of salt also stimulated the production of bacteriocin by *Lactobacillus pentosus* B96 (Delgado *et al.* 2005) and *Lactobacillus amylovorus* DCE 471 (Neysens *et al.* 2003). In addition, stress caused by the presence of competing Gram-positive organisms was found to induce the synthesis of plantaricin by *Lb. plantarum*, with the bacteriocin itself also able to promote its own production (Maldonado *et al.* 2004). More recently, nitrite was found to reduce significantly the production of bacteriocin by *Lb. curvatus* CWBI-B28 in pork meat (Kouakou *et al.* 2009), which completely abolished this strain's antilisterial activity. Therefore, it is extremely important to understand the factors affecting bacteriocin production by a particular strain before its successful application in foods to reduce or eliminate *Ls. monocytogenes* (Castro *et al.* 2011).

Aymerich *et al.* (2000) examined the influence of additives found in dry-fermented sausages on the production of enterocin by *Ent. faecium* CTC 492. They found that all the ingredients tested (salt, nitrate, nitrite and black pepper), with the exception of nitrate, reduced the amounts of enterocin A and B produced. However, natural fermentation has been used for a very long time as the typical method to produce dry-fermented sausages. Thus, the use of indigenous bacteria present in specific products as their own starter cultures has been suggested, since they are highly adapted to that environment (Villani *et al.* 2007) and may generate significant amounts of bacteriocin (Työppönen *et al.* 2003b).

Albano *et al.* (2007) screened the antimicrobial capacity of 226 LAB strains isolated from a typical naturally fermented Portuguese sausage called 'Alheiras'. These bacteria were tested against pathogens such as *Ls. monocytogenes*, *Staph. aureus*, *E. coli* O157:H7, *Salmonella typhimurium* and *Salmonella enteriditis*. From the initial LAB strains, 14 were found to have antimicrobial effect against *Ls. monocytogenes*, 4 against *Staph. aureus*, but none was able to affect the growth of the Gram-negative bacteria. However, only two strains showed antibacterial activity related to bacteriocin production and they were both identified as *Pe. pentosaceus*. This bacteriocin was recently identified as bacPPK34 and has shown inhibitory activity against *Ls. monocytogenes in vitro* (Abrams *et al.* 2011). Moreover, lactobacilli and pediococci strains isolated from suçuk (typical Turkish fermented sausage) also showed inhibitory activity against *Ls. monocytogenes*, and the utilization of such LAB as starter cultures specifically for suçuk manufacture was recommended (Çon *et al.* 2001; Cosansu *et al.* 2010). A similar study has shown that use of the autochthonous *Pe. acidilactici* MCH14, which produces pediocin PA-1, reduced by ~4 log cfu/g the population of *Ls. monocytogenes* in dry-fermented sausage. This reduction was two times higher than was found for a dry-fermented sausage produced with a non-bacteriocinogenic *Pe. acidilactici* as starter culture (Nieto-Lozano *et al.* 2010).

The synthesis of sakacin K by *Lb. sakei* CTC 494 was found to be temperature and pH dependent (Leroy and De Vuyst, 1999). This bacterium was isolated from Spanish dry-fermented sausages and diverse characteristics of bacteriocin production were evaluated in a model simulating the conditions of the fermenting process. Interestingly, the production of sakacin K was found to achieve the best rates at temperatures between 20 °C and 25 °C and pH 5.0. These values coincide with the typical fermentation temperature and pH of the Spanish sausage. Therefore, it is expected that *Lb. sakei* CTC 494 might produce high concentrations of sakacin K if used as starter culture for these products. This combination could be highly favourable, since sakacin K has been demonstrated to present antilisterial activity (Hugas *et al.* 1995).

Some bacteriocinogenic LAB have been introduced to the market as protective cultures for meat products. For example, Bovamine® Meat Cultures, developed by Texas Tech University (Lubbock, TX, USA), is composed of a mixture of *Lactobacillus acidophilis, Lc. lactis* and *Pe. acidilactici* and is active against *Salmonella* and *E. coli*. Moreover, HOLDBAC® Listeria (*Lactobacillus rhamnosus* and *Propionibacterium freudenreichii*) from Danisco (Copenhagen, Denmark) was specially designed for fermented meat and dairy products to reduce/eliminate *Ls. monocytogenes* from these foods. Chr. Hansen (Hørsholm, Denmark) has released a family of products based on protective cultures, known as SafePro®, to avoid the growth of spoilage and pathogenic bacteria in fresh and fermented meat products. These products contain at least one bacteriocinogenic LAB, such as *Pe. acidilactici, Lb. curvatus* or *Lb. sakei.*

Inhibition of mycotoxinogenic moulds

Another potential hazard found in fermented meats is caused by the growth of mycotoxinogenic fungi. Smoking usually confers protection against the growth of moulds on the surface of dry-fermented meat products (Holley and Patel 2005). However, traditional products, such as Italian sausages and hams, do not receive this type of treatment and are susceptible targets for the growth of undesirable mycotoxinogenic strains. Some of the most common species found in fermented meat products are *Penicillium nordicum, Penicillium verrucosum* and *Aspergillus ochraceus*, which are ochratoxin A (OTA) producers (Larsen *et al.* 2001). Therefore, Italy has imposed a limit of 1 μg/kg of OTA in dry-cured meat products.

More recently, some studies have shown that indigenous yeasts can be used as starter cultures in dry-cured meat products to avoid the growth of *Pen. nordicum* (Virgili *et al.* 2012; Andrade *et al.* 2014).

Production of bioactive peptides by starter culture in fermented meat products

Potential use of starter cultures to improve the healthiness of meat products

The health effects reported for fermented products are thought to be partly associated with low molecular weight peptides liberated from food proteins due to microbial proteolytic activities during fermentation. The complex proteome of meat is known to harbour numerous peptide sequences within its primary structures, and the extensive proteolytic activities during microbial fermentation can facilitate the liberation of these peptides in the final meat product. Nowadays, there are extensive reports or clinical evidence on the health benefits of meat products fermented with starter cultures: the complex endogenous proteolytic systems of meat and starter cultures are known to generate several peptides that can be explored for bioactivity relevant to health promotion.

Proteolytic systems during meat processing

Peptide sequences within the protein structure can be released by endogenous meat and starter culture proteases during ageing or fermentation. Endogenous meat proteases such as calpains, cathepsins, proteasomes and caspases (cysteine-dependent aspartate-directed proteases) have been reported to participate in hydrolytic post-mortem modification of meat ultrastructure during ageing and storage (Sentandreu *et al.* 2003; Huang *et al.* 2011). Moreover, several microbial populations (e.g. *Lactobacillus, Leuconostoc, Pediococcus, Enterococcus* etc.) are present during meat processing (Danilović *et al.* 2011). *Lactobacillu*s spp. are known to possess a comprehensive proteolytic system comprising endopeptidases, aminopeptidases, tripeptidases, dipeptidates and proline-specific peptidases (Savijoki *et al.* 2006). These proteases are mostly metallopeptidases, but a number of the enzymes function as cysteine-type and serine peptidases. Among the several lactobacilli identified, *Lb. sakei* appears to be the most commonly utilized starter culture for meat fermentation (Flores and Toldrá 2011). Therefore, extensive knowledge of its proteolytic system could enhance its use for the production of beneficial compounds during meat processing. Environmental (especially stress-related) changes may have an impact on starter culture gene expression and related proteolytic systems (Hüfner *et al.* 2007). For instance, whole-genome DNA microarray analysis indicated that several genes involved in peptide and amino acid metabolism and transport are differentially upregulated in *Lb. sakei* in the presence of both meat sarcoplasmic and myofibrillar proteins (Xu *et al.* 2014). In particular, proteomic analysis revealed that meat sarcoplasmic (not myofibrillar) proteins induced overexpression of two cysteine-type dipeptidases in *Lb. sakei*, and this resulted in sarcoplasmic protein degradation (Fadda *et al.* 2010a). This is particularly important as meat is not considered a competitive carbon source, and the ability of starter cultures to initiate proteolysis during fermentation depends on the microbial response to environmental changes. The proteases identified in *Lb. sakei* possess distinct cleavage sites with optimum activity near neutral pH and 37–55 °C (Flores and Toldrá 2011). For instance, *Lb. sakei* produces prolyl dipeptidylpeptidase that can cleave meat proteins to liberate X-Pro-type peptides from the N terminus (Sanz and Toldrá 2001). This activity is particularly important in processing meat collagen considering its high proline content. There is evidence that the proteolytic systems of mixed cultures can also interact to provide more extensive proteolysis (Tremonte *et al.* 2010).

Potential bioactivity of peptides released during meat fermentation

Apart from their contribution to the flavour of fermented meat, peptides released during proteolysis can play key roles as bioactive ingredients for health promotion. These peptides are different from histidyl dipeptides, carnosine (β-alanyl-L-histidine) and anserine (N-β-alanyl-1-methyl-L-histidine), which are known to exhibit strong antioxidative activities. Meat proteome contains sarcoplasmic and myofibrillar proteins, which can serve as precursors of peptides. Peptides have been investigated since the 2000s as functional ingredients for health promotion (Udenigwe and Aluko 2012). Although milk proteins remain the most investigated sources of bioactive peptides, meat proteins are gaining particular attention due to the presence of several cryptic untapped peptide sequences within their diverse structure (Udenigwe and Howard 2013). The major bioactivity reported for peptides includes the inhibition of angiotensin-converting enzyme (ACE), which catalyses the blood pressure–regulating pathway (for antihypertensive effects). Moreover, some histidine-containing peptides derived from chicken essence were found to exhibit antioxidative properties (Wu *et al.* 2005), whereas others from beef sarcoplasmic proteins inhibited ACE activity and exhibited antimicrobial activities against pathogenic bacteria (Jang *et al.* 2008).

Several other meat protein hydrolysates and peptides derived from exogenous proteases are known to demonstrate anticancer, antithrombotic and lipid-lowering activities and potential for the treatment of osteoporosis (see the review by Udenigwe and Howard 2013), although there is no indication that these particular components can be produced during meat fermentation. However, Arihara and Ohata (2008) reported that fermentation of raw sausages and dry-cured ham increased the ACE-inhibitory activity of the product extracts, and that the activity of LAB on porcine skeletal muscle proteins produced antihypertensive effects. Moreover, 90-day ripening of fermented sausage resulted in a 17% degree of protein hydrolysis, with a concomitant two- to threefold increase in antioxidative and ACE-inhibitory activities (Vaštag *et al.* 2010). These observations suggest that peptides released in the fermented meat could be the principal bioactive candidates.

The proteolytic activity of LAB is thought to be weak during meat fermentation; however, the starter cultures can provide the acidic environment required for the activity of endogenous meat proteases (Arihara and Ohata 2008). This can facilitate the breakdown of meat proteins into polypeptide products that can be further cleaved by the microbial peptidases (Figure 16.1). This is supported by a recent study demonstrating that the digestibility and release of peptides from myofibrillar and sarcoplasmic proteins in dry-cured ham after simulated gastric digestion were reported to be dependent on maturation time (Paolella *et al.* 2015). Broncano *et al.* (2012) reported that the strongest free radical scavenging (antioxidative) fraction from fermented chorizo sausage contained not only small peptides, but also free amino acids and natural dipeptides, and that peptides

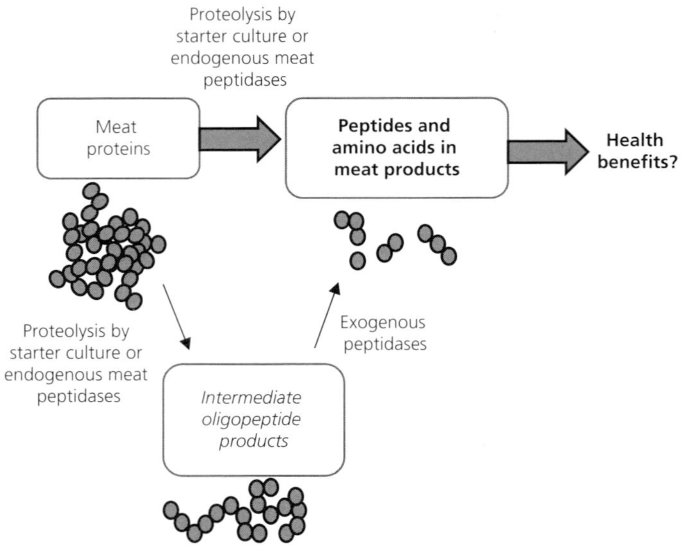

Figure 16.1 Starter culture, exogenous and endogenous meat proteases hydrolyse meat proteins to liberate peptides with potential bioactivity.

are not the major components of the fermented products. The authors suggested that the peptides may have been lost due to extensive hydrolysis during the ripening of chorizo, especially if assimilated by the starter cultures. Besides the use of starter cultures, the addition of exogenous proteases could enhance the peptide profile and healthiness of fermented meat products. This was demonstrated for fermented sausage extract, which exhibited increased radical scavenging and lipid peroxidation inhibitory activities, reducing power and oxidative stability when supplemented with exogenous microbial proteases (Broncano *et al.* 2011).

Probiotic starter cultures for fermented meat products

In addition to the sensorial and safety benefits provided by starter cultures in fermented meat products, the use of bacterial starters as functional microbiota has recently gained more importance as a result of the growing concern of consumers regarding food health and safety issues.

Among the different types of functional food, probiotics represent a large share of the functional food market, being used mainly in dairy beverages, cereal products, infant feeding formulas, fruit juices and ice creams (Santos *et al.* 2003; De Vuyst *et al.* 2008). In the meat industry, the addition of probiotics to products could promote health benefits and contribute to the increase in their consumption (Lücke 2000; De Vuyst *et al.* 2008).

Although the concept of including probiotics in meat products is not entirely new, only a few manufacturers consider the use of meat products as vehicles for probiotics (De Vuyst *et al.* 2008; Rivera-Espinoza and Gallardo-Navarro 2010). Salami containing three intestinal LAB (*Lactobacillus acidophilus, Lb. casei* and *Bifidobacterium* spp.) were produced by a German company in 1998. In the same year, a meat spread containing an intestinal LAB (*Lb. rhamnosus* FERM P-15120) was produced by a Japanese company (Arihara 2006; Toldrá and Reig 2011).

Among meat products, fermented sausages are suitable for incorporating probiotic bacteria, since mild or no heat treatment is usually required by dry-fermented meat products, thus providing suitable conditions for their survival (Arihara 2006; Ammor and Mayo 2007; Khan et 2011). The sausage has to be designed in order to keep the number and viability of probiotic strains in the optimum range. Thus, reduction in pH (<5.0), extended ripening (>1 month), dry or excessive heating has to be avoided if the beneficial effects of the probiotic are to be maintained (De Vuyst *et al.* 2008; Khan *et al.* 2011).

Thus, the incorporation of probiotic bacteria into these products represents a technological challenge, because probiotics are sensitive to curing salts, spices and other ingredients used in the formulation of the fermented sausages (Erkkilä *et al.* 2001). Furthermore, this inclusion requires the use of microorganisms able to persist in the fermentation process and remain viable in order to survive the stomach pH and exert beneficial effects in the intestines (Lücke 2000).

Additionally, the processing of probiotic meat products implies taking into account the appropriateness of the probiotic culture to the target consumer, the intestinal functionality expected for the probiotic species, the rate of survival of the probiotic during food processing and the need for maintenance in the probiotic product of same sensory attributes characterizing the standard product (Andersen 1998; Lücke 2000; Ferreira *et al.* 2003). Many studies report the successful addition of these LAB in fermented meat sausages (Table 16.3).

Several studies suggest the selection of probiotic properties in LAB from commercial starter cultures traditionally used in fermented meat products, since they are already adapted to grow in these conditions (Lücke 2000; Hammes and Hertel 1998; Maragkoudakis *et al.* 2009). Among the LAB used as commercial starters in fermented meat products, *Lb. brevis, Lb. plantarum, Lb. fermentum* and *Pe. pentosaceus* have been characterized as probiotics (Silvi *et al.* 2003; Klingberg *et al.* 2005). Strains of *Lb. sakei* and *Pe. acidilactici* have also been proposed as potential probiotics in meat products, due to their survival under acid conditions and high concentrations of bile (Erkkilä and Petäjä 2000). Probiotic cultures can also be selected from the naturally presented LAB in fermented meat products (Papamanoli *et al.* 2003; Villani *et al.* 2005; Klingberg *et al.* 2005; Rebucci *et al.* 2007; De Vuyst *et al.* 2008).

Lactobacillus rhamnosus CTC1679 was used as a probiotic culture in Fuet (a Spanish low-acid fermented sausage), where it was able to grow, survive and dominate (levels of log 8 cfu/g) the endogenous LAB of the product. The probiotic

Table 16.3 Species of lactic acid bacteria (LAB) used in fermented meat products as probiotics.

LAB Species	Reference
Lb. plantarum, Lb. casei	Khan *et al.* (2011)
Lb. paracasei, Lb. casei, Lb. rhamnosus	Macedo *et al.* (2008)
Pe. acidilactici PA-2, *Lb. sakei* Lb3	De Vuyst *et al.* (2008)
Lb. casei, Lb. paracasei	Rebucci *et al.* (2007)
Lb. plantarum, Lb. paracasei	Pennacchia *et al.* (2004)
Lb. gasseri, Lb. rhamnosus, Lb. paracasei subsp. *paracasei, Lb. casei, Bif. lactis*	Erkkilä *et al.* (2001)
Lb. rhamnosus LC-705, *Lb. rhamnosus* VTT-97800, *Lb. rhamnosus* VTT, *Lb. rhamnosus* GG	Erkkilä and Petäjä (2000)
Lb. gasseri	Arihara *et al.* (1998)
Lb. rhamnosus, Lb. paracasei	Sameshima *et al.* (1998)
Lb. casei LC-01, *Bif. lactis* Bb-12	Andersen (1998)

Notes: Bif. = Bifidobacterium; Lb. = Lactobacillus; Pe. = Pediococcus.

culture also prevented the growth of *Ls. monocytogenes* throughout the ripening process and eliminated *Salmonella enterica* (Rubio *et al.* 2014b). The antimicrobial activity towards *Ls. monocytogens* could be attributed to a decrease of pH and water activity rather than a bacteriocin-related effect, because *Lb. rhamnosus* CTC1679 showed acid-related antagonistic activity *in vitro* against *Ls. monocytogenes* (Rubio *et al.* 2014b).

Criteria for the selection of probiotic cultures for meat products

The criteria for a microbial culture to be considered probiotic are stomach acidity resistance, lysozyme and bile resistance and the ability to colonize the human intestinal tract using mechanisms of adhesion or binding to intestinal cells (Lücke 2000; Pidcock *et al.* 2002; Papamanoli *et al.* 2003; De Vuyst *et al.* 2008). Other authors have also included the ability to tolerate pancreatic enzymes as a required characteristic of probiotic cultures (Ruiz-Moyano *et al.* 2008).

Additionally to the criteria already described, the probiotic bacteria need to have GRAS status (Pan *et al.* 2009). Currently, this concept also includes antibiotic resistance evaluated by the Qualified Prediction Security Programme suggested by EFSA (European Food Safety Authority). The ability of probiotic bacteria used in meat products to resist some antibiotics can be genetically transmitted to other bacteria. Scientific studies report genetic determinants for bacterial resistance to chloramphenicol, erythromycin and tetracycline (Ammor and Mayo 2007). Normally, lactic bacteria are sensitive to penicillin G, ampicillin, tetracycline, erythromycin, chloramphenicol and aminoglycosides, quinolones and glycopeptides (Maragkoudakis *et al.* 2009). Thus, the selection of probiotic cultures for meat products implies confirmation of the absence of antibiotic resistance transferable genes in selected strains (Ammor and Mayo 2007).

However, among the criteria for the selection of probiotic cultures, the main condition to be evaluated is the ability of strains to promote beneficial effects and to prevent diseases in the host through probiotic–host interactions (FAO-WHO 2006). These effects on human health may occur in three different ways according to the specificity of the strain: antagonist action against other microorganisms in the same environment (by nutrient competition, bacteriocin production or competitive exclusion); barrier effect on the intestinal mucosa; and boosting of the immune system (De Vuyst *et al.* 2008; Pan *et al.* 2009).

Technological characteristics

To be included in fermented meat products, probiotic bacteria need to maintain their viability against the adverse conditions generated in their production: low pH (<5.0), high salt content (2–3%), high nitrite content (around 120 ppm) and low water activity (<0.85). The probiotics should also be able to grow fast during fermentation, be easily cultivated on an industrial scale, resist freezing and lyophilization processes, provide a longer shelf life to the product as well as contribute to the sensory quality of the final product (Erkkilä *et al.* 2001).

Probiotic cultures could be added to fermented sausages as part of the starter culture or as an additional culture incorporated during mass mixing (Figure 16.2);

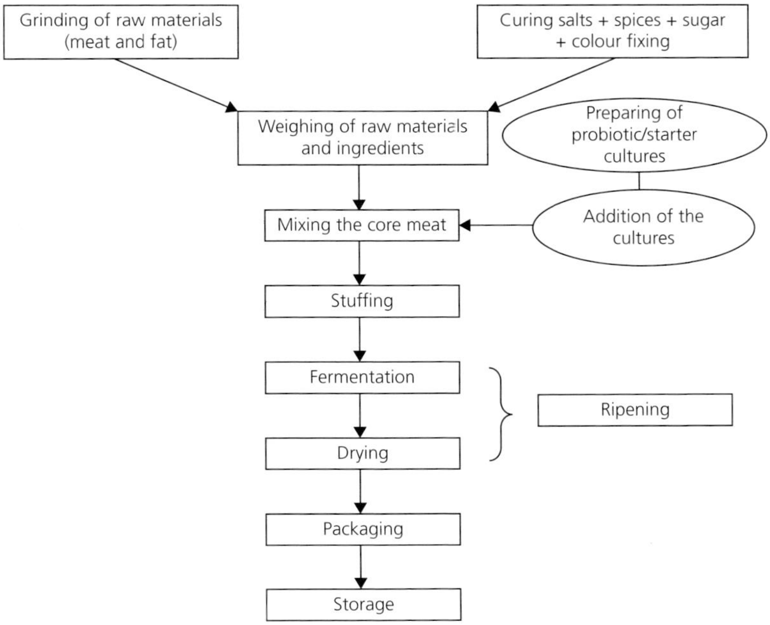

Figure 16.2 Basic flowchart of the processing of fermented dry sausage with the addition of probiotic cultures. Source: Macedo *et al.* (2012).

they may be added to the core meat as liquid inoculum, in high concentrations or lyophilized. However, the addition of lyophilized culture could delay fermentation and reduce culture viability in the final product. These effects can be reduced with the culture microencapsulation prior to lyophilization. This procedure is also indicated when probiotic strains are inhibited by ingredients of the sausage composition (Leroy *et al.* 2006; Muthukumarasamy and Holley 2007).

Muthukumarasamy and Holley (2007) tested the microencapsulation of *Lb. reuteri* ATCC 55730 in alginate for use in fermented meat products and found no adverse effect on the sensory quality of the product. Despite the microcapsules being visible to the naked eye, they were detected as fat particles by the panelists due to their size and colour similarity.

Rivera-Espinoza and Gallardo-Navarro (2010) encapsulated *Bifidobacterium longum* and *Lb. reuteri* in alginate to increase the survival of probiotics in fermented meat.

Recently, Sidira *et al.* (2014) and Sidira *et al.* (2015a) investigated the effect of immobilized probiotic *Lb. casei* ATCC 393 on wheat grains on spoilage and pathogenic microorganisms and on the evolution of flavour compounds in probiotic dry-fermented sausages during ripening, respectively. Authors found a more prominent resistance to spoilage in samples containing immobilized cells and the sausages containing high amounts of the immobilized starter culture showed an improved profile of aroma-related compounds at the end of the ripening process. It is important to confirm that the LAB used as probiotics are able to cause a fast decrease in pH values during fermentation of sausages to contribute to prevention of the accumulation of biogenic amines, which are harmful to health (Ammor and Mayo 2007). Together with fast acidification of the meat mixture, the use of cultures resistant to curing salts is another important condition for the production of probiotic sausages (Papamanoli *et al.* 2003).

Physiological characteristics

Tolerance to acidity and bile are two fundamental properties that indicate the ability of a probiotic microorganism to survive through the gastrointestinal tract, resisting the acidic conditions of the stomach and the bile salts in the initial portion of the small intestine (Erkkilä and Petäjä 2000; Annuk *et al.* 2003).

Due to the sensitivity of most bacteria to the low pH of the stomach, probiotic bacteria have to be ingested with food, because it acts as a buffer on the high acidity of the stomach, allowing the survival of the bacteria during gastric transit (Collado and Sanz 2006). Meat, as well as milk, has buffer characteristics in an acid environment and can thereby protect the probiotic from the adverse environment of the stomach (Työppönen *et al.* 2003b).

Taking into account the pH conditions of the stomach and the digestion time, probiotic bacteria ingested with food must be capable of resisting a pH value of 3 for a period of 2–4 hours to allow their survival during gastric transit. Macedo *et al.* (2005) found that *Lactobacillus paracasei* used in probiotic salami was able to resist and grow in a medium at pH 3, showing a 20% increase in the initial number of cells during 4 hours of exposure to this acid condition.

Pennacchia *et al.* (2004) tested the resistance of *Lactobacillus* isolated from 10 different types of salami to low pH. The authors found that among 14 LAB surviving at pH 2.5 for 3 hours, 5 isolates affered to the *Lb. casei* group. Bile resistance is also required as a characteristic of a probiotic culture. According to Erkkilä and Petäjä (2000) and Pennacchia *et al.* (2004), its concentration in the human intestinal tract is 0.3%, thus this is the critical concentration used to select probiotic bacteria. Papamanoli *et al.* (2003) consider a strain tolerant to bile salts when its cellular population decreases from 10^6–10^7 cfu/mL to 10^5 cfu/mL in a 4–hour period.

Meat has also been reported to protect bacteria against bile. During meat sausage processing, *Lactobacillus* added to the batter is encapsulated by the matrix consisting of meat and fat. Due to the protection exerted by the food, the survival of *Lactobacillus in vivo* during transit through the stomach and intestine appears to be higher than that observed by the *in vitro* exposure of the microorganisms to low pH and bile salt (Erkkilä and Petäjä 2000; Työppönen *et al.* 2003b).

Biogenic amines are produced by the microbial decarboxylation of amino acids present in meat products, either by naturally occurring microorganisms or from the starter culture. In fermented meat products, biogenic amine producers have a favourable environment due to the high protein content and the intense proteolytic activity occurring during ripening. However, some strains of *Lactobacillus* are able to produce amino acid decarboxylase that prevents the accumulation of biogenic amines in the product. Consequently, the selection of probiotic bacteria to be used in fermented meat products must also be based on their ability to oxidate biogenic amines formed in the product and/or to prevent the formation of new amines. In fermented meat products, amine oxidase activity was determined in strains of *Lb. casei* and *Lb. plantarum* (Leroy *et al.* 2006; Ammor and Mayo 2007).

Ergönül and Kundakçi (2011) observed low biogenic amine contents in a Turkish fermented sausage manufactured by using three different probiotic starter culture combinations (*Lb. casei, Lb. acidophilus* or their combination).

The capabilities of probiotics to dominate the meat environment and to inhibit the growth of pathogenic bacteria in fermented meat products are also desirable characteristics in potential probiotic LAB and were reported in several studies.

Probiotic *Lb. plantarum* 299V, used as a starter culture in Spanish fermented sausages and inoculated at log 5 cfu/g, achieved and maintained high counts until the end of ripening and storage (log 8 cfu/g), co-dominating (60%) with the endogenous microbiota and producing functional sausages with a satisfactory overall sensory quality (Rubio *et al.* 2013).

Mahoney and Henriksson (2003) tested the inhibition of colonization and virulence of *Ls. monocytogenes* in the intestinal tract of rats by the consumption of fermented meat products with the addition of starter cultures, probiotic cultures and *Ls. monocytogenes*. The results showed that the starter culture consisting of *Pe. pentosaceus* and *Staph. xylosus*, and the probiotic culture consisting of *Lb. acidophilus*, *Lb. paracasei* and *Bifidobacterium* sp., were able to inhibit the growth of *Ls. monocytogenes* during its passage through the gastrointestinal tract.

Health-promoting effects associated with the consumption of probiotic meat products

It is recognized that the estimated number of viable cells of probiotic bacteria to be ingested to exert beneficial effects and temporary colonization of the intestine is about 10^9–10^{10} cfu/g of product, in accordance with the counts of 10^6–10^8 viable cells found in 1 g of faeces. Therefore, in a fermented meat product containing 10^8 cfu/g, the minimum daily consumption might be 10–100 g of product (Penna 2002; Työppönen *et al.* 2003b). Rivera-Espinoza and Gallardo-Navarro (2010) recommend a concentration of viable probiotic cells of at least 10^8–10^9 cfu/g of product to obtain the physiological effects associated with the use of probiotic food.

Most research involving probiotics in meat products focuses on the survival of these species in the matrix and on their potential influence on the technological and sensory characteristics of the final product. Few studies report the effects of consumption of these products on host health (De Vuyst *et al.* 2008). This is mainly due to the fact that *in vivo* tests are expensive, and require more time for experimentation and approval by ethics committees (Pan *et al.* 2009).

Jahreis *et al.* (2002) evaluated the effect of daily consumption of 50 g of probiotic salami containing *Lb. paracasei* LTH 2579 on the immunity system and blood triglycerides and cholesterol levels of healthy volunteers for a period of a few weeks, obtaining moderately satisfactory results. Although an effect was observed on the immunity of the host, a non-significant effect was observed on plasmatic lipid levels.

In laboratory animals probiotic administration has been shown to decrease the blood cholesterol level and increase the feed conversion rate (Alkhalf *et al.* 2010). *Lb. plantarum* administration was reported to increase CD-8 and CD-4 lymphocytes in lab rats (Mao *et al.* 1996).

More recently, Rubio *et al.* (2014a) carried out a human intervention study to assess the ability of the probiotic strain *Lb. rhamnosus* CTC1679 to survive through the gastrointestinal tract (GIT). This study consisted of 5 healthy volunteers who consumed 25 g of reduced-fat and sodium low-acid fermented sausage (Fuet) every day for 21 days. *Lb. rhamnosus* CTC1679 was used as a starter culture in the production of Fuet. These authors found that *Lb. rhamnosus* CTC1679 produced a transient colonization of the human gastrointestinal tract and persisted during the period of ingesting Fuet. Even after 3 days of non-consumption, the strain was still recovered in the faeces of all the volunteers.

In addition to these studies, probiotic bacteria have been assessed for nutritional enhancement of foods and production of nutraceutical compounds in meat products. Ammor and Mayo (2007) describe studies related to the high production of folate by a genetically modified *Lb. plantarum*. Likewise, the production of conjugated linoleic acid (CLA), which has anticancer, antiobesity, antidiabetic and antiatherogenic properties as well as a stimulating effect on the immune response, has been reported for some probiotic bacteria. Thus, the property of some probiotic bacteria to produce micronutrients and nutraceutical compounds may allow *in situ* fortification of meat products, making them more nutritious and healthy.

The combined effect of the addition of probiotics and other active ingredients such as dietary fibre to meat products has also been studied. Sayas-Barberá *et al.* (2012) reported that the addition of *Lb. casei* CECT 475 to a traditional Spanish dry-cured sausage (*Longaniza de Pascua*) accelerates the curing process, whereas the incorporation of 1% orange fibre promotes the growth and survival of lactobacilli and micrococci, enhancing the microbial quality and safety of the sausages.

Genetics and biotechnology of starter cultures used in fermented meat products

The use of more competitive, well-adapted starter cultures with high metabolic capacities can beneficially affect the quality and safety of fermented meat products while preserving their traditional characteristics (Leroy *et al.* 2006). However, the isolation and selection of starter cultures to be used in meat fermentation present a considerable challenge in standardizing and managing the quality of dry-fermented sausages. The genetic potential of starter culture strains intended to be applied as meat starter cultures should be considered to guarantee safety and technological properties. In this context, genetic tools have been applied to develop multifunctional starter culture strains containing all desirable metabolic features to achieve flavour, texture and safety in fermented meat. Some of these genetic technologies include genetic engineering, microarray analysis, transcriptomics and proteomics.

In genetic engineering, the recombinant DNA technique has been applied to produce appropriate starter cultures possessing the technical and metabolic characteristics for specific fermentation requirements (e.g. accelerated acid production, health-promoting properties, overproduction of bacteriocins or particular enzymes necessary for degradation of undesirable factors; Holzapfel 2002). Genome analysis has revealed survival strategies as well as metabolic properties enabling starter cultures to compete effectively in the meat environment (Chaillou *et al.* 2005). The transcriptional and functional examination of starter cultures in meat matrix has indicated that stress-tolerance genes are induced to support the ability of these organisms to compete and grow in this harsh fermentation environment (e.g. Hüfner *et al.* 2007). Proteomic approaches have focused on the functionality of detected genes contributing to the connection between genome sequences and their biological role, in which protein expression, molecular responses and growth mechanisms of LAB strains were investigated in the presence of meat proteins (Fadda *et al.* 2010b). Thus, the application of genomic tools has exponentially increased knowledge of meat starter cultures, opening new opportunities for better performance and application of these microorganisms in industrial settings. However, new methods of analysis, such as metabolomics and metagenomics, should be further employed to investigate characteristics and select starter cultures in fermented meat.

The isolation of strains able to carry out fermentation processes is extremely important in the selection of starters (Cocolin *et al.* 2011). For many years, culture-dependent methods based on conventional culturing followed by phenotypical and/or genotypical (e.g. sequencing, genetic fingerprinting) identification of a selected subset of purified isolates have been used to identify the members of the LAB communities associated with fermented food products (Rantsiou and Cocolin 2008). More recently, modern molecular approaches for the analysis of nucleic acids have been applied during the traditional fermentation of sausages to allow better detection, identification and characterization of microorganisms (Flores and Toldrá 2011). For instance, denaturing gradient gel electrophoresis (DGGE) has been employed as a culture-independent approach that allows profiling of microbial populations and follows their dynamics during fermentation and ripening (Cocolin *et al.* 2011). Aquilanti *et al.* (2007) used the DGGE technique to investigate the ecology of the bacteria and yeast of the traditional Italian *Ciauscolo* salami and found that, among the bacteria, *Lb. curvatus, Lb. plantarum* and *Staph. xylosus* were the prevalent species, while *D. hansenii* was the prevalent species among the yeasts. Moreover, a combination of both culture-dependent and culture-independent methods has been proved favourable for a better description of LAB populations involved in the production of fermented meat products (Nguyen *et al.* 2013).

Conclusion

Fermented sausages already satisfy some current food consumption trends: ready-to-eat, long shelf life and versatility of use (individually or garnishing other plates). More recently, demand for safe and functional meat products has arisen from the growing consumer concern about safer and healthier foods. Thus, bacteriocin or other antimicrobial metabolite-producing starter cultures may be used to eliminate foodborne pathogens and extend the shelf life of fermented sausages. Many peptides with antihypertensive and antioxidant effects can be generated by starter cultures during the meat fermentation process and play key roles as bioactive ingredients for health promotion. The functional value of fermented meat products can also be improved with the use of probiotic starter cultures that promote beneficial effects on human health. In addition, probiotic bacteria have been assessed for nutritional enhancement or production of nutraceutical compounds in meat products. Folate and conjugated linoleic acid production has been reported for some probiotic bacteria, which may allow *in situ* fortification of meat products, making them more nutritious and healthy. Thus, the use of starter cultures able to produce compounds that can improve safety, nutritional and functional value is a great opportunity for the development of innovative meat products.

References

Abrams, D., Barbosa, J., Albano, H. and Teixeira, P. (2011) Characterization of bacPPK34 a bacteriocin produced by *Pediococcus pentosaceus* strain K34 isolated from 'Alheira'. *Food Control*, **22**, 940–946.

Ahmad, S. and Amer, B. (2013) Sensory quality of fermented sausages as influenced by different combined cultures of lactic acid bacteria fermentation during refrigerated storage. *Food Processing and Technology*, **4**, 2–8.

Albano, H., Oliveira, M., Aroso, R. and Teixeira, P. (2007) Antilisterial activity of lactic acid bacteria isolated from 'Alheiras' (traditional Portuguese fermented sausages), *in situ* assays. *Meat Science*, **76**, 796–800.

Al-Holy, M.A., Al-Nabulsi, A., Osaili, T.M., Ayyash, M.M. and Shaker, R.R. (2012) Inactivation of *Listeria innocua* in brined white cheese by acombination of nisin and heat. *Food Control*, **23**, 48–53.

Alkhalf, A., Alhaj, M. and Al-Homidan, I. (2010) Influence of probiotic supplementation on immune response of broiler chicks. *Egypt Poultry Science*, **30**, 271–280.

Ammor, S., Tauveron, G., Dufour, E. and Chevallier, I. (2006) Antibacterial activity of lactic acid bacteria against spoilage and pathogenic bacteria isolated from the same meat small-scale facility 1—Screening and characterization of the antibacterial compounds. *Food Control*, **17**, 454–461.

Ammor, M. and Mayo, B. (2007) Selection criteria for lactic acid bacteria to be used as functional starter cultures in dry sausage production: An update. *Meat Science*, **76**, 138–146.

Ananou, S., Garriga, M., Hugas, M. *et al.* (2005a) Control of *Listeria monocytogenes* in model sausages by enterocin AS-48. *International Journal of Food Microbiology*, **103**, 179–190.

Ananou, S., Maqueda, M., Martínéz-Bueno, M., Gálvez, A. and Valdivia, E. (2005b) Control of *Staphylococcus aureus* in sausages by enterocin AS-48. *Meat Science*, **71**, 549–556.

Andersen, L. (1998) Fermented dry sausages produced with the admixture of probiotic cultures. In *Proceedings of the 44th International Commitment of Meat Science and Technology* (pp. 826–827), Institut de Recerca i Tecnologia Agroalimentaries, Barcelona.

Andrade, M.J., Córdoba, J.J., Casado, E.M., Córdoba, M.G. and Rodríguez, M. (2010) Effect of selected strains of *Debaryomyces hansenii* on the volatile compound production of dry fermented sausage 'salchichón'. *Meat Science*, **85**, 256–264.

Andrade, M.J., Thorsen, L., Rodríguez, A., Córdoba, J.J. and Jespersen, L. (2014). Inhibition of ochratoxigenic moulds by *Debaryomyces hansenii* strains for biopreservation of dry-cured meat products. *International Journal of Food Microbiology*, **170**, 70–77.

Annuk, H., Shchepetova, J., Kullisaar, T., Songisepp, E., Zilmer, M. and Mikelsaar, M. (2003) Characterization of intestinal lactobacilli as putative probiotic candidates. *Journal of Applied Microbiology*, **94**, 403–412.

Aquilanti, L., Santarelli, S., Silvestri, G., Osimani, A., Petruzzelli, A. and Clementi, F. (2007) The microbial ecology of a typical Italian salami during its natural fermentation. *International Journal of Food Microbiology*, **120**, 136–145.

Arihara, K. (2006) Strategies for designing novel functional meat products. *Meat Science*, **74**, 219–229.

Arihara, K. and Ohata, M. (2008) Bioactive compounds in meat. In F. Toldrá (ed.), *Meat Biotechnology* (pp. 231–249), New York: Springer.

Arihara, K., Ota, H., Itohm M. *et al.* (1998) *Lactobacillus acidophilus* group lactic acid bacteria applied to meat fermentation. *Journal of Food Science*, **63**, 544–547.

Aro, J.M.A., Nyam-Osor, P., Tsuji, K., Shimada, K., Fukushima, M. and Sekikawa, M. (2010) The effect of starter cultures on proteolytic changes and amino acid content in fermented sausages. *Food Chemistry*, **119**, 279–285.

Arqués, J.L., Rodríguez, E., Nuñes, M. and Medina M. (2008) Antimicrobial activity of nisin, reuterin, and the lactoperoxidase system on *Listeria monocytogenes* and *Staphylococcus aureus* in cuajada, a semisolid dairy product manufactured in Spain. *Journal of Dairy Science*, **91**, 70–75.

Aymerich, T., Artigas, M.G., Garriga, M., Monfort, J.M. and Hugas, M. (2000) Effect of sausage ingredients and additives on the production of enterocin A and B by *Enterococcus faecium* CTC492. Optimization of *in vitro* production and anti-listerial effect in dry fermented sausages. *Journal of Applied Microbiology*, **88**, 686–694.

Barrière, C., Centeno, D., Lebert, A., Leroy-Sétrin, S., Berdagué, J.L. and Talon, R. (2001) Roles of superoxide dismutase and catalase of *Staphylococcus xylosus* in the inhibition of linoleic acid oxidatíon. *Microbiology Letters*, **201**, 181–185.

Benkerroum, N., Daoudi, A., Hamraoui, T. *et al.* (2005) Lyophilized preparations of bacteriogenic *Lactobacillus curvatus* and *Lactococcus lactis* subsp. *Lactis* as potential protective adjuncts to control *Listeria monocytogenes* in dry-fermented sausages. *Journal of Applied Microbiology*, **98**, 56–63.

Bonomo, M.G., Ricciardi, A., Zotta, T., Parente, E. and Salzano, G. (2008) Molecular and technological characterization of lactic acid bacteria from traditional fermented sausages of Basilicata region (Southern Italy). *Meat Science*, **80**, 1238–1248.

Bonomo, M.G., Ricciardi, A., Zotta, T. and Salzano, G. (2009) Technological and safety characterization of coagulase-negative staphylococci from traditionally fermented sausages of Basilicata region (Southern Italy). *Meat Science*, **83**, 15–23.

Borch, E., Berg, H. and Holst, O. (1999) Heterolactic fermentation by a homofermentative *Lactobacillus sp.* during glucose limitation in anaerobic continuous culture with complete cell recycle. *Journal of Applied Bacteriology*, **71**, 265–269.

Broncano, J.M., Otte, J., Petrón, M.J. and Timón, M.L. (2012) Isolation and identification of low molecular weight antioxidant compounds from fermented 'chorizo' sausages. *Meat Science*, **90**, 494–501.

Broncano, J.M., Timón, M.L., Parra, V. and Petrón, M.J. (2011) Use of proteases to improve oxidative stability of fermented sausages by increasing low molecular weight compounds with antioxidant activity. *Food Research International*, **44**, 2655–2659.

Casaburi, A., Aristoy, M.C., Cavella, S. *et al.* (2007) Biochemical and sensory characteristics of traditional fermented sausages of Vallo di Diano (Southern Italy) as affected by the use of starter cultures. *Meat Science*, **76**, 295–307.

Casaburi, A., Monaco, R.D., Cavella, S., Toldrá, F., Ercolini, D. and Villani, F. (2008) Proteolytic and lipolytic starter cultures and their effect on traditional fermented sausages ripening and sensory traits. *Food Microbiology*, **25**, 335–347.

Casquete, R., Benito, M.J., Martín, A., Ruiz-Moyano, S., Córdoba, J.J. and Córdoba, M.G. (2011) Role of an autochthonous starter culture and the protease EPg222 on the sensory and safety properties of a traditional Iberian dry-fermented sausage 'salchichón'. *Food Microbiology*, **28**, 1432–1440.

Castellano, P., Aristoy, M.C., Sentandreu, M.A., Vignolo, G. and Toldrá, F. (2012) *Lactobacillus sakei* CRL1862 improves safety and protein hydrolysis in meat systems. *Journal of Applied Microbiology*, **113**, 1407–1416.

Castellano, P., Belfiore, C., Fadda, S. and Vignolo, G. (2008) A review of bacteriocinogenic lactic acid bacteria used as bioprotective cultures in fresh meat produced in Argentina. *Meat Science*, **79**, 483–499.

Castro, M.P., Palavecino, N.Z., Herman, C., Garro, O.A. and Campos, C.A. (2011) Lactic acid bacteria isolated from artisanal dry sausages: Characterization of antibacterial compounds and study of the factors affecting bacteriocin production. *Meat Science*, **87**, 321–329.

Chaillou, S., Champomier-Vergès, M.C., Cornet, M. *et al.* (2005) The complete genome sequence of the meat-borne lactic acid bacterium *Lactobacillus sakei* 23K. *Nature Biotechnology*, **23**, 1527–1533.

Cocolin, L., Dolci, P. and Rantsiou, K. (2011) Biodiversity and dynamics of meat fermentations: The contribution of molecular methods for a better comprehension of a complex ecosystem. *Meat Science*, **89**, 296–302.

Collado, M.C. and Sanz, Y. (2006) Method for direct selection of potencially probiotic *Bifidobacterium* strains from human feces based on their acid-adaptation ability. *Journal of Microbiological Methods*, **66**, 560–563.

Coloretti, F., Tabaneli, G., Chiavari, C. and Montanari, C. (2014) Effect of wine addition on microbiological characteristics, volatile molecule profiles and biogenic amine contents in fermented sausages. *Meat Science*, **96**, 1395–1402.

Çon, A.H., Gökalp, H.Y. and Kaya, M. (2001) Antagonistic effect on *Listeria monocytogenes* and *L. innocua* of a bacteriocin-like metabolite produced by lactic acid bacteria isolated from sucuk. *Meat Science*, **59**, 437–441.

Cosansu, S., Geomaras, I., Ayhan, K. and Sofos, J.N. (2010) Control of *Listeria monocytogenes* by bacteriocin-producing *Pediococcus acidilactici* 13 and its antimicrobial substance in a dry fermented sausage sucuk and in turkey breast. *Journal of Food Nutrition Research*, **49**, 206–214.

Danilović, B., Joković, N., Petrović, L. and Savic, D. (2011) The characterisation of lactic acid bacteria during the fermentation of an artisan Serbian sausage (Petrovská Klobása). *Meat Science*, **88**, 668–674.

De Vuyst, L., Falony, G. and Leroy, F. (2008) Probiotics in fermented sausages. *Meat Science*, **80**, 75–78.

Delgado, A., Brito, D., Peres, C. and El Bakali, N.B. (2005) Bacteriocin production by *Lactobacillus pentosus* B96 can be expressed as a function of temperature and NaCl concentration. *Food Microbiology*, **22**, 721–726.

Di Cagno, R., Lòpez, C.C., Tofalo, R. *et al.* (2008) Comparison of the compositional, microbiological, biochemical and volatile profile characteristics of three Italian PDO fermented sausages. *Meat Science*, **79**, 224–235.

Diaz, O., Fernandez, M., Garcia de Fernando, G.D., de la Hoz, L. and Ordoñez, J.A. (1997) Proteolysis in dry fermented sausages: The effect of selected exogenous proteases. *Meat Science*, **46**, 115–128.

El-Ziney, M.G. and Jakobsen, M. (2009) Effectiveness of reuterin alone and in combination with nisin or other food contact surfaces sanitizers and cleaners for disinfection of stainless steel surfaces contaminated with *Escherichia coli* and *Listeria innocua*. *International Journal of Food, Agriculture and Environment*, **7**, 145–149.

Ergönül, B. and Kundakçi, A. (2011) Microbiological attributes and biogenic amine content of probiotic Turkish fermented sausage. *Journal für Verbraucherschutz und Lebensmittelsicherheit*, **6**, 49–56.

Erkkilä, S. and Petäjä, E. (2000) Screening of commercial meat starter cultures at low pH and in the presence of bile salts for potential probiotic use. *Meat Science*, **55**, 297–300.

Erkkilä, S., Suihko, M.L., Eerola, S., Petäjä, E. and Mattila-Sandholm, T. (2001) Dry sausage fermented by *Lactobacillus rhamnosus* strains. *International Journal of Food Microbiology*, **64**, 205–210.

Essid, I. and Hassouna, M. (2013) Effect of inoculation of selected *Staphylococcus xylosus* and *Lactobacillus plantarum* strains on biochemical, microbiological and textural characteristics of a Tunisian dry fermented sausage. *Food Control*, **32**, 707–714.

Fadda, S., Anglade, P., Baraige, F. *et al.* (2010a) Adaptive response of *Lactobacillus sakei* 23K during growth in the presence of meat extracts: A proteomic approach. *International Journal of Food Microbiology*, **142**, 36–43.

Fadda, S., López, C. and Vignolo, G. (2010b) Role of lactic acid bacteria during meat conditioning and fermentation: Peptides generated as sensorial and hygienic biomarkers. *Meat Science*, **86**, 66–79.

FAO-WHO (2006) Probiotics in Food: Health and Nutritional Properties and Guidelines for Evaluation. FAO Food and Nutrition paper 85. http://www.fao.org/3/a-a0512e.pdf (accessed 23 February 2015).

Farber, J. and Peterkin, P. (1991) *Listeria monocytogenes*: A food pathogen. *Microbiology Reviews*, **55**, 476–511.

Ferreira, C.L.L.F. (2003) Grupo de bactérias láticas – Caracterização e aplicação tecnológica de bactérias probióticas. In C. Ferreira (ed.), *Prebióticos e probióticos: atualização e prospecção* (pp. 7–33), Viçosa: Célia L.L.F. Ferreira.

Fernández, M., Ordóñez, J.A., Bruna, J.M. and de la Hoz, L. (2000) Accelerated ripening of dry fermented sausages. *Trends in Food Science and Technology*, **11**, 201–209.

Fiorentini, A.M., Sawitzki, M.C., Bertol, T.M. *et al.* (2009) Phenotypic and molecular characterization of *Staphylococcus xylosus*: Technological potential for use in fermented sausage. *Brazilian Archives of Biology and Technology*, **52**, 737–746.

Flores, M. and Toldrá, F. (2011) Microbial enzymatic activities for improved fermented meats. *Trends in Food Science and Technology*, **22**, 81–90.

Fonseca, S., Cachaldora, A., Gómez, M. and Carballo, J. (2013) Effect of different autochthonous starter cultures on the volatile compounds profile and sensory properties of Galician chorizo, a traditional Spanish dry fermented sausage. *Food Control*, **33**, 6–14.

Gálvez, A., Abriouel, H., López, R.L. and Ben Omar, N. (2007) Bacteriocin-based strategies for food biopreservation. *International Journal of Food Microbiology*, **120**, 51–70.

Gøtterup, J., Olsen, K., Knøchel, S., Tjener, K., Stahnke, L.H. and Møller, J.K.S. (2008) Colour formation in fermented sausages by meat-associated staphylococci with different nitrite- and nitrate-reductase activities. *Meat Science*, **78**, 492–501.

Hammes, W.P., Bantleon, A. and Min, S. (1990) Lactic acid bacteria in meat fermentation. *FEMS Microbiology Review*, **87**, 165–174.

Hammes, W.P. and Hertel, C. (1998) New developments in meat starter cultures. *Meat Science*, **49**, 125–138.

Herranz, C., Casaus, P., Mukhopadhyay, S. and Cintas, L.M. (2001) *Enterococcus faecium* P21: A strain occurring naturally in dry-fermented sausages producing the class II bacteriocins enterocin A and enterocin B. *Food Microbiology*, **18**, 115–131.

Hinkens, J., Faith, N., Lorang, T. and Luchansky, J.B. (1996) Validation of pepperoni processes for control of *Escherichia coli* O157:H7. *Journal of Food Protection*, **59**, 1260–1266.

Holley, R.A. and Patel, D. (2005) Improvement in shelf-life and safety of perishable foods by plant essential oils and smoke antimicrobials. *Food Microbiology*, **22**, 273–292.

Holzapfel, W.H. (2002) Appropriate starter culture technologies for small-scale fermentation in developing countries. *International Journal of Food Microbiology*, **75**, 197–212.

Huang, F., Huang, M., Xu, X. and Zhou, G. (2011) Influence of heat on protein degradation, ultrastructure and eating quality indicators of pork. *Journal of the Science of Food and Agriculture*, **91**, 443–448.

Hüfner, E., Markieton, T., Chaillou, S., Crutz-Le Coq, A.M., Zagorec, M. and Hertel, C. (2007) Identification of *Lactobacillus sakei* genes induced during meat fermentation and their role in survival and growth. *Applied and Environmental Microbiology*, **73**, 2522–2531.

Hugas, M., Garriga, M., Aymerich, M.T. and Monfort, J.M. (1995) Inhibition of *Listeria* in dry fermented sausages by the bacteriocinogenic *Lactobacillus sake* CTC494. *Journal of Applied Microbiology*, **79**, 322–330.

Hughes, M.C., Kerry, J.P., Arendt, E.K., Kenneally, P.M., McSweeney, P.L. and O'Neill, E.E. (2002) Characterization of proteolysis during the ripening of semi-dry fermented sausages. *Meat Science*, **62**, 205–216.

Jahreis, G., Vogelsang, H., Kiessling, G., Schubert, R., Bunte, C. and Hammes, W.P. (2002) Influence of probiotic sausage (*Lactobacillus paracasei*) on blood lipids and immunological parameters of health volunteers. *Food Research International*, **35**, 133–138.

Jang, A., Jo, C., Kang, K.S. and Lee, M. (2008) Antimicrobial and human cancer cell cytotoxic effect of synthetic angiotensin-converting enzyme (ACE) inhibitory peptides. *Food Chemistry*, **107**, 327–336.

Khan, M.I., Arshad, M.S., Anjum, F.M., Sameen, A., Aneeq-ur-Rehman, Gill, W.T. (2011) Meat as a functional food with special reference to probiotic sausages. *Food Research International*, **44**, 3125–3133.

Kingcha, Y., Tosukhowong, A., Zendo, T. *et al.* (2012) Anti-listeria activity of *Pediococcus pentosaceus* BCC 3772 and application as starter culture for Nham, a traditional fermented pork sausage. *Food Control*, **25**, 190–196.

Klingberg, T.D., Axelsson, L., Naterstad, K., Elsser, D. and Budde, B.B. (2005) Identification of potential probiotic starter cultures for Scandinavian-type fermented sausage. *International Journal of Food Microbiology*, **105**, 419–431.

Kouakou, P., Ghalfi, H., Destain, J. and Thonart, P. (2009) Effect of curing sodium nitrite additive and natural meat fat on growth control of *Listeria monocytogenes* by the bacteriocin-producing *Lactobacillus curvatus* strain CWBI-B28. *Food Microbiology*, **26**, 623–628.

Kuleasan, H. and Çakmakçi, M.L. (2002) Effect of reuterin produced by *Lactobacillus reuteri* on the surface of sausages to inhibit the growth of *Listeria monocytogenes* and *Salmonella* ssp. *Nahrug/Food*, **46**, 408–410.

Landeta, G., Curiel, J.A., Carrascosa, A.V., Muñoz, R. and de las Rivas, B. (2013) Characterization of coagulase-negative staphylococci isolated from Spanish dry cured meat products. *Meat Science*, **93**, 387–396.

Langa, S., Landete, J.M., Martín-Cabrejas, I. and Medina, M. (2013) *In situ* reuterin production by *Lactobacillus reuteri* in dairy products. *Food Control*, **33**, 200–206.

Larsen, T.O., Svendsen, A. and Smedsgaard, J. (2001) Biochemical characterization of ochratoxin A-producing strains of the genus *Penicillium*. *Applied and Environmental Microbiology*, **67**, 3630–3635.

Leroy, F. and De Vuyst, L. (1999) Temperature and pH conditions that prevail during fermentation of sausages are optimal for production of the antilisterial bacteriocin sakacin K. *Applied and Environmental Microbiology*, **65**, 974–981.

Leroy, F., Geyzenb, A., Janssensa, M. and Scholliers, P. (2013) Meat fermentation at the crossroads of innovation and tradition: A historical outlook. *Trends in Food Science and Technology*, **31**, 130–137.

Leroy, F., Verluyten, J. and De Vuyst, L. (2006) Functional meat starter cultures for improved sausage fermentation. *International Journal of Food Microbiology*, **106**, 270–285.

Li, P., Kong, B., Chen, Q., Zheng, D. and Liu, N. (2013) Formation and identification of nitrosylmyoglobin by *Staphylococcus xylosus* in raw meat batters: A potential solution for nitrite substitution in meat products. *Meat Science*, **93**, 67–72.

Liepe, H.U. (1983) Starter cultures in meat production. In H.J. Rehm and G. Reed (eds), *Biotechnology, Food, and Feed Production with Microorganisms* (pp. 400–424), Weinheim: Verlag Chemie.

Lücke, F.K. (2000) Utilization of microbes to process and preserve meat. *Meat Science*, **56**, 105–115.

Macedo, R.E.F, Pflanzer, S.B. and Gomes, C.L. (2012) Probiotic meat products. In E.C. Rigobelo (ed.), *Probiotic in Animals* (pp. 85–102), Rijeka: InTech.

Macedo, R.E.F., Pflanzer, S.B., Terra, N.N. and Freitas, R.J.S. (2005) Características de culturas lácticas probióticas para uso em produtos cárneos fermentados: Sensibilidade aos sais de cura e uso de antibióticos para contagem seletiva. *Boletim do CEPPA*, **23**, 123–134.

Macedo, R.E.F., Pflanzer, S.B., Terra, N.N. and Freitas, R.J.S. (2008) Desenvolvimento de embutido fermentado por *Lactobacillus* probióticos: Características de qualidade. *Ciência e Tecnologia de Alimentos*, **28**, 509–519.

Mahoney, M. and Henriksson, A. (2003) The effect of processed meat and meat starter cultures on gastrointestinal colonization and virulence of *Listeria monocytogenes* in mice. *International Journal of Food Microbiology*, **84**, 255–261.

Makarova, K., Slesarev, A., Wolf, Y. *et al.* (2006) Comparative genomics of the lactic acid bacteria. *Proceedings of the National Academy of Sciences USA*, **103**, 15611–15616.

Maldonado, A., Ruiz-Barba, J.L. and Jimenez-Diaz, R. (2004) Production of plantaricin NC8 by *Lactobacillus plantarum* NC8 is induced in the presence of different types of gram-positive bacteria. *Archives of Microbiology*, **181**, 8–16.

Mao, Y., Nobaek, S., Kasravi, B. *et al.* (1996) The effects of *Lactobacillus* strains and oat fiber on methotrexate-induced enterocolitis in rats. *Gastroenterology*, **111**, 334–344.

Maragkoudakis, P.A., Mountzouris, K.C., Psyrras, D. *et al.* (2009) Functional properties of novel protective lactic acid bacteria and application in raw chicken meat against *Listeria monocytogenes* and *Salmonella enteritidis*. *International Journal of Food Microbiology*, **130**, 219–226.

Martín, A., Córdoba, J.J., Benito, M.J., Aranda, E. and Asensio, M.A. (2003) Effect of *Penicillium chrysogenum* and *Debaryomyces hansenii* on the volatile compounds during controlled ripening of pork loins. *International Journal of Food Microbiology*, **84**, 327–338.

Mauriello, G., Casaburi, A., Blaiotta, G. and Villani, F. (2004) Isolation and technological properties of coagulase negative staphylococci from fermented sausages of Southern Italy. *Meat Science*, **67**, 149–158.

Montville, T.J., Winkowski, K. and Ludescher, R.D. (1995) Models and mechanisms for bacteriocin action and application. *International Dairy Journal*, **5**, 797–814.

Muthukumarasamy, P. and Holley, R.A. (2007) Survival of *Escherichia coli* O157:H7 in dry fermented sausages containing micro-encapsulated probiotic lactic acid bacteria. *Food Microbiology*, **24**, 82–88.

Nedelcheva, P., Denkova, Z., Denev, P., Slavcheva, A. and Krastanov, A. (2010) Probiotic strain *Lactobacillus plantarum* NBIMCC 2415 with antioxidant activity as a starter culture in the production of dried fermented meat products. *Biotechnology and Biotechnological Equipment*, **24**, 1624–1630.

Neysens, P., Messens, W. and De Vuyst, L. (2003) Effect of sodium chloride on growth and bacteriocin production by *Lactobacillus amylovorus* DCE 471. *International Journal of Food Microbiology*, **88**, 29–39.

Nguyen, D.T.L., Hoorde, K.V., Cnockaert, M. *et al.* (2013) A culture-dependent and -independent approach for the identification of lactic acid bacteria associated with the production of nem chua, a Vietnamese fermented meat product. *Food Research International*, **50**, 232–240.

Nieto-Lozano, J.C., Reguera-Useros, J.I., Paláez-Martínez, M.C., Sacristán-Pérez-Minayo, G., Gutiérrez-Fernández, Á.J. and Hardisson de la Torre, A. (2010) The effect of the pediocin PA-1 produced by *Pediococcus acidilactici* against *Listeria monocytogenes* and *Clostridium perfringens* in Spanish dry-fermented sausages and frankfurters. *Food Control*, **21**, 679–685.

Palanichamy, A., Jayas, D.S. and Holley, R.A. (2008) Predicting survival of *Escherichia coli* O157:H7 in dry fermented sausage using artificial neural networks. *Journal of Food Protection*, **71**, 6–12.

Pan, X., Chen, F., Wu, T. and Zhao, Z. (2009) The acid, bile tolerance and antimicrobial property of *Lactobacillus acidophilus* NIT. *Food Control*, **20**, 598–602.

Paolella, S., Falavigna, C., Faccini, A. *et al.* (2015) Effect of dry-cured ham maturation time on simulated gastrointestinal digestion: Characterization of the released peptide fraction. *Food Research International*, **67**, 136–144.

Papamanoli, E., Tzanetakis, N., Litopoulou-Tzanetaki, E. and Kotzekidou, P. (2003) Characterization of lactic acid bacteria isolated from a Greek dry-fermented sausage in respect of their technological and probiotic properties. *Meat Science*, **65**, 859–867.

Penna, A.L.B. (2002) Probióticos and Saúde. In *Proceedings of the XVIII Congresso Brasileiro de Ciência e Tecnologia de Alimentos* (pp. 4045–4046), Porto Alegre: SBCTA.

Pennacchia, C., Ercolini, D., Blaiotta, G., Pepe, O., Mauriello, G. and Villani, F. (2004) Selection of *Lactobacillus* strains from fermented sausages for their potential use as probiotics. *Meat Science*, **67**, 309–317.

Perrone, G., Samson, R.A., Frisvad, J.C. *et al.* (2015) *Penicillium salami*, a new species occurring during seasoning of dry-cured meat. *International Journal of Food Microbiology*, **193**, 91–98.

Pidcock, K., Heard, G.M. and Henriksson, A. (2002) Application of nontraditional meat starter cultures in production of Hungarian salami. *International Journal of Food Microbiology*, **76**, 75–81.

Rantsiou, K. and Cocolin, L. (2008) Fermented meat products. In C. Luca and E. Danilo (eds), *Molecular Techniques in the Microbial Ecology of Fermented Foods* (pp. 91–118), New York: Springer.

Rebucci, R., Sangalli, L., Fava, M., Bersani, C., Cantoni, C. and A. Baldi, A. (2007) Evaluation of functional aspects in *Lactobacillus* strains isolated from dry fermented sausages. *Journal of Food Quality*, **30**, 187–201.

Resch, M., Nagel, V. and Hertel, C. (2008) Antibiotic resistance of coagulase-negative staphylococci associated with food and used in starter cultures. *International Journal of Food Microbiology*, **127**, 99–104.

Ricke, S.C., Diaz, I.Z. and Keeton, J.T. (2007) Fermented meat, poultry, and fish products. In M.P. Doyle and L.R. Beuchat (eds), *Food Microbiology: Fundamentals and Frontiers* (pp. 795–815), Washington, DC: ASM Press.

Rivera-Espinoza, Y. and Gallardo-Navarro, Y. (2010) Non-dairy probiotic products. *Food Microbiology*, **27**, 1–11.

Rubio, R., Aymerich, T., Bover-Cid, S. and Garriga, M. (2013) Probiotic strains *Lactobacillus plantarum* 299 V and *Lactobacillus rhamnosus* GG as starter cultures for fermented sausages. *LWT – Food Science Technology*, **54**, 51–56.

Rubio, R., Jofré, A., Martín, B., Aymerich, T. and Garriga, M. (2014a) Characterization of lactic acid bacteria isolated from infant faeces as potential probiotic starter cultures for fermented sausages. *Food Microbiology*, **38**, 303–311.

Rubio, R., Martín, B., Aymerich, T. and Garriga, M. (2014b) The potential probiotic *Lactobacillus rhamnosus* CTC1679 survives the passage through the gastrointestinal tract and its use as starter culture results in safe nutritionally enhanced fermented sausages. *International Journal of Food Microbiology*, **186**, 55–60.

Ruiz-Moyano, S., Martín, A., Benito, M.J., Nevado, F.P. and de Guía Córdoba, M. (2008) Screening of lactic acid bacteria and bifidobacteria for potential probiotic use in Iberian dry fermented sausage. *Meat Science*, **80**, 715–721.

Sameshima, T., Magome, C., Takeshita, K., Arihara, K., Itoh, M. and Kondo, Y. (1998) Effect of intestinal *Lactobacillus* starter cultures on the behaviour of *Staphylococcus aureus* in fermented sausage. *International Journal of Food Microbiology*, **41**, 1–7.

Santos, F.L., Ferreira, C.L.L.F. and Costa, N.M.B. (2003) Modulação da colesterolemia por meio de prebióticos e probióticos. In C.L.L.F. Ferreira (ed.), *Prébióticos e Probióticos: Atualização e Prospecção* (pp. 61–78), Viçosa: Rubio.

Sanz, Y. and Toldrá, F. (2001) Purification and characterization of an X-prolyl-dipeptidyl peptidase from *Lactobacillus sakei*. *Applied and Environmental Microbiology*, **67**, 1815–1820.

Savijoki, K., Ingmer, H. and Varmanen, P. (2006) Proteolytic systems of lactic acid bacteria. *Applied Microbiology and Biotechnology*, **71**, 394–406.

Sayas-Barberá, E., Viuda-Martos, M., Fernández-López, F., Perez-Alvarez, J.A. and Sendra, E. (2012) Combined use of a probiotic culture and citrus fiber in a traditional sausage '*Longaniza de pascua*'. *Food Control*, **27**, 343–350.

Sentandreu, M.A., Stoeva, S., Aristoy, M.A., Laib, K., Voelter, W. and Toldrá, E. (2003) Identification of small peptides generated in Spanish dry cured ham. *Journal of Food Science*, **68**, 64–69.

Shaefer, L., Auchtung, T.A., Hermans, K.E., Whitehead, D., Borhan, B. and Britton, R.A. (2010) The antimicrobial compound reuterin (3-hydroxypropionaldehyde) induces oxidative stress via interaction with thiol groups. *Microbiology*, **156**, 1589–1599.

Sidira, M., Kandylis, P., Kanellaki, M. and Kourkoutas, Y. (2015a) Effect of immobilized *Lactobacillus casei* on volatile compounds of heat treated probiotic dry-fermented sausages. *Food Chemistry*, **178**, 201–207.

Sidira, M., Kandylis, P., Kanellaki, M. and Kourkoutas, Y. (2015b) Effect of immobilized *Lactobacillus casei* on the evolution of flavor compounds in probiotic dry-fermented sausages during ripening. *Meat Science*, **100**, 41–51.

Sidira, M., Karapetsas, A., Galanis, A., Kanellaki, M. and Kourkoutas, Y. (2014) Effective survival of immobilized Lactobacillus casei during ripening and heat treatment of probiotic dry-fermented sausages and investigation of microbial dynamics. *Meat Science*, **96**, 948–955.

Silvi, S., Oripianesi, M.C. and Cresci, A. (2003) Functional foods, gut microbiota and healthy ageing. Isolation and identification of *Lactobacillus* and *Bifidobacterium* strains from faecal samples of elderly subjects for possible probiotic use in functional foods. *Journal of Food Engineering*, **56**, 195–200.

Sunesen, L.O. and Stahnke, L.H. (2003) Mould starter cultures for dry sausages: Selection, application and effects. *Meat Science*, **65**, 935–948.

Toldrá, F., Nip, W.K. and Hui, H.Y. (2007) Dry-fermented sausage: An overview. In F. Toldrá (ed.) *Handbook of Fermented Meat and Poultry* (pp. 321–325), Ames, IA: Wiley-Blackwell.

Toldrá, F. and Reig, M. (2011) Innovations for healthier processed meats. *Trends in Food Science and Technology*, **22**, 517–522.

Tremonte, P., Reale, A., Di Renzo, T. *et al.* (2010). Interactions between *Lactobacillus sakei* and CNC (*Staphylococcus xylosus* and *Kocuria varians*) and their influence on proteolytic activity. *Letters in Applied Microbiology*, **51**, 586–594.

Työppönen, S., Markkula, A., Petäjä, E. and Mattila-Sandholm, T. (2003a) Survival of *Listeria monocytogenes* in North European type dry sausages fermented by bioprotective meat starter cultures. *Food Control*, **14**, 181–185.

Työppönen, S., Petäjä, E. and Mattila-Sandholm, T. (2003b) Bioprotectives and probiotics for dry sausages. *International Journal of Food Microbiology*, **83**, 233–244.

Udenigwe, C.C. and Aluko, R.E. (2012) Food protein derived bioactive peptides: Production, processing, and potential health benefits. *Journal of Food Science*, **77**, R11–R24.

Udenigwe, C.C. and Howard, A. (2013) Meat proteome as source of functional biopeptides. *Food Research International*, **54**, 1021–1032.

Vaštag, Ž., Popović, L., Popović, S., Petrović, L. and Peričin, D. (2010) Antioxidant and angiotensin-I converting enzyme inhibitory activity in the water-soluble protein extract from petrovac sausage (Petrovská Kolbasá). *Food Control*, **21**, 1298–1302.

Villani, F., Casaburi, A., Pennacchia, C., Filosa, L., Russo, F. and Ercolini, D. (2007) Microbial ecology of the soppressata of Vallo di Diano, a traditional dry fermented sausage from southern Italy, and *in vitro* and *in situ* selection of autochthonous starter cultures. *Applied and Environmental Microbiology*, **73**, 5453–5463.

Villani, F., Mauriello, G., Pepe, O., Blaiotta, G., Ercolini, D. and Casaburi, A. (2005) Technological and probiotic characteristics of *Lactobacillus* and coagulase negative *Staphylococcus* strains as starter for fermented sausage manufacture. *Italian Journal of Animal Science*, **4**, 498–502.

Virgili, R., Simoncini, N., Toscani, T., Camardo Leggieri, M., Formenti, S. and Battilani, P. (2012) Biocontrol of Penicillium nordicum growth and ochratoxin A production by native yeasts of dry cured ham. *Toxins (Basel)*, **4**, 68–82.

Wu, H.C., Pan, B.S., Chang, C.L. and Shiau, C.Y. (2005) Low-molecular-weight Peptides as Related to Antioxidant Properties of Chicken Essence. *Journal of Food and Drug Analysis*, **13**, 176–183.

Xu, H.Q., Gao, L., Jiang, Y.S. *et al.* (2014) Transcriptome response of *Lactobacillus sakei* to meat protein environment. *Journal of Basic Microbiology*, **55**(4), 490–499.

Yang, R. and Ray, B. (1994) Factors influencing production of bacteriocins by lactic acid bacteria. *Food Microbiology*, **11**, 281–291.

Zhang, X., Kong, B. and Xiong, Y. (2007) Production of cured meat color in nitrite-free Harbin red sausage by *Lactobacillus fermentum* fermentation. *Meat Science*, **77**, 593–598.

CHAPTER 17

Fermentation of fish-based products: A special focus on traditional Japanese products

Takashi Kuda

Department of Food Science and Technology, Tokyo University of Marine Science and Technology, Japan

Unlike farm products and stock farm products, fishery products are difficult to produce in a standardized manner, for various reasons. First, the quantity of fish caught varies widely each season; in addition, fish rapidly undergo autolysis, degradation and spoilage after death (Ghaly *et al.* 2010). Historically, the most important problems related to fishery products involved ways of preserving fish and preventing fish degradation. Processed marine products such as dried fish, salt-preserved fish, fish sausage, smoked fish and canned fish were thus developed based on the need to prevent putrefaction. For example, canned fish and fish sausages are heat sterilized to destroy microorganisms and then packed into sealed containers for protection against microbial contamination from the external environment. On the other hand, salt preservation, drying and soaking in vinegar inhibit the growth of microorganisms through the use of high salt levels, low moisture and low pH (Kuda *et al.* 2002, 2013b).

Currently, various traditional fermented fish foods have appeared in the Japanese mass media, including newspapers, television programmes and comics. Various unique fermented foods have been introduced. For example, the characteristic strong smells of *funazushi* in the Biwako Lake region (Fujii *et al.* 2011) and *kusaya* in the Izu Islands (Fujii *et al.* 1992) are famous. Salted and fermented puffer fish ovaries with rice bran, called *fugu no mako zuke*, are exclusively made in the Ishikawa Prefecture (Kobayashi *et al.* 1995). The detoxification mechanism remains unclear. Fish sauce products (i.e. *nam pla*, *nook mam* and *patis*) are made from salted and fermented fish, not only in Japan but also in other countries in Southeast Asia. These Asian fish sauces are now being imported into and utilized in the United States and Western countries.

Traditional fermented fish products are also made in Europe and Africa. For example, *surströmming*, which is marketed in Sweden, is made from herrings

---

Starter Cultures in Food Production, First Edition. Edited by Barbara Speranza, Antonio Bevilacqua, Maria Rosaria Corbo and Milena Sinigaglia.

and emits a strong smell after spontaneous fermentation in cans without sterilization (Kobayashi *et al.* 2000b). *Feseekh* is made in Africa from mullet or other fish by salting and fermentation (Mohamed *et al.* 2009). Several studies on Southeast and East Asian cultures and their traditional foods since the 1950s were reported by Shinoda (1969). These reports showed the spread and distribution of *narezushi*, salted and fermented fish with cooked rice. Ishige and Ruddle indicated in their report, entitled 'Research on fish sauce and *narezushi*' (1990), that the distribution of *narezushi* and fish sauce products is correlated with the paddy-rice culture in Monsoon Asia.

Traditional fermented fish foods can be roughly classified as follows: the salted and ripening type; and the pickled (lactic acid fermented) type. The former type is made through a ripening process using autolytic enzymes: fish guts pickled in salt called *shiokara* in Japan and fish sauces are examples of these products. The latter type is made through a fermentation process with lactic acid bacteria (LAB), such as *narezushi* made in Japan and fish *kimchi* made in Korea. Researchers have performed different systematic studies on microorganisms in both types of fermented fish products. In recent years, new genera and species have been isolated and identified using molecular techniques (Kobayashi *et al.* 2000a; Satomi *et al.* 1997, 1998, 2004). This chapter presents various traditional fermented fish products, particularly those made in the Hokuriku region of Japan, which is located in the central part of the country, facing Eurasia (China and Korea; Figure 17.1).

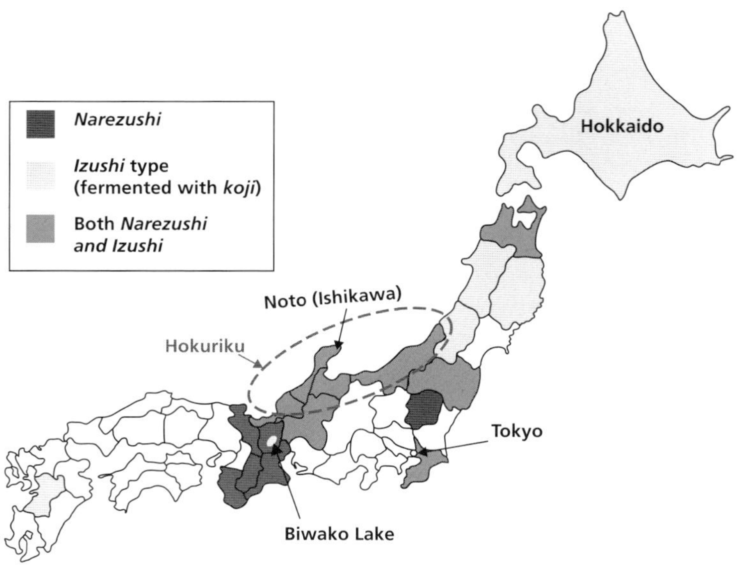

Figure 17.1 Distribution of *narezushi* and *izushi* in Japan.

Salted and fermented fish products

Fish sauce

In Noto in Ishikawa Prefecture (Figure 17.1), traditional fish sauce is prepared from whole sardines or squid guts, called *ishiru*, *ishiri*, *yoshiru* or *yoshiri* (Figure 17.2; Enomoto *et al.* 2009). *Ishiru* is made from raw materials that undergo a spontaneous ripening process in high-salt conditions from November of each year to May of the next year. To prepare fish sauce using squid guts, salt (18% w/w) is added to the fish material and stirred if necessary; then the mixture is aged for approximately 2 years. To prepare fish sauce using whole sardines, 20% salt is added to the raw materials, stirred if necessary and aged for approximately 6–12 months. Halophilic LAB, *Tetragenococcus*, occasionally increase during salting and ripening (Kuda *et al.* 2014a). *Tetragenococcus* is detected in other salted and fermented foods, such as *miso* and soy sauce (Ohata *et al.* 2011). Some of the *miso* and soy-sauce brewers use *Tetragenococcus halophilus* as a starter. The exuded and accumulated liquid in the lower layer of the fermentation tank is collected using a drainage system, subjected to deproteinization by boiling and used as the finished product. Fish sauce made from sardines or squid is also known as *nam pla* and *nuoc mam* in Thailand and Vietnam, respectively. Compared with soy sauce, *ishiru* contains high amounts of nitrogen and salt (Michihata *et al.* 2000). *Ishiru* made from squid has a total nitrogen content of 1.7–2.5 g/100 mL and a salinity level of 14.8–26.5 g/100 mL. *Ishiru* made from sardines has a total nitrogen content of 1.4–2.2 g/100 mL and a salinity level of 25.7–27.3 g/100 mL.

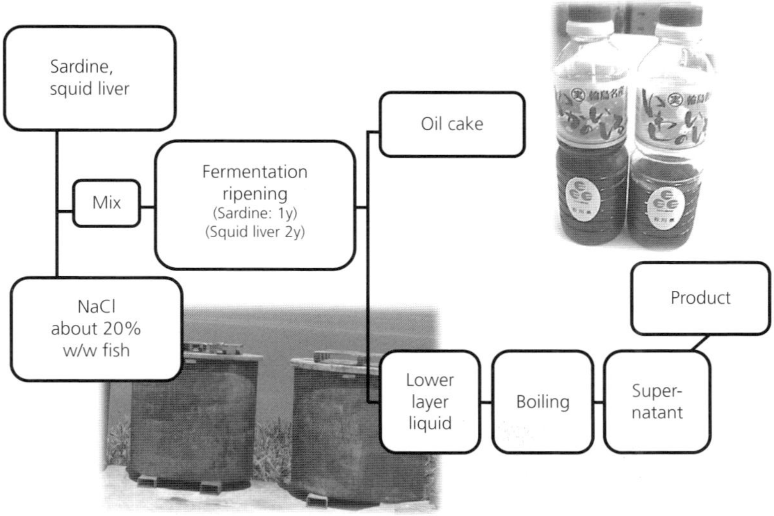

Figure 17.2 Recipe for *ishiru*.

High amounts of various free amino acids, such as alanine (Ala), glutaminic acid (Glu), glycine (Gly), lysin (Lys) and valine (Val), and oligopeptides consisting of Glu, Gly, aspartic acid (Asp) and lysine (Lys) are also present. Major organic acids in fish sauce include lactic acid, pyroglutamic acid, acetic acid, formic acid, succinic acid and malic acid. It has been suggested that these ingredients are responsible for generating the *ishiru*-specific delicate *umami* taste. Volatile compounds from lipids, such as aldehydes and ketones (Michihata *et al.* 2002) and polyamines (Michihata *et al.* 2006), may affect the taste and flavour of *ishiru*.

Shiokara: Salted and fermented fish with their guts

Shiokara are basically manufactured using the following process. Salt – generally at least 10% (w/w) – is added to fish and/or guts to prevent putrefaction. During ripening, their taste is generated by the autolytic enzymes in the raw materials. It has been suggested that this specific taste is mainly due to the increase in free amino acids. There are various types of *shiokara*, including those from squid, bonito (*shuto*), sea urchin, sweetfish (*uruka*), sea cucumber (*konowata*) and salmon (*mefun*). This section describes the preparation of squid *shiokara*, which is the most common *shiokara* and is therefore produced in large quantities.

The production of squid *shiokara* is shown in Figure 17.3. The guts are first carefully removed to prevent the squid ink bag from bursting. The mantle and leg are then washed with water. After draining the water, the meat is chopped and placed in a large barrel. The liver is salted, minced and ripened in another barrel. The mixture is then thoroughly stirred every morning and evening. In the traditional recipe, the amount of salt added is over 10% (less than 20%) of the amount of meat, and that of the liver is around 3–10%. The rate of ripening of the squid liver depends on the amount of salt and temperature used in the process.

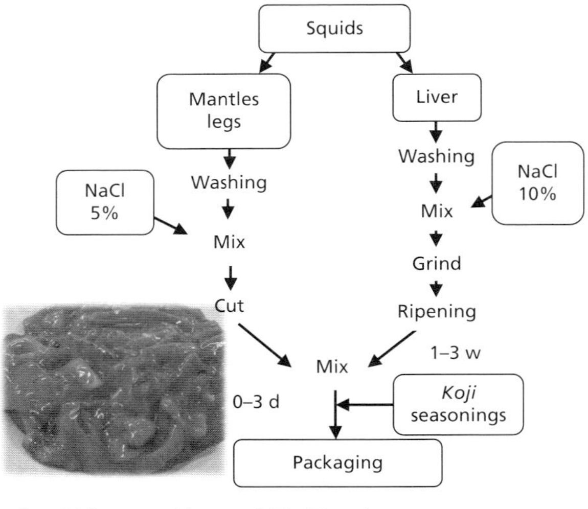

Figure 17.3 Recipe for *shiokara*, squid guts pickled in salt.

During ripening of the *shiokara*, the levels of amino acids, organic acids and volatile basic nitrogens increase (Fujii *et al.* 1994). For example, the levels of amino acids, such as Glu, leucine (Leu), Lys and Asp, increase during ripening: Glu levels increase from 53 mg/100 g during the early preparation stage and reach approximately 600–700 mg/100 g by the time of consumption. When squid *shiokara* is aged at 10 °C, it is ready for consumption within 1–2 weeks if pickled in 10% salt or within 1 month if pickled in 13% salt.

Since 1975, low-salt (approximately 3–7% salt) squid *shiokara* has gained significant attention over the traditional *shiokara* with a salt concentration of 10% and higher. Traditional *shiokara* usually has a reduced risk of contamination by food spoilage bacteria. On the other hand, low-salt *shiokara* is usually prepared within a shorter time period to prevent bacterial growth, thus resulting in a less intense *umami* taste due to the shortened ripening (auto-digestion) period. Instead, low-salt *shiokara* is mixed with seasonings and requires low temperature, low pH, low water activity and various types of additives to preserve its quality and safety.

Most food-poisoning and putrefying bacteria cannot grow in traditional *shiokara* because of its high salt content. *Vibrio parahaemolyticus*, which can generally tolerate environments containing 2–8% salt, cannot grow under water activity <0.93; that is, conditions of >10% salt. *Staphylococcus aureus* has comparatively strong resistance against salt and can grow even in environments containing ≥10% salt. However, in squid *shiokara*, bacteria other than *Staphylococcus* genus may exist, with no congeneric *Staph. aureus* detected.

In September 2007, a massive food-poisoning incident occurred in Miyagi Prefecture, involving *V. parahaemolyticus* present in 'squid *shiokara*' and affecting a total of 620 patients (Kato *et al.* 2008). *V. parahaemolyticus* is widely distributed in coastal seawater during the summer and its growth rate is high. Although food-poisoning bacteria are usually unable to grow in traditional *shiokara*, the salt concentration of the contaminated fish guts in this food-poisoning incident was only approximately 2%.

Fish *nukazuke*

Nukazuke fish products (Figure 17.4) are commonly called '*heshiko*' in the Fukui Prefecture and '*konka-zuke*' in the Ishikawa Prefecture. Regions primarily recognized for *nukazuke* production include Obama, Fukui Prefecture, and Mikawa, Ishikawa Prefecture. Even in these high-salt (15–20%) food items, LAB may be active during fermentation and ripening, and the main lactic acid–producing bacterium is *Tetragenococcus*, which is also utilized in the preparation of *miso* and soy sauce (Kuda *et al.* 2002, 2007, 2012, 2014a). In fact, the degree of lactic acid fermentation varies according to the factory, barrel and fish species: for example, some products show a small amount of LAB and low lactic acid levels, whereas others produce more than 2% (w/w) of lactic acid (Michihata *et al.* 2000). The authors have observed that for a product containing salt close to 20% (w/w)

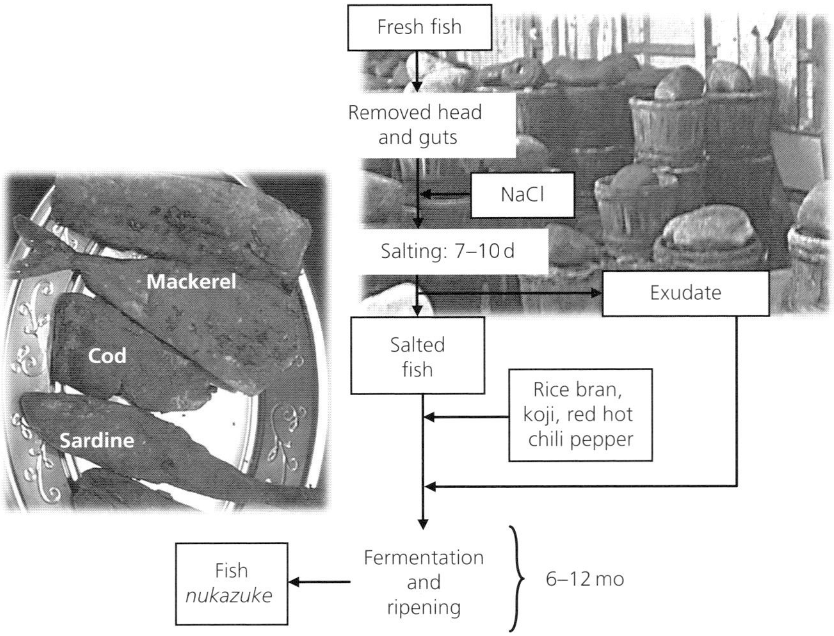

Figure 17.4 Recipe for common fish *nukazuke*.

the number of LAB is generally small, and that the lactic acid level is relatively low, with the levels of free amino acid and amines also being low (Kuda *et al.* 2002). Isolated *Tetragenococcus* grows well in a medium with 15% (w/v) salt, whereas it does not grow well in a medium with 20% (w/v) salt. However, these traditional foods have no reference or regulatory standard for the number of LAB, lactic acid level, salinity or pH value.

The accumulation of histamine (Hm) is occasionally a problem in *nukazuke* fishery products and fish sauces. For example, the cumulative amount may often be more than 1100 mg/kg and 1500 mg/L in sardine *nukazuke* and sardine *ishiru*, respectively (Kuda and Miyawaki 2010; Kuda *et al.* 2012). Hm will be generated immediately by *Morganella morganii*, *Photobacterium phosphoreum* and other Hm-generating bacteria when the raw fish are stored at an inadequate temperature (Torido *et al.* 2014). Some of these bacteria survive in a low pH environment through the action of histidine decarboxylase. In salted foods, *Tetragenococcus* is involved in generating Hm (Kuda *et al.* 2012). Because the actual consumption of high-salt-content foods at one time is very low, these authors think that the presence of these bacteria should not be a major problem. However, from the viewpoint of strict regulatory values established in the United States and Canada, and consumer preference for safety, technologies for practical microbiological control regarding decreased Hm in these fermented foods should be developed.

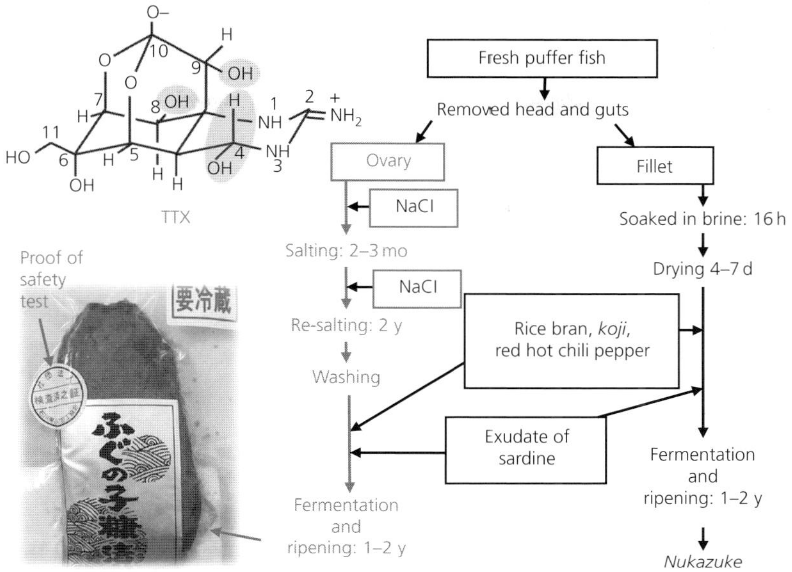

Figure 17.5 Recipe for *nukazuke* using ovaries of puffer fish.

Nukazuke from the ovaries of puffer fish is made in limited regions in the Ishikawa Prefecture (Kobayashi *et al.* 1995). The ovaries, which contain a deadly poison, are salted for at least 1 year and soaked in rice bran for 2 years (Figure 17.5). Although researchers have challenged this mysterious detoxification mechanism for more than 30 years, no clear answer has been found. Unlike the common fish *nukazuke*, the puffer ovary *nukazuke* requires resalting several times during the salting process. By resalting, a significant amount of water-soluble *fugu* poison (tetrodotoxin, TTX) may be removed. In addition, this poison is further diluted by soaking the ovaries in rice bran. However, microbiology researchers are continuously attracted to this mysterious mechanism and consider that some kind of microbial action, including conversion and binding, influences the amount of TTX that remains in these organs.

The Association of Puffer Fish Processing in the Ishikawa Prefecture strictly requires that a *fugu* poison test be conducted per production lot, and a certificate proving that a product passed this test is attached to every product distributed in the market.

Narezushi

Japanese *narezushi* culture

In Southeast and East (Monsoon) Asia, traditional fermented foods are prepared from fish, salt and cooked rice (called *narezushi* in Japan), paddy rice. The old-type *narezushi* is pickled in only salt, fish and rice (Shinoda 1969; Ishige and

Ruddle 1990). Currently, in Japan the old-type *narezushi* is made in the Kinki region, particularly around Biwa Lake (Figure 17.1). *Funazushi* made from crucian carp is the most famous *narezushi*. Furthermore, various other *narezushi* are made from sweetfish, dace, three-lips, freshwater minnow, trout, loach and salmon. On the other hand, in northeast regions of Japan the *izushi*-type *narezushi* is produced, utilizing fermentation with *koji*: *Aspergillus oryzae* malted rice (Kuda and Yano 2010). In the Hokuriku region located between the old-type *narezushi* and the *izushi*-type making regions, both food cultures prevail. It is recognized that *izushi* originates in the Hokuriku region, although it is famous as a traditional food in Aomori and Hokkaido, Japan (Aida *et al.* 1999). It has also been associated with *shihhe* (or *sikhye*), made in Korea. Although malt is used to promote fermentation for *shihhe*, it can also be regarded as an *izushi*-type product.

In the Ishikawa Prefecture, *izushi*-type fermented foods, including *kaburazushi* and *daikonzushi*, remain in the Kaga district, which is an urbanized region. On the other hand, old-type *narezushi* is made in Noto (Kuda *et al.* 2009), which is a rural area that follows traditional cultures. *Ishiru* is made in Noto (Michihata *et al.* 2000), as previously mentioned. These facts correspond to the findings obtained by Ishige and Ruddle (1990) indicating that rice paddy cultivation is deeply associated with fish sauce and *narezushi*.

Narezushi made in Noto

The process of preparing *narezushi* in Noto is as follows (Figure 17.6; Kuda *et al.* 2009; Kuda and Yano 2010): the gills, guts and eyes of the fresh fish are removed; the fish is salted; the fish is soaked in diluted vinegar; the prepared fish and cooked rice are put in a barrel; and the barrel is allowed to ripen at an ambient temperature. The following criteria may greatly vary according to the manufacturer: salinity for salting (18–33%), duration of soaking in vinegar (ranging from a few seconds to 5 hours), quantity of cooked rice (ranging from half to the same amount as the fish used in the recipe) and duration of ripening (ranging from 6 weeks to 1 or more years).

It is believed that the original *narezushi* (*hinezushi* in Noto) was not soaked in vinegar but took at least 1 year to ripen. However, currently *narezushi* is eaten within 2–4 months after initiation of ripening. The procedures are as follows. Rice is spread on the bottom of a barrel. Thin vinegar and liquor are used for hand wetting. The fish are placed in the barrel to cover all of the rice layer, then the layer is sealed together with a slight sprinkling of red pepper and Japanese pepper leaves. After the filling, which results in several layers of rice and fish, rice and Japanese pepper leaves are spread across the top layer. Then, stones approximately 1.5–4 times heavier than the weight of the materials are placed on the lid and the entire set-up is stored in a cool place to allow ripening.

LAB, such as lactobacilli and lactococci, are predominant in *aji no susu* (*narezushi* made from horse mackerel), which is made in each family or is commercially

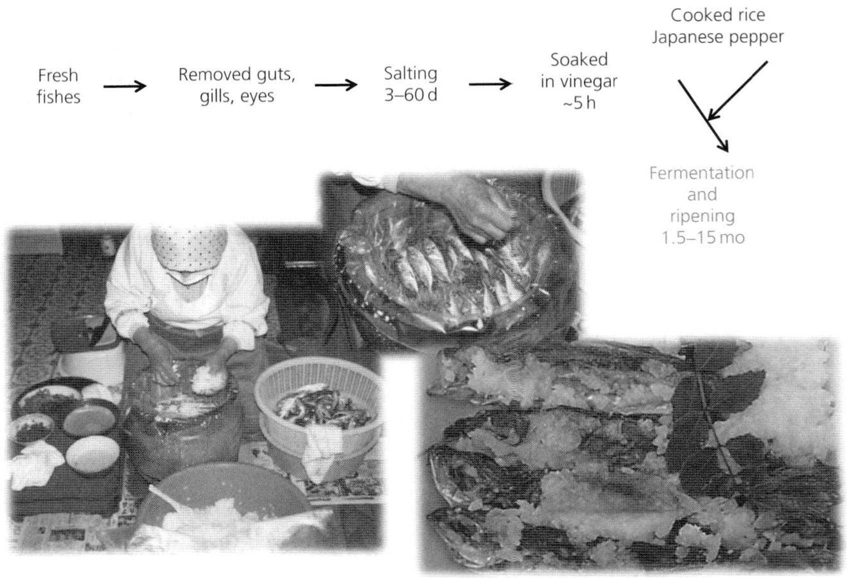

Figure 17.6 Recipe for *narezushi* in Noto.

available. Most of the samples (1.5–6 months of ripening) contained 8 log colony forming units (cfu) per gram LAB, indicating that *aji no susu* is a typical lactic acid–fermented food. In 1-year fermented products, the amount of LAB decreases to approximately 6 log cfu/g. Moulds occur on the lid of some barrels. These fungi might have played a role in the development of the unique taste of *aji no susu*.

Lactobacillus plantarum, which possesses high fermentative ability, is the predominant bacterium isolated from *narezushi*, whereas *Lactobacillus rennini* strains are also recovered. *Lb. rennini* has been isolated from foods with a moderately high salt content, such as cheese (Chenoll *et al.* 2006). *Tetragenococcus* strains have also been isolated from some *narezushi* products. Among yeasts, *Debaryomyces hansenii* is commonly isolated from pickles (Tominaga 2004), while *Wickerhamomyces anomalus* (synonym *Pichia anomala*), known as the causative organism of acetate ester (Rojas *et al.* 2003), has also been detected. On the other hand, when genomes were directly extracted from samples and analysed by denaturing gradient gel electrophoresis (DGGE) and pyrosequencing methods, strains of *Lactobacillus acidipiscis* or *Latobacillus versmoldensis* (generally not detected by the described culture method) resulted as dominant strains (An *et al.* 2010; Koyanagi *et al.* 2011).

The pH of *aji no susu* is within the range of 3.9–4.3, with the main organic acid lactic acid (3.1–8.6 g/100 g). In addition, volatile short-chain fatty acids, such as acetic acid (0.1–0.6 g/100 g), were detected. The concentration of lactic acid in *narezushi* (*aji no susu*) made from horse mackerel and fermented for 1 year was higher than that in *funazushi* and *saba* (mackerel) *narezushi* made in the

Fukui Prefecture. Butyric acid, which is considered the main causative factor for the characteristic odour of *funazushi* (Kasahara 1979), was hardly detected. Currently, Japanese *sushi* is prepared by placing a slice of fresh fish on hand-shaped vinegar rice. It is considered that *narezushi* is one origin of the current *sushi*.

The aromatic compounds in *narezushi* made from horse mackerel were analysed by a Headspace GC/MS system (Katsuyama *et al.* 2009): the content was higher and the aroma constituents were greater in the fish part compared to those in the rice part. The amounts of acetic acid (4100–6500 ng equivalent to BHT/g) and 1-penten-3-ol (980–2900 ng equivalent to BHT/g: dried small sardine smell) were significantly high. From the value of the Odour Unit, which was converted from the standard threshold in humans, the ingredients that are strongly felt as an 'odour' in *narezushi* were estimated to be aldehydes such as hexanal (raw smell), methylbutanal (almond smell), pentanal (pungent smell) and 1-octen-3-ol. These aldehydes and alcohols are reported as deterioration factors during the storage and processing of meat, fish and milk. Long (12-month) fermented *narezushi* showed a higher odour content; it was estimated that 2-nonenal and butyric acid might be the main aroma constituents. In recent years, 2-nonenal has been associated with the distinctive body odours of an old man, which are expressed as orris, oil smell and raw smell.

Generally, acetic acid, propionic acid, butyric acid and amines are regarded as specific aroma compounds of *funazushi* and other *narezushi*. However, lipid metabolites and alcohols are the major odour compounds, as previously described. Although the major factor of production of these aromas is oxidation, alcohols (i.e. 1-octen-3-ol and 1-penten-3-ol), able to emit a mushroom smell and a mould smell, are regarded as metabolites of fungi and bacteria.

Kaburazushi and *daikonzushi* (*izushi*-type products)

Kaburazushi and *daikonzushi* are known as winter-special products in the Hokuriku region. *Kaburazushi* is prepared using two steps (Kuda *et al.* 1998, 2010): cut turnips and yellowtails are salted for 2–3 days; salted yellowtails are sliced and inserted into the cut turnips; these materials are fermented with *koji* for several days (main fermentation). An image of *Lb. plantarum* isolated from *kaburazushi* by atomic force microscopy is shown in Figure 17.7. The preparation of *daikonzushi* is similar to that of *kaburazushi*. Salted Japanese radishes and dried and rehydrated herrings are fermented with *koji*. *Kaburazushi* is a special food to pray the big catch and safety, and has been produced in regions near fishing ports in Kanazawa since the Edo era (1603–1867 CE). In the past, turnips and yellowtails were expensive and harvested in a limited season and the quantity of crops was small. Thus, *kaburazushi* is considered a special food item in ordinary households. On the other hand, *daikonzushi* made from Japanese radishes and herrings might be affordable and easy to obtain.

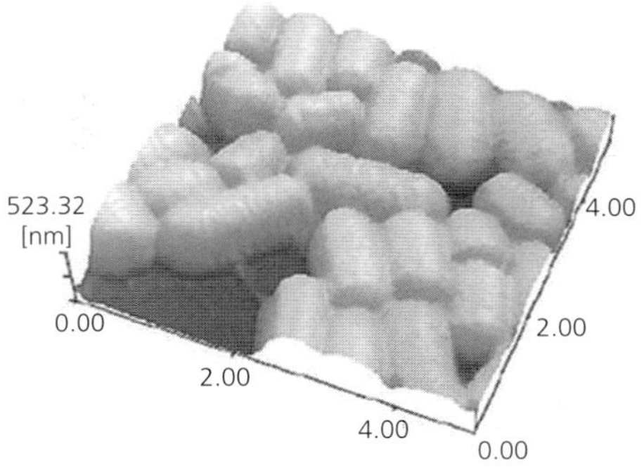

Figure 17.7 Image of *Lactobacillus plantarum* isolated from *kaburazushi* by atomic force microscopy.

The materials and recipe for *daikonzushi* are similar to those for other *izushi*-type products made in the Tohoku region and Hokkaido. Originally, old-type *narezushi* was mainly prepared from freshwater fish in inland or mountain areas. *Izushi*-type products are mainly made from marine fish in plains and urban areas.

We examined the changes in microbial flora in *kaburazushi* and *daikonzushi* prepared for a family using its own traditional recipe (Kuda *et al.* 1998). The number of LAB slightly increased during salting and then rapidly increased during the main fermentation, reaching at least 8 log cfu/g or more. Although the lactic acid level was lower than that in old-type *narezushi*, reaching 1.5–2.0% (w/w), the level of lactic acid fermentation was the same as that of milk fermentation. It is possible that the growth of LAB is promoted by the products obtained by degradation of starch (glucose and maltose) in *koji*, which was used in the main fermentation, and by glucose and fructose in turnips and Japanese radishes. Thus, the number of saprophytes observed during the early period of fermentation decreased, and only LAB and yeasts remained after 7 days of fermentation (time to eat).

We also examined the microbial flora in commercial products. In some products, the number of LAB was fewer than 4 log cfu/g, and the level of acetic acid was higher than that of lactic acid. In addition, even in products in which at least 8 log cfu/g of LAB were detected, the quantity of lactic acid was 0.26–0.46% (w/w), lower than the value recovered in domestic products. Nowadays, since refrigerators can be used, low pH and high salt concentrations have not been required for food preservation. In addition, consumers tend to prefer milder tastes and textures. Therefore, it is considered that commercial *kaburazushi* and *daikonzushi* products have different tastes from domestic and traditional products. Similar to common vegetable pickles, *Lb. plantarum* have been isolated as the dominant strain from *kaburazushi* and *daikonzushi* (Kuda *et al.* 2010). On the other hand, a culture-independent method, pyrosequencing analysis, showed that *Lb. sakei* is dominant in *kaburazushi* (Koyanagi *et al.* 2013).

Health functionality of lactic acid–fermented fish foods

Functions of fermented fish products

Many researchers have described the third function of LAB or lactic acid–fermented foods, which are the health benefits due to normalization of the enteral environment (Mitsuoka 2014; Reuter 1997). In regions along the Biwa Lake and in Noto, there are some folk medicines with *narezushi*, with recommendations such as: 'When gastrointestinal condition is not good, take a piece of rice potion of *narezushi*.' However, reports on the functionality of LAB derived from *narezushi* are limited. Highly concentrated GABA (gamma aminobutyric acid) accumulates in some types of *narezushi*. In addition, radical-scavenging activities (antioxidative properties) have been observed for the lactate ion itself (Groussard *et al.* 2000).

Lactic acid–fermented foods made from fishery products, including *narezushi* and *kaburazushi*, might have been created as a preserved food by using antibacterials such as salt and lactic acid, while they have been utilized as meals during celebrations for hundreds of years. On the other hand, *nukazuke* and salted foods continue to be consumed as preserved foods. However, a large amount of *nukazuke* cannot be consumed at one time because of its intense salty taste; its salinity is approximately 15–20% (w/v). LAB in these traditional fermented foods, as well as the functional ingredients generated by lactic acid fermentation, attract researchers interested in explaining the functions of the fermented foods and their LAB, with their high salinity and balances of nutrients.

Isolation of lactic acid bacteria from fish products and their environment

Recently, we have tried to isolate LAB from fresh fish, fermented fish products and their environments to use as starters and/or functional compounds. We expect the isolates to have higher resistance against salt and bile than isolates from milk-fermented products. Furthermore, we think that the resistance might be related to several food functions.

Lactococcus lactis subsp. *lactis* and *Leuconostoc mesenteroides* subsp. *mesenteroides* are isolated from fish intestine ubiquitously. For example, *Lc. lactis* JS1-3, isolated from Japanese seabass, showed *p*-nitrophenol-lowering activity with acid, alkaline and salt resistance (Kuda *et al.* 2011); *Lc. lactis* BF3 and *Ln. mesenteroides* BF7, isolated from chum salmon intestine, showed soy-milk fermentation, antioxidant activities and protective effects against hydrogen peroxide toxicity on *Saccharomyces cerevisiae*, a live cell model (Kuda *et al.* 2015). *Ln. mesenteroides* 1RM3, isolated from *saba* (mackerel) *narezushi* and selected as acid and bile tolerant, had antioxidant properties (Kanno *et al.* 2012) and inhibitory effects on lipopolysaccharide-induced inflammation in RAW264.7 mouse macrophage cells and dextran sodium sulphate–induced inflammatory bowel disease in mice

(Kuda *et al.* 2014c). Furthermore, *Ln. mesenteroides* 1RM3 showed inhibitory effects on *Listeria monocytogenes* infection to Caco-2 cells and A/J mice (Nakamura *et al.* 2012).

Lb. plantarum is also isolated from fish intestine and fermented fish products. *Lb. plantarum* S-SU5 isolated from Korean rockfish intestine showed immune-promoting and anti-inflammatory effects on RAW264.7 cells and also showed a protective effect on *S. cerevisiae* (Kuda *et al.* 2014b). *Lb. plantarum* AN6 isolated from *aji* (horse mackerel) *narezushi* showed strong acid and bile resistance with some exopolysaccharides (Kuda *et al.* 2013a) and a cholesterol- and bile acid–lowering effect.

References

Aida, K., Abe, J., Sumino, S. *et al.* (1999) Various component and bacteria of home-made Izushi in Hokkaido area. *Journal for the Integrated Study of Dietary Habits*, **10**, 70–74.

An, C., Takahashi, H., Kimura, B. and Kuda T. (2010) Comparison of PCR-DGGE and PCR-SSCP analysis for bacterial flora of Japanese traditional fermented fish products, *aji-narezushi* and *iwashi-nukazuke*. *Journal of the Science of Food and Agriculture*, **90**, 1796–1801.

Chenoll, E., Macián, M. C. and Aznar, R. (2006) *Lactobacillus rennini* sp. nov., isolated from rennin and associated with cheese spoilage. *International Journal of Systematic and Evolutionary Microbiology*, **56**, 449–452.

Enomoto, T., Michihata, T. and Taniguchi, H. (2009) Physiological function of a Japanese traditional fish sauce, Ishiru. In C.T. How and J.F. Shaw (eds), *Biocatalysis and Agricultural Biotechnology* (pp. 199–206), London: CRC Press.

Fujii, T., Kuda, T. and Okuzumi, M. (1992) Anaerobic microflora of '*Kusaya*' gravy. *Nippon Suisan Gakkaishi*, **59**, 185.

Fujii, T., Matsubara, M., Itoh, Y. and Okuzum, M. (1994) Microbial contribution on ripening of squid *shiokara*. *Nippon Suisan Gakkaishi*, **60**, 265–270.

Fujii, T., Watanabe, S., Horikoshi, M., Takahashi, H. and Kimura, B. (2011) PCR-DGGE analysis of bacterial communities in *funazushi*, fermented crucian carp with rice, during fermentation. *Fisheries Science*, **77**, 151–157.

Ghaly, A.E., Dave, D., Budge, S. and Brooks, M.S. (2010) Fish spoilage mechanisms and preservation techniques: Review. *American Journal of Applied Sciences*, **7**, 859–877.

Groussard, C., Morel, I., Chevanne, M., Monnier, M., Cillard, J. and Delamarche, A. (2000) Free radical scavenging and antioxidant effects of lactate ion: An in vitro study. *Journal of Applied Physiology*, **89**, 169–175.

Ishige, N. and Ruddle, K. (1990) *Gyosho to Narezushi no Kenkyu (Research on Fish Sauce and Narezushi)*, Tokyo: Iwanami Shoten.

Kanno, T., Kuda, T., An, C., Takahashi, H. and Kimura, B. (2012) Radical scavenging capacities of *saba-narezushi*, Japanese fermented chub mackerel, and its lactic acid bacteria. *LWT – Food Science and Technology*, **47**, 25–30.

Kasahara, K. (1979) Volatile flavour components of funazushi. *Eiyo to Shokuryo*, **32**, 119–122.

Kato, H., Sasaki, H., Kobayashi, T., Sugiwara, Y., Yatsu, J. and Saito, N. (2008) An outbreak of *Vibrio parahaemolyticus* food poisoning on squid guts pickled in salt. *Annual Report of Miyago Prefectural Institute of Public Health and Environment*, **26**, 48–50.

Katsuyama, Y., Michihata, T., Yano, T. and Kuda, T. (2009) Volatile compounds of a traditional fermented food, *aji-no-susu*, in Noto. *Report of the Industrial Research Institute of Ishikawa*, **59**.

Kobayashi, T., Kimura, B. and Fujii, T. (2000a) *Haloanaerobium fermentans* sp. nov., a strictly anaerobic, fermentative halophile isolated from fermented puffer fish ovaries. *International Journal of Systematic and Evolutionary Microbiology*, **50**, 1621–1627.

Kobayashi, T., Kimura, B., and Fujii, T. (2000b) Strictly anaerobic halophiles isolated from canned Swedish fermented herrings (Surströmming). *International Journal of Food Microbiology*, **54**, 81–89.

Kobayashi, T., Okuzumi, M., and Fujii, T. (1995) Microflora of fermented puffer fish ovaries in rice-bran 'fugunoko nukazuke'. *Fisheries Science*, **61**, 291–295.

Koyanagi, T., Kiyohara, M., Matsui, H. *et al.* (2011) Pyrosequencing survey of the microbial diversity of '*narezushi*', an archetype of modern Japanese *sushi*. *Letters in Applied Microbiology*, **53**, 635–640.

Koyanagi, T., Nakagawa, A., Kiyohara, M. *et al.* (2013) Pyrosequencing analysis of microbiota in *kaburazushi*, a traditional medieval sushi in Japan. *Bioscience, Biotechnology, and Biochemistry*, **77**, 2125–2130.

Kuda, T., Izawa, Y., Ishii, S. and Kimura, B. (2012) Suppressive effect of *Tetragenococcus halophilus*, isolated from fish-*nukazuke*, on histamine accumulation in salted and fermented fish. *Food Chemistry*, **130**, 569–574.

Kuda, T., Izawa, Y., Yoshida, S. and Kimura, B. (2014a) Rapid identification of *Tetragenococcus halophilus* and *Tetragenococcus muriaticus*, important species in the production of salted and fermented foods, by matrix-assisted laser desorption ionization-time of flight mass spectrometry (MALDI-TOF MS). *Food Control*, **35**, 419–425.

Kuda, T., Kaneko, N., Yano, T. and Mori, M. (2010) Induction of superoxide anion radical scavenging capacity in Japanese white radish juice and milk by *Lactobacillus plantarum* isolated from *aji-narezushi* and *kaburazushi*. *Food Chemistry*, **120**, 517–522.

Kuda, T., Kawahara, M., Nemoto, M. and Kimura, B. (2014b) In vitro antioxidant and anti-inflammation properties of lactic acid bacteria isolated from fish intestines and fermented fish from the Sanriku *Satoumi* region in Japan. *Food Research International*, **64**, 248–255.

Kuda, T., Kyoi, D, Takahashi, H., Obama, K. and Kimura, B. (2011) Detection and isolation of *p*-nitrophenol-lowering bacteria from intestine of marine fishes caught in Japanese waters. *Marine Pollution Bulletin*, **62**, 1622–1627.

Kuda, T., Mihara, T., and Yano, T. (2007) Detection of histamine and histamine-related bacteria in fish-nukazuke, a salted and fermented fish with rice-bran, by simple colorimetric microplate assay. *Food Control*, **18**, 677–681.

Kuda, T. and Miyawaki, M. (2010) Reduction of histamine in fish sauces by rice bran *nuka*. *Food Control*, **21**, 1322–1326.

Kuda, T., Noguchi, Y., Ono, M. *et al.* (2015) In vitro evaluation of the fermentative, antioxidant, and anti-inflammation properties of *Lactococcus lactis* subsp. *lactis* BF3 and *Leuconostoc mesenteroides* subsp. *mesenteroides* BF7 isolated from *Oncorhynchus keta* intestines in Rausu, Japan. *Journal of Functional Foods*, **11**, 269–277.

Kuda, T., Okamoto, K., and Yano, T. (2002) Population of halophilic bacteria in salted fish products made in the Loochoo Islands, Okinawa and the Noto Peninsula, Ishikawa, Japan. *Fisheries Science*, **68**, 1265–1273.

Kuda, T., Shouda, A., Morimura, N. and Yokoyama, M. (1998) Microflora in kaburazushi and daikonzushi made in Kanazawa, Japan. *Nippon Suisan Gakkaishi*, **64**, 1053–1059.

Kuda, T., Tanibe, R., Mori, M. *et al.* (2009) Microbial and chemical properties of *aji-no-susu*, a traditional fermented fish with rice product in the *Noto* Peninsula, Japan. *Fisheries Science*, **75**, 1499–1506.

Kuda, T., Tomomi, K., Kawahara, M., Takahashi, H. and Kimura, B. (2014c) Inhibitory effects of *Leuconostoc mesenteroides* 1RM3 isolated from *narezushi* on lipopolysaccharide induced inflammation in RAW264.7 mouse macrophage cells and dextran sodium sulphate induced inflammatory bowel disease in mice. *Journal of Functional Foods*, **6**, 631–636.

Kuda, T. and Yano, T. (2010) Lactic fermented food with fish: Noto narezushi and Kaga kaburazushi. *Japanese Journal of Food Microbiology*, **27**, 185–195.

Kuda, T., Yazaki, T., Ono, M., Takahashi, H. and Kimura, B. (2013a) In vitro cholesterol-lowering properties of *Lactobacillus plantarum* AN6 isolated from *aji-narezushi*. *Letters in Applied Microbiology*, **57**, 187–192.

Kuda, T., Yazaki, T., Takahashi, H. and Kimura, B. (2013b) Effect of dried and vinegar flavored squid products on acid resistance of *Salmonella* Typhimurium and Staphylococcus aureus. *Food Control*, **30**, 569–574.

Michihata, T., Kato, D., Yano, T. *et al.* (2006) Contents of polyamines in ISHIRU (fish sauce). *Nippon Shokuhin Kagaku Kogaku Kaishi*, **53**, 337–343 (in Japanese with abstract in English).

Michihata, T., Sado, Y., Yano, T. and Enomoto, T. (2000) Free amino acids, oligopeptides, organic acids and nucleotides of ISHIRU (fish sauce). *Nippon Shokuhin Kagaku Kogaku Kaishi*, **47**, 241–248.

Michihata, T., Yano, T., and Enomoto, T. (2002) Volatile compounds of headspace gas in the Japanese fish sauce *ishiru*. *Bioscience, Biotechnology, and Biochemistry*, **66**, 2251–2255.

Mitsuoka, T. (2014) Development of functional foods. *Bioscience of Microbiota, Food and Health*, **33**, 117–128.

Mohamed, R., Livia, S.S., Hassan, S., Soher, E. and Ahmed-Adel, E.B. (2009) Changes in free amino acids and biogenic amines of Egyptian salted-fermented fish (Feseekh) during ripening and storage. *Food Chemistry*, **115**, 635–638.

Nakamura, S., Kuda, T., An, C., Kanno, T., Takahashi, H. and Kimura, B. (2012) Inhibitory effects of *Leuconostoc mesenteroides* 1RM3 isolated from *narezushi*, a fermented fish with rice, on *Listeria monocytogenes* infection to Caco-2 cells and A/J mice. *Anaerobe*, **18**, 19–24.

Ohata, E., Yoshida, S., Masuda, T. *et al.* (2011) *Tetragenococcus halophilus* MN45 isolated from miso Inhibits IgE production. *Food Science and Technology Research*, **17**, 129–138.

Reuter, G. (1997) Present and future of probiotics in Germany and in Central Europe. *Bioscience Microflora*, **16**, 43–51.

Rojas, V., Gil, J. V., Piñaga, F. and Manzanares, P. (2003) Acetate ester formation in wine by mixed cultures in laboratory fermentations. *International Journal of Food Microbiology*, **86**, 181–188.

Satomi, M., Kimura, B., Hayashi, M., Okuzumi, M. and Fujii, T. (2004) *Marinospirillum insulare* sp. nov., a novel halophilic helical bacterium isolated from kusaya gravy. *International Journal of Systematic Bacteriology*, **54**, 163–167.

Satomi, M., Kimura, B., Hayashi, M., Shouzen, Y., Okuzumi, M. and Fujii, T. (1998) *Marinospirillum* gen. nova, with descriptions of *Marinospirillum megaterium* sp. nov., isolated from kusaya gravy, and transfer of *Oceanospirillum minutulum* to *Marinospirillum minutulum* comb. nov. *International Journal of Systematic Bacteriology*, **48**(4), 1341–1348.

Satomi, M., Kimura, B., Mizoi, M., Sato, T. and Fujii, T. (1997) *Tetragenococcus muriaticus* sp. nov., a new moderately halophilic lactic acid bacterium isolated from fermented squid liver sauce. *International Journal of Systematic Bacteriology*, **47**, 832–836.

Shinoda, O. (1969) *Sushi no Hon (a Book of Sushi)*, Tokyo: Shibata Shoten (in Japanese).

Tominaga, T. (2004) Rapid identification of pickle yeasts by fluorescent PCR and microtemperature-gradient gel electrophoresis. *FEMS Microbiology Letters*, **238**, 43–48.

Torido, Y., Ohshima, C., Takahashi, H. and Kimura, B. (2014) Distribution of psychrophilic and mesophilic histamine-producing bacteria in retailed fish in Japan. *Food Control*, **46**, 338–342.

CHAPTER 18

Traditional alkaline fermented foods: Selection of functional *Bacillus* starter cultures for *soumbala* production

Labia Irène I. Ouoba
Consultant – Senior Research Scientist, London, UK

In Africa and worldwide, alkaline fermented foods play an important role in the diet of many people. Alkaline fermented foods are usually made from protein-rich raw materials. The main microorganisms responsible for the fermentation are *Bacillus* species, but many other secondary microorganisms such as lactic acid bacteria, staphylococci and micrococci are also involved, although their role in fermentation is not clearly defined (Parkouda *et al.* 2009). The main biochemical reaction during alkaline fermentation is proteolysis, which is associated with a release of peptides and essential amino acids in the products. Furthermore, a part of amino acids is degraded into alkaline components such as ammonia that are responsible for an increase of the final product pH to alkaline values of 8–10. Alkaline fermented foods include African products such as *soumbala*, *ugba*, *bikalga* and *ntoba mbodi* and Asian products such as *kinema*, *natto* and *thua-nao* (Parkouda *et al.* 2009).

Soumbala is an alkaline fermented food condiment very rich in protein (30–47%) that is commonly consumed in many West African countries. It is also known as *dawadawa* or *iru* (Ghana, Nigeria), *nététu* (Senegal), *kinda* (Sierra Leone) and *aftin* (Benin). *Soumbala* is usually obtained by fermenting African locust bean seeds (*Parkia Biglobosa*) and plays an economic, social and cultural role in Burkina Faso (Diawara *et al.* 1992).

Traditionally, its production comprises the following main steps (Figure 18.1): boiling of the seeds for 24–40 hours, dehulling, second boiling for 1–2 hours, fermentation for 24–72 hours (25–37 °C), air drying and moulding into balls of various sizes. *Bacillus* spp. have been reported as the main microorganisms responsible for the fermentation of African locust bean seeds into *soumbala*. The predominant species is *B. subtilis*, but other species such as *B. pumilus*, *Bacillus megaterium*, *Bacillus licheniformis*, *Bacillus sphaericus* and *B. cereus* can also be

Starter Cultures in Food Production, First Edition. Edited by Barbara Speranza, Antonio Bevilacqua, Maria Rosaria Corbo and Milena Sinigaglia.

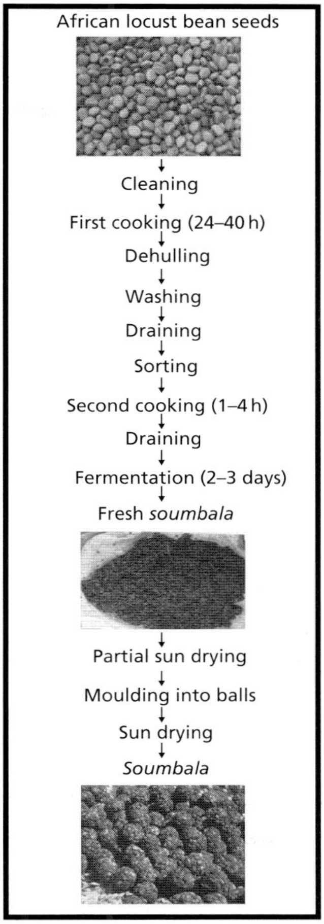

Figure 18.1 Flow diagram of *soumbala* production.

found. Secondary microorganisms include *Staphylococcus* spp., *Micrococcus* spp., *Lactobacillus* spp., *Enteroccus* spp. and *Pediococcus* spp. (Odunfa 1981; Antai and Ibrahim 1986; Diawara *et al.* 1992; N'dir *et al.* 1997; Ouoba *et al.* 2004, 2010; Parkouda *et al.* 2009). The product contains proteins (30–47%), lipids (20–43%) and carbohydrates (13–17%), and constitutes a great source of energy (464–546 Kcal/100 g; Odunfa 1986; Nordeide *et al.* 1996; Diawara *et al.* 1998). The main biochemical change during the fermentation is proteolysis, but lipolysis and degradation of carbohydrates also occur, making the condiment a rich source of essential amino and fatty acids (Odunfa 1985; Wang and Fung 1996; Ouoba *et al.* 2003a, 2003b; Azokpota *et al.* 2006). *Soumbala* is also rich in vitamins, especially vitamin B (thiamine, riboflavin and niacin), and in minerals such as calcium, iron and phosphorus (Eka 1980; Odunfa 1986). It has a characteristic strong flavour and odour attributed to the production of components such as ammonia, pyrazines, esters, acids and ketones during the fermentation (Ouoba *et al.* 2005;

Azokpota *et al.* 2008). *Soumbala* is usually added generously into sauces and soups by poor families and serves as a low-cost meat replacement because of the high concentration of protein and fat (Eka 1980).

This kind of product, along with other African fermented foods, is readily available and can be used as a cheap and accessible way to help eliminate nutritional deficiencies, assist in control of infections, decrease loss of raw materials and save energy. However, its production is still traditional and uncontrolled, without application of a quality management system or understanding of the concept of good manufacturing practice (GMP). This results in immense variability of the nutritional and sensory qualities, as well as the safety and stability of the final products. In *soumbala* and other African fermented foods, pathogenic bacteria such as *B. cereus* are commonly isolated and this constitutes a serious threat that should be addressed. Optimization of the processing conditions by for instance use of multifunctional starter cultures for controlled production of African fermented foods is essential to provide more nutritious and reliable products, with consequent improvements in the health of the African population.

In order to optimize the production of *soumbala*, various research activities have been undertaken, including a selection of *Bacillus* starter cultures for controlled fermentation. A case study on a selection based on the technical and functional properties of the isolates is described in this chapter.

Selection of multifunctional *Bacillus* starter cultures for *soumbala* production

Identification of microorganisms

The first important step in the selection of a starter culture is the identification of the microorganisms using well-defined phenotypical and genotypical methods (Ouoba *et al.* 2004).

Isolates of *Bacillus* spp were isolated from different samples of *soumbala* from different production sites in Burkina Faso. After initial screening, 17 bacteria were selected and first identified using conventional phenotypical methods (Berkeley *et al.* 1984) and API 50 CHB identification kits. Then, they were genotyped by polymerase chain reaction (PCR) amplification of the 16S–23S rDNA intergenic transcribed spacer (ITS-PCR), restriction fragment length polymorphism of the ITS-PCR (ITS-PCR RFLP), pulsed field gel electrophoresis (PFGE) and sequencing of the 968–1401 region of the 16S rDNA.

Phenotyping and genotyping allowed the identification of eight isolates of *B. subtilis* (B3, B4, B7, B9, B11, B15, B16 and B19), two of *B. pumilus* (B6 and B10), one of *Brevibacillus bortelensis* (B14) and six of *B. cereus* (B1, B2, B8, B13, B17 and B18). Using PFGE, it was possible to differentiate three groups of *B. subtilis* and two of *B. pumilus*. One group of *B. subtilis* included isolates

B3, B4, B11, B15, B16 and B19; the second group was represented by isolate B7 and the last group by isolate B9. The two isolates of *B. pumilus* form two distinct groups.

The isolates of *B. subtilis* and *B. pumilus* were selected for further investigation of their technological and functional properties.

Screening the proteolytic activity of the *Bacillus* isolates

Proteolysis is the main biochemical change occurring during the fermentation of African locust bean seeds for *soumbala* production. Therefore, the isolates of *B. subtilis* and *B. pumilus* were screened for their ability to degrade African locust bean protein (ALBP), as described by Ouoba *et al.* (2003b).

ALBP was extracted and used for this assay. The isolates were first screened by diffusion tests in casein and ALBP agars; then fermentations using a pure culture of each *Bacillus* isolate were carried out in broth media supplemented with ALBP. This was followed by the determination of the profiles of water-soluble proteins and free amino acids (FAA) by electrophoresis (SDS-PAGE) and cation exchange chromatography, respectively.

All isolates were able to degrade casein and ALBP in agar. The profile of soluble proteins changed during the fermentation time and varied according to the *Bacillus* isolate. As seen in Table 18.1, the quantity of total and essential FAA considerably increased during the first 48 hours of fermentation. Essential FAA (histidine, isoleucine, leucine, lysine, methionine, phenylalanine, threonine, tryptophan and valine) constituted 39.7–48.5% of total FAA in 48-hour fermented samples. Norleucine was the only amino acid that was not detected in all samples. In the unfermented seeds, cysteine, isoleucine, leucine, methionine, phenylalanine and tyrosine were not detected, but were released during the fermentation by most isolates. Cysteine was not detected in *B. pumilus* B6 and *B. subtilis* B7 samples and proline was absent in *B. subtilis* B7 and *B. pumilus* B6 and B10 samples. All other known amino acids – that is, alanine, aspartic acid, arginine, glutamic acid, glycine, histidine, isoleucine, lysine, ornithine, proline, serine, threonine, valine and tryptophan – were present in the products.

The study revealed that the *Bacillus* isolates studied degrade ALBP, leading to a profile of soluble proteins and FAA specific to each isolate. *Bacillus subtilis* B9 and B15 exhibited the best proteolytic activity.

Screening the lipolytic activity of the *Bacillus* isolates

As described by Ouoba *et al.* (2003b), the *Bacillus* isolates were screened for their ability to degrade African locust bean oil (ALBO). Esterase and lipase activities were investigated by diffusion tests in tributyrin and ALBO agar, respectively. Titration and gas chromatography were used to determine the amount and profile of free fatty acids (FFA) released during 72-hour degradation of ALBO by each isolate in broth media supplemented with ALBO.

Table 18.1 Quantity of free amino acids (FAA), essential FAA, free fatty acids (FFA) and non-digestible oligosaccharides obtained after incubation of African locust bean protein/oil and oligosaccharides with *B. subtilis* and *B. pumilus* isolates from *soumbala*.

Isolates	Total FAA (nmol/100 μL of sample)[1]				Total essential FAA (nmol/100 μL of sample)[1]				Total FFA (μmol/g of fat)[2]				Stachyose (mg/100 mL)[3]		Raffinose (mg/100 mL)[3]	
	0 h	24 h	48 h	72 h	0 h	24 h	48 h	72 h	0 h	24 h	48 h	72 h	0 h	48 h	0 h	48 h
B. subtilis																
B7	61.6	56.9	814.4	909.6	34.6	33.1	323.7	205.2	73.4	524.3	796.9	922.7	1000.0	410.0	1000.0	–
B9	61.6	790.2	1328.9	1407.6	34.6	406.4	606.8	569.0	73.4	377.5	440.4	524.3	1000.0	820.0	1000.0	320.0
B15	61.6	486.8	1619.1	1402.4	34.6	198.9	785.4	409.8	73.4	73.4	78.6	94.4	1000.0	150.0	1000.0	–
B. pumilus																
B6	61.6	8.8	747.4	1489.6	34.6	7.2	308.0	569.7	73.4	608.2	692.1	650.1	1000.0	990.0	1000.0	930
B10	61.6	338.2	1104.9	1359.9	34.6	137.1	472.2	508.2	73.4	566.2	713.0	755.0	1000.0	990.0	1000.0	900

Notes: *B.* = *Bacillus*.
[1] Ouoba *et al.* (2003b).
[2] Ouoba *et al.* (2003a).
[3] Ouoba *et al.* (2007a).

Esterase and lipase activity was isolate dependent (Table 18.1). A significant increase of the quantity of FFA was observed mainly after 24-hour incubation of ALBO with *B. subtilis* B7 and B9 and as well as the two isolates of *B. pumilus*. With the *B. subtilis* B15 group of isolates, a very slight increase in FFA between 48 and 72 hours was observed, indicating poor lipolytic activity. Palmitic (C16), stearic (C18), oleic (C18:1), linoleic (C18:2), arachidic (C20), behenic (C22), lignoceric (C24), palmitoleic (C16:1), linolenic (C18:3) and gadoleic (C20:1) were detected in *B. subtilis* B7 and *B. pumilus* B6 and B10 samples in variable quantities according to the isolate. For *B. subtilis* B9 and B15, lignoceric (C24), pamitoleic (C16:1), linolenic (C18:3) and gadoleic (C20:1) acids were not detected. In general, the essential FFA linoleic acid (C18:2) was predominant in all samples.

The lipolytic activity of the *Bacillus* isolates from *soumbala* was also isolate dependent. *B. subtilis* B7 and *B. pumilus* B6 and B10 were the best lipolytic isolates.

Screening the ability of the *Bacillus* isolates to degrade carbohydrates

African locust bean seeds contain non-digestible oligosaccharides such as stachyose and raffinose. Consequently, the *Bacillus* isolates were investigated for the ability to degrade these oligosaccharides as well as various polysaccharides (Ouoba *et al.* 2007a).

The ability of the isolates to degrade galactomannan polysaccharides from locust bean (*Ceratonia siliqua*) and guar (*Cyamopsis tetragonoloba*) was screened by inoculating each isolate into gels made of the polysaccharides and observing the rate of liquefaction. Furthermore, their ability to degrade arabinogalactan (the main polysaccharide in African locust bean seeds) and the non-digestible oligosaccharides stachyose and raffinose was screened by inoculating each isolate in broth media supplemented with the pre-cited carbohydrates, followed by detection of the residual sugars using high performance anion exchange chromatography, pulsed amperometric detection (HPAEC-PAD).

All isolates were able to liquefy the galactomannan gels within 5 hours of incubation. Degradation of arabinogalactan, stachyose and raffinose varied according to the isolate and the carbohydrate (Table 18.1). The rate of degradation of stachyose was as follows: 85% degradation by *B. subtilis* B15, 59% by *B. subtilis* B7, 18% by *B. subtilis* B9 and 1% by the two isolates of *B. pumilus*. Raffinose was not detected after 48 hours' incubation with *B. subtilis* B7 and B15 (100% degradation). The rate of degradation of raffinose for *B. subtilis* B9 and *B. pumilus* B6 and B10 was 68%, 10% and 7%, respectively. Melibiose, fructose and traces of galactose were detected as residual sugars from the degradation of the oligosaccharides.

Overall, the degradation of the oligosaccharides by the *B. pumilus* isolates was very poor. The *B. subtilis* isolates, especially B7 and B15, were very much more efficient in reducing the level of non-digestible oligosaccharides.

Screening the antimicrobial activity of the *Bacillus* isolates

The aim of the study was to investigate the ability of the *Bacillus* isolates to inhibit and inactivate indicator microorganisms such as *Micrococcus luteus, Staphyloccocus aureus, B. cereus, Enterococus facium, Listeria monocytogenes, Escherichia coli, Salmonella typhimurium, Shigella dysenteriae, Yersinia enterocolitica, Aspergillus ochraceus* and *Penicillium roqueforti* (Ouoba *et al.* 2007b).

Agar diffusion tests were first used to screen the inhibition capacity of the *Bacillus* isolates: the antimicrobial activity of both living cells and culture supernatants was investigated. This was followed by inactivation tests in broth media and during the fermentation of African locust bean seeds, where *Bacillus* and indicators were inoculated at a ratio of 1:1.

The antimicrobial activity of the *Bacillus* isolates was variable according to the *Bacillus* isolate and the indicator microorganism being screened (Table 18.2). *Bacillus subtilis* B7 and B15 exhibited the best activity and were used for the inactivation experiments. Supernatants of pure cultures of *B. subtilis* B7 and B15 inhibited one isolate of *B. cereus*, one of *Staph. aureus* and one of *E. coli*, and caused abnormal germination of *A. ochraceus*. However, supernatants of mixed cultures of *B. subtilis* and indicators inhibited all the indicators screened, suggesting that the antimicrobial power of the *B. subtilis* isolates is increased by the presence of the indicator. When supernatants were treated with protease, no inhibition was observed, indicating that the antimicrobial factor may be a protein. During the fermentation of the African locust bean seeds, all indicators were inactivated by the *Bacillus* isolates.

The ability of the *Bacillus* isolates from *soumbala* to inhibit and inactivate Gram-positive and Gram-negative potential pathogenic bacteria as well as ochratoxin A–producing fungi constitutes an important asset for the production of safe products. Additionally, this confers a functional property on the product.

Screening the tolerance of the *Bacillus* isolates to bile and acid

To exert potential functional properties in humans, the *Bacillus* isolates need to survive in the acidic condition of the stomach and tolerate the presence of bile in the small intestine. Therefore, the isolates were first screened for their ability to survive in Brain Heart Infusion adjusted to pH 2.5 and also supplemented with oxgall bile (0.3%). Furthermore, the bile tolerance of the *Bacillus* isolates after pre-exposure to a low pH was investigated. The procedures and media described by Compaoré *et al.* (2013) were used. The initial inoculation concentration was 10^5 colony forming units (cfu) per mL, and broth media adjusted to pH 7 and not supplemented with bile salt were used as controls.

As seen in Table 18.3, all the *Bacillus* isolates studied were able to survive at pH 2.5 after 4 hours of incubation at 37 °C, suggesting that they may be able to survive in the acidic condition of the stomach. For most isolates, there was no sensitive variation in viable cell numbers from 0 to 4 hours. However, a drop in

Table 18.2 Inhibition of indicator bacteria by *B. subtilis* and *B. pumilus* isolates from *soumbala*.

Indicator bacteria	Inhibition (24 h)				
	B. subtilis			*B. pumilus*	
	B7	B9	B15	B6	B10
M. luteus A1	++	+	+	+	+
B. cereus A4	+	–	+	–	–
B. cereus A10	++	++	+	++	++
Staph. aureus A11	–	–	+	+	+
Staph. aureus A12	++(p)	–	+++	+	+
Ent. facium A19	–	–	–	+	+
Ls. monocytogenes A18	++	+	+	+	+
E. coli A13	++	–	+(p)	–	++(p)
Sal. enteriditis A14	+	–	+	–	–
Sal. typhimurium A15	++	–	+	–	+
Sh. flexneri A16	+	–	–	–	–
Sh. dysenteria A17	++	–	+	–	++
Y. enterocolitica A21	++(p)	–	++	–	+

Notes: *B.* = *Bacillus*; *E.*=*Escherichia*; *Ent.*=*Enterococcus*; *Ls.* = *Listeria*; *M.* = *Micrococcus*; *Sal.* = *Salmonella*; *Sh.* = *Shigella*; *Staph.* = *Staphilococcus*; *Y.* = *Yersinia*.
Key: – No inhibition; + inhibition 1–3 mm; ++ inhibition 3–5 mm; +++ inhibition >5 mm; p partial
Source: Ouoba *et al.* (2008b).

the number of viable cells to 10^2 cfu/mL was observed for *B. pumilus* B6 after 4 hours' incubation and no cells were detected after 24 hours.

Interestingly, the vegetative cells of all *Bacillus* isolates showed excellent resistance to 0.3% bile (Table 18.3). They were able to survive and even grow in the media supplemented with bile. After pre-exposure to an acid condition for 4 hours, all *Bacillus* isolates were able to grow in the presence of 0.3% bile, with viable cell counts reaching up to 10^8 cfu/mL after 24 hours' incubation at 37 °C.

From the results, it can be concluded that most *Bacillus* isolates have the potential to survive the conditions in the stomach and the intestine. However, *B. pumilus* B6 may not be a primary probiotic candidate due to its weak tolerance to pH 2.5.

Screening the ability of the *Bacillus* isolates to produce toxins

An important requirement for starter and probiotic cultures is that they are safe. To screen the safety of the *Bacillus* isolates, their ability to produce toxins was investigated (Ouoba *et al.* 2008b).

The *Bacillus* isolates were screened for their potential to produce haemolytic and non-haemolytic enterotoxins using the *B. cereus* enterotoxin reverse passive latex agglutination test kit (BCETRPLA) and *Bacillus* diarrhoeal enterotoxin

Table 18.3 Survival in low pH and tolerance to bile of *B. subtilis* and *B. pumulis* isolates from *soumbala*.

Isolates	Viable count (cfu/mL)									
	pH7			pH2.5		Bile		Bile/pre-exposure to low pH		
	0 h	4 h	24 h	4 h	24 h	4 h	24 h	0 h	4 h/pH 2.5	24 h/bile
B. subtilis										
B7	1.9×10^5	5.4×10^6	6.3×10^8	1.2×10^5	1.1×10^5	1.1×10^6	3.9×10^7	1.0×10^5	1.1×10^5	2.9×10^8
B9	1.3×10^5	1.4×10^6	4.0×10^8	7.6×10^4	6.0×10^3	8.0×10^4	4.9×10^7	1.2×10^5	1.4×10^5	2.9×10^7
B15	1.1×10^5	4.6×10^6	2.4×10^8	2.3×10^5	1.5×10^5	1.9×10^5	1.4×10^8	8.2×10^4	3.4×10^4	3.1×10^8
B. pumilus										
B6	3.0×10^5	2.1×10^7	5.0×10^9	1.0×10^2	ND	5.0×10^6	5.0×10^9	1.0×10^5	2.8×10^2	1.4×10^7
B10	2.2×10^5	3.4×10^7	9.4×10^9	9.0×10^4	1.3×10^4	2.3×10^6	4.4×10^8	1.0×10^5	5.0×10^4	6.5×10^7

Notes: *B.* = *Bacillus*; cfu = colony forming units; ND = not detected.

visual immunoassay (BDEVIA). Furthermore, the presence of genes responsible for the production of haemolysin BL (Hbl A, Hbl C, Hbl D), non-haemolytic enterotoxin (NheA, NheB, NheC), cytotoxin K (CytK) and emetic (EM1) was screened by PCR using single-pair and multiplex primers.

The results demonstrated that the potential starter cultures of *B. subtilis* (B7, B9 and B15) and *B. pumilus* (B6 and B10) do not produce the enterotoxins screened for, nor do they host any of the toxin genes investigated. The isolates are thus safe with regard to the production of such toxins.

Screening the resistance of the *Bacillus* isolates to antimicrobials

Another safety requirement for multifunctional starter cultures is the absence of transferable antimicrobial resistance genes in the microorganisms. The *Bacillus* isolates were screened for their susceptibility to 25 antimicrobials (Table 18.4) according the Clinical and Laboratory Standard Institute (CLSI, Wayne, PA, USA) standards. The minimal inhibitory concentrations (MIC) obtained were analysed using breakpoints described by the European Food Safety Agency (EFSA 2008) for *Bacillus*. For antimicrobials not included in the EFSA list, breakpoints used by Compaoré *et al.* (2013) were applied. Furthermore, the presence of genes encoding resistance to chloramphenicol: catpIP501; erythromycin: erm(A), erm(B), erm(C); gentamycin: aac(6′)aph(2″), aac(3″)II, aac(3″)IV, ant(2″)-I; kanamycin: (aph(3″)-I, aph(3″)-III, ant(2″)-I; spectinomycin: (aadA); streptomycin: aadA, aadE, StrA, StrB, ant(6)-Ia; and tetracycline: tet(S), tet(O) tet(K), tet(M) tet(L), tet(Q) was investigated by PCR using specific primers and positive control bacteria, as reported by Ouoba *et al.* (2008a) and Compaoré *et al.* (2013).

As seen in Table 18.4, all bacteria screened were resistant to bacitracin and flavomycin and susceptible to ampicilin, avilamycin, ciprofloxacin, daptomycin, florfenicol, linezolid, penicillin, salynomycin, tigecycline, trimethoprim, vancomycin and trimethoprim + sulphamethoxazol. For ceftiofur, chloramphenicol, erythromycin, gentamycin, kanamycin, spectinomycin, streptomycin, sulphamethoxazol, synercid, tiamulin and tetracycline, the susceptibility was variable according to the isolate and the antimicrobial.

None of the resistance genes screened was detected in the *Bacillus* isolates, suggesting that the resistance observed for some antimicrobials may be intrinsic.

The isolates are not likely to introduce antimicrobial resistance and are safe regarding their ability to host and transfer the resistance genes investigated.

Controlled fermentations and sensory evaluation

The ability of the isolates of *B. subtilis* B7, B9 and B15 and *B. pumilus* B10 to ferment African locust bean seeds into *soumbala* were screened during controlled fermentations. This was done by using a pure culture of each *B. subtilis* isolate and mixed cultures composed of BT: *B. subtilis* B7, B9 and B15; and BP: *B. subtilis* B7, B9, B15 and *B. pumilus* B10. Simultaneously, traditional fermentations using the same batch of cooked and unfermented seeds were conducted for comparison

Table 18.4 Minimal inhibitory concentrations (MIC) and antimicrobial susceptibility of *B. subtilis* and *B. pumilus* isolates from *soumbala*.

Antimicrobials	MIC(μg/mL)/susceptibility									
	B. subtilis						*B. pumilus*			
	B7		B9		B15		B6		B10	
Ampicilin	<2	S	<2	S	<2	S	<2	S	<2S	S
Avilamycin	<2	S	<2	S	<2	S	<2	S	<2	S
Bacitracin	>64	R	>64	R	>64	R	>64	R	>64	R
Chloramphenicol	4	S	<2	S	<2	S	16	I	8 S	S
Ceftiofur	0.5	S	0.25	S	<0.12	S	8	R	16	R
Ciprofloxacin	0.25	S	<0.12	S	<0.12	S	0.5	S	0.5	S
Daptomycin	4	S	2	S	4	S	4	S	4	S
Erythromycin	<0.12	S	<0.12	S	<0.12	S	0.25	S	<0.12	S
Flavomycin	>32	R	>32	R	>32	R	>32	R	>32	R
Florfenicol	8	S	<1	S	<1	S	2	S	2	S
Gentamycin	≤2	S	≤2	S	≤2	S	≤2	S	4	I
Kanamycin	4	S	4	S	4	S	≤2	S	32	R
Linezolid	2	S	<1	S	<1	S	<1	S	2	S
Penicillin	<0.06	S	<0.06	S	<0.06	S	0.12	S	<0.06	S
Salynomycin	4	S	<2	S	<2	S	<2	S	<2	S
Spectinomycin	64	S	32	S	64	S	128	R	256	R
Streptomycin	16	I	4	S	4	S	16	I	8	S
Sulphamethoxazol	512	R	64	S	512	R	512	R	512	R
Synercid	4	R	<0.5	S	4	R	2	S	2	S
Tetracycline	16	R	<0.5	S	8	I	<0.5	S	<0.5	S
Tiamulin	>32	R	<0.25	S	>32	R	>32	R	>32	R
Tigecycline	0.06	S	0.25	S	0.03	S	0.03	S	0.03	S
TrimethopRim	<1	S	<1	S	<1	S	<1	S	4	S
TMP+SMX	<0.25	S	<0.25	S	<0.25	S	<0.25	S	<0.25	S
Vancomycin	<2	S	<2	S	<2	S	<2	S	<2S	S

Notes: I = intermediary; R = resistant; S = susceptible; TMP + SMX = trimethoprim/sulfamethoxazole.

(Ouoba *et al.* 2005). During the fermentation course, the pH and aroma component profile of the samples as well the microbial growth were monitored. The fermentation trials were followed by sensory analyses of the *soumbala* samples as well as rice dishes prepared with each type of *soumbala*. This was performed by 10 panellists composed of producers (5) and regular consumers (5) of *soumbala* who was requested to score the samples from 6 to 1 (6 = like extremely; 1 = dislike extremely), based on overall acceptability according to appearance, colour, odour and taste.

As expected, a pH increase was observed from 6.0 to 7.8–8.1 after 48 hours of fermentation. The organoleptic characteristics of the *soumbala* samples were variable according to the starter culture used and the type of fermentation (controlled or traditional). The characteristics (appearance, colour, odour and taste) were as follows: B7: very slimy, red-brown, mild odour and taste; B9: individual, dark brown, very strong odour and taste; B15: slimy, yellow-brown, middling odour and taste; BT: middling slimy, brown, strong odour and taste; BP: individual, brown, very strong odour and taste; traditional sample: individual, light brown, strong odour and taste. The scores were panellist and sample dependent.

Statistical analyses revealed that overall, the samples produced with the starter cultures were scored higher (score 3.3–4.4) than the traditional sample (score 2.2). It was also noticed that samples produced with the pure and mixed cultures of *B. subtilis* (score 3.8–4.4) were preferred to those produced with the mixed culture of *B. subtilis* and *B. pumilus* (score 3.3).

Conclusion

The *Bacillus* isolates exhibited different performance according to the technological and functional property screened. *Bacillus subtilis* B9 and B15 were the best proteolytic isolates, *B. subtilis* B7 and *B. pumilus* B6 and B10 the best lipolytic isolates, and *B. subtilis* B7 and B15 the best in the degradation of carbohydrates as well as in the inhibition and inactivation of the indicators of pathogenic microorganisms. All bacteria were tolerant to acid and bile, with however a lower tolerance to acid for *B. pumilus* B6.

It is therefore clear that the best *soumbala* product in terms of nutritional quality, safety and functional properties would be obtained with a mixed-starter culture, preferably composed of all three *B. subtilis* isolates.

The selected starter cultures were introduced in a *soumbala* production unit supported by the Department of Food Technology of the National Research Centre in Burkina Faso (DTA/IRSAT/CNRST) and are commonly used for research purposes at the department and collaborator institutions in Europe.

References

Antai, S.P. and Ibrahim, M.H. (1986) Microorganisms associated with African locust bean (*Parkia filicoidea welw*) fermentation for 'dawadawa' production. *Journal of Applied Bacteriology*, **61**,145–148.

Azokpota, P., Hounhouigan, D.J., Annan, N.T., Nago, M.C. and Jakobsen, M. (2008) Diversity of volatile compounds of afitin, iru and sonru, three fermented food condiments from Benin. *World Journal of Microbiology and Biotechnology*, **24**, 879–885.

Azokpota, P., Hounhouigan, D.J. and Nago, M.C. (2006) Microbiological and chemical changes during the fermentation of African locust bean *Parkia biglobosa*, to produce afitin, iru and sonru, three traditional condiments produced in Benin. *International Journal of Food Microbiology*, **107**, 304–309.

Berkeley, R.C.W., Logan, N.A., Shute, L.A. and Capey, A.C. (1984) Identification of *Bacillus* species. In T. Bergan (ed.), *Methods in Microbiology* (pp. 292–328), London: Academic Press.

Compaoré, C.S., Jensen, L.B., Diawara, B., Ouedraogo, G.A., Jakobsen, M. and Ouoba, L.I.I. (2013) Resistance to antimicrobials and acid and bile tolerance of *Bacillus* spp. isolated from *Bikalga*, fermented seeds of *Hibiscus sabdariffa*. *African Journal of Food Science*, **7**, 408–414.

Diawara, B., Sawadogo, L., Amoa Amwa, W.K.A. and Jakobsen, M. (1998) HACCP system for traditional African fermented foods: *Soumbala*. Taastrup: WAITRO Danish Technological Institute.

Diawara, B., Sawadogo, L. and Kabore, I.Z. (1992) Contribution à l'étude des procédés traditionnels de fabrication du '*soumbala*' au Burkina Faso. Aspects biochimiques, microbiologiques et technologiques. *Sciences et Techniques*, **20**, 5–14.

EFSA (2008) Technical guidance prepared by the Panel on Additives and Products or Substances used in Animal Feed (FEEDAP) on the update of criteria used in the assessment of bacterial resistance to antibiotics of human or veterinary importance. *EFSA Journal*, **732**, 1–15.

Eka, O.U. (1980) Effects of fermentation on the nutrient status of locust bean. *Food Chemistry*, **5**, 303–308.

N'dir, B., Gningue, R.D., Keita, N'D.G. *et al.* (1997) Caractéristiques microbiologiques et organoleptiques du nététu du commerce. *Cahiers Agricultures*, **6**, 299–304.

Nordeide, M.B., Hatløy, A., Følling, M., Lied, E. and Oshaug, A. (1996) Nutrient composition and nutritional importance of green leaves and wild food resources in an agricultural district, Koutiala, in Southern Mali. *International Journal of Food Sciences and Nutrition*, **47**, 455–468.

Odunfa, S.A. (1981) Microorganisms associated with African locust bean (*Parkia filicoidea*) during 'iru' preparation. *Qualitas Plantarum/Plant foods for Human Nutrition*, **32**, 3–10.

Odunfa, S.A. (1985) Biochemical changes during 'iru' fermentation. *Journal of Food Technology*, **20**, 295–303.

Odunfa, S.A. (1986) Dawadawa. In N.R. Reddy, M.D. Pierson and D.K. Salunkhe (eds), *Legumes Based Fermented Foods* (pp.173–189), Boca Raton, FL: CRC Press.

Ouoba, L.I.I., Cantor, M.D., Diawara, B., Traore, A.S. and Jakobsen, M. (2003a) Degradation of African locust bean oil by *Bacillus subtilis* and *Bacillus pumilus* isolated from soumbala, a fermented African locust bean condiment. *Journal of Applied Microbiology*, **95**, 868–873.

Ouoba, L.I.I., Diawara, B., Amoa-Awua, W.K., Traoré, A.S. and Møller, P.L. (2004) Genotyping of starter cultures of *Bacillus subtilis* and *Bacillus pumilus* for fermentation of African locust bean (*Parkia biglobosa*) to produce Soumbala. *International Journal of Food Microbiology*, **90**, 197–205.

Ouoba, L.I.I., Diawara, B., Annan, N.T., Poll, L. and Jakobsen, M. (2005) Volatile compounds of Soumbala, a fermented African locust bean (*Parkia biglobosa*) food condiment. *Journal of Applied Microbiology*, **99**, 1413–1421.

Ouoba, L.I.I., Diawara, B., Christensen, T., Dalgaard Mikkelsen, J. and Jakobsen, M. (2007a) Degradation of polysaccharides and non-digestible oligosaccharides by *Bacillus subtilis* and *Bacillus pumilus* isolated from Soumbala, a fermented African locust bean (*Parkia biglobosa*) food additive. *European Food Research and Technology*, **224**, 689–694.

Ouoba, L.I.I., Diawara, B., Jespersen, L. and Jakobsen, M. (2007b) Antimicrobial activity of *Bacillus subtilis* and *Bacillus pumilus* during the fermentation of African locust bean (Parkia biglobosa) for Soumbala production. *Journal of Applied Microbiology*, **102**, 963–970.

Ouoba, L.I.I., Lei, V. and Jensen, L.B. (2008a) Resistance of potential probiotic lactic acid bacteria and bifidobacteria of African and European origin to antimicrobials: Determination and transferability of the resistance genes to other bacteria. *International Journal of Food Microbiology*, **121**, 217–224.

Ouoba, L.I.I, Nyanga-Koumou, C.A.G., Parkouda, C. *et al.* (2010) Genotypic diversity of lactic acid bacteria isolated from African traditional alkaline fermented foods. *Journal of Applied Microbiology*, **108**, 2019–2029.

Ouoba, L.I.I., Rechinger, K.B., Barkholt, V., Diawara, B., Traore, A.S. and Jakobsen, M. (2003b) Degradation of proteins during the fermentation of African locust bean (*Parkia biglobosa*) by strains of *Bacillus subtilis* and *Bacillus pumilus* for production of Soumbala. *Journal of Applied Microbiology*, **94**, 396–402.

Ouoba L.I.I., Thorsen, L. and Varnam, A.H. (2008b) Enterotoxins and emetic toxins production by *Bacillus cereus* and other species of *Bacillus* isolated from Soumbala and Bikalga, African alkaline fermented food condiments. *International Journal of Food Microbiology*, **124**, 224–230.

Parkouda, C., Nielsen, D.S., Azokpota, P. *et al.* (2009) The microbiology of alkaline-fermentation of indigenous seeds used as food condiments in Africa and Asia. *Critical Reviews in Microbiology*, **35**, 139–156.

Wang, J. and Fung, D.Y.C. (1996) Alkaline fermented foods: A review with emphasis on pidan fermentation. *Critical Reviews in Microbiology*, **22**, 101–138.

CHAPTER 19

Ethnic fermented foods

Marianne Perricone, Ersilia Arace, Giuseppe Calò and Milena Sinigaglia
Department of the Science of Agriculture, Food and Environment, University of Foggia, Italy

The history of fermented foods is lost in antiquity; after drying, fermentation is in fact the oldest food preservation method. In literature there are many references highlighting the importance of fermented foods and beverages to the cultural history of human society. It seems that the art of fermentation originated on the Indian subcontinent, in the settlements preceding the great Indus Valley civilization. Some documents also suggest that fermentation was known from ancient times in Egypt and the Middle East (Prajapati and Nair 2008); others confirm that fermented foods also have an important role in ancient Eastern and Western mythologies, in which writers defined them as 'gifts of the gods'. For example, in Greek mythology Dionysus was the god of wine (Bacchus according to Roman mythology) and in some classic poems, such as *The Iliad* and *The Odyssey*, there are numerous references to wine, cheese and bread. Korean and Japanese mythology also refers to the gods who provided *miso* and other Asian fermented foods. Furthermore, fermented foods are frequently mentioned in biblical passages (Old and New Testaments) and other religious tracts too.

Thus, fermented foods and beverages are indispensable components of the dietary culture of every community in the world and every culture has a variety of fermented products as part of its diet.

Ethnic fermented foods

The term 'ethnic' derives from the Latin word *ethnicus*, meaning 'pagan'; that is, a characteristic of a population subgroup sharing a common cultural or national heritage or a popular tradition or religion. On the other hand, the term 'fermentation' derives from the Latin word *fervere*, meaning 'to boil'.

Since antiquity, indigenous people have developed ethnic fermented foods as a response to the harsh conditions of the environment in which they live; for this reason, different types of substrates and fermenting organisms have been

Starter Cultures in Food Production, First Edition. Edited by Barbara Speranza, Antonio Bevilacqua, Maria Rosaria Corbo and Milena Sinigaglia.

employed for the production of various products and they change from place to place. Although sometimes several ethnic foods seem to be similar, they are unique and distinguishable due to their geographical origin, environmental factors (religion and beliefs), food presence and availability of a plant or animal source (Das and Deka 2012; Tamang *et al.* 2012).

Ethnic fermented foods and beverages are an integral part of dietary culture; they may constitute a staple meal, or could be consumed as an accompaniment to the meal, as snacks, drinks, desserts, condiments or ingredients of cooked dishes.

Table 19.1 lists some of most common ethnic fermented foods and beverages eaten in the world.

Scientific knowledge of fermentation and ethnic fermented foods has recently increased and the scientific community has highlighted the importance and the advantages of these foods: they are superior in digestibility and nutritional value compared to their unfermented counterparts and, in addition, the fermentation technique improves their organoleptic qualities. For this reason, some ethnic fermented products are being manufactured on a large commercial scale and are popular around the world. Although some fermented foods and beverages are industrially produced, many of them continue to be produced in households for daily domestic consumption. Today some tribes still produce fermented beverages for medical purposes through native skills and by propagating knowledge orally, generation by generation.

Cereal-based fermented foods and beverages

Cereals and legumes are important in the human diet, especially for the world's vegetarian population, as they contain high amounts of carbohydrates and proteins. Ancient peoples used techniques of hydrolysing starch and proteins in these products to improve their digestibility and organoleptic properties. In particular, the Asian population was the first to use fermentation of plant proteins to produce meat-like flavours, while Indonesians exploited fermentation methods to obtain a meat-like texture in vegetable products. Even Egyptians made wheat bread leavened with yeasts, whereas Indians discovered methods to leaven a mixture of cereals and legumes (Prajapati and Nair 2008).

Cereals are considered one of the most important sources of dietary carbohydrates, proteins, vitamins, minerals and fibres for people all over the globe. For cereal processing, a great number of technologies (e.g. cooking, sprouting and milling) are used, but fermentation remains the best choice; in fact, it is commonly performed in most parts of the world. Although cereals are deficient in some basic components (essential amino acids), fermentation represents the most simple and economic way to improve their nutritional value, sensorial properties (taste, flavour and texture) and functional qualities. For this reason, large amounts of cereals are processed into foods and beverages by fermentation.

Table 19.1 Most common ethnic fermented foods and beverages in the world.

Product	Substrate	Country	Microrganisms	Product use
		CEREAL FERMENTED FOODS		
Atole	Maize	Southern Mexico	LAB	Porridge based on maize dough
Banku	Maize or maize and cassava	Ghana	LAB, moulds	Dough as staple
Bhattejaanr	Rice	India, Sikkim	*Ha. anomala*, *Mu. rouxianus*	Sweet-sour alcoholic paste
Boza	Wheat, millet, maize and other cereals	Albania, Turkey, Bulgaria and Romania	*Lactobacillus* spp., *S. cerevisiae*, *Leuconostoc* spp.	Thick, sweet, slightly sour beverage
Burukutu	Sorghum	Nigeria, Benin and Ghana	*S. cerevisiae*, *S. chavelieri*, *Ln. mesenteroides*, *C.* spp., *Acetobacter* spp.	Alcoholic beverage of vinegar-like flavour
Dosa	Rice and bengal gram	India	*Ln. mesenteroides*, *Strep. faecalis*, *C. glabrata*, *Trich. pullulans*	Griddled cake for breakfast or snack food
Hamanatto	Wheat and soybeans	Japan	*A. oryzae*, *Streptococcus* spp., *Pediococcus* spp.	Raisin-like, soft, flavouring agent for meat and fish, eaten as snack
Idli	Rice grits and black gram	South India and Sri Lanka	*Ln. mesenteroides*, *Strep. faecalis*, *C.* spp., *Trich. pullulans*	Steamed cake for breakfast food
Ogi	Maize, sorghum, millet	West Africa	*Lb. plantarum*, *S. cerevisiae*, *C. mycoderma*, *Corynebacterium* spp., *Aerobacter* spp., *Rhodotorula* spp., *Cephalosporium* spp., *Fusarium* spp., *Aspergillus* spp., *Penicillium* spp.	Paste as staple. For breakfast or weaning food for babies
Pozol	Maize	Southeastem Mexico	*Lc. lactis*, *Strep. suis*, *Lb. plantarum*, *Lb. casei*, *Lb. alimentarium*, *Lb. delbruekii*, *Clostridium* spp.	Spongy dough formed into balls
Sake	Rice	Japan and China	*A. oryzae*, yeast, LAB	Alcoholic beverage
Tarhana	Parboiled wheat meal and yogurt	Turkey	LAB	Solid powder, dried seasoning for soups
		FERMENTED MILK		
Kefir	*Kefir* grains	Caucasian mountains, Eastern Europe, Russia and Southwest Asia	*Lb. acidophilus*, *Lb. kefiri*, *Lb. parakefir*, *Lb. casei*, *Lb. delbrueckii*, *S. cerevisiae*, *K. marxianus* var. *lactis*, *C. kefiri*, *P. fermentans*, *T. delbrueckii*, *S. unisporus*, *Ya. lipolytica*	Milk beverage

Koumiss	Mare's milk	Central Asia	*C. pararugosa*, *De. anomala*, *Geotrichum* spp., *I. orientalis*, *Ka. unispora*, *K. marxianus*, *P. deserticola*, *P. fermentans*, *P. manshurica*, *P. membranaefaciens*, *T. delbrueckii*, *K. marxianus*, *Ka. unispora*, *S. cerevisiae*	Alcoholic beverage
Viili	Cow's milk	Finland	*Lc. lactis* subsp. *lactis*, *Ln. mesenteroides* subsp. *cremoris*, *G. candidum*, yeasts	Fermented milk
Yogurt	Cow's milk	Worldwide	*Lb. delbrueckii* subsp. *bulgaricus*, *Strep. thermophilus*	Fermented milk
			FERMENTED FISH	
Balao-balao	Shrimps	Philippines	*Leuconostoc* spp., *Lb.* spp., *Pediococcus* spp., *Streptococcus* spp.	Fish paste
Budu	Carp, herrings and sardines, mackerel and anchovies	Malaysia/Thailand	*Pe. halophilus*, *Staphylococcus* spp., *Halobacterium* spp., *Halococcus* spp.	Fish sauce
Patis	Salmon, mackerel, ponyfish, sardines and anchovies	Philippines	*Pe. halophilus*, *Micrococcus* spp., *Halobacterium* spp., *Halococcus* spp., *Bacillus* spp.	Fish sauce
			FERMENTE MEAT	
Jinhua	Pork	Eastern China	*Staphylococcus* spp., LAB, yeasts, *Penicillium* spp.	Ham
Nham	Red pork	Thailand	LAB	Sausage
Sai-krork-prieo	Pork	Thailand	LAB	Sausage
			OTHER FERMENTED FOODS	
Bukalga	Kartade red sorrel (*Hibiscus sabradiffa*)	West Africa	*B. subtilis*	Fermented vegetable food
Ogiri	Melon (*Citrullus vulgaris*)	Southwestern Nigeria	*Bacillus* spp. (predominant) *Proteus* spp., *Pediococcus* spp.	Fermented vegetable food
Ogiri-igbo	Castor oil seed (*Ricinus communis*)	Southeastern Nigeria	*B. subtilis*, *B. megaterium*, *B. firmus*	Fermented vegetable food
Palm wine	Palm	West Africa	*S. cerevisiae*, *Schiz. pombe*, *Lb. plantarum*, *Ln. mesenteroides*	Alcoholic beverage

Notes: *A.* = *Aspergillus*; *B.* = *Bacillus*; *C.* = *Candida*; *De.* = *Dekkera*; *G.* = *Geotrichum*; *Ha.* = *Hansenula*; *I.* = *Issatchenkia*; *K.* = *Kluyveromyces*; *Ka.* = *Kazachstania*; LAB = lactic acid bacteria; *Lb.* = *Lactobacillus*; *Lc.* = *Lactococcus*; *Ln.* = *Leuconostoc*; *Mu* = *Mucor*; *Pe.* = *Pediococcus*; *S.* = *Saccharomyces*; *Schiz.* = *Schizosaccharomyces*; *Strep.* = *Streptococcus*; *Trich.* = *Trichosporon*; *Ya.* = *Yarrowia*.

Sources: Blandino *et al.* (2003); Guyot (2010, 2012).

Raw grain materials and/or type of fermentation are the main criteria for classifying cereal-based fermented foods (Coda *et al.* 2012). The preparation of many indigenous cereal-based foods, normally obtained by natural fermentation, involves mixed cultures of bacteria, yeasts and/or fungi. The most common fermenting bacteria are lactic acid bacteria (LAB; *Leuconostoc* spp., *Lactobacillus* spp., *Streptococcus* spp., *Pediococcus* spp. and *Micrococcus* spp.); the yeasts most frequently found are of the *genera Saccharomyces* and *Candida*, although *Zygosaccharomyces*, *Geotrichum* and *Torulopsis* have also been identified in some foods (Blandino *et al.* 2003).

However, it is well known that the type of bacterial flora developed in each fermented food depends on its water activity, pH, salt concentration, temperature and the composition of the food matrix.

In most Asian countries, rice is fermented by using mixed cultures in alcoholic or food beverages, whereas in Europe, America and Australia other cereals (mostly wheat, rye, barley or maize) are fermented by natural fermentation or by adding commercial baker's yeast into leavened sourdoughs, forming dough breads or loaves, which are usually baked or steamed. In Africa, fermented cereal foods are traditionally used as a staple food as well as complementary and weaning foods for infants and young children. This wide range of cereal-based fermented foods and related processes confirms how humans are able to find ways to produce foods in different contexts (Guyot 2012).

In cereal-based fermented drinks produced around the world, two major categories are notable: alcoholic food beverages and non-alcoholic foods. Alcoholic fermented beverages can be classified into wines and beers, while the great majority of non-alcoholic fermentations involve lactic acid fermentations (Blandino *et al.* 2003). Most of these products are produced in Africa and Asia and a number of them utilize cereals in combination with legumes, thus improving the overall protein quality of the fermented product. Some examples of ethnic fermented beverages and foods are reported in this chapter.

Boza

Boza is a traditional cereal-based non-alcoholic fermented beverage, which originated in Mesopotamia 8000–9000 years ago and is produced in different parts of the world from different cereals such as millet, maize, wheat and rice. It is very popular and widely consumed by people of all ages in Bulgaria, Albania, Turkey and Romania, due to its pleasant taste, flavour and nutritional properties.

It is a slightly acid and highly viscous beverage with a pale yellow colour, characterized by spontaneous fermentation; it is usually stored at 4 °C and has a short shelf life (1–2 weeks).

There is very little information concerning the taxonomy of *boza*-fermenting microflora. They mainly consist of yeasts and LAB, although the latter are always predominant in the microbial association with a ratio of LAB:yeasts of

2:4 (Altay *et al.* 2013). Moreover, the LAB that are often isolated have been identified as *Lactobacillus plantarum, Lactobacillus acidophilus, Lactobacillus fermentum, Lactobacillus coprophilus, Lactobacillus brevis, Lactococcus* spp., *Pediococcus* spp., *Leuconostoc raffinolactis* and *Leuconostoc mesenteroides*; while the yeasts mainly isolated are *Saccharomyces cerevisiae, Candida tropicalis, Candida glabrata, Geotricum penicillatum* and *Geotricum candidum, Torulaspora* spp. and *Pichia fermentans* (Blandino *et al.* 2003; Prado *et al.* 2008; Nyanzi and Jooste 2012; Altay *et al.* 2013). During *boza* production and storage, the growth of LAB leads to an increase in free short-chain fatty acid concentration, mainly represented by acetic acid, which helps the product get its characteristic taste and reduces the growth of harmful bacteria such as *Escherichia coli, Pseudomonas aeruginosa* and *Enterococcus faecalis* (Caputo *et al.* 2012).

The interaction among microorganisms during the process is uncontrolled, leading to variations in the quality and stability of the product; on the other hand, the fermentation time represents an important factor affecting its physicochemical properties (Altay *et al.* 2013). In addition, the compositional variations of *boza* could also depend on the utilization of different types and amounts of cereals and cereal products as raw material.

Altay *et al.* (2013) highlight that this beverage is also regarded a rich source of probiotic bacteria, because some bacteriocins and antimicrobial metabolite-producing LAB (*Lb. plantarum, Lactobacillus pentosus, Lactobacillus rhamnosus, Lactobacillus paracaesi, Lactococcus lactis* subsp. *lactis* and *Pediococcus pentosaceus*) have been isolated (Todorov *et al.* 2008).

LAB produce antimicrobial substances (mainly lactic acid) that are considered one of the factors in the unique taste of *boza* and increase its acidity, providing a preservative effect. On the other hand, metabolites of the alcohol fermentation of yeasts affect its odour and mouthfeel.

Thus, by using LAB and yeast isolates and appropriately selected starter cultures, controlled fermentation can be carried out; the selection of the proper strains with probiotic and antimicrobial properties may additionally enhance the functional properties of *boza*.

Figure 19.1 shows the flowchart of *boza* production. Its manufacture is divided into six main stages: preparation of raw materials, boiling, filtration and cooling, sugar addition, fermentation and bottling. The selection of raw materials is an important stage that affects the degree of fermentability, viscosity and dry matter content; then the cereal grains are cleaned, milled to the size of 300–800 μm and water is added at a ratio of 1:5 (cereal:water). Boiling is carried out for 1–2 hours until homogeneous pulp formation is observed, hence the mixture is filtered to remove bran, hull and other foreign materials. Afterwards, the product is cooled by stirring to prevent the formation of a thin layer on the surface; sugar/saccharose powder (20–25%) is added to the mix as a substrate for LAB and yeasts. Previously fermented *boza* can be used as a starter culture. Fermentations (both lactic and alcoholic) occur at 30 °C for 24 hours. Finally, the *boza* is cooled and

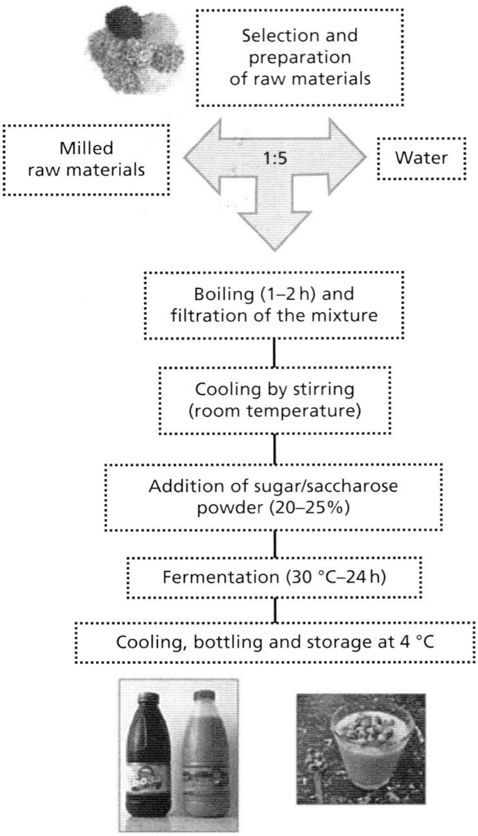

Figure 19.1 *Boza* production flowchart.

bottled in plastic or glass containers, and stored at 4 °C before being consumed. A significant increase in glucose content is observed during fermentation, while the pH, viscosity, free amino nitrogen content and dry matter decrease (Zorba *et al.* 2003; Yegin and Uren 2008).

Ogi

In the western states of Nigeria *ogi* is a fermented food obtained from white maize, although sorghum or millet is also used as a raw material. *Ogi* is considered the most traditional food for weaning infants in West Africa and the major breakfast cereal for adults; moreover, it is consumed as a porridge (*pap*) or as a gel (*agidi*) by a very large number of Nigerians. In the West African coastal region, this product is also called *eko, agidi, kamu, akamu, koko* and *furah*, depending on the substrate used and the form in which it is eaten. *Ogi* from sorghum is known as '*ogi-baba*', while '*ogi-gero*' is prepared from millet; in East Africa it is known as *uji* (Blandino *et al.* 2003).

Ogi has a smooth texture and a sour flavour similar to yogurt, as well as a characteristic aroma differentiating it from other known cereal-based fermented products. The colour depends on the cereal grain used: cream-white for maize, reddish brown for sorghum and grey for millet.

The principal microorganisms involved in fermentation are LAB, yeasts and moulds. Although *Lb. plantarum* is the predominant microorganism, other bacteria are involved in fermentation: for example, *Corynebacterium* spp. hydrolyses the corn starch, while *Saccharomyces* spp. and *Candida* spp. contribute to flavour development (Blandino *et al.* 2003).

For the preparation of *ogi*, the cereal grains are washed and steeped in earthenware, plastic or enamel pots to sediment the filtrate during 1–3 days; then, the fermented grains are wet milled and wet sieved to obtain the *ogi* slurry (Omemu *et al.* 2007; Figure 19.2).

After fermentation, the pH of *ogi* is 3.0–4.0 depending on fermentation time and LAB presence. It was also observed that the use of *Lb. brevis* alone rapidly reduced the pH to 3.0 in 48 hours, compared to the sterile slurry fermented by *S. cerevisiae*. Some studies underlined that the use of starter cultures

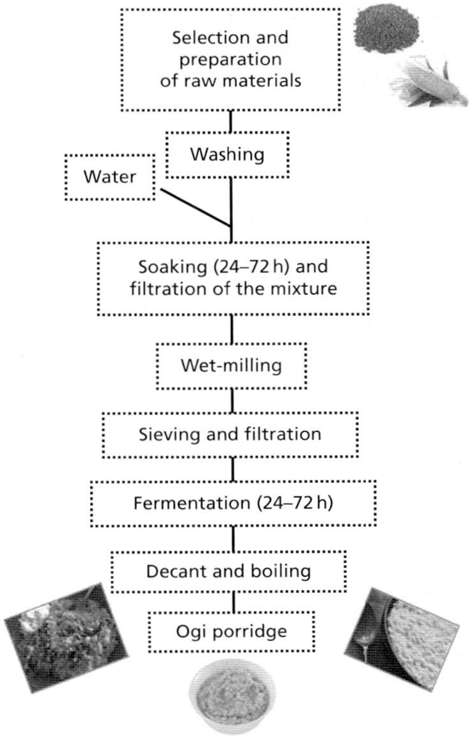

Figure 19.2 *Ogi* production flowchart.

(such as *Lb. brevis*) resulted in a rapid drop of the pH of the food matrix. The rapid pH decrease may significantly increase LAB and the organic acid concentration, which is indicative of the antipathogenic and/or prophylactic and therapeutic potential of *ogi* or other fermented cereal beverages (Nyanzi and Jooste, 2012; Oyedeji *et al.* 2013).

The nutritional qualities of *ogi* have also been studied; it has been highlighted that during fermentation phosphorous is released from phytate and the niacin and riboflavin contents increase. In general, most of the nutrients available in the original cereal grains (20–50%) are lost through *ogi* production (wet milling and wet sieving); specifically, amino acid analysis of *ogi* and its raw materials indicates substantial losses in lysine and tryptophan. To avoid these losses, lysine and methionine excreted to mutants of *Lactobacillus* spp. and yeasts have been used to fortify *ogi* (Blandino *et al.* 2003).

Idli

Idli is a traditional cereal-legume fermented food in Southeast Asian countries and is made of a blend of rice and dehulled black gram; it is very popular in India and Sri Lanka in particular. It is famous for its soft, spongy texture, desirable sour taste and characteristic aroma, as well as being a low-calorie, starchy and nutritious food, which is consumed as breakfast or a snack. Steamed *idli* contains about 3.4% protein, 20.3% carbohydrate and 70% water (Blandino *et al.* 2003; Nisha *et al.* 2005).

Idli fermentation is spontaneous; the microrganisms present in the raw materials as well as the environmental conditions determine the nature of the microflora involved, even if the main microorganisms are LAB: *Ln. mesenteroides, Lactobacillus delbrueckii* and *Pediococcus cerevisiae*. In particular, *Ln. mesenteroides* and *Streptococcus faecalis* are considered the microorganisms responsible for batter leavening and acid production. Furthermore, the yeasts *G. candidum, Torulopsis holmii, Torulopsis candida* and *Trichosporon pullulans* have also been involved in fermentation. They contribute to the leavening of the batter and flavour formation: the LAB reduce the pH from 6 to about 4.2, which favours yeast growth, while the yeasts contribute to starch degradation and gas formation, as well as to the accumulation of vitamin B and free amino acids (Nout 2009).

Idli preparation in the conventional manner takes at least 18 hours (Figure 19.3). Traditionally, rice and black gram in various proportions are soaked and ground, adding water in a mortar and pestle to obtain a batter with the desired consistency. In general, the rice and black gram are soaked and ground separately; the rice is coarsely ground and the black gram is finely ground. Then the rice and the black gram batters are mixed together (2:1 ratio) with the addition of a little salt and the fermentation takes place overnight at room temperature (about 30 °C). Finally, the fermented batter is steamed for 5–8 min (Blandino *et al.* 2003; Rekha and Vijayalakshmi 2011).

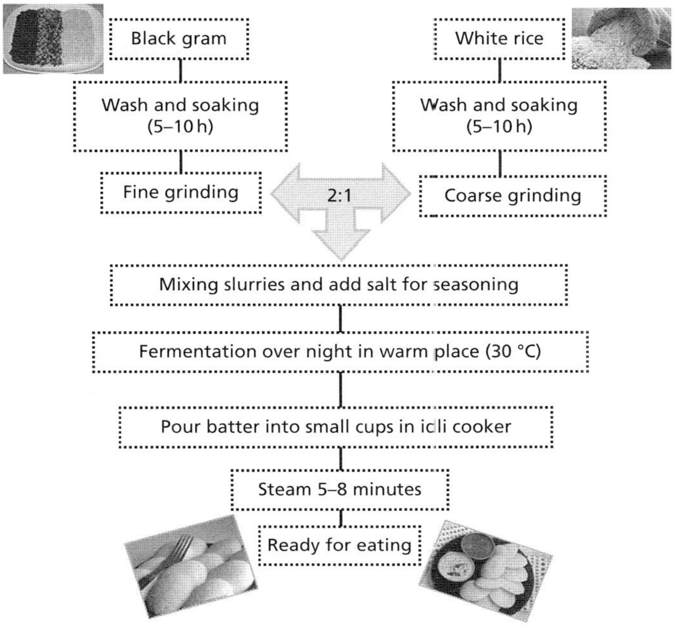

Figure 19.3 *Idli* production flowchart.

Fermentation of ingredients is essential in determining the quality of the final product. It is a better source of protein, calories and vitamins, especially B-complex vitamins, compared to the unfermented raw materials (Rekha and Vijayalakshmi 2011).

The biochemical changes occurring during natural fermentation include an increase in non-protein nitrogen, total acids, soluble solids, methionine and cystine and a decrease in sugars, pH and soluble nitrogen. During natural fermentation, water-soluble B vitamins and vitamin C increase, whereas phytate phosphorous (able to interfere with the absorption of both calcium and iron) decreases significantly. Some authors report that rice and legume batter fermentation may reduce phytate and tannin concentration. These changes during fermentation are highly significant for the nutritional propierties: for example, an increase in methionine, a limiting essential amino acid in legumes, greatly improves the protein value; thereafter, the vitamin content of *idli* batter is 0.76 mg/100 g for folic acid and 0.59 mg/100 g for riboflavin and thiamine (Ghosh and Chattopadhyay 2011).

In recent times, fermented legume and cereal products are becoming popular in developed countries due to their nutritive value and organoleptic characteristics, and for this reason they are considered health foods. The fortification of popularly consumed staple foods, such as cereals, with inexpensive plant protein sources, notably legumes, is increasingly exploited in these countries.

Consequently, research attention has recently been aimed at studying these traditional foods as potential functional foods. For example, Rekha and Vijayalakshmi (2011) studied the effect of the fortification of *idli* batter with okara (insoluble parts of the soybean that remain after the production of soy milk and tofu) on the natural fermentation rate, physico-chemical parameters, microbial growth, texture and sensorial attributes.

The fortification of *okara* to *idli* batter has a beneficial effect in terms of higher gas production and leavening during natural fermentation, resulting in a soft and spongy final product. Moreover, a reduction of fermentation time from 14 to 10 hours was observed, with great commercial significance for large-scale *idli* production. Thus, for the growing consumption of *idli* as a breakfast food, numerous studies have been carried out to develop its production on a large scale (Blandino *et al.* 2003).

Fermented milk

Fermented milk products are dairy foods that have been fermented with some LAB strains such as *Lactobacillus* spp., *Lactococcus* spp. and *Leuconostoc* spp. Today approximately 400 different products deriving from the fermentation of milk are consumed around the world.

The actual role of milk and dairy products in the diet of human communities varies greatly in different regions: tropical countries have not traditionally been milk consumers, unlike northern regions like Europe (especially Scandinavia) and North America (Mayo *et al.* 2010). The high nutritive nature of milk has encouraged its use for human consumption from the beginning of human societies; it probably started when the first animals were domesticated, about 10,000–15,000 years ago, most likely under Sumerian, Babylonian, Egyptian or Indian civilizations (Chandan 2013). These products resulted in the conservation of valuable nutrients that would otherwise deteriorate rapidly in the high temperatures of some countries in South Asia and the Middle East; indeed, the fermentation process increases the shelf life of the product, allowing the consumption of milk constituents over a longer period. Concomitantly, the conversion of milk to fermented milks results in the generation of a distinctive viscous consistency, smooth texture and unmistakable flavour. Thereafter, fermentation conserves the vital nutrients of the milk and modifies certain milk constituents, enhancing their nutritional status (Chandan 2013).

Milk can be fermented into hundreds of different products. Chandan (2013) highlights that the great diversity of fermented milks probably depends on factors such as the different kinds of milk obtained from a variety of domesticated animals; the use of variable microflora as starters; the addition of sugar, condiments, grains, fruits and nuts to create a variety of flavours and textures; and the application of additional preservation methods such as freezing, concentrating and drying.

The milk of various mammals exhibits significant differences in total solids, fat, mineral and protein content; these factors have a profound effect on the rheological characteristics of fermented milks, leading to bodies and textures ranging from drinkable fluid to firm curd. The fermentation of milk from buffalo, sheep and yak produces a well-defined custard-like body and firm curd, while milk from other animals tends to generate a soft gel consistency.

Cow's milk is used for the production of fermented milks, including yogurt, in the majority of countries around the world. In the Indian subcontinent, buffalo milk or mixtures of buffalo and cow's milk are used to produce *dahi* (fermented milk) using mixed mesophilic cultures. In several Asian countries, buffalo milk is the base for making yogurt using thermophilic cultures, whereas sheep's, goat's and camel's milk are the starting materials for fermenting milks in several Middle Eastern countries.

Milk is the normal habitat of LAB, which can cause spontaneous souring. Depending on the type of LAB that may result from environmental sources (air, utensils, milking equipment, milkers, cows, feed etc.), the sour milk has a characteristic flavour and texture. When spontaneous fermentation occurs, the growth of LAB promotes a rapid lowering of pH (acidification), which inhibits spoiling microorganisms and pathogens. Bacterial growth plays a further part in the digestion of milk constituents, increasing their bioavailability and the formation of desirable flavour compounds (Chandan 2013).

The *genera* of LAB generally affecting milk fermentation are *Lactococcus*, *Lactobacillus*, *Leuconostoc*, *Pediococcus*, *Propionibacterium* and *Bifidobacterium*. Growth of LAB to high cell densities during fermentation modifies milk constituents (protein and fats) through their proteolytic and lipolytic complex systems, changing the final rheological and sensorial characteristics of the fermented products (Mayo *et al.* 2010).

Fermented milks may be mixed with water to make a refreshing beverage such as liquid and spoonable yogurt; the addition of salt, sugar, spices or fruits to enhance the taste is also reported.

To extend their shelf life, fermented milks and yogurts may be sun dried to concentrate them or spray dried to get a powder form. For example, in Lebanon concentrated yogurt is salted, compressed into balls, sun dried and preserved in oil; another way to preserve yogurt is to smoke it and dip it in oil. In Iran, Iraq, Lebanon, Syria and Turkey, the concentrated yogurt is mixed with wheat products and sun dried to get *kishk*, while in North America and countries around the world frozen yogurt is available (Chandan 2013).

Dairy product manufacture involves mainly lactic acid fermentation. In mixed lactic acid–alcohol fermentation (as in *kefir* and *koumiss*), in addition to LAB yeasts are also associated to form alcohol.

To enhance health appeal, the current trend is the addition of prebiotics and/or probiotics to conventional yogurt. In many countries yogurt and fermented milks with probiotic cultures are available; these are made with defined cultures that have been scientifically documented to display certain health benefits (Chandan 2013).

Kefir

Kefir is a fermented milk beverage produced by the action of bacteria and yeasts existing in symbiotic association in *kefir* grains. More than two decades ago, Kemp (1984) described *kefir* as the 'champagne' of cultured dairy products and suggested that it might be an alternative drink to soft drinks.

The traditional production of *kefir* originated in the Caucasian mountains and from the late 19th century it spread to other parts of the world (Eastern Europe, Russia and Southwest Asia); nowadays, an increase in its consumption in many countries has been reported, due to its unique sensorial properties and long history associated with beneficial effects on human health. The word *kefir* is derived from the Turkish word *keyif*, which means 'feeling good' after its ingestion (de Oliveira Leite *et al.* 2013). This fermented product looks like a thick liquid dairy beverage containing carbonic acid and ethyl alcohol; *kefir* grains are cauliflower-like florets of white to yellowish-white colour and have a slimy but firm texture.

The large number of microorganisms present in *kefir* and their microbial interactions, the possible bioactive compounds resulting from microbial metabolism and the benefits associated with the use of this beverage confer on it the status of a natural probiotic (de Oliveira Leite *et al.* 2013).

There are several types of *kefir* (e.g. milk and water kefir); in particular, it can be produced from cow's, sheep's or goat's milk through fermentation with *kefir* grains (or culture). These grains are composed of microorganisms immobilized on a polysaccharide and protein matrix, where several species of bacteria (acetic acid bacteria and LAB such as *Lb. acidophilus*, *Lactobacillus kefiri*, *Lactobacillus arakefir*, *Lb. casei*, *Lb. delbrueckii* etc.) and yeasts (*S. cerevisiae*, *Kluiveromyces marxianus* var. *lactis*, *Candida kefiri*, *P. fermentans*, *T. delbrueckii*, *Saccharomyces unisporus* and *Yarrovia lipolytica*) coexist in symbiotic association. This microbial population provides the synthesis of bioactive metabolites, which are essential for grain growth and the inhibition of pathogens and contaminant microorganisms (Pogačić et al. 2013).

Kefir grains are supposed to have developed spontaneously after storing milk in animal-based containers, such as pelt-made recipients and/or intestinal and bladder vessels. Grain formation may have happened several times in history and at different geographical locations.

At present, a great variety of *kefir* grains exists, showing different microbial communities and having grain-specific distinctive sensorial properties. There are two main ways to produce this ethnic product: the artisanal process and the commercial process (Russian method or pure cultures; Figure 19.4).

Traditional artisanal production involves milk inoculation with a variable amount of grains and fermentation over a period of between 18 and 24 hours at

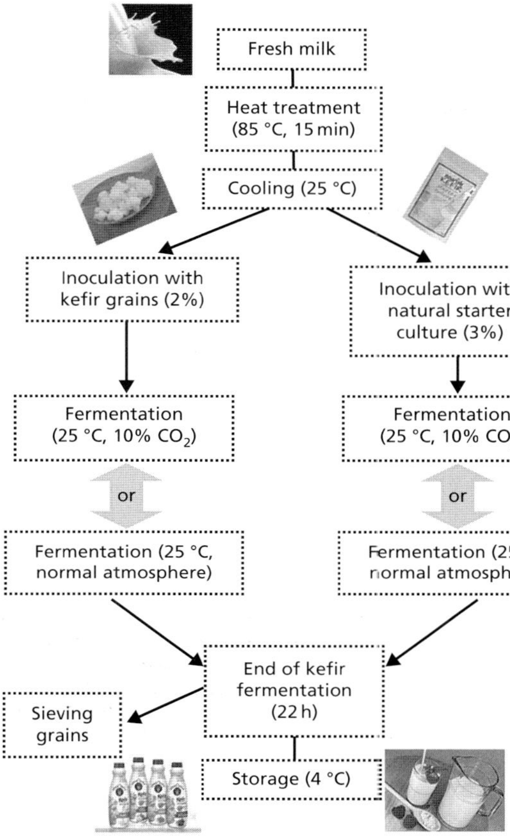

Figure 19.4 *Kefir* production flowchart.

20–25 °C; at the end of fermentation the grains are sieved and can be used for a new fermentation or kept (1–7 days) in fresh milk, while the *kefir* beverage is stored at 4 °C and is ready for consumption. The second method, known as the 'Russian method' and used for the production of *kefir* on a larger scale, employs a process of fermentation in series, from the percolate resulting from the first fermentation of the grains (fermented without the grains or mother culture). Different techniques can be used in the industrial process, but all are based on the same principle. The milk is inoculated with pure cultures isolated from *kefir* grains and commercial cultures. The maturation phase can be performed or not, maintaining the product at 8–10 °C for up to 24 hours to allow microorganism (primarily yeast) growth and to contribute to the specific flavour of the product. Although the commercial beverage is available in many countries, not all the properties of traditional *kefir* are always present. For this reason, de Oliveira Leite *et al.* (2013) highlight that the use of commercial cultures can standardize the commercial production of *kefir*, if the selection of species and strains of yeasts and bacteria is carried out accurately and carefully, thus allowing the production of a '*kefir*-type' beverage with an acceptable flavour and good conservation properties. Moreover, the commercial beverage may have a commercial life of up to 28 days, while it is recommended that *kefir* produced with grains be consumed at between 3 and 12 days; on the other hand, the '*kefir*-type' beverage may not offer the same therapeutic and probiotic properties as are present in the traditional one (Rattray and O'Connell 2011).

The main products of *kefir* fermentation are lactic acid, ethanol and carbon dioxide (CO_2), which confer viscosity, acidity and a low alcohol content; minor components include diacetyl, acetaldehyde, ethyl and amino acids, contributing to the flavour composition. In addition, a wide variety of bioactive compounds such as other organic acids, hydrogen peroxide (H_2O_2), bioactive peptides, exopolysaccharides (kefiran) and bacteriocins may act independently or together to produce the various health benefits attributed to *kefir* consumption (de Oliveira Leite *et al.* 2013).

Kefir is a traditional example of the coexistence of bacteria and yeasts. The importance of this symbiotic relationship seems clear, since it is necessary to produce compounds beneficial to health; existing scientific studies demonstrate the health benefits reported empirically by historical *kefir* consumption. *Kefir* is rich in vitamins B12 and B1, but also contains vitamins E and K, minerals and essential amino acids. Finally, the complete proteins in *kefir* are partially digested and therefore more easily utilized in the gut (Farnworth and Mainville 2008).

Both *kefir* and yogurt are cultured milk products, but they contain different type of beneficial bacteria: yogurt contains transient beneficial bacteria able to keep the digestive system clean and provide food for the friendly bacteria that reside there, whereas *kefir* bacteria could really colonize the intestinal tract. On the other hand, *kefir* contains several major strains of friendly bacteria not commonly found in yogurt, which dominate, control and eliminate pathogenic yeasts; they are able to penetrate the mucosal lining where unhealthy yeasts and bacteria reside and clean the intestine. Hence, the body becomes more efficient at resisting such pathogens and intestinal parasites. Therefore, *kefir*'s active yeast and bacteria provide more nutritive value than yogurt by keeping the colon environment clean and healthy.

Fermented fish

Fermentation of fish products has been practised for many years in many parts of the world, because fish undergoes rapid spoilage, especially in the presence of high temperature and high humidity. In particular, there have been reports of fermented fish products as far back as ancient Greece, while the trade in *garum*, a traditional fermented sauce, began in the Roman era (Salampessy *et al.* 2010).

Traditional fish fermentation is a low-cost preservation method; moreover, it is often combined with the addition of salt or drying to reduce water activity and eliminate proteolytic and putrefying microorganisms.

Fermented products from fish or shellfish are proteineous products obtained through natural hydrolysis by endogenous enzymes and microorganisms. There are a wide variety of fermented fish products, including liquid products such fish sauces and fish paste, and they have been consumed since ancient times. Because of poor roads and lack of other methods of transport, the provision of fresh fish to potential inland consumers was impossible, and this encouraged fermentation as a preservation technique. In Southeast Asian countries such as Thailand, Cambodia, Malaysia, the Philippines and Indonesia, the use of fermentation as a preservation method for fish has been of great value since the earliest times. In the countries of Northern Europe, fermented fish products are used mainly as condiments, whereas in Southeast Asia various fish products are regarded as staples.

At present, different fermented fish sauces exist. However, fish pastes are more popular than fish sauces, and are consumed as condiments. In general, the fermentation time for paste is shorter than that for sauces. In Southern India and Sri Lanka, pickled or Colombo curd fish have been known for many years. In this food, fish and salt in a 3:1 ratio are mixed in concrete tanks, dried tamarind fruit is added and the mixture is pickled.

In general, fish pastes and sauces contain relatively high levels of salt: about 15–25%. Fish sauces are produced on a large scale and exported to European, North American and other markets; in economic terms, they are the most important of all fermented fish products.

Budu

The term 'fish sauce' refers to a clear, brown liquid hydrolysate from salted fish. Fish sauce is known as *nam pla* in Thailand, *patis* in Philippines, *ngapi* in Burma, *kecap ikan* in Indonesia, *budu* in Malaysia, *nuoc-mam* in Vietnam and Cambodia, *ngan pya ye* in Myanmar and *ishiru* or *shottsuru* in Japan; each product is unique in terms of taste, organoleptic features and characteristics. The worldwide popularity of fish sauce has been rising with the growing popularity of ethnic foods, such as Vietnamese and Thai dishes.

The processing method for fish sauce can be different in each country, but the basic principle is quite similar. The main ingredients of fish sauce are fish and salt. First of all the fish is washed and mixed with salt (in a ratio of 1:1 to 1:5), then the mixture is left at room temperature for a variable period (5–24 months). During this fermentation process, the fish undergoes microbial and enzymatic hydrolysis. Although the traditional process of fish fermentation can take up to 24 months to complete, there are some attempts to accelerate the fermentation process without affecting the characteristic flavour and nutritional quality of fish sauce, such as increasing temperature, adding antibacterial agents instead of salt and using plant protease or soya sauce (Hariono *et al.* 2005).

The complex ripening process consists of chemical and biochemical reactions that change the characteristics of the fish tissue and thus its sensorial properties. Ripening is believed to be caused mainly by enzymatic actions, which split macromolecules such as protein and fat in the fish musculature into low molecular weight compounds, for example peptides, amino acids and free fatty acids. The texture of the salted fish becomes softer and more tender during the ripening phase and a pleasant, typical taste is formed.

Furthermore, the natural microflora plays an important role in the production of fish sauces, enhancing the degradation of fish proteins and developing flavour and aroma. These bacteria can also be classified into proteolytic enzyme-producing bacteria and flavour- or aroma-producing bacteria. The microflora of various traditional fish sauces shows little differences corresponding to the different species of fish used, the environment where the fish is caught, the microflora of salt used for fermentation and cross-contamination from workers' hands or tools.

During fermentation, the microflora changes as a response to pH: in the first 10 days the number of microorganisms increases rapidly, while after 20 days pH decreases, consequently reducing the number of microorganisms. *Lactobacillus* spp., *Streptococcus* spp. and *Pediococcus* spp. become the predominant species. The decrease of pH may be attributed to the production of lactic acid by LAB. Moreover, the high salt concentration of fish sauce means that only halophilic and halotolerant bacteria are able to grow.

Budu is an indigenous food produced and consumed mainly by people in the east coast states of west Malaysia and the southern provinces of Thailand. This fish product is a sauce normally used as a condiment and seasoning, as well as being an important source of cheap protein in the diet of those residing in rural areas. Since *budu* sauces have a high salt concentration, the growth of pathogenic or spoilage bacteria (*Aspergillus chevalieri*, *Escherichia* spp., *Serratia* spp., *Pseudomonas* spp. and *Clostridium* spp.) is controlled and the salt results in the typical *umami* taste and aroma. The types of fish commonly used in *budu* manufacturing are *Stolephorus* spp., *Sardinella* spp. and *Decaterus macrosoma*.

Budu is brown or dark brown in colour and is a mixture of fermented fish and salt (3:1 or 2.5:1). Like fish sauces in other countries, it has a unique taste that is developed during the fermentation process, in which enzymes and some halotolerant and halophilic microorganisms provoke the hydrolysis of the fish proteins and the production of free amino acids, peptides and ammonia. *Budu* is a very good source of protein and contains a high number of essential amino acids (Rosma *et al.* 2009).

The fish and salt mixture is fermented in closed tanks at 30–40 °C for several months (6–12). After fermentation, palm sugar, tamarind, monosodium glutamate and flavouring compounds are added to the mixture. which is then boiled, filtered and packaged (Figure 19.5).

There are three main contributing odours: cheesy, meaty and ammoniacal notes. These notes are formed by different classes of volatile compounds.

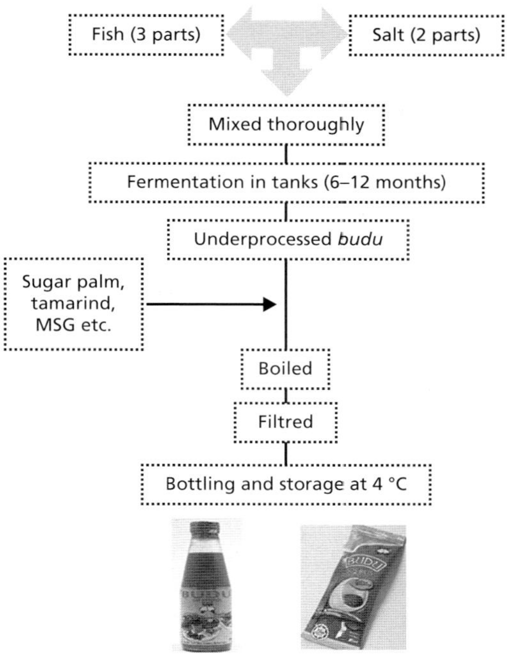

Figure 19.5 *Budu* production flowchart.

In the alcohol group, ethanol is the most important compound in the different types of *budu* since its concentration is higher, due to the action of microorganisms during fermentation. However, fatty acids in fish may reach alcohol concentration through the secondary decomposition of hydroperoxides (Mohamed *et al.* 2012).

There are 12 registered *budu* producers in Kelantan and Terengganu, Malaysia, but there are also many unregistered backyard-scale producers (Rosma *et al.* 2009).

Hentak

Fermented fish pastes are common products that have a special place in ethnic cuisine; they are used mainly as a flavouring agent or in condiment preparations. A widespread product in South and Southeast Asian countries is *hentak*, also known as *ngari* and *tungtap* in India, *bagoong* in the Philippines, *terasi* in Indonesia, *belacan* in Malaysia, *ngapi* in Myanmar, *kapi* in Thailand and many other names specific to the surrounding countries. Fish pastes are made from various species of freshwater and marine fish as well as shrimps. For example, *hentak* is made from a mixture of sun-dried fish (*Esomus danricus*) powder and the petioles of aroid plants (Alocasia macrorhiza), *tungtap* is prepared by mixing salt and dried fish (*Danio* spp.), whereas anchovy, sardine, herring and small shrimp are used in *bagoong* preparation.

The fish is allowed to ferment for several months until it develops the characteristic flavour and salty aroma. Generally, small shrimps are mixed with salt to give a final salt concentration of 10–15%, or even higher to get a better product (Salampessy *et al.* 2010).

As in fish sauce, the microflora plays an important role in the preparation of fermented fish pastes, contributing to the degradation of proteins and the development of flavour and aroma. In general, LAB are the predominant microorganisms, *Bacillus* spp., *Micrococcus* spp. and *Moraxella* spp., and some high-salt-tolerant species are also present.

Fermented meat

Generally, fermented meats are presented in the form of a sausage; in fact, this name derives from the latin word *salsus*, which means salted. It has been stated that the ancient Babylonians and people from ancient China prepared and consumed meat sausages as far back as 1500 BCE, while Greek literature after Homer's time makes frequent mention of sausages or *oryae*. Fermented dry sausages probably had their origin in Italy about 250 years ago.

The widespread use of fermented meat occurred mainly because meat plays an important role as a high-protein food in most cultures and societies; moreover, the consumption of fermented meat originated from the achievement of the desired flavour and taste, but also from the need to extend the shelf life of these products. Southeast and East Asia, in particular China, have a rich tradition in the further processing of meat dating back several thousands of years, while Europe is also famous for a great variety of processed meat products. The tradition of meat processing in Asia, especially in China, is older than in Europe; in fact, the manufacture of raw fermented ham is known in China (*Jinhua* ham) and this product is similar to the European Parma (Italy) or Serrano-type (Spain) raw fermented hams. Generally, however, other traditional Chinese or Asian processed meat products are completely different from the European ones. The typical characteristic of different products is the utilization of sugar as an ingredient for the following reasons: achievement of a moderately sweet taste, lowering of the water activity (a_w) and, finally, enhancement of fermentation processes. The most popular product available throughout East and Southeast Asia is Chinese sausage, locally called *kunchiang*, *yuen chang* or *lup-cheong*.

Other regions took a different approach. In Africa, for example, the requirement for a longer shelf life was met by drying the meat through sun drying, as for *biltong* or *kilishi*; moreover, a harsh meat processing method used in West Africa is the hot smoking of large pieces of fresh meat with the aim of preserving it, as the smoking reduces the moisture content of meat and the smoke substances add a preservation effect. However, meat treated in this way has a reduced sensorial quality. Apart from drying and smoking, the only traditional meat products popular in Africa are mixtures of meat with vegetables and spices.

Jinhua ham

Jinhua ham is a dry-cured ham, named for the city of its production in the Zhejiang province of eastern China by the first emperor of the southern Song dynasty. It is traditionally produced using the hind legs of a breed of Chinese pig; this breed is quick to mature and has excellent meat quality and a thin skin. It is highly regarded in Chinese cuisine and may be eaten as part of a cold meat dish or used in producing soup stock, imparting its unique *umami* flavour to any dish (Zhou and Zhao 2012).

The art of making special *Jinhua* ham is passed down through generations by masters to apprentices through performance and verbal instruction.

As shown in Figure 19.6, this traditional ham is produced through natural curing methods dependent on the climate and weather conditions; production begins in winter (when the temperature drops below 10 °C) and the entire process takes approximately 8–10 months to complete. First of all fresh hind legs from Jinhua pigs are selected, trimmed, shaped and salted. Salting is done at a temperature between 5 and 10 °C, because lower temperatures reduce salt penetration and high temperatures promote bacterial growth. The salting process is repeated 5–7 times and lasts about 1 month. The exact salting process is a secret that only *Jinhua* ham masters know. After the leg is well salted and partially dried, it is soaked in water for 4–6 hours and then scrubbed; the hams are successively soaked for another 16–18 hours, trimmed to the desired bamboo-leaf shape, de-hooved, branded with an iron and then hung up to dry in the sun. Sun drying is terminated when the hams begin to drip liquefied fat, which usually takes a week's worth of sun. The dried hams are hung in a low-temperature room (15 °C and 55–57% humidity) and allowed to dry, cure and develop aromas over 6–8 months. During this period the hams ferment through moulding, and the proteins and fats are hydrolysed through endogenous enzymes. This improves the flavour by creating free amino acids and other compounds. Finally, the ripened ham is brushed clean of mould and dust, and a thin layer of vegetable oil is applied to soften the ham and prevent excessive oxidation of the fat. The dried hams are piled on top of one another and allowed to ripen further for 2–3 months, which enables the flavours to stabilize and intensify (Rentfrow *et al.* 2012; Zhou and Zhao 2012).

The taste of *Jinhua* ham is due to the high concentration of free amino acids and nucleotides in the ham (glutamic acid and 5′-inosinic acid) and fermentation is due largely to the moulds and some yeasts present on the ham. Furthermore, the fragrant compounds consist of aldehydes, sulfur compounds (methanethiol and dimethyl disulfide) and branched alkenes, which derive from the breakdown and rearrangement of amino acids and fatty acids (Rentfrow *et al.* 2012).

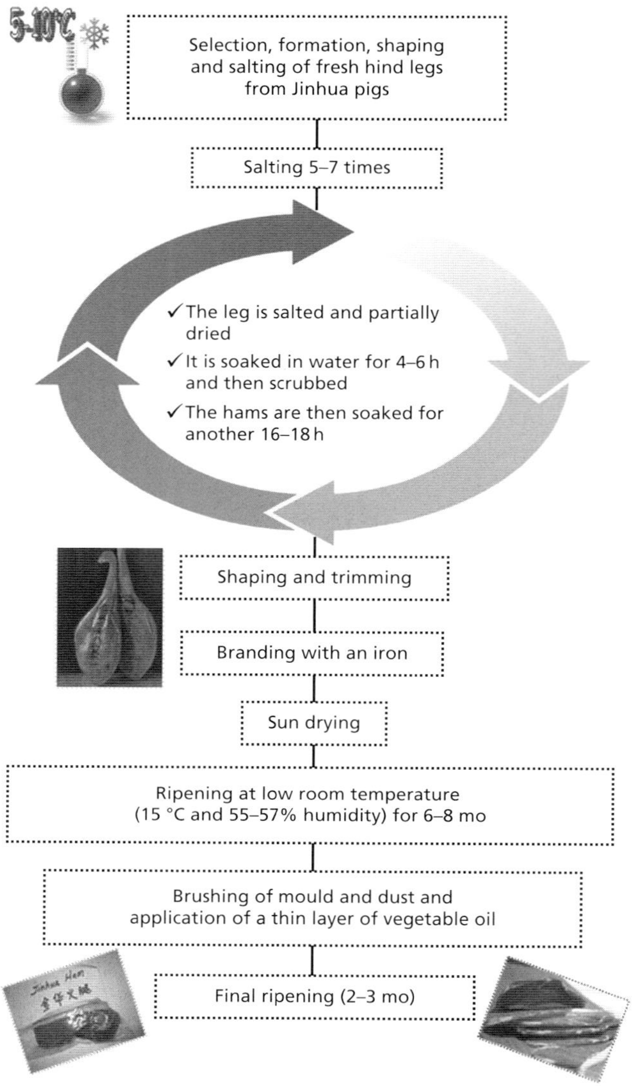

Figure 19.6 *Jinhua* ham production flowchart.

References

Altay, F., Karbancıoglu-Güler, F., Daskaya-Dikmen, C. and Heperkan, D. (2013) A review on traditional Turkish fermented non-alcoholic beverages: Microbiota, fermentation process and quality characteristics. *International Journal of Food Microbiology*, **167**, 44–56.

Blandino, A., Al-Aseeri, M.E., Pandiella, S.S. and Webb, C. (2003) Cereal based fermented foods and beverages. *Food Research International*, **36**, 527–543.

Caputo, L., Quintieri, L., Baruzzi, F., Borcakli, M. and Morea, M. (2012) Molecular and phenotypic characterization of *Pichia fermentants* strains found among boza yeasts. *Food Research International*, **48**, 755–762.

Chandan R.C. (2013) History and consumption trends. In R.C. Chandan and A. Kilara (eds), *Manufacturing Yogurt and Fermented Milks* (pp. 1-20), Chichester: John Wiley & Sons.

Coda, R., Lanera, A., Trani, A., Gobbetti, M. and Di Cagno, R. (2012) Yogurt-like beverages made of a mixture of cereals, soy and grape must: Microbiology, texture, nutritional and sensory properties. *International Journal of Food Microbiology*, **155**, 120–127.

Das, A.J. and Deka, S.C. (2012) Fermented foods and beverages of the North-East India. *International Food Research Journal*, **19**, 377–392.

de Oliveira Leite, A.M., Miguel, M.A.L., Peixoto, R.S., Rosado, A.S., Silva, J.T. and Paschoalin, V.M. (2013) Microbiological, technological and therapeutic properties of kefir: A natural probiotic beverage. *Brazilian Journal of Microbiology*, **44**, 341–349.

Farnworth, E.R. and Mainville, I. (2008) Kefir: A fermented milk product. In E.R. Farnworth (ed.), *Handbook of Fermented Functional Foods* (2nd edn, pp. 89–127), Boca Raton, FL: CRC Press.

Ghosh, D. and Chattopadhyay, P. (2011) Preparation of idli batter, its properties and nutritional improvement during fermentation. *Journal of Food Science and Technology*, **48**, 610–615.

Guyot, J.P. (2010) Fermented cereal products. In J.P. Tamang and K. Kailasapathy (eds), *Fermented Foods and Beverages of the World* (pp. 247–261), Boca Raton, FL: CRC Press.

Guyot, J.P. (2012) Cereal-based fermented foods in developing countries: Ancient foods for modern research. *International Journal of Food Science and Technology*, **47**, 1109–1114.

Hariono, I., Yeap, S.E., Kok, T.N. and Ang, G.T. (2005) Use of koji and protease in fish sauce fermentation. *Singapore Journal of Primary Industries*, **32**, 19–29.

Kemp, N. (1984) Kefir, the champagne of cultured dairy products. *Cultured Dairy Products Journal*, **20**, 29–30.

Mayo, B., Ammor, M.A., Delgado, S. and Alegría, Á. (2010) Fermented milk products. In J.P. Tamang and K. Kailasapathy (eds), *Fermented Foods and Beverages of the World* (pp. 263–288), Boca Raton, FL: CRC Press.

Mohamed, H.N., Man, Y.C., Mustafa, S. and Manap, Y.A. (2012) Tentative identification of volatile flavor compounds in commercial budu, a malaysian fish sauce, using GC-MS. *Molecules*, **17**, 5062–5080.

Nisha, P., Ananthanarayan, L. and Rekha, S.S. (2005) Effect of stabilizers on stabilization of Idli (traditional south Indian food) batter during storage. *Food Hydrocolloids*, **19**, 179–186.

Nout, M.J.R. (2009) Rich nutrition from the poorest-cereal fermentations in Africa and Asia. *Food Microbiology*, **26**, 685–692.

Nyanzi, R. and Jooste, P.J. (2012) Cereal based functional foods. In E.C. Rigobelo (ed.), *Probiotics* (pp. 161–196), Rijeka: InTech.

Omemu, A.M., Oyewole, O.B. and Bankole, M.O. (2007) Significance of yeasts in the fermentation of maize for ogi production. *Food Microbiology*, **24**, 571–576.

Oyedeji, O., Ogunbanwo, S.T. and Onilude, A.A. (2013) Predominant lactic acid bacteria involved in the traditional fermentation of fufu and ogi, two Nigerian fermented food products. *Food and Nutrition Sciences*, **4**, 40–46.

Pogačić, T., Šinko, S., Zamberlin, Š. and Samaržija, D. (2013) Microbiota of kefir grains. *Mljekarstvo*, **63**, 3–14.

Prado, F.C., Parada, J.L., Pandey, A. and Soccol, C.R. (2008) Trends in non-dairy prebiotic beverages. *Food Research International*, **41**, 11–23.

Prajapati, J.B. and Nair, B.M. (2008) The history of fermented foods. In E.R. Farnworth (ed.), *Handbook of Fermented Functional Foods* (2nd edn, pp. 1–24), Boca Raton, FL: CRC Press.

Rattray, F.P. and O'Connell, M.J. (2011) Fermented Milks Kefir, in *Encyclopedia of Dairy Sciences*, 2nd edn (ed. J.W. Fukay), Academic Press, San Diego, pp. 518-524.

Rekha, C.R. and Vijayalakshmi, G. (2011) Accelerated fermentation of 'idli' batter using soy residue okara. *Journal of Food Science and Technology*, **48**, 329–334.

Rentfrow, G., Chaplin, R. and Suman, S.P. (2012) Technology of dry-cured ham production: Science enhancing art. *Animal Frontiers*, **2**, 26–31.

Rosma, A., Afiza, T.S., Wan Nadiah, W.A., Liong, M.T. and Gulam, R.R.A. (2009) Short communication microbiological, histamine and 3-MCPD contents of Malaysian unprocessed 'budu'. *International Food Research Journal*, **16**, 589–594.

Salampessy, J., Kailasapathy, K. and Thapa, N. (2010) Fermented fish products. In J.P. Tamang and K. Kailasapathy (eds), *Fermented Foods and Beverages of the World* (pp. 289–307), Boca Raton, FL: CRC Press.

Tamang, J.P., Tamang, N., Thapa, S. and Kharel, N. (2012) Microorganisms and nutritional value of ethnic fermented foods and alcoholic beverages of North East India. *Indian Journal of Traditional Knowledge*, **11**, 7–25.

Todorov, S.D., Botes, M., Guigas, C. *et al.* (2008) Boza, a natural source of probiotic lactic acid bacteria. *Journal of Applied Microbiology*, **104**, 465–477.

Yegin, S.R. and Uren, A. (2008) Biogenic amine content of boza: A traditional cereal-based, fermented Turkish beverage. *Food Chemistry*, **111**, 983–987.

Zhou, G.H. and Zhao, G.T. (2012) History and heritage of Jinhua ham. *Animal Frontiers*, **2**, 62–67.

Zorba, M., Hancioglu, O., Genc, M. and Ova, G. (2003) The use of starter culture in the fermentation of Boza, a traditional Turkish beverage. *Process Biochemistry*, **38**, 1405-1411.

Index

Notes: Page numbers in *italics* indicate figures and those in **bold** denote tables

Starter Cultures in Food Production, First Edition. Edited by Barbara Speranza, Antonio Bevilacqua, Maria Rosaria Corbo and Milena Sinigaglia.